Understanding Aquatic Science

Understanding Aquatic Science

Edited by Jeremy Harper

SYRAWOOD
PUBLISHING HOUSE
New York

Published by Syrawood Publishing House,
750 Third Avenue, 9th Floor,
New York, NY 10017, USA
www.syrawoodpublishinghouse.com

Understanding Aquatic Science
Edited by Jeremy Harper

International Standard Book Number: 978-1-68286-442-5 (Hardback)

Cataloging-in-publication Data

Understanding aquatic science / edited by Jeremy Harper.
p. cm.
Includes bibliographical references and index.
ISBN 978-1-68286-442-5
1. Aquatic sciences. 2. Marine sciences. 3. Aquatic biology. 4. Aquatic ecology. I. Harper, Jeremy.
QH90 .U53 2017
551.46--dc23

Printed in China.

TABLE OF CONTENTS

PREFACE

Aquatic science is an emerging field of study which is multidisciplinary in nature. It is the study of aquatic systems. It closely studies both freshwater and marine systems. This book explores the various factors causing degeneration to the aquatic life and ecosystem. Aquatic science is a discipline that integrates several fields such as biogeochemistry, oceanography, marine biology, hydrology etc. The topics including in this book are of utmost significance and bound to provide incredible insights to readers. It includes some of the vital pieces of work being conducted across the world, on various aspects related to aquatic science. Coherent flow of topics, student-friendly language and extensive use of examples make this book an invaluable source of knowledge.

The information shared in this book is based on empirical researches made by veterans in this field of study. The elaborative information provided in this book will help the readers further their scope of knowledge leading to advancements in this field.

Finally, I would like to thank my fellow researchers who gave constructive feedback and my family members who supported me at every step of my research.

Editor

1

Response of Rocky Reef Top Predators (Serranidae: Epinephelinae) in and Around Marine Protected Areas in the Western Mediterranean Sea

Carlos Werner Hackradt[1]*, José Antonio García-Charton[1], Mireille Harmelin-Vivien[2], Ángel Pérez-Ruzafa[1], Laurence Le Diréach[3], Just Bayle-Sempere[4], Eric Charbonnel[5], Denis Ody[6], Olga Reñones[7], Pablo Sanchez-Jerez[4], Carlos Valle[4]

1 Departamento de Ecología e Hidrología, Universidad de Murcia, Murcia, Spain, **2** Aix-Marseille Université, Université du Sud Toulon-Var, CNRS/INSU/IRD, Institut Méditerranéen d'Océanologie (MIO), Marseille, France, **3** GIS Posidonie, Institut Méditerranéen d'Océanologie (MIO), Marseille, France, **4** Departamento de Ciencias del Mar y Biología Aplicada, Universidad de Alicante, Alicante, Spain, **5** Parc Marin de la Côte Bleue, Observatoire, Carry-le-Rouet, France, **6** WWF-France, Marseille, France, **7** Instituto Español de Oceanografía, Centro Oceanográfico de Baleares, Palma de Mallorca, Spain

Abstract

Groupers species are extremely vulnerable to overfishing and many species are threatened worldwide. In recent decades, Mediterranean groupers experienced dramatic population declines. Marine protected areas (MPAs) can protect populations inside their boundaries and provide individuals to adjacent fishing areas through the process of spillover and larval export. This study aims to evaluate the effectiveness of six marine reserves in the Western Mediterranean Sea to protect the populations of three species of grouper, *Epinephelus marginatus*, *Epinephelus costae* and *Mycteroperca rubra*, and to understand in which circumstances MPAs are able to export biomass to neighbouring areas. All the studied MPAs, except one where no grouper was observed, were able to maintain high abundance, biomass and mean weight of groupers. Size classes were more evenly distributed inside than outside MPAs. In two reserves, biomass gradients could be detected through the boundaries of the reserve as an indication of spillover. In some cases, habitat structure appeared to exert a great influence on grouper abundance, biomass and mean individual weight, influencing the gradient shape. Because groupers are generally sedentary animals with a small home range, we suggest that biomass gradients could only occur where groupers attain sufficient abundance inside MPA limits, indicating a strongly density-dependent process.

Editor: Athanassios C. Tsikliras, Aristotle University of Thessaloniki, Greece

Funding: This work was developed in the framework of the EC Project BIOMEX (QLRT-2001-0891). The authors express thanks to the National Council for Technology and Science Development (CNPq-MCT, Brazil, for Brazilian acronym) from the Ministry of Science of Brazil for giving personal financial support to CWH (Proc: 200630/2008-3), and to the Fundación BBVA (Foundation of Bilbao-Viscaya Bank, Spain) for additional support. The funders had no role in study design, data collection and analysis, decision to publish, or preparation of the manuscript.

Competing Interests: The authors confirm that Just Bayle-Sempere is a PLOS ONE Editorial Board member and this does not alter their adherence to PLOS ONE Editorial policies and criteria.

* E-mail: hackradtcw@gmail.com

Introduction

Several marine species are seriously threatened by an array of anthropogenic actions [1], from which fishing is likely the main human activity seriously affecting fish population abundance and size structure, and causing marine biodiversity loss. Worm et al. [2] stated that although increasing efforts to restore marine ecosystems and rebuild fisheries are under way, most fish stocks worldwide still require rebuilding. Lower exploitation rates are needed to reverse the collapse of vulnerable species, such as high trophic level species, which are likely to cause upheavals in the global ecosystem through the loss of particular functions played by these key species (e.g. reduction of natural predation) [2].

Recent studies, based on meta-analyses and reviews, show that MPAs can reverse most deleterious effects of fisheries on the marine environment [3–8] provided that they are properly managed [9,10]. The notable effects of marine reserves are an increase in abundance and an enlargement of the average size of individuals of the target species inside the boundaries of the protected area, so that greater abundance and size theoretically imply an increase in reproductive potential [3–5,7–8]. Eggs and larvae from restored spawning stocks inside MPAs could then be exported by currents to adjacent fishing areas ([11], but see [12]). On the other hand, because of increased density inside the MPA, adults and juveniles fishes from target species may emigrate from inside the protected locations to outside where the density is lower ("spillover", [13]). An indirect method to estimate the magnitude and importance of such export of larvae, juveniles and adults fishes from MPA to neighbouring areas is to look for the likely existence of gradients of biomass of target species across MPA limits, under the rationale that, if spillover occurs, there would be more fishes near than far away from the MPA [6,14].

This research strategy has been used in several studies in the Mediterranean (e.g., [15–18]) and worldwide (e.g., [19–20]). It has been hypothesized that the shape of biomass gradient for a given fish population responds to the distribution of the fishing pressure outside the reserve, and to the flux of individuals over the reserve boundary, which in turn would depend on the extent to which the system's carrying capacity is reached by the population [6]. The instantaneous population growth rate of a species would affect the speed of recovery of the population after cessation of fishing activity and the ability of the species to maintain abundances close to the carrying capacity inside the integral reserve; even more importantly, growth rate likely determines the fishing mortality that the population can support without collapsing [6]. For its part, flux of adults through MPA limits will depend on movement patterns, home range and spatial use of the species concerned [4,6,21].

On the other hand, habitat structure is one of the factors explaining the small-scale spatial variability of fish assemblage [22–23], and spatial variations in habitat structure is likely to affect the strength and even the occurrence of biomass increase within MPA boundaries and spillover [14,16,21,24], by influencing resource availability (food or refuge against predators or fishing), and ultimately affecting population growth and mobility.

Among the most common species affected by fishing pressure are top predators [25–27], and especially groupers [28,29]. Groupers (Perciformes: Serranidae: Epinephelinae) are emblematic species around the world, as they are of great importance for both recreational and artisanal fisheries [30,31]. Many species of Epinephelinae are overexploited, and about 25% of the species are under some level of threat [31]. The high susceptibility of grouper species to overfishing and habitat loss is likely due to their biology and life style, which promote a synergetic effect with anthropogenic activities. High site fidelity, high longevity, late maturity, formation of spawning aggregations, slow growth rate and low resilience (5 to 14 years to minimum population doubling time) are some of the characteristics that determine a high to very high level of vulnerability of these species [31].

The aim of this study is to evaluate the efficiency of protection measures to promote the recovery of populations of three groupers species both within and around MPAs in the Western Mediterranean Sea. The hypotheses to be tested are whether, and to what extent, there are higher grouper abundances inside than outside marine reserves, and whether biomass gradients can be found across the boundaries of the studied MPAs, suggesting spillover to neighbouring areas. Moreover, the present study aims at exploring the possible interference of the spatial distribution of structural habitat on grouper abundance within MPAs, and the occurrence of spillover towards adjacent areas.

Materials and Methods

Data acquisition for this work was made by visual censuses only; no animals were collected or manipulated. Research permissions were provided by "Ministerio de Medio Ambiente, Medio Rural y Marino", "Servicio de Pesca y Acuicultura – Comunidad Autónoma de la Región de Murcia", and "Departament de Medi Ambient y Habitatge – Generalitat de Catalunya" for Spanish MPAs, and "Ministère de l'Écologie, du Développement durable et de l'Énergie" for MPAs located in France.

Study Area

The work was conducted from July to October of 2003 and 2004 in six MPAs spread over the Western Mediterranean Sea (Fig. 1): the natural marine reserve of Cerbère-Banyuls (hereafter called Banyuls) and the Carry-le-Rouet (Carry) marine park in France, and the National Park of Cabrera (Cabrera), and the marine reserves of Medes islands (Medes), Tabarca island (Tabarca) and Cabo de Palos – Hormigas islands (Cabo de Palos) in Spain. All MPAs (except Carry) include one or more no-take zones surrounded by buffer zones, where some uses are permitted (usually recreational diving and some kind of artisanal fishing). Carry, is formed exclusively by a no-take zone, although it belongs to a larger conservation unit in the region. Common criteria used to select the MPAs included in this study were that they were established for more than 10 years, and presented a high level of enforcement. All marine reserves involved in this study are similar regarding the composition and constitution of the seabed, presenting *Posidonia oceanica* (Posidoniaceae) meadows and rocky bottoms.

Sampling Design and Data Acquisition

Seven to nine sectors, separated by 1000's of metres, were positioned at increasing distances from the core of each MPA. In each sector, three zones were haphazardly located at a scale of 100s of metres. Finally, six transects (replicates) separated by 10s of metres were sampled in each zone. In three MPAs (Banyuls, Carry and Cabo de Palos), gradients in fish parameters were studied along three sectors located inside the MPA (one inside the no-take zone and two within the buffer zone (herafter NTZ and BZ respectively) and six sectors in unprotected areas (UP), where fishing is allowed: three in one direction and three in the opposite direction (Table 1). In the three MPAs located on islands (Cabrera, Medes and Tabarca) only one direction (northward) could be studied due to the absence of suitable rocky habitats southward (Table 1). In Cabrera, three sectors were located within the NTZs, three sectors inside the BZs, and three within UP, outside the MPA. For Medes, three sectors were located within the NTZ, one in the BZ, and three outside the MPA (Table 1). In Tabarca, the sampling was performed within: one sector inside the NTZ, three sectors in the BZ, and the rest outside the reserve (Table 1). In this MPA, the sampling was done on two different bottom types: rocky bottoms and *Posidonia oceanica* meadows, respectively named Tabarca – rocky and Tabarca - *Posidonia* hereafter (see Table 1 in Harmelin-Vivien et al. [16] for further information).

Three species of groupers were assessed during this study, *Epinephelus marginatus* (Lowe, 1834) (dusky grouper), *E. costae* (Steindachner, 1878) (goldbloch grouper) and *Mycteroperca rubra* (Bloch, 1793) (mottled grouper). According to IUCN red list, *E. marginatus* is defined as an "endangered" species, while *E. costae* is categorized as "data deficient" and *M. rubra* as "least concern". The abundance of these species was assessed by visual census in 25×5 m transect belts located at 6–12 m depth and parallel to the

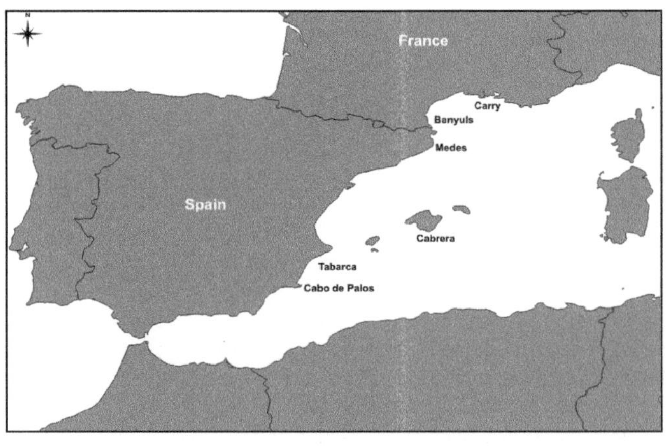

Figure 1. Location of the Mediterranean marine protected areas (MPAs) studied. 1: Carry-le-Rouet, 2: Banyuls, 3: Medes, 4: Cabrera, 5: Tabarca, 6: Cabo de Palos.

Table 1. Summary of spatial sampling effort applied in each studied MPAs, depending on MPA management plan and natural habitat distribution, showing the number of sectors sampled in each existing level of protection.

MPAs	Sectors			Zones	Transects
	NTZ	BZ	UP		
Marine Park of Carry-le-Rouet	3	0	6*	27	162
Natural Marine Reserve of Cerbère-Banyuls	1	2	6*	27	162
Marine Reserve of Medes Island	3	1	3†	21	126
National Park of Cabrera Island	3	3	3†	27	162
Marine Reserve of Tabarca Island	3	1	3†	21	126
Marine Reserve of Cabo de Palos-Hormigas Islands	1	2	6*	27	162

In each sector, 3 random zones and 6 replicas were always defined in each MPA. NTZ: No-take zone; BZ: Buffer zone; UP: unprotected area; Zones: total number of zones randomly located; Transects: total number of replicates.
*Three sectors were located to the South and other 3 to the North.
†All sectors located to the North direction.

coast. As *P. oceanica* beds covered large areas around Tabarca island, seagrass beds were surveyed in this habitat in 50×5 m transects, as fishes were more dispersed. In each transect, fishes were identified and the size of each individual was recorded within 10-cm size classes, so that fish weight could be estimated from length-weight relationship found in the literature. Within each transect the following descriptors of structural habitat were also registered: number of rocky boulders (classified as small, medium and large) and percentage of cover by different types of substrate (namely rock, sand, *Posidonia* and pebbles), for further details on habitat data acquisition, see García-Charton et al. [24] and Harmelin-Vivien et al. [16].

Data Analysis

The effect of protection on groupers was evaluated separately for each MPA, because differences in sampling design precluded making a unique analysis with all MPAs together. Distance-based permutational multivariate analysis of variance (PERMANOVA, [32]) based on Bray-Curtis dissimilarities [33] on log-transformed data was run in univariate mode, using abundance, biomass and individual mean weight of the three target species of groupers as response variables. In general, the field experimental design for each MPA consisted of three factors: Location (factor L, fixed, comparing two or three levels of protection in each MPA, depending on the MPA, considering NTZ and BZ as compared to the unprotected –UP– locations), Sector (factor S, 3 levels, random, nested in L) and Zone (factor Z, 3 levels, random, nested in S). Only in the case of Cabrera all NTZ, BZ and UP levels were compared. For Medes and Tabarca, one sector (in BZ) was excluded from the analyses to get a balanced design, so that only the comparison between NTZ and UP was taken into account. In the case of MPAs where 6 sectors were surveyed outside the protected location against 3 sectors within it (Carry, Banyuls and Cabo de Palos), an asymmetrical design was considered, for which the Location term was partitioned into two portions: the one degree-of-freedom contrast Protected (P) (including NTZ and BZ locations) *vs.* UPs locations, and the variability between UPs. The overall mean squares of the terms S(L) and Z(S(L)) were similarly partitioned into S(P *vs* UPs) and S(UPs), and Z(S(P *vs* UPs)) and Z(S(UPs)), respectively. Nevertheless, as no groupers were recorded in Carry, and in Banyuls they were observed exclusively in the NTZ (see results), only the case of Cabo de Palos was explored using this asymmetrical design. In order to minimize the effect of habitat variability on data and exploring only the effect on species protection, we used the environmental data as covariables. For all analysis 9999 permutations were applied under a full model, using PRIMER v.6 package.

Trends and significance of gradients of grouper biomass across MPAs boundaries were calculated using linear correlation with the distance from MPA limits at the scale of zones. Negative correlations would indicate that biomass decreases from the core of MPA to distant fished zones. The border of the NTZ was defined as zero, so negative distances indicate zones inside the NTZ, and positive distances zones outside the NTZ. To explore the actual shape of grouper biomass gradients across MPA borders, generalized additive modelling (GAM) were applied using gam v.1.06.2 in R statistic package. GAM is known to be useful when the actual relationship between the variables is unknown and expected to be of a complex form, not easily fitted by standard linear or non-linear models [34]. Distance to the boundary of the no-take zone was introduced as a continuous smooth variable modelled non-parametrically using a loess smoother (lo(Distance)). We applied to model Gaussian variance and identity link functions, both based on Hastie and Tibshirani [34] and Venables and Ripley [35]. The gradients were tested for both sides of MPA (South and North), or as a unique gradient, depending on the study case.

The likely influence of habitat structure on gradient shape was further assessed by performing multiple linear regression analyses to measure the strength of the relationship between the whole set of habitat variables (including their quadratic and cubic terms) and each species' population parameter (abundance, biomass and individual mean weight). Prior to analyses, the extreme and influential cases were detected and removed by carrying out analysis of residuals [36–37]. Then, the residuals of these analyses were used as a dependent variable in linear correlations and GAMs with the distance from MPA limits [16]. If habitat quality is equal both inside and outside the MPA, or does not influence fish biomass, we hypothesize that the extraction of habitat influence would not affect the shape of biomass gradient as depicted by GAMs (Fig. 2a and b). If habitat quality is better inside than outside the MPA (i.e. it promotes higher fish biomass within the MPA because structural habitat provides either enhanced food or/and refuge resources as compared to surrounding areas), the shape of biomass gradient across MPA limits would be smoothed once extracted the influence of habitat from raw data (Fig. 2c).

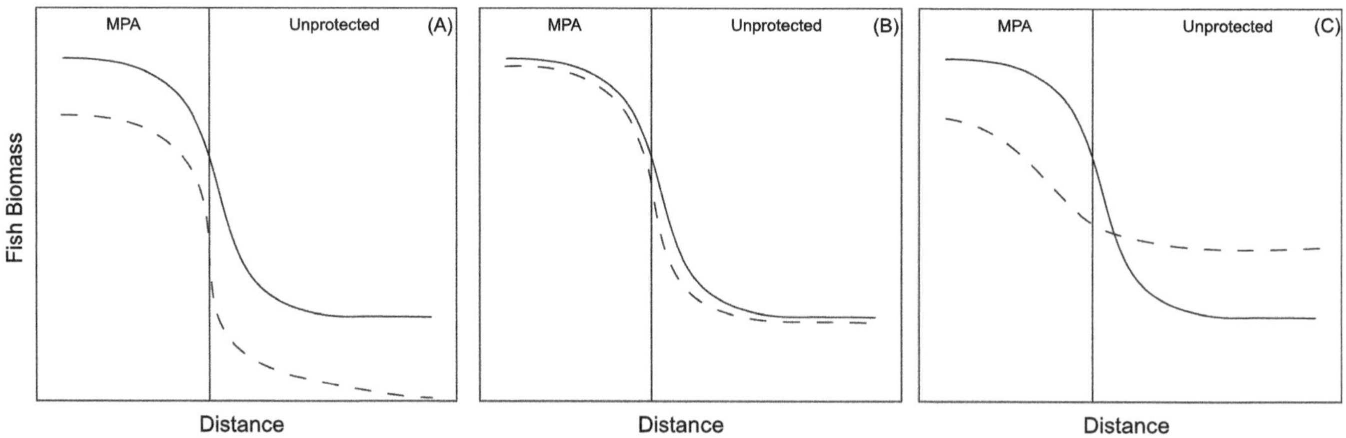

Figure 2. Hypotheses for the shape of gradient of fish biomass across MPA boundaries with raw data (solid line) and residual data after extracting habitat influence (dotted line) where habitat quality (a) is equally good or (b) has no influence both inside and outside the MPA, and (c) where habitat quality is better within the MPA. The vertical line indicates the limit of the MPA, with protected area to the left and fishes zones to the right.

Results

Effects of Protection on Grouper Abundance and Size Structure

Epinephelus marginatus was recorded in all MPAs, except Carry. In Banyuls and Medes, this grouper was censused only inside the NTZ, and in Tabarca – *Posidonia* this species was censused both in NTZ and BZ areas, but not outside the MPA. The two other species, *E. costae* and *M. rubra*, were recorded only in Cabrera, Tabarca – rocky and Cabo de Palos MPAs. Thus, the effect of protection levels and nested spatial factors on the abundance and biomass of all grouper species will be explored by PERMANOVA only for the latter three case studies (Fig. 3). Cabrera and Cabo de Palos MPAs presented the highest grouper abundance amongst all reserves, while the lowest abundance values were recorded in the northernmost MPAs where these species were observed (Banyuls and Medes).

Further to the fact that in Banyuls, Medes and Tabarca – *Posidonia E. marginatus* was observed only inside the protected areas, the abundance, biomass and mean weight of this species were significantly higher within the protected areas than in unprotected ones in the three MPAs analysed, as shown by the statistically significant effect of the fixed factor Location (Table 2, Fig. 3). In Cabrera and Cabo de Palos, the values of the analysed parameters in the NTZ and BZ areas were similar, and presented significantly higher values than in unprotected areas (Table 2, Fig. 3). In the case of Tabarca (both in rocky bottoms and *Posidonia* meadows) the NTZ showed greater values of the three parameters analysed compared to both BZ and UP areas (Fig. 3). A significant medium-scale (among sectors) variability was evidenced in Cabo de Palos, mostly in the unprotected locations, and a fine-scale (among zones) variability was detected for all dependent variables in Tabarca – rocky and Cabo de Palos (Table 1).

No significant effect of the factor Location was found in any MPA analysed for *E. costae* and *M. rubra* (Tables 3 and 4), despite these species were more abundant inside the protected area of Cabrera and Cabo de Palos (Fig. 3). This result was likely due to the high spatial variability at several scales, including among replicates. At Cabo de Palos a significant spatial variability among sectors within the protected location in abundance, biomass and mean weight of *E. costae* was evident (Table 3). Significant variability among zones within sectors was also found in all MPAs analyzed for *E. costae*, and in Tabarca – rocky and Cabo de Palos for *M. rubra* (Table 4).

The size structure of *E. marginatus* populations was more evenly distributed within Banyuls and Medes marine reserves, with the presence of large individuals, than inside the other MPAs, where individuals larger than 60 cm were scarce (Fig. 4). In all MPAs, larger individuals, if censused, were only seen within MPAs limits, while small ones were abundant outside MPA in Cabrera and Cabo de Palos (Fig. 4). Small individuals (<40 cm) of *E. costae* were also more frequently seen outside MPAs, while large individuals (>60 cm) of *M. rubra* occurred exclusively within MPA limits (Fig. 4).

Influence of Habitat

Habitat structure appeared to exert a great influence on the abundance, biomass and mean individual weight of *E. marginatus* in Tabarca (both in rocky and *Posidonia* bottoms) and Cabo de Palos (Table 4). Abundance of *E. marginatus* in Banyuls and of *E. costae* in Cabrera, Tabarca rocky and Cabo de Palos was positively correlated to habitat structure. Habitat exerted also a significant influence on the mean weight of *E. costae* individuals in Tabarca – rocky and Cabo de Palos, and also on biomass of *M. rubra* in Cabo de Palos (Table 5). In the cases where it resulted to be significant, the influence of habitat on the parameters examined attained on average around 33% of total variability (range 14–53%) (Table 5).

Gradients of Biomass Export

Raw values of grouper biomass decreased with increasing distance from the boundary of MPAs, for Cabrera, Tabarca – rocky (except for *E. costae* in Tabarca – rocky), Cabo de Palos – North and Cabo de Palos – South, and these relationships were statistically significant in 8 out of the 11 species by MPA studied cases (Table 6). When residuals of biomass data obtained from multiple linear regressions were used as dependent variables instead of raw data, the number of significant negative correlations with distance decreased. Remarkably, the non-significant negative relationship of *E. costae* raw biomass southwards of Cabo de Palos became a significant positive relationship when using residuals (Table 6). These shifts when using residuals instead of raw data as dependent variables suggested a high habitat effect superimposed to that of protection measures.

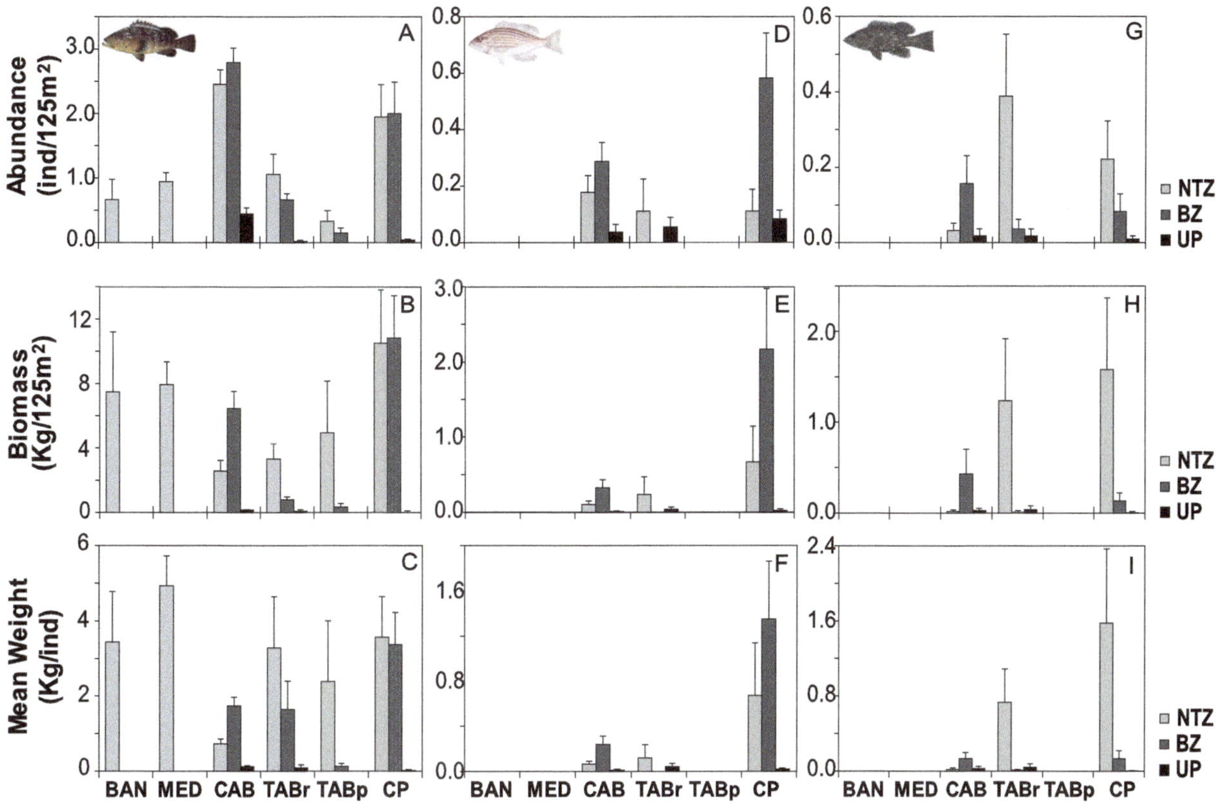

Figure 3. Abundance, biomass and mean weight of the three species of groupers, *E. marginatus, E. costae* and *M. rubra* according to the tree levels of protection (No-take zone: NTZ, Buffer zone: BZ and unprotected: UP) in all MPAs studied (BAN: Banyuls; MED: Medes; CAB: Cabrera; TAB-r: Tabarca Rocky habitat; TAB-p: Tabarca *Posidonia* habitat; CP: Cabo de Palos), Carry excepted as no groupers were recorded there.

Results of general additive models (GAM) on fish raw biomass as a function of distance to reserve boundaries yielded significant non-linear relationships in 8 out of 12 studied crossings between MPAs and species, these significant models explain 16–58% of total data variability (Table 7). When using residual biomass resulting from multiple linear regressions with habitat variables, the number of significant non-linear relationships with distance dropped to 5, and the range of percentage variability explained by the models decreased to 10–39% (Table 7). Graphical representation of GAMs evidenced three major patterns: the biomass of grouper species decreased i) constantly between integral reserve, buffer zone and outside (Fig. 5 b, c, g); ii) rapidly between the integral reserve and the buffer zone (Fig. 5 f, i, l); or iii) at the edge of the MPA, in the fishing area (Fig. 5 a, d, h, j, k). No difference was observed among raw and residuals data pattern for Cabrera, which was the expected pattern when habitat structure does not influence grouper abundances either inside or outside the MPA (Fig. 2). However, for the curves constructed with residuals of biomass data from Tabarca and Cabo de Palos, slopes were less pronounced and shapes smoother, following what it was hypothesized when habitat quality is better inside than outside the MPA. The actual shape of the observed biomass gradients suggested in these cases that there was no export, or it occurred at a very short distance (<1000 m).

Discussion

Reserve Effect on Grouper Populations

In this work, we showed that grouper species, and especially *E. marginatus*, are extremely favoured by protection measures in Western Mediterranean MPAs. Average abundance, biomass and individual weight of groupers are higher within the protected areas than immediately outside. In some MPAs (Banyuls, Medes, Tabarca – *Posidonia*), dusky groupers were observed only inside the MPA, indicating the occurrence of a high fishing effort just at the edge of the protected areas [17,38]. Therefore, at present abundant grouper populations (≥ 1 indiv.125 m^{-2}) are found only within MPAs in the western Mediterranean, highlighting the heavy impact of fishing on the coastal fish populations and the importance of MPAs to maintain epinephelid populations. A noticeable recovery in grouper abundance as a response to fisheries closure has been already documented in the Mediterranean Sea [39–41,24], corroborating that these species respond more or less rapidly to protection [4,28,42] even in small reserves [29]. Thus, MPAs are very effective to counteract the rapid depletion of predatory fish observed worldwide [43]. Because high biomass of top predators can be considered the natural state of marine reef fish communities, as demonstrated by studying remote reefs [26,44], MPAs would serve to recover pristine fish community structures [25]. The key questions are how much protection time is required until local carrying capacity is attained [42], and which is the maximum value of biomass of apex predator [26] for unfished Mediterranean locations?

Table 2. Permutational multivariate analysis of variance (PERMANOVA) results for abundance, biomass and mean weight of *E. marginatus* found in the locations studied on Western Mediterranean Sea where groupers are present both inside and outside the MPA.

Marine Reserve			Abundance			Biomass			Mean Weight		
	Source	df	MS	PF	p	MS	PF	p	MS	PF	p
CAB	L	2	9960.3	14.43	**0.006**	29525	22.05	**0.001**	26915	23.99	**0.001**
	S(L)	6	660.0.	1.92	0.120	1278.3	1.13	0.374	1070.8	1.01	0.454
	Z(S(L))	18	344.9	1.53	0.073	1137	1.18	0.277	1065.5	1.16	0.310
	Res	128	224.6			965.7			922.0		
TAB-r	L	1	3174.3	92.15	**0.001**	21930	271.77	**0.0001**	22613	275.57	**0.0001**
	S(L)	4	28.8	0.10	0.981	67.4	0.035	0.998	68.6	0.003	0.998
	Z(S(L))	12	290	2.83	**0.002**	2032.2	2.97	**0.002**	2190.7	3.15	**0.0005**
	Res	83	102.3			682.2			695.8		
CP	L	2	2436.2	4.09	0.059	11668	5.03	**0.037**	11267	5.11	**0.030**
	P *vs* UPs	1	4816	6.83	**0.044**	22996	8.79	**0.020**	22204	8.97	**0.025**
	UPs	1	0.002	0.0005	0.984	3.4	0.008	0.984	3.4	0.008	0.984
	S(L)	6	623.6	3.35	**0.009**	2419.7	2.71	**0.027**	2298.1	2.66	**0.031**
	S(P *vs* UPs)	4	905.9	3.44	**0.031**	3327.9	2.67	0.067	3141.9	2.61	0.069
	S(UPs)	4	59.9	4.25	**0.023**	450.5	3.95	**0.017**	450.5	3.95	**0.021**
	Z(S(L))	18	186.7	1.45	0.108	895.8	1.69	**0.042**	868.4	1.69	**0.046**
	Z(S(P *vs* UPs))	12	270.5	2.20	**0.010**	1282.3	2.50	**0.006**	1241.3	2.49	**0.005**
	Z(S(UPs))	12	14	0.60	0.859	113.59	0.67	0.807	113.6	0.67	0.802
	Res	128	129.1			531.2			515.2		

L: locality, S: sector, Z: zone, P: protected, UP: unprotected, res: residuals, df: degree-of-freedom, MS: mean square, PF: pseudo-F, p: significance level (bold characters are used for significant results). CAB – Cabrera; TAB-r – Tabarca rocky habitat; CP – Cabo de Palos.

Table 3. Permutational multivariate analysis of variance (PERMANOVA) results for abundance, biomass and mean weight of *E. costae* found in the locations studied on western Mediterranean Sea where groupers are present both inside and outside the MPA.

Marine Reserve			Abundance			Biomass			Mean Weight		
	Source	**df**	**MS**	**PF**	**p**	**MS**	**PF**	**p**	**MS**	**PF**	**p**
CAB	**L**	2	428.8	2.85	0.129	3034.1	3.52	0.082	3015.3	3.61	0.083
	S(L)	6	144.3	0.92	0.507	826.9	0.73	0.640	800.4	0.71	0.652
	Z(S(L))	18	157.6	2.12	**0.010**	1140.9	2.24	**0.005**	1133.4	2.26	**0.005**
	Res	135	74.4			509.0			501.6		
TAB-r	**L**	1	13.5	0.24	0.750	143.5	0.34	0.693	175	0.36	0.681
	S(L)	4	69.6	1.38	0.29	476.9	1.16	0.398	542.1	1.14	0.406
	Z(S(L))	12	52.5	2.18	**0.019**	428.7	2.59	**0.006**	495.9	2.68	**0.006**
	Res	83	24.1			165.8			185.3		
CP	**L**	2	170	0.26	0.798	1443.5	0.35	0.733	1444	0.36	0.732
	P*vs*UPs	1	40.3	0.05	0.869	363.31	0.07	0.832	352.8	0.07	0.826
	UPs	1	150.9	4.82	0.089	1257.4	4.89	0.084	1262.7	4.90	0.084
	S(L)	6	680.7	6.09	**0.0001**	4343.7	5.86	**0.0001**	4260.5	5.85	**0.0001**
	S(P*vs*UPs)	4	1041.5	7.87	**0.0001**	6643.4	7.54	**0.0001**	6516.9	7.53	**0.0001**
	S(UPs)	4	31	0.60	0.684	256.1	0.75	0.585	256.4	0.76	0.577
	Z(S(L))	18	112.4	1.80	**0.030**	744.2	1.69	**0.044**	731.8	1.68	**0.048**
	Z(S(P*vs*UPs))	12	136	2.16	**0.015**	904	2.03	**0.026**	888.1	2.02	**0.025**
	Z(S(UPs))	12	51.4	1.03	0.431	342.5	0.94	0.525	338.3	0.94	0.522
	Res	128	62.5			439.8			435.5		

L: locality, S: sector, Z: zone, P: protected, UP: unprotected, res: residuals, df: degree-of-freedom, MS: mean square, PF: pseudo-F, p: significance level (bold characters are used for significant results). CI – Cabrera; TAB-r – Tabarca rocky habitat; CP – Cabo de Palos.

Table 4. Permutational multivariate analysis of variance (PERMANOVA) results for abundance, biomass and mean weight of *M. rubra* found in the locations studied on Western Mediterranean Sea where groupers are present both inside and outside the MPA.

Marine Reserve			Abundance			Biomass			Mean Weight		
	Source	**df**	**MS**	**PF**	**p**	**MS**	**PF**	**p**	**MS**	**PF**	**P**
CAB	**L**	2	1.2	0.11	0.897	14.3	0.13	0.909	14.2	0.13	0.903
	S(L)	6	17.9	1.31	0.290	153.9	1.40	0.251	153.8	1.39	0.259
	Z(S(L))	18	13.6	0.83	0.685	110.0	0.83	0.685	109.9	0.82	0.683
	Res	135	16.3			131.8			131.8		
TAB-r	**L**	1	18.6	0.46	0.576	168.8	0.43	0.596	166	0.42	0.602
	S(L)	4	54.2	0.67	0.624	497.5	0.77	0.580	507	0.76	0.574
	Z(S(L))	12	83.9	2.17	**0.016**	674.9	2.27	**0.014**	689.4	2.30	**0.014**
	Res	83	38.6			297.1			299.2		
CP	**L**	2	44.2	1.35	0.325	441.9	1.32	0.330	441.9	1.32	0.341
	P *vs* UPs	1	87.3	2.06	0.218	873.1	2.03	0.208	873.1	2.03	0.214
	UPs	1	1.3	0.38	0.597	11.2	0.38	0.598	11.2	0.38	0.598
	S(L)	6	32.7	0.88	0.570	334.8	0.92	0.550	334.8	0.92	0.553
	S(P *vs* UPs)	4	46.6	0.91	0.544	481.0	0.95	0.521	481.3	0.95	0.535
	S(UPs)	4	3.6	0.57	0.744	29.6	0.57	0.736	29.6	0.57	0.746
	Z(S(L))	18	37.3	1.35	0.164	366.4	1.40	0.131	366.4	1.40	0.135
	Z(S(P *vs* UPs))	12	52.8	2.00	**0.031**	522.1	2.11	**0.021**	522.1	2.11	**0.017**
	Z(S(UPs))	12	6.3	1.15	0.327	52.3	1.15	0.331	52.3	1.15	0.324
	Res	128	27.8			261.9			261.9		

L: locality, S: sector, Z: zone, P: protected, UP: unprotected, res: residuals, df: degree-of-freedom, MS: mean square, PF: pseudo-F, p: significance level (bold characters are used for significant results). CI – Cabrera; TAB-r – Tabarca rocky habitat; CP – Cabo de Palos.

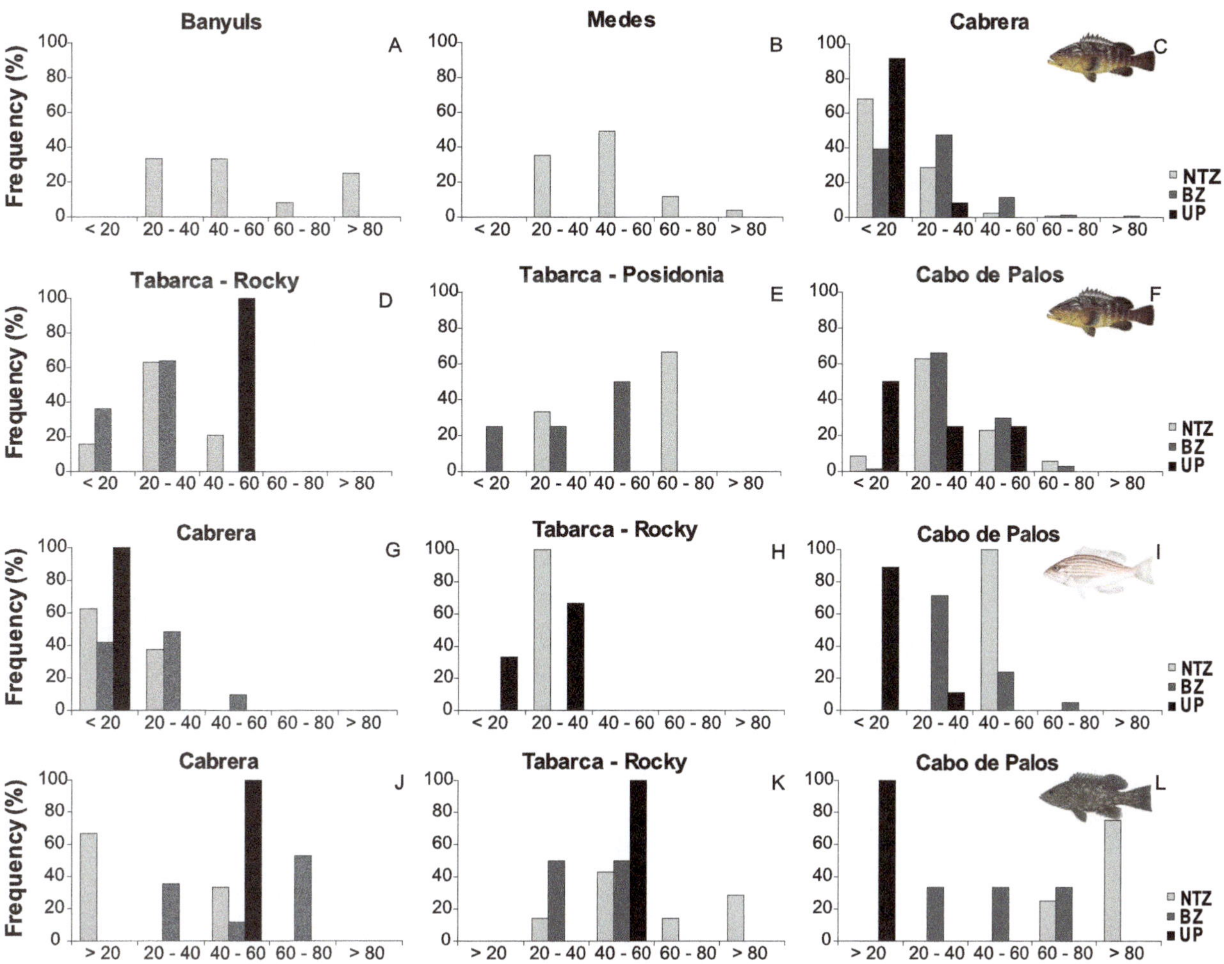

Figure 4. Distribution of frequencies of groupers size classes in each studied MPA (a–f, *E. marginatus*; g–i, *E. costae*; j–l, *M. rubra*).

Regarding individual size of groupers, almost all size classes are represented within the studied MPA, with larger individuals generally restricted to inside MPA limits. This indicates that grouper populations protected from fisheries are well established and are constituted from both mature and juvenile individuals. The high abundance, large size and stable size class distribution can allow the reproduction of groupers inside MPAs [45–47]. Due to the high site fidelity and strong territorial behaviour displayed by mature dusky grouper [48], younger individuals must swim larger distances in order to find food, shelter and constitute their own territory. During this wandering search, juveniles may establish themselves outside MPAs as a density-dependent response to high competition for territory inside the reserve [49]. Both mechanisms, egg and larvae exportation from the restored spawning biomass and density-dependent movement of juveniles, can be reflected by an increased observation of younger dusky grouper outside the limits of marine reserves [50].

Our study shows important geographical differences in species composition, abundance and size structure of grouper populations. Grouper species were generally more frequent and abundant in the southernmost MPAs (Cabo de Palos, Tabarca and Cabrera), where the three species were found. Moreover, in the northernmost reserves, only dusky groupers (in Banyuls and Medes) or even no groupers at all (in Carry) were recorded. Dusky groupers showed also much lower abundances in northern MPAs than in southern ones. Additionally, the higher mean biomass and individual weight values recorded for this species in Banyuls and Medes indicated that *E. marginatus* populations were composed of larger and older individuals than in the other MPAs, which could be due to the older ages of these two marine reserves (dating from 1974 and 1983, respectively), compared to the southernmost MPAs (from 1986 for Tabarca to 1995 for Cabo de Palos). On the other hand, environmental conditions could also shed light on the spatial differences observed. Duration of the spawning activity and the spawning survivorship (recruitment success) are limited by temperature thresholds, and can largely influence the differential population parameters observed [46–47]. The geographic pattern observed in this study corroborates the thermal affinity of grouper species, especially *E. costae* and *M. rubra*. Moreover, in our data juveniles were virtually absent from the northern MPAs despite the recovery of the adult population and years of protection, although this pattern could be reverting during the last decade due to seawater warming [50]. Thus, latitudinal/oceanographic effects could be limiting northern populations.

Biomass Export of Groupers

Identifying life-stages that are critical to the population dynamics of a threatened species along its distribution range is essential to effectively design MPA networks in order to achieve conservation objectives. Based on the estimation of biomass gradients, significant evidences of spillover of groupers from inside to outside MPA borders taking the effect of habitat into account

Table 5. Summary of results of multiple linear regressions (adjusted R^2 and significance level) of mean abundance (Ab), biomass (Bm) and mean individual weight (IW) of the three species of groupers studied against linear, quadratic and cubic terms of all habitat characteristics measured in the transects for each MPA.

		Banyuls	Medes	Cabrera	Tabarca – rocky	Tabarca – *Posidonia*	Cabo de Palos
E. marginatus	**Ab**	0.210 (*)	0.117 (ns)	0.062 (ns)	0.296 (**)	0.525 (***)	0.331 (***)
	Bm	0.176 (ns)	0.048 (ns)	0.086 (ns)	0.350 (***)	0.473 (***)	0.359 (***)
	IW	0.083 (ns)	0.066 (ns)	0.112 (ns)	0.323 (**)	0.451 (***)	0.374 (***)
E. costae	**Ab**			0.139 (*)	0.257 (*)		0.253 (**)
	Bm			0.060 (ns)	0.173 (ns)		0.191 (ns)
	IW			0.084 (ns)	0.300 (**)		0.211 (*)
M. rubra	**Ab**			0.049 (ns)	0.397 (***)		0.210 (*)
	Bm			0.048 (ns)	0.366 (***)		0.159 (ns)
	IW			0.044 (ns)	0.378 (***)		0.159 (ns)

(ns: not significant; *: $p<0.05$; **: $p<0.01$; ***: $p<0.001$).

Table 6. Results of linear correlations performed on raw data and residuals after extracting habitat effects on the biomass of the three studied grouper species (significant *p*-value in bold).

	E. marginatus				*E. costae*				*M. rubra*			
	Raw		**Res**		**Raw**		**Res**		**Raw**		**Res**	
	r	p	r	p	r	p	r	p	r	p	r	p
Cabrera	−0.483	**0.000**	−0.450	**0.000**	−0.140	**0.027**	−0.109	0.084	−0.056	0.374	−0.074	0.239
Tabarca-rocky	−0.413	**0.000**	−0.158	0.077	0.012	0.890	−0.071	0.429	−0.076	0.397	−0.021	0.814
C. Palos North	−0.516	**0.000**	−0.181	0.061	−0.216	**0.025**	0.047	0.627	−0.276	**0.004**	−0.057	0.556
C. Palos South	−0.573	**0.000**	−0.085	0.380	−0.156	0.107	0.226	**0.019**	−0.316	**0.001**	−0.121	0.212

Table 7. Analysis of deviance for the generalised additive models (GAMs) fitted with raw data and residuals of multiple linear regressions with habitat variables on biomass of the three studied grouper species.

		Term	Res df	F	p	R^2
E. marginatus						
Cabrera	**Raw**	lo (distance)	246	3.73	0.005	0.28
	Res	lo (distance)	246	3.80	0.004	0.25
Tabarca - rocky	**Raw**	lo (distance)	120	6.87	<0.001	0.32
	Res	lo (distance)	120	1.48	0.211	0.07
C Palos North	**Raw**	lo (distance)	102	13.64	<0.001	0.54
	Res	lo (distance)	102	2.37	0.05	0.12
C Palos South	**Raw**	lo (distance)	102	12.00	<0.001	0.51
	Res	lo (distance)	102	3.25	0.02	0.10
E. costae						
Cabrera	**Raw**	lo (distance)	246	1.20	0.31	0.04
	Res	lo (distance)	246	1.11	0.35	0.03
Tabarca - rocky	**Raw**	lo (distance)	120	2.42	0.053	0.07
	Res	lo (distance)	120	1.15	0.33	0.03
C Palos North	**Raw**	lo (distance)	102	29.12	<0.001	0.58
	Res	lo (distance)	102	14.68	<0.001	0.39
C Palos South	**Raw**	lo (distance)	102	20.61	<0.001	0.40
	Res	lo (distance)	102	10.97	<0.001	0.29
M. rubra						
Cabrera	**Raw**	lo (distance)	246	1.55	0.18	0.03
	Res	lo (distance)	246	1.35	0.25	0.03
Tabarca - rocky	**Raw**	lo (distance)	120	2.09	0.08	0.07
	Res	lo (distance)	120	0.70	0.59	0.02
C Palos North	**Raw**	lo (distance)	102	2.37	0.05	0.16
	Res	lo (distance)	102	1.47	0.21	0.06
C Palos South	**Raw**	lo (distance)	102	2.37	0.05	0.16
	Res	lo (distance)	102	1.47	0.21	0.06

were detected in only two of the five marine reserves analyzed, Cabo de Palos and Cabrera. In Cabo de Palos, an area with low artisanal fishing effort directed to grouper species [17], the spatial scale of grouper spillover seemed to be of less than 1000 m, a value lower than that estimated by Harmelin-Vivien et al. [16] for the whole visually-censused fish assemblage (3000 m). This observation corroborates the hypothesis that biomass gradients are sharper (and thus spillover lower) for low-mobility or high-catchability species, provided that fishing pressure outside the MPA is high compared to that exerted in the BZ, as it is clearly the case in Cabo de Palos [17,38]. It is important to highlight that groupers are targeted by recreational fishing (mainly spearfishing) as heavily as by professional fishing [51], and spearfishing is highly practiced around the Cabo de Palos marine reserve (Carlos W. Hackradt, *personal observation*). In Cabrera, the closer unprotected sites are located at about 10 km from the MPA with unsuitable habitats for juvenile groupers in between. Therefore, grouper biomass export from Cabrera MPA is rather likely due to egg and larval dispersal, as shown by Crec'hriou et al. [11]. In Banyuls, Medes and Tabarca, no spillover of groupers outside the reserves could be detected. In Cabo de Palos and Cabrera, grouper abundances were the highest among the six MPAs studied, and there was no difference in grouper abundance and biomass between NTZs and BZs, where artisanal fishing occurred, while in Tabarca higher abundances were recorded in the NTZ than in the BZ. We hypothesise that the importance of biomass gradients (and thus the occurrence of fish spillover) is likely to be a function of fish density inside MPAs [20]: when groupers are abundant inside MPA, biomass gradients from inside to outside are to be evidenced, as a result of movement of individuals (usually small-sized ones); when, in turn, there is a low number of individuals within an MPA, gradients could be detected from the NTZ to the BZ, but not from the BZ to outside the MPA. Therefore, because the efficiency of an MPA for a target species will depend on its total size compared to the home range of this species [21], the carrying capacity, zonation and size of the MPA should be taken into consideration when estimating its potential for spillover. Although several studies establish that Mediterranean groupers have high site fidelity and small home range [48,52], further studies about the mobility and home range of these species inside and outside Mediterranean MPAs are still needed, taking into account fish densities and sizes.

Habitat Characteristics and MPA Designs

The present study emphasizes the importance of considering explicitly habitat structure when evaluating biomass exportation

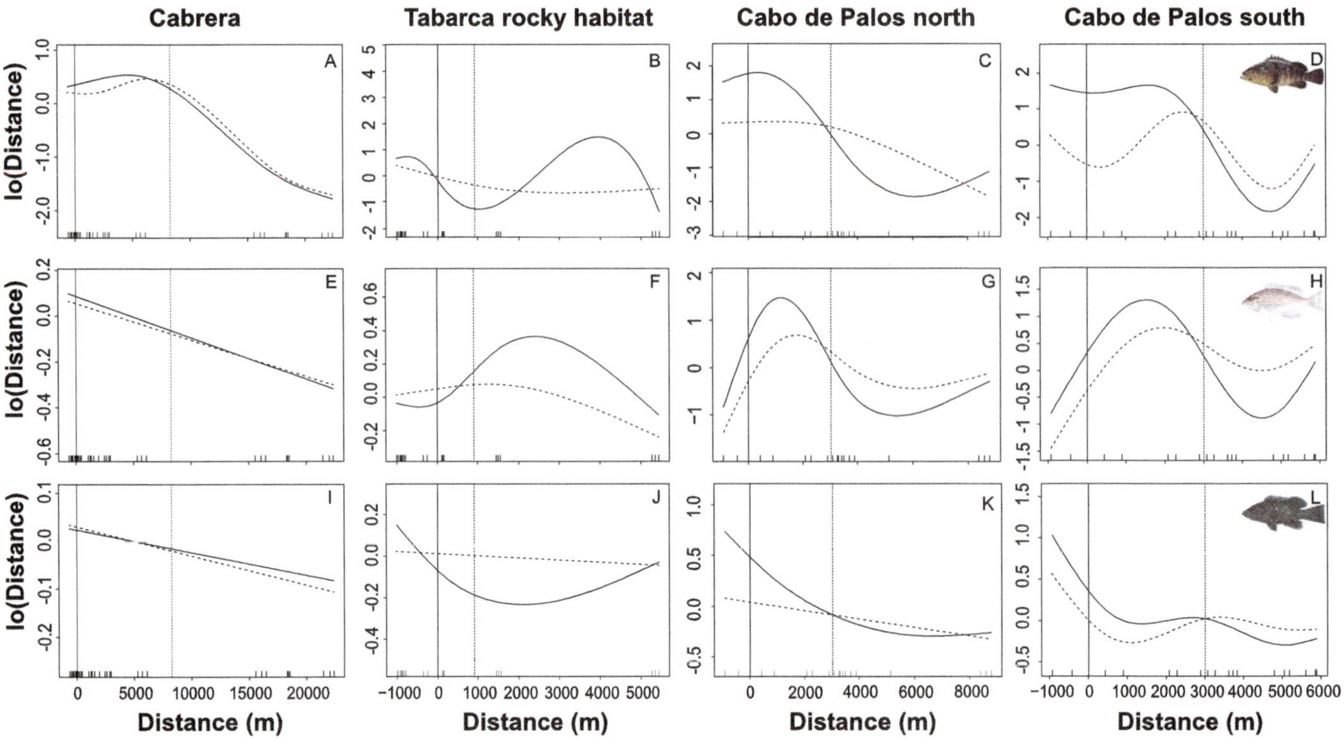

Figure 5. General additive model fitted for raw data (solid line) and residuals (dashed line) of biomass ($Log_{[x+1]}$ transformed) of *E. marginatus, E. costae* and *M. rubra* as a function of the smooth variable distance (in m) from the boundaries of integral reserve. The y-axis is scaled so zero corresponds to the mean in log scale. The solid vertical line indicates the limit of the no-take zone (NTZ) and the dashed vertical line indicates the limit of the buffer zone (BZ).

patterns [14,16,24]. The influence of habitat structure on the shape of the curve depicting the relationship between the response variable and the geographical distance to MPA limits tells much about the relative importance of structural habitat to favour or, at contrary, to hinder spillover. In those MPAs where habitat structure appears to exert a great influence on grouper abundance, biomass and mean individual weight (Tabarca – rocky and Cabo de Palos), extracting the habitat effect smoothes the shape of biomass gradient, although it does not affect the estimated spillover distance. According to the theoretical model proposed here, this shift is likely due to the fact that habitat structure is of higher quality within than outside MPAs, which is a quite normal situation in the Mediterranean [24]. In these situations, structural habitat, by providing additional food and refuge resources, would act as an "attractant" for groupers, hence boosting the carrying capacity of protected sites and, consequently, reducing the strength of spillover to neighbouring, unprotected sites. However, Cabo de Palos – Islas Hormigas marine reserve can be considered a singularity within the Mediterranean context regarding the habitat structure: the offshore steep and structurally complex rocky shoals, with significant water motion and currents, are not typical of most Mediterranean rocky reefs [53]. The fact that, despite these particular features, some spillover can be detected highlights that this process is possible even for highly sedentary fish species, once a minimum density has been attained. Moreover, spillover is likely to be favoured by the existence of habitat continuity from inside to outside the MPA [17,38,54]. In this work spillover has not been detected when exploring continuous habitats (*Posidonia oceanica* meadows), contrarily to what has been observed in discontinuous (rocky bottoms) habitats around Tabarca MPA. On the other hand Forcada et al. [54] working in the same location, observed decreasing total abundance gradients in both habitats studied, *P. oceanica* meadow and rocky substrate, and independently of their different continuity through the reserve boundaries. Although groupers can be observed in seagrass meadows, they primarily live in rocky bottoms, which could explain the fact that no groupers were observed in *P. oceanica* outside the reserve, and therefore the absence of spillover in continuous habitats.

Further from oceanographic and hydrological factors (see above), low recruitment level of groupers in northernmost locations could be explained in part by the specific habitat features existing in each MPA, given that spatial variability in reef fish recruitment at the scale of locations (thousands of metres apart) as well as sites (separated by hundreds of metres) are likely a reflection of habitat differences at these spatial scales [55]. Juvenile groupers show a preference for cavities and recesses in shallow areas from near-surface bottoms to 10 m depth [56]. Therefore, the capacity of marine reserves to protect early fish life stages will rely on the extent to which these crucial habitat characteristics are included within the protected areas [57].

These results provide novel insights into the debate about the consequences of establishing a buffer zone when designing an MPA. Some studies argue that BZ's can have detrimental effects on the protection of fish species [3], and that only no-take marine reserves should be created, as partial protection is an ineffective conservation strategy [58]. The present study found that the success of BZ to protect grouper species will depend on both the ecological and management conditions established in each particular MPA. The ecological role of a partially closed area to protect commercially important fish species would depend on a variety of aspects, such as the difference in habitat quality between both zones, the heterogeneity and spatial continuity of structural habitat between both areas of different protection levels and with unprotected ones, the intensity of fishing pressure allowed in the buffer zone compared to neighbouring unprotected areas, the relative size of both no-take and buffer zones, and the mobility and home range of the species involved [3–4]. On the other hand, it has been demonstrated that partial protection of coastal areas together with an adaptive co-management plan that involves fishers, scientists, and managers may benefit fishing communities and reduce overfishing [59].

In summary, the evidence presented here shows that MPAs are an essential tool to protect overexploited populations and threatened species, such as Mediterranean groupers, provided that they are adequately enforced and managed [10,58,60]. Moreover, MPAs are able to export fish biomass to neighbouring areas, even in the case of very sedentary species, but only if they are appropriately designed in terms of reserve location, size, zoning, and management. Therefore, spillover by eggs release and larval dispersion could be the key process to ensure the connectivity among distant sites for sedentary and strongly habitat-linked species such as groupers. Further studies on grouper population mobility, connectivity, habitat preferences, demography and carrying capacity are urgently needed in order to establish ecological criteria to optimize MPA design.

Acknowledgments

We thank the support of the coordinator of the BIOMEX project, Serge Planes, and all researchers involved in project activities, as well as the staffs of MPAs investigated. Also we would like to express our gratitude to Fabiana C. Félix-Hackradt for comments and assistance on manuscript preparation and to Katie Hogg for the English review.

Author Contributions

Conceived and designed the experiments: APR MHV LLD JBS EC DO OR PSL CV JAGC. Performed the experiments: MHV LLD JBS EC DO OR PSL CV JAGC. Analyzed the data: CWH JAGC. Contributed reagents/materials/analysis tools: CWH APR. Wrote the paper: CWH.

References

1. Lotze HK, Lenihan HS, Bourque BJ, Bradbury RH, Cooke RG, et al. (2006) Depletion, degradation, and recovery potential of estuaries and coastal seas. Science 312(5781): 1806–1809.
2. Worm B, Hilborn R, Baum JK, Branch TA, Collie JS, et al. (2009) Rebuilding Global Fisheries. Science 325: 578–585 DOI:10.1126/science.1173146.
3. Claudet J, Osenberg CW, Benedetti-Cecchi L, Domenici P, García-Charton JA, et al. (2008) Marine reserves: size and age do matter. Ecology Letters 11: 481–489.
4. Claudet J, Osenberg CW, Domenici P, Badalamenti F, Milazzo M, et al. (2010) Marine reserves: Fish life history and ecological traits matter. Ecological Applications 20(3): 830–839.
5. García-Charton JA, Pérez-Ruzafa A, Marcos C, Claudet J, Badalamenti F, et al. (2008) Effectiveness of European Atlanto-Mediterranean MPAs: Do they accomplish the expected effects on populations, communities and ecosystems? Journal for Nature Conservation 16: 193–221.
6. Pérez-Ruzafa A, Martín E, Marcos C, Zamarro JM, Stobart B, et al. (2008) Modeling spatial and temporal scales for spillover and biomass exportation from MPAs and their potential for fisheries enhancement. Journal of Nature Conservation 16: 234–255.
7. Lester SE, Halpern B, Grorud-Colvert K, Lubchenco J, Ruttenberg BI, et al. (2009) Biological effects within no-take marine reserves: a global synthesis. Marine Ecology Progress Series 384: 33–46.
8. Fenberg PB, Caselle JE, Claudet J, Clemence M, Gaines SD, et al. (2012) The science of European marine reserves: Status, efficacy, and future needs. Marine Policy 36(5): 1012–1021.
9. Samoilys MA, Martin-Smith KM, Giles BG, Cabrera B, Anticamara JA, et al. (2007) Effectiveness of five small Philippines' coral reef reserves for fish populations depends on site-specific factors, particularly enforcement history. Biological Conservation 136: 584–601.
10. Guidetti P, Milazzo M, Bussotti S, Molinari A, Murenu M, et al. (2008) Italian marine reserves effectiveness Does enforcement matter? Biological Conservation 141: 699–709.
11. Crec'hriou R, Alemany F, Roussel E, Chassanite A, Marinaro JY, et al. (2010) Fisheries replenishment of early life taxa: potential export of fish eggs and larvae from a temperate marine protected area. Fisheries Oceanography 19: 135–150.
12. Pelc RA, Warner RR, Gaines SD, Paris CD (2010) Detecting larval export from marine reserves. Proceedings of the National Academy of Sciences 107: 18266–18271.
13. Rowley RJ (1994) Case studies and reviews: marine reserves in fisheries management. Aquatic Conservation: Marine and Freshwater Ecosystems 4: 233–254.
14. Chapman MR, Kramer DL (1999) Gradients of coral reef fish density and size across the Barbados Marine Reserve boundary: effects of reserve protection and habitat characteristics. Marine Ecology Progress Series 181: 81–96.
15. Guidetti P (2006) Marine reserves reestablish lost predatory interactions and cause community changes in rocky reefs. Ecological Applications 16(3): 963–976.
16. Harmelin-Vivien M, Direach L, Bale-Sempere J. Charbonnel E, García-Charton JA, et al. (2008) Gradients of abundance and biomass in six Mediterranean marine protected areas: Evidence of fish spillover? Biological Conservation 141: 1829–1839.
17. Goñi R, Adlerstein S, Alvarez-Berastegui, Forcada A, Reñones O, et al. (2008) Spillover from six western Mediterranean marine protected areas: evidence from artisanal fisheries. Marine Ecology Progress Series 366: 159–174.
18. Stobart B, Warwick R, González C, Mallol S, Díaz D, et al. (2009) Long-term and spillover effects of a marine protected area on an exploited fish community. Marine Ecology Progress Series 384, 47–60.
19. Russ GR, Alcala AC, Maypa AP (2003) Spillover from marine reserves: the case of *Naso vlaminii* at Apo Island, the Philippines. Marine Ecology Progress Series 264, 15–20.
20. Amargós FP, Sansón GG, Castillo AJ, Fernández AZ, Blanco FM, et al. (2010) An experiment of fish spillover from a marine reserve in Cuba. Environmental Biology of Fishes 87: 363–372.
21. Kramer DL, Chapman MR (1999) Implications of fish home range size and relocation for marine reserve function. Environmental Biology of Fishes 55: 65–79.
22. García-Charton JA, Perez-Ruzafa A (2001) Spatial pattern and the habitat structure of a Mediterranean rocky reef fish local assemblage. Marine Biology 138: 917–934.
23. Hackradt CW, Félix-Hackradt FC, García-Charton JA (2011) Influence of habitat structure on fish assemblage of an artificial reef in southern Brazil. Marine Environmental Research 72: 235–247.
24. García-Charton JA, Pérez-Ruzafa A, Sánchez-Jerez P, Bayle-Sempere JT, Reñones O, et al. (2004) Multi-scale spatial heterogeneity, habitat structure, and the effect of marine reserves on Western Mediterranean rocky reef fish assemblages. Marine Biology 144: 161–18882.
25. McClanahan TR, Graham NAJ, Calnan JM, MacNeil MA (2007) Toward pristine biomass: Reef fish recovery in coral reef marine protected areas in Kenya. Ecological Applications 17: 1055–1067.
26. Sandin SA, Smith JE, De Martini EE, Dinsdale EA, Donner SD, et al. (2008) Baselines and degradation of coral reefs in the Northern Line Islands. PLoS ONE 3(2), e1548.
27. Stallings CD (2009) Fishery-independent data reveal negative effect of human population density on Caribbean predatory fish communities. PLoS ONE 4(5): e5333.
28. Chiappone M, Sluka R, Sealey KS (2000) Groupers (Pisces: Serranidae) in fished and protected areas of the Florida Keys, Bahamas and northern Caribbean. Marine Ecology Progress Series 198: 261–272.
29. Unsworth RKF, Powell A, Hukom F, Smith DJ (2007) The ecology of Indo-Pacific grouper (Serranidae) species and the effect of small scale no-take area on grouper assemblage, abundance and size frequency distribution. Marine Biology 152; 243–254.
30. Coll J, Linde M, García-Rubies A, Riera F, Grau AM (2004) Spear fishing in the Balearic Islands (west central Mediterranean): species affected and catch evolution during the period 1975–2001. Fisheries Research 70: 97–111.
31. Sadovy de Mitcheson Y, Craig MT, Bertoncini AA, Karpenter KE, Cheung WWL, et al. (2012) Fishing groupers towards extinction: a global assessment of threats and extinction risks in a billion dollar fishery. Fish and Fisheries 14: 119–136.
32. Anderson MJ (2001) A new method for non-parametric multivariate analysis of variance. Austral Ecology 26: 32–46.
33. Bray JR, Curtis JT (1957) An ordination of the upland forest communities of southern Wisconsin. Ecological Monographs 27: 325–349.
34. Hastie T, Tibshirani R (1990) Generalized additive models. 1st edition. London: Chapman and Hall. 335p.
35. Venables WN, Ripley BD (2000) Modern Applied Statistics with S. New York: Springer-Verlag. 495p.
36. McCullagh P, Nelder JA (1989) Generalized Linear Models. New York: Chapman and Hall. 532p.

37. García-Charton JA, Pérez-Ruzafa A (1998) Correlation between habitat structure and a rocky reef fish assemblage in the Southwest Mediterranean. P.S.Z.N.: Marine Ecology 19: 111–128.
38. Stenzenmüller V, Maynou F, Bernard G, Cadiou G, Camilleri M, et al. (2008) Spatial assessment of fishing effort around European marine reserves: Implications for successful fisheries management. Marine Pollution Bulletin 56: 2018–2026.
39. Harmelin JG, Bachet F, Garcia F (1995) Mediterranean marine reserves fish indices as test of protection efficiency. P.S.N.Z. Marine Ecology 16: 233–250.
40. Reñones O, Goñi R, Pozo M, Deudero S, Moranta J (1999) Effects of protection on the demographic structure and abundance of *Epinephelus marginatus* (Lowe, 1834). Evidence from Cabrera Archipelago National Park (West-central Mediterranean). Marine Life 9: 45–53.
41. Lenfant P, Louisy P, Licari ML (2003) Recensement des mérous bruns (*Epinephelus marginatus*) de la réserve naturelle de Cérbere-Banyuls (France, Méditerranée) effectué en septiembre 2001, aprés 17 années de protection. Cybium 27: 27–36.
42. Russ G, Alcala A (2010) Decadal-scale rebuilding of predator biomass in Philippine marine reserves. Oecologia 163(4): 1103–1106.
43. Myers RA, Worm B (2003) Rapid worldwide depletion of predatory fish communities. Nature 423: 280–283.
44. Jackson JBC (2008) Ecological extinction and evolution in the brave new ocean. Proceedings of the National Academy of Science 105: 11458–11465.
45. Zabala M, Garcia-Rubes A, Loisy P, Sala E (1997) Spawning behavior of the Mediterranean dusky grouper, *Epinephelus marginatus* (Lowe, 1834) (Pisces, Serranidae) in the Medes Islands Marine Reserve (NW Mediterranean, Spain). Scientia Marina 61: 65–77.
46. Hereu B, Diaz D, Pasqual J, Zabala M, Sala E (2006) Temporal predictability of spawning of the dusky grouper, *Epinephelus marginatus*, in relation to environmental factors. Marine Ecology Progress Series 325: 187–194.
47. Reñones O, Grau A, Mas X, Riera F, Saborido-Rey F (2010) Reproductive pattern of an exploited dusky grouper *Epinephelus marginatus* (Lowe 1834) (Pisces: Serranidae) population in the western Mediterranean. Scientia Marina 74: 523–537.
48. Pastor J, Verdoit-Jarraya M, Astruch P, Dalias N, Pasqual JSN, et al. (2009) Acoustic telemetry survey of the dusky grouper (*Epinephelus marginatus*) in the Marine Reserve of Cerbére-Banyuls: information on the territoriality of this emblematic species. Comptes Rendus Biologies 332: 732–740.
49. Sánchez-Lizaso JL, Goñi R, Reñones O, García-Chartón JA, Galzin R, et al. (2000) Density dependence in marine protected populations: A review. Environmental Conservation 27: 144–158.
50. Bodilis P, Ganteaume A, Francour P (2003) Presence of 1 year-old dusky groupers along the French Mediterranean coast. Journal of Fish Biology 62: 242–246.
51. Lloret J, Zaragoza N, Caballero D, Riera V (2008) Biological and socioeconomic implications of recreational boat fishing for the management of fishery resources in the marine reserve of Cap de Creus (NW Mediterranean). Fisheries Research 91: 252–259.
52. Lembo G, Spedicato MT, Okland F, Carbonara P, Fleming IA, et al. (2002) A wireless communication system for determining site fidelity of juvenile dusky groupers *Epinephelus marginatus* (Lowe, 1834) using coded acoustic transmitters. Hidrobiologia 483: 249–257.
53. Sala E, Ballesteros E, Dendrinos P, Di Franco A, Ferretti F, et al. (2012) The structure of Mediterranean rocky reef ecosystems across environmental and human gradients, and conservation implications. PLOS ONE 7(2): e32742.
54. Forcada A, Valle C, Bonhomme P, Criquet G, Cadiou G, et al. (2009) Effects of habitat on spillover from marine protected areas to artisanal fisheries. Marine Ecology Progress Series 379, 197–211.
55. Félix-Hackradt FC, Hackradt CW, Treviño-Otón J, Pérez-Ruzafa Á, García-Charton JA (2014) Habitat use and ontogenetic shifts of fish life stages at rocky reefs in South-western Mediterranean Sea. Journal of Sea Research 88, 67–77.
56. La Mesa G, Louisy P, Vacchi M (2002) Assessment of microhabitat preferences in juvenile dusky grouper (*Epinephelus marginatus*) by visual sampling. Marine Biology 140, 175–185.
57. Planes S, Galzin R, García-Rubies A, Goñi R, Harmelin JG, et al. (2000) Effects of marine protected areas on recruitment processes with special reference to Mediterranean littoral ecosystems. Environmental Conservation 27, 126–143.
58. Di Franco A, Bussotti S, Navone A, Panzalis P, Guidetti P (2009). Evaluating effects of total and partial restrictions to fishing on Mediterranean rocky-reef fish assemblages. Marine Ecology Progress Series 387, 275–285.
59. Guidetti P, Claudet J (2010) Co-management practices enhance fisheries in marine protected areas. Conservation Biology 24(1): 312–318.
60. Pauly D, Christensen V, Guénette S, Pitcher TJ, Sumaila UR, et al. (2002) Towards sustainability in world fisheries. Nature 418: 689–695.

2

Towards A Network of Locally Managed Marine Areas (LMMAs) in the Western Indian Ocean

Steve Rocliffe[1]*, Shawn Peabody[2], Melita Samoilys[3], Julie P. Hawkins[1]

1 Environment Department, University of York, York, United Kingdom, **2** Blue Ventures Conservation, London, United Kingdom, **3** CORDIO East Africa, Mombasa, Kenya

Abstract

In the Western Indian Ocean (WIO), local communities are increasingly assuming responsibility for inshore marine resources either on their own or through collaborative management arrangements with governments or non-state actors. In this paper, we trace the evolution and expansion of community management in the WIO and present the first ever inventory and assessment of the region's locally managed marine areas (LMMAs). We compare the key attributes of these areas to those under government stewardship and assess their relative contributions to progress towards the Convention on Biodiversity (CBD) target of 10% of marine and coastal ecological regions to be effectively conserved by 2020. We also explore the legal frameworks that underpin locally managed marine initiatives in Kenya, Madagascar, Mozambique and Tanzania to assess the potential for future expansion. A principal finding is that whilst LMMAs protect more than 11,000 square kilometres of marine resource in the WIO, they are hampered by underdeveloped local and national legal structures and enforcement mechanisms. In our recommendations to improve local management, we suggest establishing a network of LMMA practitioners in the WIO region to share experiences and best practice.

Editor: Gretchen E. Hofmann, University of California- Santa Barbara, United States of America

Funding: SR was supported by the Natural Environment Research Council and the Economic and Social Research Council. SP was supported by the John D. and Catherine T. MacArthur Foundation as part of a grant to support information sharing and networking of LMMAs in Madagascar. The funders had no role in study design, data collection and analysis, decision to publish, or preparation of the manuscript.

Competing Interests: The authors have declared that no competing interests exist.

* Email: sr588@york.ac.uk

Introduction

Despite their value to humans, marine ecosystems worldwide are threatened by a range of anthropogenic pressures, including pollution, habitat loss, climate change and overfishing [1–3]. These impacts have drained populations of culturally and economically important fish stocks and reduced structural complexity of various marine communities across a rich range of habitats, species and trophic levels [3–5].

In the Western Indian Ocean (WIO) as throughout the world, Marine Protected Areas (MPAs) have been a primary management approach in attempts to alleviate anthropogenic pressures [6]. An MPA is defined by IUCN as: "A clearly defined geographical space, recognised, dedicated and managed, through legal or other effective means, to achieve the long-term conservation of nature with associated ecosystem services and cultural values." [7]. Solid evidence from MPAs, particularly for No-take Zones (MPAs that allow no extraction), shows that protection can increase average size, diversity, abundance and biomass of species [3,8,9] and that some of this biomass can be exported beyond protected boundaries [10–12]. MPAs can also play an important role in climate change adaptation, enhancing ecosystem resilience and protecting vital ecosystem services [13,14].

In 2002, international leaders at the World Summit on Sustainable Development set the first target for the establishment of a global system of MPAs [15]. This target was formally quantified four years later, when the parties to the Convention on Biological Diversity (CBD) committed to effectively conserving 10% of each of the world's ecological regions by 2012 [16]. In 2010, the parties pushed back the deadline to 2020 and adopted Aichi Biodiversity Target 11, with a revised goal of conserving "at least 17 per cent of terrestrial and inland water, and 10 per cent of coastal and marine areas, through effectively and equitably managed, ecologically representative and well connected systems of protected areas and other effective area-based conservation measures" [17].

Initial progress towards these targets was slow: based on the rate of MPA expansion to 2008, Wood estimated that the 10 percent figure would not be achieved until 2047 [18]. In contrast, the most recent analysis [19], paints a more optimistic picture. MPA coverage has increased dramatically, quadrupling between 2002 and 2012 (*ibid.*). MPAs now cover 8.3 million km^2, 2.3% of the global ocean area and 7.9% of the continental shelf and equivalent areas (i.e. less than 200 m deep) (*ibid.*). So pronounced is the increase that the 10 percent Aichi CBD target could be reached, even before 2020 (*ibid.*). However, a few very large MPAs are largely responsible for this apparent reversal of fortunes, a trend that looks set to continue as new super-sized protected areas come online in Cook Islands and New Caledonian waters [19,20]. Further, most of these MPAs are located in uninhabited or low-population-density areas [19] and/or in developing countries where enforcement is weak to non-existent [21].

Although their popularity continues to increase, marine protected areas often fall short of their original goals and sometimes fail entirely, though published negative evaluations are rare. Inadequate long-term funding and widespread management failure have resulted in unenforceable and ineffectual "paper parks" [22]. The most recent global evaluations suggest that less than 16% of MPA managers feel they have adequate funding for effective conservation [23] and that just 15% of coral reef MPAs are effectively managed [24]. Regional evaluations have reached similar conclusions. In a recent review of marine conservation successes in the WIO, for example, Samoilys & Obura [25] only mention one example of successful government-established MPAs: those of Kenya.

Locally managed marine areas

As a result of disappointment with top-down, centralised government interventions, and facilitated by increasing recognition of the relative strength of local institutions [26–28] local communities throughout the Indo-Pacific are increasingly assuming responsibility for inshore marine resources through collaborative partnerships with governments and/or non-state actors [29–32]. In the Pacific, areas where marine resources are at least in part under community control are usually termed "locally managed marine areas" (LMMAs) [33]. In the WIO, the terms used vary more widely, encompassing not only LMMA, but also Collaborative Fisheries Management Area and Community Conservation Area, as well as local names such as *tengefu*, *hifadhi za kijamii* and *vilindo vya wenyeji* in Kenya [25,34]. Although the level of community involvement and the overall management model is context-specific to an extent, a key aspect is local control. Technical support may be provided by government agencies, private sector stakeholders or non-governmental organisations, but it is the resource users themselves who make most of the management decisions, including the location of any protected areas [35–37].

Despite its relatively recent popularity, the approach of managing and conserving marine resources at the local level is actually centuries old [28,38,39]. In many tropical nations, especially in Pacific Island countries, informal systems of community marine management were in place prior to colonialism [38,39]. Further, the tradition of customary marine tenure (CMT) – the right to control access to local fishing grounds – in Pacific Island countries provided an ideal socio-cultural platform on which modern day LMMAs could evolve [40]. However, although there are sacred coastal sites that are protected for spiritual reasons in places like Kenya and Tanzania [41,42] as well as several taboos around fishing [43], there is no tradition of CMT and this may partly explain why the establishment of LMMAs is a more recent phenomenon in the region.

In the Pacific, more than 500 communities in 15 countries manage 12,000 km^2 of coastal resources, 1,000 km^2 of which constitutes full no-take protection [33]. In the Western Indian Ocean, little is known about the status or extent of LMMAs. In this paper, we present the first inventory of LMMAs in the WIO and assess them in terms of geography, numbers, size and governance structures. We compare the key attributes of these areas to those under government stewardship and evaluate potential contributions to international biodiversity commitments. To determine prospects for future LMMA expansion, we also explore the legal frameworks that underpin locally managed marine initiatives in Madagascar, Kenya, Mozambique and Tanzania. Finally, we make recommendations for improving local marine management, including the establishment of a regional network of practitioners to facilitate the sharing of experiences and best practice.

Methods

Locally managed marine areas: a definition

Following Govan et al [44], we define a locally managed marine area as *"An area of nearshore waters and coastal resources that is largely or wholly managed at a local level by the coastal communities, land-owning groups, partner organizations, and/or collaborative government representatives who reside or are based in the immediate area."* Under this definition, LMMAs are managed for sustainable use rather than for conservation per se [24]. Many LMMAs employ a combination of management techniques, including periodic closures, gear restrictions, species specific reserves and permanent fully protected (closed) no-take zones [45,46].

This wide variety of approaches and focus on sustainable use has lead some to question whether LMMAs should qualify as protected areas and thereby count towards international biodiversity targets [47]. For the IUCN, for example, "only those areas where the main objective is conserving nature can be considered protected areas; this can include many areas with other goals as well, at the same level, but in the case of conflict, nature conservation will be the priority" [7].

Others [44,48], notably the CBD, are perhaps mindful of the informal nature of many LMMAs and less concerned about the need for an overarching conservation objective. By assuming that LMMAs can help WIO nations to meet their CBD obligations, this second approach is the one we adopt here.

Based on categories suggested by Sen and Raakjaer Nielsen [49] we developed a typology that classifies sites in the WIO along a four-point spectrum according to the extent to which resource management is shared between government and user groups.

Level 1: *Central*:. Governments or non-state actors designate and manage the area. No mechanisms exist for dialogue with users and decisions are taken by resource managers.

Level 2: *Consultative*:. Governments or partner organisations designate and manage the area. Whilst mechanisms exist for dialogue with users, in practice, most decisions are taken by resource managers.

Level 3: *Cooperative*:. Local communities and governments or non-state actors cooperate together as equal partners in decision making.

Level 4: *Local*:. In this type of arrangement, government has delegated management authority to local communities. The remit of government or partner organisations is largely restricted to providing advice and endorsing management decisions made by local communities.

For this paper, we classified level 3 and 4 sites as LMMAs and level 1 and 2 sites as MPAs. For the sake of clarity, we refer to the four levels collectively as Marine Managed Areas (herein MMAs) [44].

Study Site

The Western Indian Ocean region refers to the African coastal states of Somalia, Kenya, Tanzania, Mozambique and South Africa, together with the Indian Ocean island states of Comoros, Madagascar, Mauritius and Seychelles, as well as the two French overseas departments Mayotte and Réunion (Figure 1). Basic geographic and socio-economic information for the region is summarised in Table 1.

The mainland WIO area stretches for 13,000 km along the coast from Somalia in the north to South Africa in the south. The island states consist of more than 400 islets and islands with a combined coastline of 6,360 km. The region is ecologically and socio-economically diverse. Overall species composition is rich,

Figure 1. LMMAS and MPAs in the Western Indian Ocean.

exceeding 11,000 species of marine flora and fauna, 60–70% of which are endemic to the Indo-Pacific ocean [50,51]. There are at least 369 species of coral, 10 mangrove and 12 seagrass, 2,200 coastal fishes, 3,000 molluscs, 450 crabs, 400 echinoderms and five of the world's seven marine turtle species [50–52].

Table 1. Geographic scope, population and socio-economic characteristics of Western Indian Ocean (WIO) nations.

Country	Area (km^2)[1]	Coastline (km)	Pop. (m) 2010	% Coastal pop. 2000[2]	Pop. density km^2	GDP 2010 (US$bn)	GNI per cap. 2010 (US$)[3]	HDI* 2011
Comoros	1,860	340	0.73	100	395	0.54	750	0.433
Kenya	569,140	536	40.51	7.7	71.2	31.41	790	0.509
Madagascar	581,540	4,828	20.71	50.5	35.6	8.72	430	0.48
Mayotte[4]	374	-	0.19	100	-	-	-	-
Mauritius	2,030	372	1.28	100	631.1	9.73	7,750	0.728
Mozambique	786,380	2,470	23.39	55.9	29.7	9.59	440	0.322
Réunion[5]	2,513	-	0.83	100	331.7	-	-	-
Seychelles	460	491	0.09	100	188.1	0.94	9,760	0.773
Somalia	627,340	3,025	9.33	62.7	14.9	-	-	-
South Africa[6]	94,361	-	10.27	37.3	108.8	57.46	6,090	0.619
Tanzania	885,800	1,424	44.84	19.8	50.6	23.06	530	0.466

Source: The World Bank [120], except for coastal population and HDI, which are provided respectively by CIESIN [121] and UNDP [122].
Note: South Africa only includes KwaZulu Natal as the Province that borders the WIO.
*HDI = Human development index (http://hdr.undp.org/en/statistics/indices/hpi/).
[1]Excludes area under inland water bodies, national claims to continental shelf, and exclusive economic zones.
[2]Percentage of population living within 100 km of a coastline.
[3]Gross national income, calculated using World Bank Atlas Method.
[4]Figures for Mayotte are calculated from INSEE [123]. Population estimate from 2009.
[5]Figures for Réunion are calculated from INSEE [124].
[6]South Africa only includes KwaZulu Natal, the Province that borders the WIO. Area and population figures from Statistics South Africa [125]; GDP figures calculated from Statistics South Africa [126].

Table 2. Mean area of MPAs in the Western Indian Ocean and numbers existing by decadal creation date.

Year created	Number of MPAs created	Average size (km^2)
1965–1974	7	49.1
1975–1984	16	36.5
1985–1994	10	25.3
1995–2004	26	292.4
2005–2014	15	8920.8

The WIO region has a population of around 152 million people, of which approximately 48.3 million (31.8%) live within 100 km of the coast (Table 1). Population density is diverse, ranging from 15 people per square kilometre in Somalia to 395 in Comoros and 631 in Mauritius. Economically, excluding the French territories (Mayotte, Réunion), GNI per capita totals in the island nations are significantly higher than on the mainland, with figures from the Seychelles more than 12 times those of Kenya and more than 18 times those of Tanzania, two of the largest economies on the mainland. The higher level of socio-economic development in Mauritius and the Seychelles is also underscored by human development index scores, with Seychelles ranked 52 in the world and Mauritius 77–91 places higher than Kenya, which is ranked third in the WIO region.

Data Compilation

To gather data on level 1 and 2 sites, we synthesised a list of MPAs in the eleven territories under consideration from the academic literature, government agencies, non-government and intergovernmental organisations' reports, and from the World Database on Protected Areas. We captured information on IUCN category, size and age.

To calculate threat levels to marine resources and MPA effectiveness, we undertook a region-specific re-analysis of global spatial data from Burke et al. [24]'s *Reefs at Risk Revisited,* thus confining ourselves to coral reef habitats. We extracted data on reef extent, reef threats, MPA extent and MPA effectiveness for each nation (one province for South Africa) in the WIO. The MPA effectiveness data for the Western Indian Ocean in Burke et al. [24] cover 59 coral reef MPAs, 66% of the MPAs we documented in this analysis. Because effectiveness was determined through a rapid review using scores from regional experts rather than from field practitioners, there may be a sampling bias toward better-known sites, with a potentially higher proportion of ecologically effective sites than would be found overall (*ibid.*). Where one or more of the authors of this study had in-depth and more recent experience of one of the sites, these ratings were updated where necessary ($n = 6$) to give a more accurate picture. All spatial modelling was performed with the ArcGIS™ 10.1 Geographic Information System software, and the ArcGIS™ Spatial Analyst extension.

To assess contributions towards international biodiversity commitments, we measured the progress of each country, except France, towards achieving the Convention on Biological Diversity's (CBD) target of 10% of marine and coastal ecological regions to be effectively conserved by 2020 [17]. Progress was assessed by calculating the percentage coverage by MPAs and LMMAs of the continental shelf to 200 m depth after Wells et al. (2007) [53].

The list of level 3 and 4 sites, the LMMAs, was based on information gathered from LMMA workshops at the Seventh WIOMSA (Western Indian Ocean Marine Science Association) Scientific Symposium, held in Mombasa, Kenya in October 2011 and the Madagascar LMMA Forum, held in June 2012 in Madagascar. The initial workshop outputs were supplemented by an extensive electronic search for published literature using several electronic databases (including Web of Science, ScienceDirect, EconLit, WorldFish Library Catalog and CAB Direct). To capture potential LMMA sites documented in grey literature, we used Google.com and Google Scholar and examined the first 100 hits from each of our searches. Criteria for inclusion of an LMMA in the final list were that its area under management had been formalised through some form of legislation, usually a by-law. We combined the outputs of the workshop and literature searches to create our inventory of a total of 136 MMAs in the Western Indian Ocean and liaised with key individuals and officials to determine or verify locations, sizes and governance structures of these areas.

Results and Discussion

MPAs in the Western Indian Ocean

Formally recognised MPAs have been around in the WIO since 1965, when the Ilhas da Inhaca e dos Portugueses Faunal Reserve (now part of the Ponta do Ouro Partial Marine Reserve) was gazetted in Mozambique [54]. Both Madagascar and Kenya followed suit over the next three years, establishing Nosy Tanikely, the Malindi and Watumu Marine National Parks and the Malindi-Watumu Marine National Reserve [54]. Today, all WIO countries have gazetted MPAs, except for Somalia, where the lack of a central administration has made it very difficult to practice conservation [55,56].

The 74 MPAs identified have a total coverage of 133,273 km^2. Early protected areas tended to be smaller than their contemporary counterparts and were often designed to protect a specific habitat such as a turtle-nesting beach [53]. Indeed, the smallest MPA in the WIO, the 0.01 km^2 Cousin Island Special Reserve in the Seychelles dates from 1975. Table 2 shows how the average size of protected areas in the region increased by a factor of six from 49.1 km^2 between 1965–1974 to 292.4 km^2 between 1995–2004 as the emphasis shifted to larger, zoned multiple-use sites such as the Quirimbas and Bazaruto Archipelago National Parks in Mozambique. The two largest MPAs in the region are also among the newest: the Marine Park of the Glorieuses, designated in February 2012, and the neighbouring Marine Park of Mayotte, designated two years previously (Personal communication, Pascale Chabanet). The combined area of these new parks alone is more than 100,000 km^2, constituting more than 84% of the total MPA coverage in the WIO region. This trend towards larger marine protected areas is continuing, with the newly designated Primeiras and Segundas MPA in Mozambique covering over 10,000 km^2 [57]. Although declarations of large MPAs represent a step

Table 3. Extent of MPA coverage and effectiveness for coral reef MPAs in the Western Indian Ocean.

Region/Country	Continental shelf (km^2)[1]	MPAs (km^2) in shelf area	MPA coverage (%)[2]	Effective MPAs (%)[3]	Reef Area (km^2)	Reef area as % of global	Overfished reefs (%)[4]
Global	24,285,959.0			15.0	249,713.0	100.0	55.0
Indian Ocean				29.0	31,543.0	13.0	60.0
Comoros	1,415.7	404.0	28.5	0.0	399.0	0.2	100.0
Kenya	8,460.1	835.3	9.9	55.6	620.0	0.2	93.8
Madagascar	96,652.8	2603.3	2.7	100.0	3,934.0	1.6	86.9
Mauritius	27,372.8	139.2	0.5	28.6	976.3	0.4	62.6
Mayotte	1,100.0	1,100.0	100.0	0.0	643.8	0.3	78.1
Mozambique	73,299.7	14,551.3	19.9	0.0	2,435.5	1.0	78.5
Réunion	965.1	35.0	3.6	0.0	27.5	0.0	100.0
Seychelles	31,478.7	201.8	0.6	30.0	1,904.3	0.8	9.6
Somalia	40,391.7	0.0	0.0	NA	546.5	0.2	100.0
South Africa[6]	16,093.8	960.9	6.0	50.0	5.0	0.0	100.0
Tanzania[7]	8,951.6	1,160.8	13.0	0.0	2,089.0	0.8	94.7
Tanzania –Zanzibar	8,951.6	1,000.5	11.2	25.0	922.0	0.4	99.7
WIO*	**315,133.6**	**22,185.8**	**7.0**	**29.6**	**14,502.8**	**5.8**	**76.4**

[1]Source: World Resources Institute [127].
[2]Percentage of continental shelf within marine protected areas.
[3]Percentage of coral reef MPAs judged to be effective in a rapid appraisal by regional experts. Developed from Burke et al. [24]. Both the Burke analysis and this one use the same definition of an MPA.
[4]Percentage of reefs at medium or high risk from overfishing and destructive fishing. Developed from Burke et al. [24].
[5]At 68,332 km^2, Mayotte Marine Park is much larger than its continental shelf area. Until 31 March 2011, Mayotte was not an overseas department of France and separate figures for continental shelf were not attainable. The figure used as a conservative proxy is the size of its lagoon [123].
[6]South Africa only includes KwaZulu Natal, the Province that borders the WIO. As continental shelf values for the province weren't available, a conservative proxy of 10% of South Africa's total shelf area was used.
[7]Figures for continental shelf of the mainland and Zanzibar were unavailable, so an arbitrary split of 50% of Tanzania's total shelf was used.
*WIO values are totals.

Table 4. Extent of MPA (level 1 and 2) and LMMA (level 3 and 4) coverage and current progress towards achieving international biodiversity conservation targets.

Country*	No. of MPAs	MPA coverage (%)[1]	No of LMMAs	LMMA av. Size (km^2)	LMMA coverage (%)[2]	LMMA + MPA coverage (%)[3]
Comoros	1	28.5	0	NA	0.0	28.5
Kenya	9	9.9	14	7.83	1.3	11.2
Madagascar	8	2.7	34	195.16	6.9	9.6
Mauritius	13	0.5	0	NA	0.0	0.5
Mayotte	1	100.0	0		0.0	100.0
Mozambique	6	19.9	1	18	0.0	19.9
Réunion	1	3.6	0	NA	0.0	3.6
Seychelles	14	0.6	0	NA	0.0	0.6
Somalia	0	0.0	0	NA	0.0	0.0
South Africa[4]	4	6.0	0	NA	0.0	6.0
Tanzania	10	13.0	12	341.38	45.8	58.7
Tanzania -- Zanzibar	3	11.2	1	470	5.3	16.4
WIO**	**69**	**7.0**	**66**	**182.73**	**3.8**	**10.9**

[1]Percentage of continental shelf within marine protected areas (level 1 and 2).
[2]Percentage of continental shelf within locally managed marine areas (level 3 and 4).
[3]Percentage of continental shelf within marine protected areas and locally managed marine areas.
[4]South Africa only includes KwaZulu Natal, the Province that borders the WIO.
*Excludes Îles Éparses.
**All are regional totals, except LMMA av. Size (km^2) which is a mean.

forward in marine biodiversity conservation, knowledge of the effectiveness of these areas is needed to properly assess achievements in addressing the Aichi CBD targets.

Effectiveness of protection of MPAs

Even with the large increase in coverage afforded by the Mayotte and Glorieuses reserves, the 74 MPAs (outlined in Table S1) protect just 7.3% of the continental shelf in the region (Table 3).

Seventy six percent of reefs in the WIO are at risk from local threats, with half (49.7%) rated at high or very high risk. The problem is most acute in the mainland coasts of Somalia and Tanzania and the islands of Réunion and Comoros, where more than 90% of reefs are threatened. The largest single threat is overfishing, which affects 72% of coral reefs, particularly on the densely populated coastlines of southern Kenya and Tanzania (Table 3). Watershed-based pollution is a problem in places like Madagascar, where widespread deforestation has caused extensive erosion and siltation in coastal areas [24]. Dynamite fishing is also an issue, primarily on the mainland Tanzania coast where it has occurred for decades [25,58], but it also occurs in Mohéli Marine Park in Comoros (MS pers. obs. 2010).

Only 29.6% of reef-related MPAs assessed in the region were found to be effective. Whilst the effectiveness of MPAs within the WIO appears to range from 0–100% effective, it is important to recognise that artefacts occur at both extremes. For example, Madagascar's MPAs receive a 100% rating because only one of the seven MPAs was actually appraised, while Mayotte's score of 0% predates the establishment of the new marine park in 2010 (Personal communication, Pascale Chabanet).

LMMAs in the Western Indian Ocean

Four of the eleven nations under consideration have active LMMA projects (Table S2). Unlike the region's legislated MPAs, the WIO's LMMAs are newer endeavours. Of the 62 sites identified, 60 (96.8%) were established after the year 2000, in line with the passing of legislation to decentralise marine resources management in Kenya, Tanzania, Mozambique and Madagascar.

The 62 LMMAs for which we obtained reliable estimates of size amount to 11,329 km^2 in total. These varied across four orders of magnitude: the largest, at 1,966.7 km^2, is Madagascar's Ankivonjy; the smallest, at 0.118 km^2, Mkwakwani/Tradewinds in Kenya. Mean LMMA size was 183 km^2, with a quarter of sites smaller than 2.12 km^2 and with a median size of 20.75 km^2. However, these figures obscure important differences between countries, highlighted in Table 4. For example, Kenya's 14 nascent LMMAs protect a total of 109.6 km^2 of marine resource, 37 times less than the 4,096.5 km^2 under management in Tanzania and 61 times less than Madagascar's LMMA coverage of 6,635.3 km^2.

Combined coverage and progress towards international targets

In the Western Indian Ocean, MMAs cover a combined 34,321.4 km^2 of the continental shelf (10.9% –Table 4). Assuming that percentage of the shelf is an acceptable proxy for the Aichi Biodiversity Target 11 target of 10% of marine and coastal areas protected by 2020, then Comoros, Kenya, and Tanzania have already achieved the target (with respective figures of 28.5%, 11.2% and 37.6%). Mozambique (19.9%) has also achieved the target, primarily due to recent designation of Primeiras and Segundas MPA, whilst Madagascar is on course to do so, should the Barren Isles LMMA be established as scheduled in 2014. In total, LMMAs in the WIO cover 11,329.4 km^2, 3.6% of the region's continental shelf. The differences between LMMA and MPA coverage are particularly pronounced in mainland Tanzania and Madagascar, where LMMAs cover 3.5 and 2.6 times more area than MPAs, respectively.

Locally managed marine areas in the WIO: legal context

Locally Managed Marine Areas exist primarily in Kenya, Madagascar, Mozambique and Tanzania. To determine potential for future expansion, here we assess the co-management systems in these four countries by exploring legal context, quantifying successes and identifying barriers to replication (Table 5).

Kenya. Before the recent introduction of Beach Management Units (BMUs) under Legal Notice 402 of the Fisheries Act, marine resource management took a centralised, top-down approach in Kenya [59]. BMUs are a co-management tool for small-scale fisheries initially developed to improve inland fisheries on Lake Victoria [60]. A BMU is an association of fishers, fish traders/mongers, boat owners, fish processors and other fishery stakeholders centred on a coastal landing site and formally led by an executive committee of stakeholders [59,61]. With the support and permission of officials from the Department of Fisheries, these BMUs are able to devise and enforce by-laws to govern their fishery, allowing them to delineate its boundaries and for example, exclude non-registered fishers or boats from the area [59,62]. There are presently around 60 BMUs along the Kenyan coastline [61]. Many of these exist only in name, and are yet to formalise their areas of jurisdiction or develop their by-laws [61,62].

The 2007 BMU Regulations provide a legislative framework to establish Locally Managed Marine Areas in Kenya. Kenya's first LMMA dates from 2006, when the community of Kuruwitu on the central Kenyan coast established the Kuruwitu Community

Table 5. Key features of LMMA initiatives in the Western Indian Oceans.

Country	Formal LMMAs	LMMA success[1]	LMMA potential[2]	Key local-level institutions	Key enabling legislation	Local name for LMMAs
Comoros	No	-	Low-Medium	Village fishing associations	-	-
Kenya	Yes	Medium	High	Beach Management Units (BMUs)	Beach Management Unit Regulations 2007	Community Conservation Areas, *tengefu*, Local Marine Management Areas (also LMMAs)
Madagascar	Yes	High	High	Village and multi-village level fishing associations. Village councils (Fokontany), Communes	Gestion Locale Sécurisée (GELOSE), *dina,* Décret d'Application No 848-05	LMMA, Community Managed Protected Area
Mauritius	No	-	Low	-	-	-
Mayotte	No	-	Low	-	-	-
Mozambique	Yes	Low	Medium	Fishing Community Councils (CCPs – Conselho Comunitário de Pescas) and Co-management Committees (CCG – Comité de Co-Gestão)	2003 Regulation on Marine Fisheries	-
Réunion	No	-	Low	-	-	-
Seychelles	No	-	Medium-low	Praslin Fishers Association	-	-
Somalia	No	-	Low	-	-	-
South Africa	No	-	Medium-high	Local Subsistence Co-Management Committees	Policy for the Small Scale Fisheries Sector in South Africa	Small Scale Fishing Community Area *
Tanzania	Yes	High	High	Beach Management Units (BMUs)	2003 Fisheries Act and its principal Regulations of 2009	Collaborative Management areas, Collaborative Fisheries Management Areas (CFMAs)
Tanzania -- Zanzibar	Yes	Medium	Medium-low	Village Fisheries Committee (VFC), Village Conservation Committee (VCC)	Environmental Management for Sustainable Development Act 1996, The Marine Conservation Unit Regulations**	-

[1]Level of success in establishing LMMAs to date.
[2]Potential to establish more LMMAs in future.
*Forthcoming. Communities will be able to apply several control measures within this area, including quotas and gear restrictions, as well as closed seasons and areas.
**In draft, awaiting finalization.

Managed Conservation Area [63,64]. Through a local umbrella organisation, the Kuruwitu Community Welfare Association, residents designated a 0.29 km^2 no-take zone [62,63]. Since establishment, live hard coral cover within the LMMA has increased by an estimated 30%, whilst fish numbers have grown by 200% [62].

Predating the 2007 BMU Regulations, Kuruwitu lacked legislative support and depended on acceptance from nearby communities and support from the East African Wildlife Society [63,65]. Nonetheless, the reserve's success has attracted interest from other fishing communities along the Kenyan coast, and it is likely that Kuruwitu, along with community exchange visits to the Collaborative Management Areas in Tanga, northern Tanzania, helped to catalyse the development of the 2007 BMU regulations and the designation of further LMMAs [25,62,63]. At present, there are 14 operational LMMAs in Kenya, covering 110 km^2 (Table S2).

This rapid increase in LMMA numbers suggests that local communities perceive them to be beneficial, and has occurred in spite of a lengthy development process involving consultations, community surveys, mapping and management plan creation [45]. However, several issues remain, including insufficient capacity for effective monitoring, control and surveillance in local communities, lack of funding and alternative livelihoods, conflicts of interest between stakeholders, legislative overlap and conflicting mandates, and a poor understanding among local communities of the legal procedures involved in designating an LMMA [34,63,66]. A further challenge relates to lack of land ownership. In Kuruwitu, for example, the Association hopes to establish a community-run eco-lodge to accommodate visiting tourists, but has so far found it impossible to obtain rights to coastal land on which to construct it [62,63].

Recent reforms to land policy offer some hope [66]. If these can be combined with guidelines to standardise LMMA establishment, as well as with education and awareness programmes at the local level, then Kenya will be well-placed to lead the LMMA revolution along the coasts of mainland East Africa [63].

Tanzania – mainland. In mainland Tanzania, co-management of marine resources dates back to the mid 1990s, when a collaborative approach was initiated in the coastal waters of the Tanga region [67]. The Tanga Coastal Zone Conservation and Development Programme operated with donor funding from 1994 until 2005 and has continued since as the Tanga Coastal Zone Resources Center, a District and Regional government initiative [68]. Under this Programme six collaborative management areas were established between 1997 and 2001 covering a total of 1,604 km^2 (Table S2). They have legal recognition in the form of a by-law, as well as formal endorsement from the Director of Fisheries [25,69].

Each of the Tanga LMMAs has a no-take-zone collaboratively policed by fisheries officers and local communities. Ecological monitoring since 1999 showed that these closures – the first on the East African coast to be established and actively managed by local fishing communities – had higher densities of fish and invertebrates, leading to positive impacts on local livelihoods, at least until 2004 [70,71]. Since then, dynamite fishing, which was almost completely eradicated in the region between 1998 and 2004, has returned [25].

Mainland Tanzania also has coastal Beach Management Units, established by the 2003 Fisheries Act and its principal Regulations of 2009 [72,73]. As in Kenya, BMUs empower communities to manage local fisheries resources, giving them the rights to restrict certain gears and control access through licencing [74]. Co-management of fisheries resources has spread rapidly in Tanzania, and there are currently 179 coastal BMUS, of which 68 have management plans and 39 have legal recognition through by-laws [75,76].

BMUs are increasingly establishing Collaborative Fisheries Management Areas (CFMAs) as a higher-level mechanism to manage their shared resources [76]. CFMAs, a type of LMMA, can protect the fishing grounds of an individual BMU or, more commonly, the shared resources of several [77]. Typically, BMUs with legal recognition first consult with neighbouring Units to determine the boundaries of the CFMA, before establishing a Co-ordination Committee. The Committee, which is composed of representatives from each BMU within the proposed area, synthesises management proposals from individual Units into a draft CFMA management plan, and acts as a networking mechanism for BMUs [77,78]. Once the draft plan is approved by each participating BMU, the CFMA can be given legal recognition in the form of a District Council by-law. At present, there are six CFMAs in mainland Tanzania, all established with the assistance of the WWF as part of a programme in the Rufiji, Mafia and Kilwa Districts of central Tanzania [34,75]. The six areas cover a total 2,498 km^2 across 21 BMUs, 2.5% of which (61.2 km^2) has no-take protection, initially for a 2-year period [78]. The programme is so far showing some promise: incidences of illegal dynamite fishing and seine netting have decreased, and there is a perception among resource users that fish abundance is starting to recover [79].

Tanzania – Zanzibar. BMUs (and CFMAs) do not exist in Zanzibar, a semi-autonomous region of the United Republic of Tanzania with separate environmental law and policy. The most comparable institution is the Village Fisheries Committee (VFC), a local-level organisation comprising 10 elected members [34]. VFCs were formed in all coastal fishing villages when devolution of marine resource management began in 1994 [80]. VFC jurisdiction depends on village boundaries and distance covered by local fishers. Responsibility for enforcement of regulations is shared between the Committee and the Department of Fisheries and VFCs can draw-up by-laws to manage resource use [80,81].

In Zanzibar, the legal basis for MPA establishment is provided by the Environmental Management for Sustainable Development Act 1996 [82]. The Act recognises the need to involve local communities in MPAs, enabling co-management arrangements analogous to LMMAs to develop at Misali Island and Menai Bay, albeit with a degree of State oversight [53,81,83].

In recent years, however, it appears that conservation of marine resources has become more centralised, less transparent and less participatory [84]. For example, Misali Island, a community-initiated attempt to resist tourism development, was subsumed into the larger Pemba Channel Conservation Area in 2005, with a consequential decline in community involvement [50,83–85]. Further, whilst the forthcoming Marine Conservation Unit regulations [86] include provisions to promote community involvement in marine resource management, the overall approach is top-down in nature [84]. These regulations will need a degree of revision if local management is to flourish in Zanzibar.

More broadly, both mainland Tanzania's BMUs and Zanzibar's VFCs lack capacity in many crucial areas including conflict resolution, financial management, project planning and marine ecology [34,75,79]. There are clear cultural, legal, political and institutional similarities between Kenya and Tanzania, so the efforts that are presently underway in both countries to address capacity constraints and promote sustainable financing would likely benefit from sharing experiences and resources.

Madagascar. One of the traditional values recovered following Madagascar's independence in 1960 was the social code. In rural communities, this social code – known as the *dina* – is a community law, generally communicated through oral tradition, though written down in some cases [87]. In 1996, the Malagasy Government introduced the Gestion Locale Sécurisée (GELOSE), a legal framework designed to integrate the *dina* with governmental laws to enable community-based management of natural resources [87].

Seven years later, at the fifth World Parks Congress in Durban, South Africa, the Malagasy president recognised the need to protect the country's unique natural assets and committed to the Durban Vision, a national conservation plan to triple the amount of protected area coverage [88]. This was codified into law shortly afterwards as a new decree (Décret d'Application No 848-05) for the existing Code des Aires Protégées [88]. The decree set up a System of Protected Areas of Madagascar, which simplified and redefined the legal process used in protected area creation [89]. Under this more flexible model, community organisations, NGOs and the private sector are permitted to manage protected areas, in addition to the parastatal protected areas agency Madagascar National Parks [90]. Since then, several LMMAs have been established along the coast by NGOs working with local communities (Table S2).

The Velondriake Community Managed Protected Area in southwest Madagascar is the country's oldest LMMA [91]. Velondriake spans nearly 1,000 km^2 of coral reefs, mangroves, lagoons, beaches and sea grass beds, making it one of the largest marine managed areas in Madagascar [91,92]. Home to around 7,500 semi-nomadic Vezo, Velondriake unites 25 coastal villages in the co-management of local marine resources [92,93]. It is legally recognised as an IUCN category V MPA and was granted definitive protected status by inter-ministerial decree in late 2012 [92]. Velondriake began as an initiative to improve the sustainability of the octopus fishery, but had since expanded to include aquaculture, temporary closures and the designation of eight permanent no-take marine reserves totalling 0.8 km^2 [92,93].

The initiative is largely guided and managed by local communities, with technical and financial support provided by the British NGO Blue Ventures. Resource use and access rights within the area are governed by a legally recognised *dina* rather than the GELOSE framework [94]. The *dina* bans destructive fishing practices including beach seining and poison fishing, regulates temporary and permanent closures and grants conflict resolution and enforcement powers to local communities, allowing them to impose fines and utilise the regional court system in cases where conflict resolution is unsuccessful [92,93].

Velondriake's perceived success has triggered widespread replication of the LMMA approach. Over the last 7 years, 34 LMMAs have been established along Madagascar's northern, western and southern coasts. Taken together, these initiatives presently cover 6.9% of the seabed, 6,635.3 km^2. In 2014, the Barren Islands is expected to add a further 4,290 km^2 to the total. This scaling up is unparalleled in the Western Indian Ocean, yet it has been achieved at low cost, without financial support from central government [93]. With severe constraints continuing to inhibit the country's capacity for environmental governance, Madagascar's LMMAs may offer an encouraging and locally acceptable solution to the challenges of marine resource management [93].

Mozambique. In Mozambique, the concept of fisheries co-management is enshrined in the 2003 Regulation on Marine Fisheries, which establishes co-management institutions at the provincial, district and local levels, as well as banning non-artisanal fisheries within three nautical miles of the coast [95]. The Decree introduces two types of institution: Fishing Community Councils (CCPs – Conselho Comunitário de Pescas) and Co-management Committees (CCG – Comité de Co-Gestão) [95]. Both were later formally established through legislation adopted in 2007 [96,97].

CCGs are multi-stakeholder committees formed principally at the provincial or district levels, but also at the local level [97,98]. Their principal objectives include deciding closed seasons and permissible types of gear and protecting endangered marine resources as well as advising on conflict resolution among fishers, fishing licences and fee collection [95]. CCPs are community-based associations of elected community members involved in artisanal fisheries [97]. Analogous to Kenya and Tanzania's BMUs, CCPs give local stakeholders rights to establish boundaries, control access and promote the sustainable use of marine resources [95,99]. Once members have been elected and the CCP established, they can apply for formal legal recognition, which, if granted, empowers them to assume responsibility for fishing licences and enforcement, functions otherwise administered at the district level [97,100].

CCP adoption has largely been driven through several donor-led artisanal fisheries programmes, especially in the Sofala Bank area, where the National Institute for the Development of Small-Scale Fisheries and the International Fund for Agricultural Development (IFAD) have worked with local communities to establish 65 Councils along a 950-kilometre stretch of coastline fronting the provinces of Sofala, Zambezia and Nampula [100,101]. Largely as a result of these initiatives, there are presently at least 156 CCPs along the coastline of Mozambique [102]. Taken together, it is estimated that these Councils provide a degree of representation to almost all coastal fishing communities in the country [103]. However, very few existing councils (approx. 20) are officially recognised by the Ministry of Fisheries [100,101], whilst many are unaware of their rights and responsibilities, and frequently lack the human resources, technical capacity and financial support necessary for effective management and enforcement [96,104]. In a study of compliance with centrally declared fisheries controls in the Sofala Bank area, for example, Wilson [104] found that fewer than 10% of inspected nets complied with minimum mesh size regulations, whilst none of the fishers interviewed ceased fishing or reduced effort during the closed season.

Similar technical and financial constraints have also plagued the higher level Co-management Committees. Outside of areas where they have direct support, very few CCGs are functioning, and of those that are, even fewer are endowed with the resources and awareness of community-level rights and obligations they need [96,98]. So far, there has been little monitoring of CCG operations at the central level and mechanisms for co-ordination across all levels of governance are practically non-existent [98].

As a result of these issues, this study was only able to identify one example of a functioning LMMA in Mozambique: the Vamizi Marine Sanctuary. This no-take reserve is managed by the Vamizi CCP, with technical and financial support from a partnership between an eco-lodge on Vamizi Island and WWF [105]. The partnership is helping to build community capacity for effective marine resource management by training members in reef and fish monitoring, developing alternative livelihood projects and providing environmental education [105]. The Council has delineated the boundaries of the Sanctuary with marker buoys and receives a $3 fee for each dive boat that enters (Personal Communication, Isabel da Silva).

Vamizi's success suggests if capacity constraints were addressed, more LMMAs could be established. Closed areas have already been trialled in other CCPs, and research suggests that these initiatives, together with restrictions on gear, length and species, may enjoy broad community support in Mozambique, particularly in the Pemba region [106,107]. To this end, a new phase of the IFAD-financed ProPESCA initiative is aiming to strengthen the capacity of the CCPs and CCGs, whilst the IUCN is considering an intervention to support the creation of community no-take zones [96,100].

Conclusions

This analysis shows that although MPAs in the Western Indian Ocean cover 133,273 km^2, only 7.0% of the region's continental shelf is protected by them. Less than 30% of reef-related MPAs in the WIO were found to offer effective ecological protection, though this compares favourably with global figures, where 15% were graded effective [24].

We found active LMMA initiatives in four countries that have passed legislation to decentralise marine resource management, with good potential for scaling up these initiatives in Kenya, mainland Tanzania and Madagascar, and lower potential in Zanzibar and Mozambique (Table 5). Due to underdeveloped legal structures supportive of local management in the other seven countries, no other formal LMMA initiatives were found to be in place, though the Seychelles [108,109], Comoros [110] and especially South Africa [111] have all acknowledged the potential of devolved management.

At 11,329.4 km^2, LMMAs cover 3.6% of the region's continental shelf, with particularly pronounced differences between LMMA and MPA coverage in mainland Tanzania and Madagascar, where LMMAs cover 3.5 and 2.6 times more area than MPAs respectively. Assuming that percentage of continental shelf covered by LMMAs and MPAs is an acceptable proxy for the CBD's 10% 2020 coverage targets, Comoros, Kenya, Mozambique and Tanzania have already achieved the target, whilst Madagascar is on course to do so.

Three caveats apply here. First, LMMA data were based on outputs from a workshop at the Seventh WIOMSA Scientific Symposium in Kenya in October 2011 and the Madagascar LMMA Forum in June 2012 [112] in addition to information gathered from a search of the published and grey literature and first-hand knowledge (MS, SP). Because many LMMAs, especially in Kenya and Mozambique, are small-scale, informal arrangements, information about them can be difficult to source. Accordingly, calculations of LMMA coverage should be considered as conservative estimates.

Secondly, the CBD 2020 targets call for 10% of the world's ecological regions to be effectively conserved. For the same reasons that the coverage estimates are conservative, the extent to which the region's LMMAs can be considered effective conservers of resources is largely unknown. Globally, empirical evidence that co-management arrangements achieve ecological [80,113] and social [32,36,114] goals is scarce and inconclusive, though this is likely to be due in part to the recently established nature of many LMMA initiatives. For example, in an examination of 42 marine co-management systems in Kenya, Tanzania, Madagascar, Indonesia and Papua New Guinea, Cinner et al. [115] found that although 88% of resource users surveyed reported high levels of compliance and 54% perceived a benefit to their livelihoods, co-management could also create a degree of social inequality by favouring wealthier users. They additionally found that fish biomass in co-managed areas was generally greater than in areas without local management, though substantially lower than in no-take marine reserves in the same countries [115].

Thirdly, although this analysis suggests that the LMMAs are increasingly numerous and could help WIO nations to meet international biodiversity commitments, we are not necessarily advocating for physically larger LMMAs with greater institutional complexity. Although the perceived success of the Velondriake Community Managed Protected Area in Madagascar provides evidence that large LMMAs with multiple stakeholders can function effectively, [92,93] other studies have reached differing conclusions. In an examination of ten fisheries cooperatives in Mexico, for example, McCay et al (2014) found that few stakeholders and a small spatial scale were critical success factors in community-based management of the commons [116].

Evidence from a recent global review of Territorial Use Rights for Fisheries (TURFs) by Auriemma et al (2014) is more equivocal [117]. LMMAs and TURFs, the latter defined by the study as "an area in which individuals or communities are given some level of exclusive access to marine resources within a defined boundary", overlap in many key areas because managed access rights are often implemented by communities within LMMAs [117]. The analysis, which drew on 103 case studies in 29 countries to test, among others, the hypothesis that larger TURFs are less successful due to increasing difficulties of enforcement, found no overall effect [117]. The question of whether bigger LMMAs are better is thus largely unresolved and would benefit from additional research.

From the country-specific analysis of LMMA implementation we present here, it is clear that a lack of organisational capacity, skills and money can all compromise the effectiveness of locally managed marine areas. And where areas are successful, further challenges may arise. For example, as biomass and fish numbers increase, so too may poaching, leading to further strain on resources devoted to enforcement [118,119].

Over the short-term, these issues may be best addressed through the establishment of an information-exchange forum to enable LMMA practitioners to share experiences and best practice, to offer training and exchange visits, and to promote local management to other communities and governments, especially in countries that have yet to devolve marine resource management to the local level. The forum could be modelled on the Pacific LMMA network and be complemented by research initiatives to better understand under what circumstances LMMAs may achieve their social and ecological objectives. Over the longer-term, the forum could form the basis for scaling-up LMMAs in the region towards a network that is lasting, effective and representative, and one that is complementary to centralised systematic conservation efforts.

Acknowledgments

We extend sincere thanks to Dave Raffaelli for insightful suggestions on improving the manuscript.

Author Contributions

Conceived and designed the experiments: SR SP MS JH. Analyzed the data: SR SP MS. Wrote the paper: SR SP MS JH.

References

1. Halpern BS, Walbridge S, Selkoe KA, Kappel CV, Micheli F, et al. (2008) A Global Map of Human Impact on Marine Ecosystems. Science 319: 948–952. doi:10.1126/science.1149345.
2. Jackson JBC (2008) Ecological extinction and evolution in the brave new ocean. Proc Natl Acad Sci 105: 11458–11465. doi:10.1073/pnas.0802812105.
3. Lester S, Halpern B, Grorud-Colvert K, Lubchenco J, Ruttenberg B, et al. (2009) Biological effects within no-take marine reserves: a global synthesis. Mar Ecol Prog Ser 384: 33–46. doi:10.3354/meps08029.
4. Fraschetti S, Claudet J, Grorud-Colvert K (2011) Transitioning from single-sector management to ecosystem-based management: what can marine protected areas offer? In: Claudet J, editor. Marine Protected Areas: A Multidisciplinary Approach. Cambridge: Cambridge University Press. 11–34.
5. Graham NAJ, McClanahan TR, MacNeil MA, Wilson SK, Polunin NVC, et al. (2008) Climate Warming, Marine Protected Areas and the Ocean-Scale Integrity of Coral Reef Ecosystems. PLoS ONE 3: e3039. doi:10.1371/journal.pone.0003039.
6. IUCN (2004) Managing Marine Protected Areas: A Toolkit for the Western Indian Ocean. Nairobi, Kenya: IUCN Eastern African Regional Programme.
7. Day J, Dudley N, Hockings M, Holmes G, Laffoley D, et al. (2012) Guidelines for applying the IUCN Protected Area Management Categories to Marine Protected Areas. Gland, Switzerland: IUCN. 36 p.
8. Roberts CM, Hawkins J (1997) How small can a marine reserve be and still be effective? Coral Reefs 16: 150 150.
9. Russ GR, Alcala AC (1996) Marine Reserves: Rates and Patterns of Recovery and Decline of Large Predatory Fish. Ecol Appl 6: 947–961. doi:10.2307/2269497.
10. Gell F, Roberts CM (2003) Benefits beyond boundaries: the fishery effects of marine reserves. Trends Ecol Evol 18: 448–455. doi:10.1016/S0169-5347(03)00189-7.
11. Russ GR, Alcala AC, Maypa AP, Calumpong HP, White AT (2004) Marine reserve benefits local fisheries. Ecol Appl 14: 597–606.
12. Russ GR, Alcala AC (2010) Enhanced biodiversity beyond marine reserve boundaries: The cup spillith over. Ecol Appl 21: 241–250. doi:10.1890/09-1197.1.
13. Hastings J, Thomas S, Burgener V, Gjerde K, Laffoley D, et al. (2012) Safeguarding the Blue Planet: Six Strategies for Accelerating Ocean Protection. Parks 18.
14. Van Lavieren H, Klaus R (2013) An effective regional Marine Protected Area network for the ROPME Sea Area: Unrealistic vision or realistic possibility? Mar Pollut Bull. doi:10.1016/j.marpolbul.2012.09.004.
15. World Summit on Sustainable Development (2002) Report of the World Summit on Sustainable Development, Johannesburg, South Africa, 26 August-4 September 2002. UN.
16. CBD (2006) Report of the Eighth Meeting of the Parties to the Convention on Biological Diversity Curitiba, Brazil: Convention on Biological Diversity.
17. CBD (2010) Strategic Plan for Biodiversity 2011–2020 and the Aichi Targets. Report of the Tenth Meeting of the Conference of the Parties to the Convention on Biological Diversity. Nagoya, Japan: Convention on Biological Diversity.
18. Wood LJ, Fish L, Laughren J, Pauly D (2008) Assessing progress towards global marine protection targets: shortfalls in information and action. Oryx 42: 340–351.
19. Spalding MD, Meliane I, Milam A, Fitzgerald C, Hale LZ (2013) Protecting Marine Spaces: Global Targets and Changing Approaches. Ocean Yearbook 27. Leiden, Netherlands: BRILL.
20. Pala C (2013) Giant marine reserves pose vast challenges. Science 339: 640–641. doi:10.1126/science.339.6120.640.
21. Samoilys MA, Martin-Smith KM, Giles BG, Cabrera B, Anticamara JA, et al. (2007) Effectiveness of five small Philippines' coral reef reserves for fish populations depends on site-specific factors, particularly enforcement history. Biol Conserv 136: 584–601. doi:10.1016/j.biocon.2007.01.003.
22. Jennings S (2009) The role of marine protected areas in environmental management. ICES J Mar Sci 66: 16–21. doi:10.1093/icesjms/fsn163.
23. Balmford A, Gravestock P, Hockley N, McClean CJ, Roberts CM (2004) The worldwide costs of marine protected areas. Proc Natl Acad Sci U S A 101: 9694–9697. doi:10.1073/pnas.0403239101.
24. Burke L, Reytar K, Spalding M, Perry A (2011) Reefs at risk revisited. Washington, D.C: World Resources Institute. 130 p.
25. Samoilys MA, Obura DO (2011) Marine conservation successes in Eastern Africa. Mombasa, Kenya: CORDIO East Africa.
26. Christie P (2004) Marine protected areas as biological successes and social failures in Southeast Asia. American Fisheries Society Symposium. Vol. 42. 155–164.
27. McClanahan TR, Marnane MJ, Cinner JE, Kiene WE (2006) A Comparison of Marine Protected Areas and Alternative Approaches to Coral-Reef Management. Curr Biol 16: 1408–1413. doi:10.1016/j.cub.2006.05.062.
28. Cinner JE, Wamukota A, Randriamahazo H, Rabearisoa A (2009) Toward institutions for community-based management of inshore marine resources in the Western Indian Ocean. Mar Policy 33: 489–496. doi:10.1016/j.marpol.2008.11.001.
29. Alcala AC, Russ GR (2006) No-take marine reserves and reef fisheries management in the Philippines: a new people power revolution. AMBIO J Hum Environ 35: 245–254.
30. Govan H, Tawake A, Tabunakawai K (2006) Community-based marine resource management in the South Pacific. Prot Areas Programme: 63.
31. Cinner JE, Basurto X, Fidelman P, Kuange J, Lahari R, et al. (2012) Institutional designs of customary fisheries management arrangements in Indonesia, Papua New Guinea, and Mexico. Mar Policy 36: 278–285. doi:10.1016/j.marpol.2011.06.005.
32. Wamukota AW, Cinner JE, McClanahan TR (2012) Co-management of coral reef fisheries: A critical evaluation of the literature. Mar Policy 36: 481–488. doi:10.1016/j.marpol.2011.09.001.
33. Govan H (2009) Achieving the potential of locally managed marine areas in the South Pacific. SPC Tradit Mar Resour Manag Knowl Inf Bull 25: 16–25.
34. Anderson J (2012) Options to Reduce IUU Fishing in Kenya, Tanzania, Uganda and Zanzibar. Ebène, Mauritius: Indian Ocean Commission – SmartFish Program. 99 p.
35. Tawake A (2007) Scaling-up networks of locally managed marine areas (LMMAs) to island wide ecosystem management while decentralising the effort of Fiji LMMA network and its implementation from national to provincial levels.
36. Evans L, Cherrett N, Pemsl D (2011) Assessing the impact of fisheries co-management interventions in developing countries: A meta-analysis. J Environ Manage 92: 1938–1949. doi:10.1016/j.jenvman.2011.03.010.
37. Gutierrez NL, Hilborn R, Defeo O (2011) Leadership, social capital and incentives promote successful fisheries. Nature 470: 386–389. doi:10.1038/nature09689.
38. Johannes RE (1978) Traditional Marine Conservation Methods in Oceania and their Demise. Annu Rev Ecol Syst 9: 349–364.
39. Christie P, White AT (1997) Trends in development of coastal area management in tropical countries: From central to community orientation. Coast Manag 25: 155–181. doi:10.1080/08920759709362316.
40. Johannes RE (1981) Words of the lagoon: fishing and marine lore in the Palau District of Micronesia. University of California Press. 280 p.
41. Burgess ND, Clarke GP, Rodgers WA (1998) Coastal forests of eastern Africa: status, endemism patterns and their potential causes. Biol J Linn Soc 64: 337–367. doi:10.1111/j.1095-8312.1998.tb00337.x.
42. Metcalfe K, Ffrench-Constant R, Gordon I (2010) Sacred sites as hotspots for biodiversity: the Three Sisters Cave complex in coastal Kenya. Oryx 44: 118–123. doi:10.1017/S0030605309990731.
43. Glaesel H (2000) Community-Level Marine Resource Management and the Spirit Realm in Coastal Kenya. Women Nat Resour 21: 35–42.
44. Govan H, Tawake A (2009) Status and potential of locally-managed marine areas in the South Pacific: meeting nature conservation and sustainable livelihood targets through wide-spread implementation of LMMAs. New Calendonia: Coral Reef Initiatives for the Pacific.
45. Abunge C (2011) Managing finite marine resources the Kenyan way. Shara: 49.
46. Mills M, Jupiter SD, Pressey RL, Ban NC, Comley J (2011) Incorporating Effectiveness of Community-Based Management in a National Marine Gap Analysis for Fiji. Conserv Biol 25: 1155–1164. doi:10.1111/j.1523-1739.2011.01749.x.
47. Govan H, Jupiter S (2013) Can the IUCN 2008 Protected Areas Management Categories Support Pacific Island Approaches to Conservation? Parks 19: 73–80.
48. Secretariat of the Convention on Biological Diversity (2004) Technical Advice on the Establishment and Management of a National System of Marine and Coastal Protected Areas. SCBD.
49. Sen S, Raakjaer Nielsen J (1996) Fisheries co-management: a comparative analysis. Mar Policy 20: 405–418.
50. WWF (2004) Towards the Establishment of an Ecologically Representative Network of Marine Protected Areas in Kenya, Tanzania and Mozambique. Dar es Salaam, Tanzania: WWF Eastern African Marine Ecoregion Programme. 74 p.
51. UNEP/Nairobi Convention Secretariat (2009) Transboundary Diagnostic Analysis of Land-based Sources and Activities Affecting the Western Indian Ocean Coastal and Marine Environment. Nairobi, Kenya: UNEP. 378 p.
52. Obura D (2012) The Diversity and Biogeography of Western Indian Ocean Reef-Building Corals. PLoS ONE 7: e45013. doi:10.1371/journal.pone.0045013.
53. Wells S, Burgess N, Ngusaru A (2007) Towards the 2012 marine protected area targets in Eastern Africa. Ocean Coast Manag 50: 67–83. doi:10.1016/j.ocecoaman.2006.08.012.
54. UNEP-WCMC (2010) World Database on Protected Areas (WDPA) Annual Release 2010. Cambridge: UNEP-World Conservation Monitoring Centre. Available: http://www.unep-wcmc.org.
55. IUCN (2000) Progress In Implementing The Jakarta Mandate In The Eastern Africa Region. A Report to the Fifth Conference of the Parties to the Convention on Biological Diversity.

56. Barrow E, Mahler F, Mosele L, Mvoyi C, Ntahuga L, et al. (2007) Sound Natural Resource Management: The Foundation for Achieving the MDGs in Somalia. Policy Brief.
57. MPA News (2012) Mozambique announces coastal protected area. MPA News 14.
58. Wells S (2009) Dynamite fishing in northern Tanzania-pervasive, problematic and yet preventable. Mar Pollut Bull 58: 20–23.
59. Government of Kenya (2007) Legal Notice No. 402, The Fisheries (Beach Management Unit) Regulations. The Fisheries Act. Cap 378.
60. Signa D, Tuda PM, Samoilys M (2008) Social, economic and environmental impacts of beach seining in Kenya.
61. Lamprey R, Murage DL (2011) Saving our seas: Coast communities and Darwin collaborate for a new future. Swara: 41–48.
62. Nelson F (2012) Recognition and Support of ICCAs in Kenya. Recognising and Supporting Territories and Areas Conserved By Indigenous Peoples And Local Communities: Global Overview and National Case Studies. Technical Series. Montreal, Canada: Secretariat of the Convention on Biological Diversity, ICCA Consortium, Kalpavriksh, and Natural Justice.
63. Maina GW, Osuka K, Samoilys M (2011) Opportunities and challenges of community-based marine protected areas in Kenya. Mombasa, Kenya: CORDIO East Afica.
64. Griffin T (2012) The Kuruwitu Conservation and Welfare Association. People Environ WIOMSA Mag: 13–17.
65. Murage D (2012) Community Conserved Areas: Experiences from Coastal Communities in Kenya. People Environ WIOMSA Mag: 18–20.
66. Samoilys M, Osuka K, Maina GW (2011) Opportunities and challenges of current legislation for effective conservation in the Tana Delta-Pate Island coast of Kenya. CORDIO Status Report. Mombasa, Kenya: CORDIO East Africa. p. 16.
67. Verheij E, Makoloweka S, Kalombo H (2004) Collaborative coastal management improves coral reefs and fisheries in Tanga, Tanzania. Ocean Coast Manag 47: 309–320. doi:10.1016/j.ocecoaman.2004.07.002.
68. Samoilys MA, Kanyange NW (2008) Natural resource dependence, livelihoods and development: Perceptions from Kiunga, Kenya. Nairobi, Kenya: IUCN Eastern Africa Regional Office. 30 p.
69. Wells S, Makoloweka S, Samoilys M, editors (2007) Putting Adaptive Management Into Practice: Collaborative Coastal Management in Tanga, northern Tanzania. Nairobi, Kenya: IUCN Eastern Africa Regional Office. 197 p.
70. Samoilys M, Horrill C, Kalombo H, Kabamba J, Wells S (2007) Coral reefs and mangroves: maintaining ecosystem health. Putting Adaptive Management Into Practice: Collaborative Coastal Management in Tanga, northern Tanzania. Nairobi, Kenya: IUCN Eastern Africa Regional Office. 77–102.
71. Wells S, Samoilys M, Anderson J, Kalombo H, Makoloweka S (2007) Collaborative Fisheries Management in Tanga, Northern Tanzania. Fisheries Management: Progress towards sustainability. Oxford, United Kingdom. 139–165.
72. United Republic of Tanzania (2003) The Fisheries Act 2003.
73. United Republic of Tanzania (2009) The Fisheries Regulations 2009.
74. Mulyila EJ, Matsuoka T, Anraku K (2012) Sustainability of fishers' communities in tropical island fisheries from the perspectives of resource use and management: a comparative study of Pohnpei (Micronesia), Mafia (Tanzania), and Guimaras (Philippines). Fish Sci 78: 947–964. doi:10.1007/s12562-012-0500-x.
75. Sobo FS (2012) Community Participation in Fisheries Management in Tanzania. IIFET 2012 Tanzania Proceedings. Dar es Salaam, Tanzania. p. 11.
76. Otsyina RM, Benno BL, Abdallah JM (2010) Strengthening Community Capacity for Fisheries Co-Management (SCCaFCoM) in Rufiji, Mafia and Kilwa Districts: Final Evaluation Report. Dar es Salaam, Tanzania: WWF Tanzania.
77. Fisheries Development Division, WWF (2009) Guidelines for Establishing Community Based Collaborative Fisheries Management in Marine Waters of Tanzania. Dar es Salaam, Tanzania: Ministry of Livestock and Fisheries Development.
78. Mwangamilo J (2012) Collaborative Fisheries Management Initiatives. Rufiji-Mafia-Kilwa (rumaki) Seascape Programme.
79. Tanzania Natural Resource Forum (2012) Community Based Natural Resource Management (CBNRM) Stocktaking Workshop Report.
80. Cinner JE, Daw TM, McClanahan TR, Muthiga N, Abunge C, et al. (2012) Transitions toward co-management: The process of marine resource management devolution in three east African countries. Glob Environ Change. doi:10.1016/j.gloenvcha.2012.03.002.
81. Mwaipopo N (2008) The social dimensions of marine protected areas: a case study of the Mafia Island Marine Park in Tanzania. Dar es Salaam, Tanzania: International Collective in Support of Fishworkers. 54 p.
82. Government of Zanzibar (2007) The Environmental Management for Sustainable Development Act, 1996. Legal Supplement (Part 1) to the Zanzibar Government Gazette. Vol. CVI No.5743.
83. Levine A (2007) Staying afloat: State agencies, local communities, and international involvement in marine protected area management in Zanzibar, Tanzania. Conserv Soc 5: 562.
84. McLean B, Hikmany AN, Mangora M, Shalli M (2012) An Assessment of Legal and Institutional Framework for Effective Management of Marine Managed Areas in Tanzania. Zanzibar Report. Dar es Salaam, Tanzania: Marine Conservation Unit.
85. Lindhjem H (2003) Sustainable Financing of Marine Protected Areas in Zanzibar. Washington DC: World Bank.
86. Government of Zanzibar (2012) The Marine Conservation Unit Regulations. Unpublished Draft Regulations.
87. Rakotoson LR, Tanner K (2006) Community-based governance of coastal zone and marine resources in Madagascar. Ocean Coast Manag 49: 855–872. doi:10.1016/j.ocecoaman.2006.08.003.
88. Durbin J (2007) Madagascar's new system of protected areas – Implementing the "Durban Vision." London, UK: Zoological Society of London.
89. IRIN (2006) Madagascar establishes new park system to protect lemurs, benefit people. IRIN. Available: http://news.mongabay.com/2006/0117-madagascar.html.
90. Rabearivony J, Thorstrom R, de Roland LA., Rakotondratsima M, Razafimanjato G, et al. (2010) Protected area surface extension in Madagascar: Do endemism and threatened species remain useful criteria for site selection? Madag Conserv Dev 5.
91. Harris A (2007) "To live with the Sea" Development of the Velondriake Community-Managed Protected Area Network, Southwest Madagascar. Madag Conserv Dev 2.
92. Westerman K, Gardner CJ (2013) Adoption of socio-cultural norms to increase community compliance in permanent marine reserves in southwest Madagascar. Conserv Evid: 4–9.
93. Harris AR (2011) Out of sight but no longer out of mind: a climate of change for marine conservation in Madagascar. Madag Conserv Dev 6.
94. Andriamalala G, Gardner CJ (2010) L'utilisation du dina comme outil de gouvernance des ressources naturelles: Leccons tirés de Velondriake, sud-ouest de Madagascar. Trop Conserv Sci 3: 447–472.
95. Government of Mozambique (2003) Decree No. 43/2003 General Regulation on Marine Fisheries (REPMAR). Decree No 432003.
96. Swennenhuis J (2011) Strengthening community based fisheries governance in Mozambique: a roadmap developed for IUCN. Gland, Switzerland: International Union for Conservation of Nature. 38 p.
97. Government of Mozambique (2007) Ministerial Order No. 49/2007 approving the Regulation of the Co-Management Committee on Fishery Sector. 49/2007.
98. Russo de Sá J (2011) Reflexão Sobre a Gestão Participava Das Pescarias. Maputo, Mozambique: Ministry of Fisheries. 30 p.
99. Rosendo S, Brown K, Joubert A, Jiddawi N, Mechisso M (2011) A clash of values and approaches: A case study of marine protected area planning in Mozambique. Ocean Coast Manag 54: 55–65. doi:10.1016/j.ocecoaman.2010.10.009.
100. IFAD (2012) Sofala Bank Artisanal Fisheries Project (PPPBAS): Project Completion Report. Completion Report. Rome, Italy: International Fund for Agricultural Development.
101. IFAD (2010) Republic of Mozambique artisanal fisheries promotion project. Project design document.
102. National Institute for the Development of Small Scale Fisheries (2012) Co-management in the artisanal fishing sub-sector. Maputo, Mozambique: National Institute for the Development of Small Scale Fisheries (IDPPE).
103. MRAG (2010) Towards sustainable fisheries management: international examples of innovation. London: MRAG. 93 p.
104. Wilson J (2012) Beach Seining in Mozambique-Bane or Benefit? IIFET 2012 Tanzania Proceedings. Dar es Salaam, Tanzania.
105. Garnier J, Silva I, Davidson J, Hill N, Muaves L, et al. (2008) Co-Management of the Reef at Vamizi Island, Northern Mozambique. CORDIO Status Report. Vol. 2008.
106. McClanahan TR, Cinner JE, Abunge C (2013) Identifying management preferences, institutional organisational rules, and their capacity to improve fisheries management in Pemba, Mozambique. Afr J Mar Sci 35.
107. McClanahan TR, Cinner JE, Abunge C, Muthiga N (2012) Identifying management preferences, institutional organizational attributes, and their capacity to improve the management of Pemba, Mozambique fisheries.
108. Clifton J, Etienne M, Barnes DKA, Barnes RSK, Suggett DJ, et al. (2012) Marine conservation policy in Seychelles: Current constraints and prospects for improvement. Mar Policy 36: 823–831. doi:10.1016/j.marpol.2011.11.009.
109. UNDP (2010) Strengthening Seychelles' protected area system through NGO management modalities.
110. Hauzer M, Dearden P, Murray G (2013) The effectiveness of community-based governance of small-scale fisheries, Ngazidja island, Comoros. Mar Policy 38: 346–354. doi:10.1016/j.marpol.2012.06.012.
111. Government of South Africa (2012) Policy for the Small Scale Fisheries Sector in South Africa: 56.
112. Blue Ventures (2012) "Share Your Story": Madagascar's First Locally Managed Marine Area (LMMA) Forum. Workshop report. Toliara, Madagascar.
113. McClanahan TR, Castilla JC, White AT, Defeo O (2008) Healing small-scale fisheries by facilitating complex socio-ecological systems. Rev Fish Biol Fish 19: 33–47. doi:10.1007/s11160-008-9088-8.
114. Syakur A, Wibowo JT, Firmansyah F, Azam I, Linkie M (2012) Ensuring local stakeholder support for marine conservation: establishing a locally-managed marine area network in Aceh. Oryx 46: 516–524. doi:10.1017/S0030605312000166.

115. Cinner JE, McClanahan TR, MacNeil MA, Graham NAJ, Daw TM, et al. (2012) Comanagement of Coral Reef Social-Ecological Systems. Proc Natl Acad Sci. doi:10.1073/pnas.1121215109.
116. McCay BJ, Micheli F, Ponce-Díaz G, Murray G, Shester G, et al. (2014) Cooperatives, concessions, and co-management on the Pacific coast of Mexico. Mar Policy 44: 49–59. doi:10.1016/j.marpol.2013.08.001.
117. Auriemma G, Byler K, Peterson K, Yurkanin A, Costello C (2014) A global assessment of Territorial Use Rights in Fisheries to determine variability in success and design. Santa Barbara, California: Bren School of Environmental Science and Management. 132 p.
118. Guilbeaux M, Parras T, Tan W (2008) Sharing Lessons Learned in Community Marine Management across the Pacific. Workshop report. University of the South Pacific, Fiji: Locally-Managed Marine Area Network. p. 112.
119. Aalbersberg B, Tawake A, Parras T (2005) Village by village: recovering Fiji's coastal fisheries. World Resources 2005: The Wealth of the Poor: Managing Ecosystems to Fight Poverty. Washington, D.C: World Resources Institute. 144–152.
120. The World Bank (2010) World development indicators 2010. [14th ed]. World Bank.
121. CIESIN (2007) National Aggregates of Geospatial Data: Population, Landscape and Climate Estimates. Palisades, NY: Center for International Earth Science Information Network, Columbia University.
122. UNDP (2011) Human Development Report 2011: Sustainability and Equity: Towards a Better Future for All.
123. INSEE (2010) Tableau Économique de Mayotte, Édition 2010. St Denis: Institut National de la Statistique et des Études Economiques.
124. INSEE (2011) Tableau Économique de La Réunion, Édition 2011. St Denis: Institut National de la Statistique et des Études Economiques.
125. Statistics South Africa (2012) Census 2011: Census in Brief. Pretoria, South Africa: Statistics South Africa.
126. Statistics South Africa (2011) Gross domestic product. Annual estimates 2002–2010. Regional estimates 2002–2010. Third quarter 2011. Statistical Release. Pretoria, South Africa: Statistics South Africa.
127. World Resources Institute (2007) EarthTrends: Environmental Information. Washington DC: World Resources Institute.

3

Assessing the Impact of Human Activities on British Columbia's Estuaries

Carolyn K. Robb*

School of Environment and Management, Royal Roads University, Victoria, British Columbia, Canada

Abstract

The world's marine and coastal ecosystems are under threat and single-sector management efforts have failed to address those threats. Scientific consensus suggests that management should evolve to focus on ecosystems and their human, ecological, and physical components. Estuaries are recognized globally as one of the world's most productive and most threatened ecosystems and many estuarine areas in British Columbia (BC) have been lost or degraded. To help prioritize activities and areas for regional management efforts, spatial information on human activities that adversely affect BC's estuaries was compiled. Using statistical analyses, estuaries were assigned to groups facing related threats that could benefit from similar management. The results show that estuaries in the most populated marine ecosections have the highest biological importance but also the highest impacts and the lowest levels of protection. This research is timely, as it will inform ongoing marine planning, land acquisition, and stewardship efforts in BC.

Editor: Simon Thrush, University of Auckland, New Zealand

Funding: Living Oceans Society (www.livingoceans.org), an environmental non-profit organization and the author's employer, provided the computer and GIS software used in the completion of this research. No other funding assistance was received. The funders had no role in study design, data collection and analysis, decision to publish, or preparation of the manuscript.

Competing Interests: This research was completed as part of the requirements for a Master of Science in Environment and Management from Royal Roads University in Victoria, British Columbia, Canada. While a student at Royal Roads University, the author also worked as the GIS Analyst for Living Oceans Society (LOS), a marine conservation organization based in British Columbia, Canada. This does not alter the author's adherence to PLOS ONE policies on sharing data and materials.

* E-mail: ckrobb18@hotmail.com

Introduction

Estuaries are one of the world's most biologically productive ecosystems [1,2,3], providing ecosystem services valued as high as $22,000 US per hectare per year [4]. In British Columbia, Canada (BC), calculations have shown that estuarine habitats sequester a minimum of 180,000 tonnes of carbon annually [5] and Kelsey [6] estimated that 80% of all coastal wildlife use estuaries. However, anthropogenic threats have made estuaries one of the most degraded ecosystems on earth [3,7] and these threats are expected to increase in prevalence and severity as coastal populations continue to expand [7].

Many estuarine threats originate within the watersheds that drain through the estuaries [7,8]. The conversion of land for agriculture, forestry, or residential and industrial development causes habitat loss or fragmentation [7,9]. These changes in land use can trigger an increase in sediment, debris, nutrient, and pollutant levels in estuaries that smothers or entangles organisms or causes eutrophication, hypoxia, and anoxia [1,3,7,10,11]. Development that expands impervious surfaces within the watershed can increase water flow into estuaries, particularly during storm events [7], while dams impede water inputs [1,7,9]. As the climate warms, increased precipitation and more frequent storm surges are also anticipated in BC [9]. All of these hydrological changes can disrupt water filtration [7] and alter estuarine productivity and trophic structure [12,13].

Other threats originate in the marine environment. Development and industrial activities within estuaries, such as log-handling, aquaculture, dredging, and vessel traffic, can result in habitat loss or degradation, environmental contamination, shoreline erosion, the re-suspension of sequestered carbon, and the invasion of alien species [1,7,9,10,12]. Commercial and recreational fishing can alter the composition of estuarine species through the preferential removal of target species [7,12]. Shoreline armouring to stabilize and protect coastal developments limits the ability of estuaries to retreat inland as the climate warms and sea levels rise and may therefore cause habitat and shorebird losses [1,9,13,14]. Higher sea temperatures may increase the severity of eutrophication in estuarine salt marshes [15]. Lastly, the acidification of estuarine waters may inhibit the growth of shelled organisms [16] and increase the bioavailability of trace metals from pollution [17].

The continuing degradation of estuarine habitats has led to a variety of research efforts. In BC, the Pacific Estuary Conservation Program (PECP), a partnership of government and non-government organizations, has mapped the boundaries of 442 estuaries and designed regional scale projects to prioritize estuaries for conservation based on their biological importance [2]. Further research has found that 38% of BC estuaries had economic tenures within the intertidal area [18]. Merrifield *et al.* [1] recently completed a regional assessment of threats to estuaries along the west coast of the United States which face similar threats to BC estuaries [19]. The researchers found that only 16% of surveyed estuaries faced few or no anthropogenic threats while 25% of estuaries had moderate levels of all assessed threats [1]. However, a systematic investigation of human activities affecting BC estuaries has not yet been completed.

In BC, the management of estuarine areas and activities requires the coordination of several government agencies. Subtidal and intertidal areas are typically public land while upland areas can be publicly or privately owned. The BC government regulates the use of public lands by issuing tenures but the ongoing activities of some industries with tenures, such as aquaculture, are managed federally. Federal jurisdiction also pertains to all organisms and most activities within the water column, areas designated for port facilities, and activities on or beneath the seabed in offshore areas [20]. In public areas, the BC government and several federal agencies have established protected areas that, to varying degrees, shield estuarine habitat from human activities [21]. In private areas, the PECP works to secure and rehabilitate important estuarine habitats through direct land purchases [2,22].

Recent reports attribute estuary degradation to a failure of management efforts that have focused predominantly on individual industrial sectors [23,24,25] without examining the cumulative impacts of activities [24,26,27,28] or the ecosystem services on which the activities rely [29]. There is a growing consensus that a broader management focus, termed ecosystem-based management (EBM), is necessary to tackle the problems facing marine and coastal environments [23,24,26,27,30,31,32]. EBM is place-based and considers both the spatial variations of, and the interactions between, the ecological, physical, and human elements of ecosystems [24,31]. In Canada, Fisheries and Oceans Canada (DFO) has developed a framework for estuarine, coastal, and marine management following the principles of EBM [23].

To implement EBM effectively, ecosystem elements must be mapped and the impacts of human activities assessed [23,24,28,33]. This research uses geographic information systems (GIS) to investigate the spatial distribution of human activities that affect BC estuaries, incorporating threats relevant to BC that have not been considered previously. Cluster analysis is used to categorize estuaries based on the presence and magnitude of threat variables. Unlike in previous research [1], the spatial range of influence of each threat is incorporated to generate a more realistic spatial representation of each threat [33], the cluster analysis uses a non-hierarchical approach to allow the use of non-normal variables, and results are compared to assessments of each estuary's biological importance and conservation status. By identifying estuaries facing similar threats, this work gives context to previous prioritization efforts that focused solely on ecological factors [2], informs regional planning efforts, and helps prioritize limited conservation and research resources to ensure vital ecosystem services are sustained.

Methodology

Study Area

BC's estuaries, comprising less than three percent of the coastline, form the study area for this research (Figure 1). I used estuaries and associated watersheds as the basic spatial units for the analysis, referred to as "estuary-watershed systems", because they are considered the complete geographic boundaries for estuarine management [8,34].

Estuary and Watershed Data Preparation

In a desktop GIS, I overlaid spatial data for watersheds and estuaries (Table S1) to identify the watershed associated with each estuary (Figure 1). I confirmed that overlapping estuaries and watersheds shared the same primary waterway using attribute information found in both datasets. Estuaries that could not be linked to a unique watershed were removed. I added attribute information on importance for waterbirds [2] and carbon sequestration [5] to each estuary based on estuary name. I also overlaid the estuary-watershed systems with spatial data on legislated protected areas and privately conserved lands (Table S1) to determine the count and total area of conservation designations within each system. Lastly, I overlaid the systems with BC's marine ecosections, which are planning areas with similar physical and biological characteristics, to determine the average estuarine importance and conservation values by ecosection (Table 1).

Threat Data Preparation

To identify the primary anthropogenic threats facing BC estuaries, I reviewed the relevant literature, focusing on research on estuaries and estuarine habitats, such as salt marshes, seagrass beds, and sand flats, from BC and the Pacific coast of North America (Table 2). Spatial data for the primary threats (Table S2) were chosen based on their relevance to estuaries, spatial coverage, quality, resolution, and currency and used to create a comprehensive spatial layer for each threat along the BC coast. While gaps did exist (Table S3), data were available for the majority of the identified threats.

Several spatial analysis methodologies were used to compile the data (Table S2) for further statistical analysis. I classified land use data into agricultural or urban areas, combining the urban areas with railroads and road networks. I identified armoured shorelines by combining dikes and manmade shoreline segments. I selected the locations of freshwater diversions from a point dataset of obstructions. To identify areas that had been clearcut over the past decade, I extracted polygons with a stand age of 10 or less from a provincial dataset of harvest depletion. I located sources of point and nonpoint watershed pollution from data on pulp mills, mines, and areas closed due to contamination. Lastly, I calculated the percent change in modeled future precipitation using data on current (1961–1990 normal) and future (2080's CGSM A2x scenario) precipitation from ClimateBC [36] and the methodology employed by Merrifield *et al.* [1].

The compilation of marine threat data (Table S2) was guided by the best available data identified by the BC Marine Conservation Analysis (BCMCA) [37]. I obtained provincial tenure data to identify public areas leased to marine-related industrial activities and confirmed site usage with site-specific finfish aquaculture production data from DFO or by visually confirming the presence of infrastructure, such as aquaculture pens or log storage sites, using Google Earth 6.1 [38] satellite imagery. I merged the tenures to create features for shoreline development and aquaculture. I transformed the tenures for marinas, commercial wharves, and ferry terminals into points and combined them with BCMCA datasets representing marinas, anchorages, and tow boat reserves. I obtained the point locations of dredging operations and extracted data for large vessels from a gridded dataset of vessel traffic (Table S2). Smaller vessels were not well represented in this dataset so I used linear data on recreational boating routes to represent the distribution of small vessels. I combined fisheries datasets to create features for commercial and recreational fisheries. To retain the catch values available in the commercial fishing datasets, the values for all fisheries were converted to kg and summed. No high resolution datasets were available for projected ocean acidification, sea surface temperature, or sea level rise within BC's coastal zone (Table S3).

Threats within Estuaries

The maximum spatial range of influence of each activity was determined from zone values specified by Ban *et al.* [33] and Quigley and Hall [39] (Table S2). Incorporating zones more

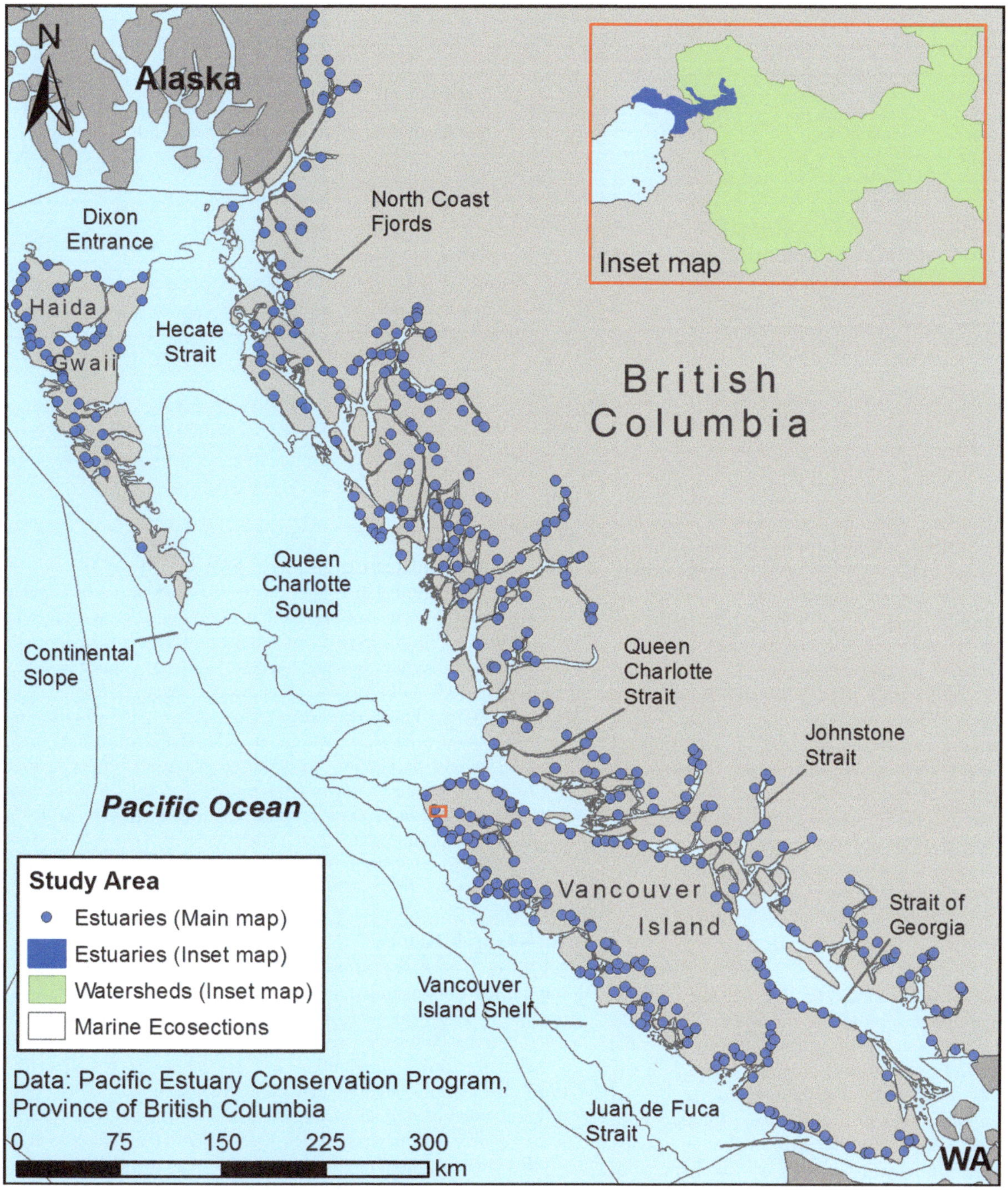

Figure 1. Study area. Description: This research investigates the presence of threats within BC estuaries, which are found along the BC coastline from the southern border with Washington to the northern border with Alaska. The inset map shows a typical estuary and watershed.

realistically represents the full spatial extent of each threat [33] and ensures that all relevant threats are considered, though the spatial footprint of associated infrastructure may not overlap directly with an estuary or watershed. I buffered each threat by the zone value to create polygons representing the larger zone of influence. Spatial ranges were not calculated for features with low data resolution [33], features where the zone of impact was not known, or features that already incorporated the zone of impact (Table S2). The spatial range was calculated for large vessel traffic, despite the low data resolution, due to the wide influence of this feature and because the original data had been fit to a coastline that did not incorporate the mapped estuaries in all areas.

To determine the impact of the threats within each estuary-watershed system, I overlaid the buffered threats with the estuaries and watersheds. To standardize for variation in the size of the systems, I calculated the proportion of each system affected by the activity. For point features, I determined the density of points within each estuary-watershed system. For a few gridded features, information on level of use was available and used to better represent their impact spatially. Because the impact of vessel

Table 1. Average conservation importance, protection level, and threat status of BC estuaries, grouped by marine ecosection.

Marine Ecosection	Importance for Waterbirds	Count of Threats	Count of Protected Areas	Proportion of Area Protected (%)	Average Watershed Area (ha)	Average Estuarine Area (ha)
Continental Slope	3.4	6.2	1.6	57.9%	4,328.3	17.0
Dixon Entrance	3.2	6.1	1.2	42.4%	6,646.5	140.0
Hecate Strait	3.5	7.0	1.3	22.4%	3,859.1	55.7
Johnstone Strait	3.6	8.9	0.8	12.8%	5,724.1	64.0
Juan de Fuca Strait	3.7	9.1	2.3	11.5%	5,019.6	25.1
North Coast Fjords	3.6	6.6	1.3	27.8%	6,514.0	192.7
Queen Charlotte Sound	3.5	5.2	1.3	20.3%	4,101.3	29.4
Queen Charlotte Strait	3.3	9.3	0.9	3.1%	5,542.8	125.4
Strait of Georgia	3.2	11.0	2.4	4.2%	7,132.2	783.3
Vancouver Island Shelf	3.4	8.9	1.0	13.3%	5,255.5	45.9

traffic, such as ship wake, can extend into an estuary from outside, I spatially joined the vessel traffic data with the estuaries and summed the number of vessel hours for all overlapping grid cells to calculate the total impact. For commercial fishing, I intersected the grid of fisheries catch data (Table S2) with the estuary polygons and summed the total catch of the overlapping grid cells proportional to the area of overlap to estimate the catch taken within each estuary. I calculated the average precipitation change for each estuary-watershed system using zonal statistics, which examined and combined the values of cells in the gridded precipitation dataset that fell within the estuary-watershed systems.

I counted the number of threats affecting each estuary and determined the proportion of the estuary and watershed area covered by those activities. The average values were compared among estuaries in different marine ecosections (Table 1) and estuaries with different conservation status and biological importance.

Statistical Analyses

I ran cluster analyses using Systat 13 [40] to group the estuaries based on the presence and magnitude of the threats (Table 3). Pairwise correlations between threats were examined to ensure the included threats were independent and considered equal in the analysis. Most of the threat variables had negatively skewed distributions and were log-transformed to minimize the impact of a few scattered higher values [28]. To determine the optimal number of clusters, I performed preliminary analyses with hierarchical clustering and the Ward method. I used non-hierarchical K-medians for the final cluster analysis because the method is resistant to non-normal variables and outliers [41]. Euclidean distance was used to determine the distance of each site from the centers of each cluster.

I mapped the results and generated spatial statistics for the clusters, including the average number of threats and the proportion of each estuary affected by each threat (Tables 3 and 4). I also analyzed the distribution of the clusters among ecosections, sites with a conservation designation, and estuaries considered biologically important (Table 4).

Results

Importance and Conservation Status of Estuaries

The average importance of estuaries for waterbirds was similar among all ecosections but priority carbon sequestration sites were only found within the Strait of Georgia and North Coast Fjords ecosections. Federally and provincially legislated protected areas overlapped 58.5% of estuary-watershed systems, predominantly in the North Coast Fjords ecosection. On average, 21% of estuarine and watershed area was covered by a legislated protected area, with the highest proportions in north coast ecosections. A further 27 estuary-watershed systems, primarily on the south coast, overlapped a privately conserved area, with an average of 1.4% of the estuarine and watershed area protected. In total, 61% of estuary-watershed systems had at least one private or legislated protected area.

Threats in Estuaries

Of 442 possible estuaries, 376 were joined to adjacent watersheds and included in the analysis. The literature review uncovered 20 threat categories and, based on the available data, 16 threat variables were created (Table S2). A spatial dataset showing the threat distribution within each estuary is available from the author. Change in future precipitation was the only threat to overlap with all estuary-watershed systems but all systems overlapped at least two threats (Figure 2). No system overlapped all of the threats, though two systems in the Strait of Georgia ecosection were affected by 15 threats. Overall, there was an average of 7.9 threats per estuary-watershed system but systems within the Strait of Georgia ecosection averaged 11.0 threats and four other ecosections, also on the south coast, had a higher than average number of threats per estuary-watershed system (Table 1).

Estuary-watershed systems with the highest importance to waterbirds had the highest counts of threats. Systems with a protected area, either legislated or privately conserved, faced an average of 7.6 threats. Estuary-watershed systems with less area protected faced a greater average number of threats and those facing the fewest threats were the most protected. Systems with a legislated protected area faced fewer average threats than those with a privately conserved area.

Table 2. The incidence of estuarine threats within the literature relevant to BC estuaries and estuarine habitats, such as salt marshes, seagrass beds, and sand flats.

	Key sources used to identify threats to estuaries and estuarine habitats						
Estuarine threats	**[9]**	**[8]**	**[19]**	**[7]**	**[1]**	**[2]**	**[35]**
Terrestrial							
Development in watershed	X	X	X	X	X	X	
Agriculture in watershed	X	X	X	X	X		
Forestry in watershed		X	X	X	X		
Watershed pollution	X	X	X	X	X	X	X
Freshwater diversions	X	X	X	X	X	X	X
Shoreline armouring in estuary	X	X	X	X	X		X
Impaired waterways		X	X	X	X	X	
Debris/litter				X		X	X
Marine							
Development of estuary shoreline	X	X	X	X	X	X	X
Port facilities in estuary	X	X		X	X		X
Dredging	X	X	X	X			X
Marine pollution				X	X	X	X
Aquaculture	X		X	X	X		X
Resource extraction				X		X	X
Introduced species	X	X	X	X		X	X
Recreation		X		X			
Increased atmospheric carbon dioxide							
Sea level rise	X	X		X	X	X	X
Sea surface temperature		X					X
Ocean acidification							X
Change in future precipitation	X	X			X		
Ecosystem focus (location)	Est* (BC)	SM (WC US)	Est (WC NA)	Est (global)	Est (WC US)	Est (BC)	SM (CA)

*Abbreviations: Est = estuary; SM = salt marsh; BC = British Columbia; WC = west coast; US = United States of America; NA = North America; CA = California.

Table 3. Average values of the threats affecting BC estuaries, grouped into the three clusters generated by the K-medians cluster analysis.

Cluster	Urban (%)	Tenures (%)	Shoreline Armouring (m/ha)	Aquaculture (%)	Ports (%)	Dredging (%)	Watershed Pollution (%)	Agriculture (%)
1	67.8%	5.4%	5.7	1.7%	61.8%	0.6%	12.6%	0.2%
2	92.1%	7.8%	9.0	11.4%	21.4%	0.5%	32.0%	0.4%
3	100.0%	17.8%	38.2	18.6%	72.4%	7.7%	87.8%	14.5%
Cluster	**Freshwater Diversion (count/ha)**	**Forestry (%)**	**Contamination Closures (%)**	**Commercial Fishing (catch in kg)**	**Recreational Fishing (%)**	**Large Vessel Traffic (hours)**	**Small Vessel Traffic (%)**	**Future Precipitation (% change)**
1	0.000010	15.4%	3.7%	890.8	19.2%	605.1	75.4%	19.2
2	0.000028	70.3%	3.1%	1,697.9	58.4%	3,650.7	67.5%	16.9
3	0.000075	52.8%	66.3%	23,345.3	37.8%	19,439.6	81.9%	15.8

Table 4. Average conservation importance, protection level, and threat status of BC estuaries, grouped into the three clusters generated by the K-medians cluster analysis.

Cluster	Number of Estuaries	Importance for Waterbirds	Priority Sites for Carbon Sequestration	Count of Threats	Count of Protected Areas	Proportion of Area Protected (%)	Average Estuarine Area (ha)	Average Watershed Area (ha)
1	180	3.6	0	6.6	1.3	34.4%	88.7	5,918.8
2	127	3.5	2	8.0	0.9	9.5%	163.6	6,034.4
3	69	3.1	6	11.1	2.1	6.7%	521.8	6,129.8

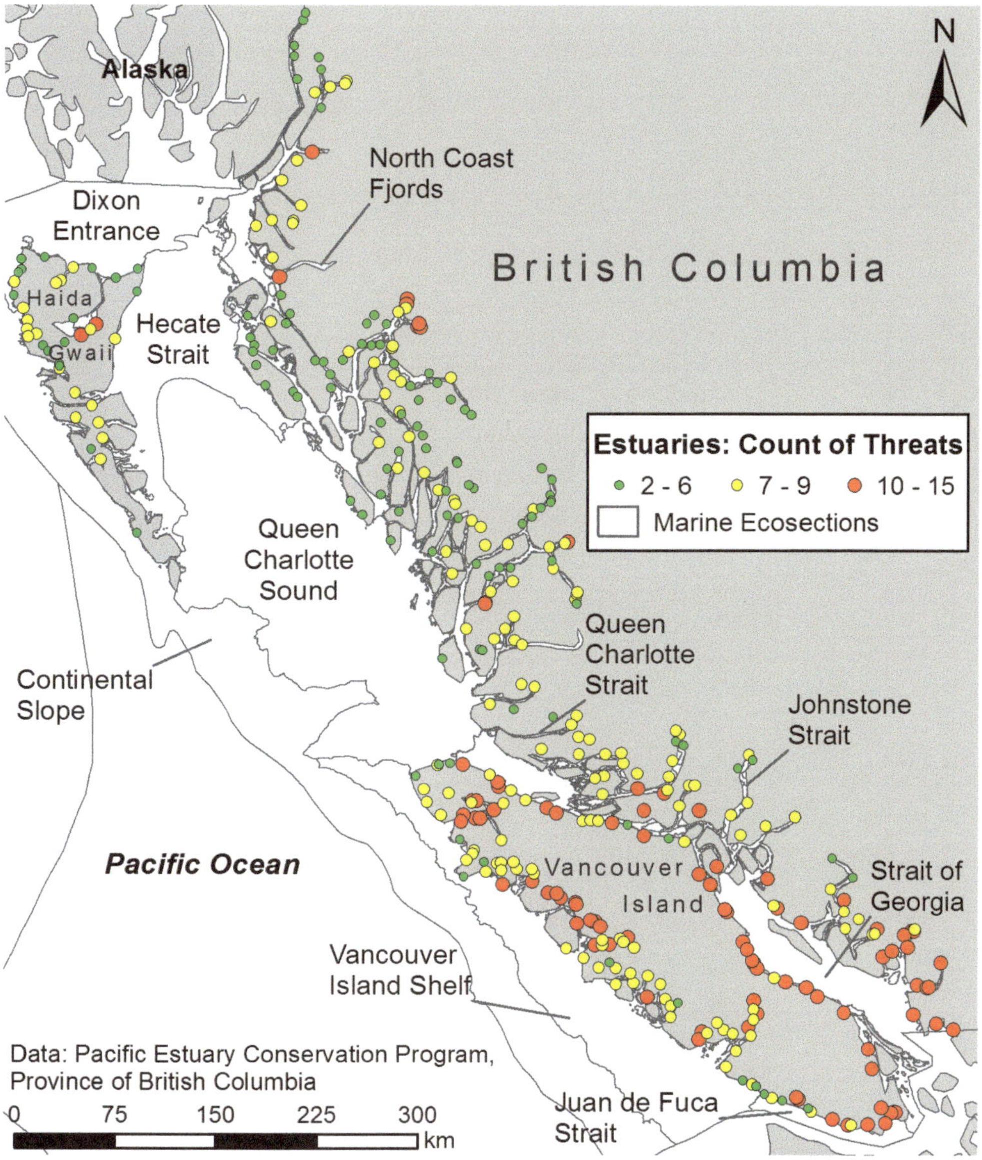

Figure 2. Count of threats by estuary-watershed system. Description: Estuary-watershed systems are shown classified based on the number of the 16 possible threats found within their bounds. The classification uses Jenks natural breaks which maximizes the variance between the classes. The number of threats faced by an estuary is greatest in the more heavily populated south coast ecosections.

Cluster Analysis

The cluster analysis revealed three optimal clusters (Figure 3). Cluster 3, representing estuaries with the greatest importance for waterbirds and carbon sequestration (Table 4), was the cluster with the greatest impact from most of the threat variables (Table 3). These 69 estuaries were found primarily in the south coast ecosections, dominated by the Strait of Georgia ecosection. Cluster 3 had the greatest average number of protected areas within each estuary, but the smallest average proportion of area protected. Cluster 2 represented 127 estuaries moderately affected by most threat variables and the highest impact from forestry and recreational fishing. The majority of estuaries in this cluster were found on the south coast in the less populated Vancouver Island Shelf ecosection. Cluster 1 faced minimal impacts from most threats but had a higher than average future change in precipitation. Cluster 1 also had the most estuaries, located predominantly within the North Coast Fjords ecosection. Based on average area conserved, these estuaries were also the most protected.

Discussion

A variety of local approaches have been important for estuary conservation, including the creation of legislated protected areas, the purchase of sites for private conservation, and the establishment of estuary management plans. A conservation designation was found within 61% of the estuary-watershed systems but systems in the north coast ecosections had greater protection than their southern counterparts (Table 1). The biased placement of legislated conservation designations in less threatened areas further from human settlements, and therefore less costly to protect, matches global trends for terrestrial protected areas [42]. Estuary-watershed systems with privately conserved areas faced a greater number of threats than those with legislated protected areas. As

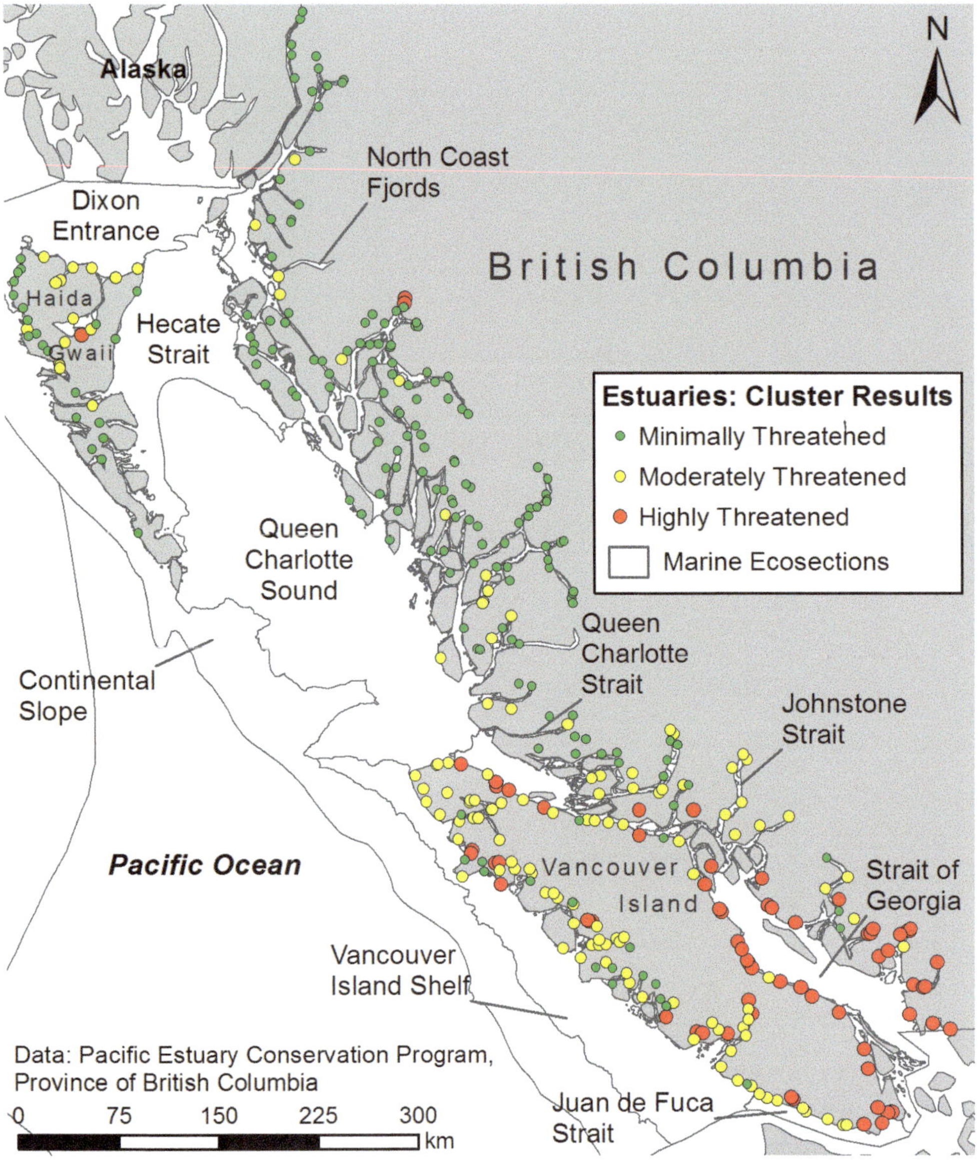

Figure 3. Distribution of K-medians cluster results. Description: Estuary-watershed systems are shown distributed into clusters that resulted from the K-medians cluster analysis based on the presence of 16 threat variables. Similar to the count of threats, the clusters highlight a north-south gradient, with highly threatened estuaries primarily observed in the south coast ecosections and minimally threatened estuaries predominantly observed in the north coast ecosections.

Brooks *et al.* [42] note, people most value "what is close to them". Privately protected areas, often negotiated by the PECP with landowners, are found primarily along the south coast, where the greatest concentration of humans, and industrial activities, is found [10,17].

In 2002, a review of estuary management plans in BC identified only nine completed plans, predominantly for estuaries on the south coast [43]. Since that time, planning has been completed for two additional estuaries but some of these plans, particularly those with limited stakeholder support, have taken up to 20 years to finalize [43] and some have not been fully implemented (K. Moore, Environment Canada, personal communication). With uneven estuarine protection along the coast and management plans for less than 2% of mapped estuaries, it is apparent that local efforts alone will not be sufficient to conserve estuarine habitats effectively into the future.

To ensure the maintenance of biodiversity, a regional approach based on the principles of EBM offers a more comprehensive, efficient, and cost-effective alternative to site-specific protection and planning [30,44]. The PECP's efforts to map and rank estuaries began the work of looking at BC's estuaries through a regional lens. However, the PECP lacked the comprehensive anthropogenic data necessary to complete the ranking [2]. Identifying clusters of estuaries that face similar threats can highlight priority activities and areas where coordinated management could achieve a higher return on investment.

Highly Threatened Estuaries

Areas with high diversity or importance and high cumulative impacts are considered conservation priorities to protect species and habitats that would otherwise be lost to habitat modification [28,42]. Identified as highly important for waterbirds and carbon sequestration, the highly threatened estuaries in Cluster 3 qualify as conservation priorities (Table 4). These estuaries were located predominantly within the heavily populated south coast ecosections (Figure 3), which correlates with results from previous research [18,33] as well as the simpler count of threats by estuary (Table 1 and Figure 2).

Protected areas that prohibit human activities are a prominent means of conserving estuaries [23] but the systems in Cluster 3 had the lowest average proportion of area protected (Table 4). Many of those estuaries are located near major urban centers, where the financial costs and logistical challenges associated with establishing and ensuring compliance of conservation designations are higher [45]. Protecting estuaries in highly populated areas will require stakeholder engagement, a component of EBM, to ensure that all relevant ecosystems and activities are considered, that priorities are established collectively, and that efforts are not duplicated [30,46]. However, only the conservation of representative sites across the coast will ensure the continued viability of estuarine ecosystems and help Canada meet its international commitment to preserve 10% of each ecological region in the marine and coastal realm [46].

Moderately Threatened Estuaries

Cluster 2 faced a moderate level of most threats (Table 3) and contained estuaries located primarily in the less populated ecosections of the south coast (Figure 3). These systems had the greatest impact from recreational fishing and forestry, suggesting that coordinating estuarine management with these two sectors might alleviate significant stresses. Regional planning is a means of engaging industries whose activities adversely affect an ecosystem from afar [47], such as the terrestrial activities like agriculture, forestry, and watershed development that affect estuaries in Clusters 2 and 3. Smith *et al.* [47] suggest that a system of local plans for highly used areas nested within regional plans might offer the most efficient solution. This suggestion meshes with recommendations from the Royal Society of Canada [48], who proposed that planning for DFO's large ocean management areas should be tied to planning for the smaller coastal management areas because the areas are physically and biologically integrated.

Minimally Threatened Estuaries

While some prioritize action in highly threatened areas, there are also calls to minimize conflict by proactively protecting areas required by few industries [49,50]. Over 40% of the estuary-watershed systems assessed, represented by Cluster 1, were minimally affected (Table 3). This result is encouraging, suggesting that many of BC's estuaries are relatively pristine. In comparison, research done for estuaries along the more heavily populated west coast of the United States found that only 16% were minimally affected [1]. The systems of Cluster 1 were located primarily in the lightly populated north coast ecosections (Figure 3). These sites faced a slightly greater average increase in future precipitation but because of the relatively low levels of other anthropogenic threats, the estuaries may be best poised to respond to the corresponding increase in water flow [51].

While the estuaries of Cluster 1 were the least affected by the combination of all threats, they did show moderate impacts from a few threats (Table 3). These estuaries are primarily located on the north coast, away from human settlements. However, small human populations do not guarantee minimal anthropogenic impacts [28] and some threats, such as forestry, shipping, and climate change, can influence even remote estuaries. Threats that impact all of the clusters are examples of regional issues that are likely managed most efficiently at a broad scale [11].

Limitations in Data and Analysis

Cluster analysis is an inherently subjective statistical methodology relying on researcher judgment to gauge the integrity of the results. A strong theoretical foundation for the analysis helps exclude irrelevant variables that might distort results [41] and the extensive literature review of anthropogenic threats (Table 2) ensured only well-researched variables were incorporated. However, the lack of fine scale or historical spatial data on marine activities and climate change is a noted impediment to conservation and research efforts [17,44].

Two specific data gaps should be noted. Observations of invasive species, one of the most important stresses in estuarine ecosystems [9,19], have not been catalogued across the BC coast (D. Buffett, Ducks Unlimited Canada, personal communication). Instead, two vectors of invasive species, aquaculture and large vessel traffic [10], were used and the results can help identify highly disturbed estuaries most susceptible to invasion [8]. Perhaps the most important data gap is projected sea level rise. The shoreline armouring feature represents estuaries with barriers preventing adaptation to higher sea levels but does not consider spatial variations in elevation. The vulnerability of BC shorelines to sea level rise has been approximated by the BC Government using coarse elevation data and the physical parameters of shoreline segments [17]. However, better projections of the impact of SLR on estuaries require high resolution elevation data that are not currently available for much of the coast [17].

Limitations in Implementation

Following the approach taken by Merrifield *et al.* [1], this analysis incorporated threats found throughout estuary-watershed systems. Estuaries link terrestrial watersheds with marine systems and the influences of the two realms both enhance [52] and threaten [3,7] the ecosystem services provided by estuaries. However, while terrestrial, coastal, and marine environments are integrated, they are often not governed harmoniously. Considering the entire land to sea continuum is a more comprehensive approach to conservation and management advocated by both government agencies [23,27] and researchers [7,8,34,47,53], despite the challenges associated with the large area and number of stakeholders involved. A failure to utilize EBM and explicitly consider threats originating across the land-sea continuum hinders effective estuary management and can undercut the success of protected areas [30]. Conservation efforts require the coordinated involvement of all relevant government bodies to facilitate successful integrated and adaptive planning [7,21,23,34]. This will be particularly relevant for the estuaries in Clusters 2 and 3, which face a suite of significant marine and terrestrial threats. In BC, improved coordination may be underway as the BC government and First Nations are developing an EBM plan that considers the cumulative impacts of human activities (www.mappocean.org) and federal agencies are collaborating on a network of marine protected areas [46].

Priorities for Management and Conservation

Land purchases such as those coordinated by the PECP play an integral role in the conservation of estuaries. To a certain extent, these purchases will likely always be done in an ad hoc manner, taking advantage of the generosity of landowners adjacent to estuaries. However, planning at a regional scale using biological and socioeconomic data can help ensure that management and conservation efforts are proactive, scientifically defensible, maximize available resources, and minimize impacts on local resource users [2,25,54]. A comprehensive network approach to protected areas can also help guarantee that protected areas fully realize their benefits [50]. To conserve estuarine ecosystem services, a network of protected areas must ensure that contributing sites include replications across the coast, maintain connectivity, and are adequately sized and protected [46].

Estuarine conservation and management would also benefit from further research into the cumulative impacts of estuarine threats. This analysis determined the specific threats facing BC's estuaries but more information is required to understand how threats interact, in particular for the highly threatened estuaries in Clusters 2 and 3. Ban *et al.* [33] began this work for BC by extrapolating expert assessments from California on the vulnerability of marine habitats to different activities [35]. However, information on the level of impact of each threat is also required and is currently only available for a few of the regional scale datasets, a shortcoming identified in other jurisdictions [1]. The BCMCA has worked with industry representatives in the marine environment to validate the accuracy of spatial data [37] and continuing that work to generate a better representation of site usage and impact is an important next step.

Austin *et al.* [9] declared climate change to be the "foremost threat to biodiversity" in BC but a lack of spatial data (Table S3) make it challenging to incorporate the relevant threats. The BC estuaries identified as priorities for carbon sequestration [5] were shown to be highly threatened (Table 4). Protecting pristine carbon sequestration sites and restoring degraded areas through an EBM process has been referred to as "one of the strongest win-win mitigation efforts known today" [32] because safeguarding carbon sequestration requires maintaining estuarine structure and function, facilitating the continued provision of the other ecosystem services that estuaries provide. Emerging information on the capacity of estuaries to sequester carbon, along with better accounting methodologies, may bring renewed impetus to improve and fund coastal management from both the public and private sectors [22,32,42,55].

Conclusions

Estuaries are an important component of BC's biodiversity and the ecosystem services they provide are well established. However, estuary degradation continues, in part due to a historical lack of integrated management and the investment required to develop and implement individual estuary management plans. Regional planning based on EBM that involves stakeholders across the entire land-sea continuum is required to prioritize sites for protection or restoration and ensure conservation benefits are realized.

By working at a broad scale and clustering estuaries based on their predominant threats, this analysis assesses human activities within BC's estuaries scientifically and reveals that estuaries deemed most important for waterfowl habitat and carbon sequestration disproportionately face the highest threats. The results also highlight a north-south gradient of threats, with minimally threatened estuaries located predominantly on the north coast and highly threatened estuaries primarily on the south coast. This information can be combined with the PECP's ecological ranking of estuaries to guide threat mitigation in marine, terrestrial, and coastal areas. Continuing to broaden our understanding of estuaries, including the cumulative impacts they face and the amount of carbon they sequester, will be essential to ensuring the sustained delivery of their vital ecosystems services.

Acknowledgments

I am grateful to the organizations that made their data available for this study, as well as the individuals who provided support and feedback, including Jonathan Moran, Kathleen Moore, Dan Buffett, Karin Bodtker, and two anonymous reviewers.

Author Contributions

Conceived and designed the experiments: CKR. Performed the experiments: CKR. Analyzed the data: CKR. Wrote the paper: CKR.

References

1. Merrifield MS, Hines E, Liu X, Beck MW (2011) Building regional threat-based networks for estuaries in the Western United States. PLoS One 6(2): 1–10.
2. Ryder JL, Kenyon JK, Buffett D, Moore K, Ceh M, et al. (2007) An integrated biophysical assessment of estuarine habitats in British Columbia to assist regional conservation planning. Delta BC: Canadian Wildlife Service, Pacific and Yukon Region.
3. Edgar GJ, Barrett NS, Braddon DJ, Last PR (2000) The conservation significance of estuaries: A classification of Tasmanian estuaries using ecological, physical and demographic attributes as a case study. Biological Conservation 92: 383–397.
4. Costanza R, d'Arge R, de Groot R, Farber S, Grasso M, et al. (2007) The value of the world's ecosystem services and natural capital. Nature 387: 253–260.
5. Campbell CR (2010) Blue carbon – British Columbia: The case for the conservation and enhancement of estuarine processes and sediments in BC. Victoria BC: Sierra Club of BC.
6. Kelsey E (1995) The Pacific Estuary Conservation Program. Victoria BC: Pacific Estuary Conservation Program.
7. Kennish MJ (2002) Environmental threats and environmental future of estuaries. Environmental Conservation 29(1): 78–107.
8. Callaway GC, Zedler JB (2009) Conserving the diverse marshes of the Pacific coast. In Batzer DP, Baldwin AH, editors. Wetland habitat of North America: Ecology and conservation concerns. London: University of California Press. 285–307.
9. Austin MA, Buffett DA, Nicolson DJ, Scudder GGE, Stevens V (2008) Taking nature's pulse: The status of biodiversity in British Columbia. Victoria BC: Biodiversity BC.
10. Environment Canada (1999) Draft Canada's National Programme of Action (NPA) for the protection of the marine environment from land-based activities. Ottawa, ON: Environment Canada.
11. Leschine TM, Ferriss BE, Bell KP, Bartz KK, MacWilliams S, et al. (2003) Challenges and strategies for better use of scientific information in the management of coastal estuaries. Estuaries 26(4B): 1189–1204.
12. Crossland CJ, Baird D, Ducrotoy JP, Lindebloom H (2005) The coastal zone – a domain of global interactions. In: Crossland, CJ, Kremer HH, Lindebloom HJ, Marshall Crossland JI, Le Tissier MDA, editors. Coastal fluxes in the anthropocene. New York: Springer-Verlag Berlin Heidelberg. 1–37.
13. Scavia D, Field JC, Boesch DF, Buddemeier RW, Burkett V, et al. (2002). Climate change impacts on U.S. coastal and marine ecosystems. Estuaries 25(2): 149–164.
14. Galbraith H, Jones B, Park R, Clough J, Herrod-Julius S, et al. (2002) Global climate change and sea level rise: Potential losses of intertidal habitat for shorebirds. Waterbirds 25(2): 173–183.
15. Nicolas D, Chaalali A, Drouineau H, Lobry J, Uriarte A, et al. (2011) Impact of global warming on European tidal estuaries: Some evidence of northward migration of estuarine fish species. Regional Environmental Change 11: 639–649.

16. Feely RA, Alin SR, Newton J, Sabine CL, Warner M, et al. (2010) The combined effects of ocean acidification, mixing, and respiration on pH and carbonate saturation in an urbanized estuary. Estuarine, Coastal and Shelf Science 88: 442–449.
17. Okey TA, Alidina HM, Lo V, Montenegro A, Jessen S (2012) Climate change impacts and vulnerabilities in Canada's Pacific marine ecosystems. Vancouver BC: CPAWS BC and WWF-Canada.
18. Ryder JL, Ceh M, Buffett D, Moore K (2006) Economic and conservation tenures in the intertidal areas of BC estuaries. In: Gilkeson L et al., editors. Alive and inseparable: British Columbia's coastal environment 2006. Victoria: Province of British Columbia. 35–41.
19. Emmett R, Llansó R, Newton J, Thom R, Hornberger CM, et al. (2000) Geographic signatures of North American West Coast estuaries. Estuaries 23(6): 765–792.
20. DFO (2008) The role of the provincial and territorial governments in the oceans sector. Ottawa ON: Fisheries and Oceans Canada.
21. Robb CK, Bodtker KM, Wright K, Lash J (2011) Commercial fisheries closures in marine protected areas on Canada's Pacific coast: The exception, not the rule. Marine Policy 35: 309–316.
22. Herr D, Pidgeon E, Laffoley D (2012) Blue carbon policy framework: Based on the discussion of the International Blue Carbon Policy Working Group. Gland Switzerland: International Union for Conservation of Nature.
23. DFO (2002) Policy and operational framework for integrated management of estuarine, coastal and marine environments in Canada. Ottawa ON: Fisheries and Oceans Canada.
24. Crowder L, Norse E (2008) Essential ecological insights for marine ecosystem-based management and marine spatial planning. Marine Policy 32: 772–778.
25. Lotze HK, Lenihan HS, Bourque BJ, Bradbury RH, Cooke RG, et al. (2006) Depletion, degradation and recovery potential of estuaries and coastal seas. Science 312: 1806–1809.
26. McLeod KL, Lubchenco J, Palumbi S, Rosenberg AA (2005) Scientific consensus statement on marine ecosystem-based management.
27. DFO (2005) Canada's oceans action plan: For present and future generations. Ottawa ON: Fisheries and Oceans Canada.
28. Halpern BS, Walbridge S, Selkoe KA, Kappel CV, Micheli F, et al. (2008) A global map of human impact on marine ecosystems. Science 319: 948–952.
29. Millennium Ecosystem Assessment (2005) Ecosystems and human well-being: A framework for assessment. Nairobi Kenya: United Nations Environment Programme.
30. Agardy T, di Sciara GN, Christie P (2011) Mind the gap: Addressing the shortcomings of marine protected areas through large scale marine spatial planning. Marine Policy 35: 226–232.
31. Douvere F (2008) The importance of marine spatial planning in advancing ecosystem-based sea use management. Marine Policy 32: 762–771.
32. Nelleman C, Corcoran E, Duarte CM, Valdés L, DeYoung C, et al. (2009) Blue carbon: A rapid response assessment. Arendal Norway: United Nations Environment Programme.
33. Ban NC, Alidina HM, Ardron JA (2010) Cumulative impact mapping: Advances, relevance and limitations to marine management and conservation, using Canada's Pacific waters as a case study. Marine Policy 34: 876–886.
34. Imperial MT, Hennessey TM (1996) The ecosystem-based approach to managing estuaries: An assessment of the National Estuary Program. Coastal Management 24: 115–139.
35. Teck SJ, Halpern BS, Kappel CV, Micheli F, Selkoe KA, et al. (2010) Using expert judgment to estimate marine ecosystem vulnerability in the California Current. Ecological Applications 20(5): 1402–1416.
36. Wang T, Hamann A, Spittlehouse DL, Murdock TQ (2012) ClimateWNA – high-resolution spatial climate data for western North America. Journal of Applied Meteorological and Climatology 51: 16–29.
37. Ban NC, Bodtker KM, Nicolson D, Robb CK, Royle K, et al. (2013) Setting the stage for marine spatial planning: Ecological and social data collation and analyses in Canada's Pacific waters. Marine Policy 39: 11–20.
38. Google (2012) Google Earth, v. 6.1. Mountainview CA: Google, Inc.
39. Quigley MP, Hall JA (1999) Recovery of macrobenthic communities after maintenance dredging in the Blyth Estuary, north-east England. Aquatic Conservation: Marine Freshwater Ecosystems 9: 63–73.
40. Systat Software (2010) Systat, v 13. Chicago IL: Systat Software, Inc.
41. Ketchen DJ Jr, Shook CL (1996) The application of cluster analysis in strategic management research: An analysis and critique. Strategic Management Journal 17: 441–458.
42. Joppa LN, Pfaff A (2009) High and far: Biases in the location of protected areas. PLoS One 4(12): 1–6.
43. Williams GL, Langer OE (2002) Review of estuary management plans in British Columbia. Vancouver BC: Fisheries and Oceans Canada.
44. Gleason MG, Merrifield MS, Cook C, Davenport A, Shaw R (2006) Assessing gaps in marine conservation in California. Frontiers in Ecology and the Environment 4(5): 249–258.
45. Balmford A, Gaston KJ, Blyth S, James A, Kapos V (2003) Global variation in terrestrial conservation costs, conservation benefits, and unmet conservation needs. Proc Natl Acad Sci U S A 100(3): 1046–1050.
46. Government of Canada (2011) National framework for Canada's network of marine protected areas. Ottawa ON: Fisheries and Oceans Canada.
47. Smith HD, Maes F, Stojanovic TA, Ballinger RC (2011) The integration of land and marine spatial planning. Journal of Coastal Conservation 15: 291–303.
48. Hutchings JA, Côté IM, Dodson JJ, Fleming IA, Jennings S, et al. (2012) Sustaining Canadian marine biodiversity: Responding to the challenges posed by climate change, fisheries, and aquaculture. Ottawa ON: Royal Society of Canada.
49. Brooks TM, Mittermeier RA, da Fonseca GAB, Gerlach J, Hoffmann M, et al. (2006) Global biodiversity conservation priorities. Science 313(5783): 58–61.
50. Jessen S, Chan K, Côté I, Dearden P, De Santo E, et al. (2011) Science-based guidelines for MPAs and MPA networks in Canada. Vancouver BC: CPAWS.
51. Baustian JJ, Mendelssohn IA, Hester MW (2012) Vegetation's importance in regulating surface elevation in a coastal salt marsh facing elevated rates of sea level rise. Global Change Biology 18: 3377–3382.
52. Barbier EB (2012) Progress and challenges in valuing coastal and marine ecosystem services. Review of Environmental Economics and Policy 6(1): 1–19.
53. Sloan NA, Vance-Borland K, Ray GC (2007) Fallen between the cracks: Conservation linking land and sea. Conservation Biology 21(4): 897–898.
54. Klein CJ, Chan A, Kircher L, Cundiff AJ, Gardner N, et al. (2008) Striking a balance between biodiversity conservation and socioeconomic viability in the design of marine protected areas. Conservation Biology 22(3): 691–700.
55. Luck GW, Chan KMA, Fay JP (2009) Protecting ecosystem services and biodiversity in the world's watersheds. Conservation Letters 2: 179–188.

4

Spatial Distribution and Conservation of Speckled Hind and Warsaw Grouper in the Atlantic Ocean off the Southeastern U.S.

Nicholas A. Farmer[1]*, Mandy Karnauskas[2]

1 National Oceanographic and Atmospheric Administration, National Marine Fisheries Service, Southeast Regional Office, St. Petersburg, Florida, United States of America,
2 National Oceanographic and Atmospheric Administration, National Marine Fisheries Service, Southeast Fisheries Science Center, Miami, Florida, United States of America

Abstract

There is broad interest in the development of efficient marine protected areas (MPAs) to reduce bycatch and end overfishing of speckled hind (*Epinephelus drummondhayi*) and warsaw grouper (*Hyporthodus nigritus*) in the Atlantic Ocean off the southeastern U.S. We assimilated decades of data from many fishery-dependent, fishery-independent, and anecdotal sources to describe the spatial distribution of these data limited stocks. A spatial classification model was developed to categorize depth-grids based on the distribution of speckled hind and warsaw grouper point observations and identified benthic habitats. Logistic regression analysis was used to develop a quantitative model to predict the spatial distribution of speckled hind and warsaw grouper as a function of depth, latitude, and habitat. Models, controlling for sampling gear effects, were selected based on AIC and 10-fold cross validation. The best-fitting model for warsaw grouper included latitude and depth to explain 10.8% of the variability in probability of detection, with a false prediction rate of 28–33%. The best-fitting model for speckled hind, per cross-validation, included latitude and depth to explain 36.8% of the variability in probability of detection, with a false prediction rate of 25–27%. The best-fitting speckled hind model, per AIC, also included habitat, but had false prediction rates up to 36%. Speckled hind and warsaw grouper habitats followed a shelf-edge hardbottom ridge from North Carolina to southeast Florida, with speckled hind more common to the north and warsaw grouper more common to the south. The proportion of habitat classifications and model-estimated stock contained within established and proposed MPAs was computed. Existing MPAs covered 10% of probable shelf-edge habitats for speckled hind and warsaw grouper, protecting 3–8% of speckled hind and 8% of warsaw grouper stocks. Proposed MPAs could add 24% more probable shelf-edge habitat, and protect an additional 14–29% of speckled hind and 20% of warsaw grouper stocks.

Editor: Craig A. Layman, North Carolina State University, United States of America

Funding: The authors have no support or funding to report.

Competing Interests: The authors have declared that no competing interests exist.

* E-mail: nick.farmer@noaa.gov

Introduction

The South Atlantic Fishery Management Council (SAFMC) manages speckled hind (*Epinephelus drummondhayi*) and warsaw grouper (*Hyporthodus nigritus*) from federal waters at the Virginia/North Carolina border through the Atlantic side of the Florida Keys. Currently, these stocks are listed as undergoing overfishing, with an unknown overfished status [1]. Stock assessments of varying degrees of resolution and rigor have indicated a declining trend for both stocks [2–10]. In the first formal stock assessment of speckled hind and warsaw grouper [11], catch curve analyses indicated that static spawning potential ratios (SPR) for warsaw grouper were between 0.2% and 6% in 1988 and 1990, and speckled hind SPR values declined from 25% in 1988 to 5% in 1999 [4–7]. SPR is the average fecundity of a recruit over its lifetime when the stock is fished divided by the average fecundity of a recruit over its lifetime when the stock is unfished; the low ratios from the most recent assessment [11] indicated the stocks were undergoing overfishing. A recent study [10] sampled 1,365 speckled hind (1977–2007) from North Carolina to central Florida and revealed trends suggesting speckled hind are overfished and undergoing overfishing, including increasing fishing mortality rate, decreasing size-at-age, and reduced numbers of mature individuals. There is a broad scientific and management interest in the development of effective and efficient regulations to reduce bycatch mortality and promote the rebuilding for these stocks.

Speckled hind and warsaw grouper have a complicated management history which makes any analysis of their distribution or current status from fishery-dependent data analytically challenging. Speckled hind and warsaw grouper regulations went from inclusion in the five grouper aggregate recreational bag limit in 1992 (56 FR 56016), to a commercial and recreational limit of one per vessel of each species with a commercial sale prohibition of these species in 1994 (59 FR 27242), to a complete harvest prohibition of both species in 2011 (75 FR 82280). In February 2009, Amendment 14 to the SAFMC's Snapper-Grouper Fishery Management Plan (S-G FMP) implemented eight deepwater marine protected areas (MPAs), in part to reduce bycatch of speckled hind and warsaw grouper. Due to continuing concerns regarding the status of these stocks, Amendment 17B established annual catch limits (ACLs) of zero pounds for speckled hind and warsaw grouper in January 2011and prohibited harvest beyond a depth of 240 ft (73.15 m) for snowy grouper, blueline tilefish, yellowedge grouper, misty grouper, queen snapper, and silk snapper in the U.S. South Atlantic. In May 2012, Regulatory Amendment 11 (Reg-11) to the S-G FMP removed the 240-ft closure to deep-water species imposed by Amendment 17B, in favor of more targeted, shelf-edge spatial protection.

To provide greater protection to these species, the SAFMC is currently developing Regulatory Amendment 17 (Reg-17), which proposes a variety of spatial closures which could reduce bycatch mortality for these stocks. A broad suite of no-take marine protected area (MPA) alternatives were developed by the SAFMC MPA Expert Working Group (EWG; Fig. 1). The analysis presented in this paper assimilates all available fishery-dependent and fishery-independent data to describe the geographic distribution of speckled hind and warsaw grouper. The relative conservation benefits of existing and proposed MPAs are also evaluated for each stock.

Methods

Data Sources

To determine locations of warsaw grouper and speckled hind encounters, observations were compiled from numerous sources (Table 1) and merged into a Geographic Information System (GIS) database. All data were used to develop a spatial classification model, and select data with underlying effort information were used to develop a more rigorous geographic distribution model. Locations of catch were provided to the highest possible resolution. Each of the data sets is described in detail below.

Since 1977, the Marine Resources Monitoring Assessment and Prediction (MARMAP) program has conducted fisheries-independent research between Cape Lookout, North Carolina, and Ft. Pierce, Florida. Gears and methodologies used have been consistent over the years to allow for long term analysis and comparisons. Sampling effort for snapper-grouper has historically been concentrated off South Carolina using various trap gears. MARMAP samples accurately identify fish to species and also collect valuable information on undersized fish. MARMAP data were aggregated by gear and set (i.e., a single trap, or a single line).

Since 2010, National Marine Fisheries Service's Southeast Fisheries Science Center (NMFS-SEFSC) has conducted fishery independent Southeast Fishery-Independent Survey video (SEFIS-V) and trap surveys (SEFIS-T). All gear-set level point observations of speckled hind and warsaw grouper from these surveys were incorporated into GIS. Additional survey data from submersible dives on continental shelf edge habitats were also incorporated (http://www.sefsc.noaa.gov/labs/beaufort/ecosystems/sefis/).

Since 2004, NMFS has conducted deep-water remote-operated vehicle (DW-ROV) surveys of the five natural bottom MPAs in the US South Atlantic between Jacksonville, FL and Cape Fear, NC. Based upon limited multibeam bathymetric maps and the local knowledge of other researchers, ROV transects were surveyed inside the MPAs and in adjacent open-to-fishing areas of similar depth and habitat type. Transects of roughly 1 km were followed and the resulting videotapes were analyzed for all detected fish and structure forming invertebrates. Observations of speckled hind and warsaw grouper were identified in a GIS map. Additional information using similar methods were collected in the Oculina Banks MPA from 2001–2005 (Oculina-ROV).

Since 1973, the Southeast Headboat Survey (HBS) has required recreational headboat captains to maintain logbooks recording trip-level information on number of anglers, trip duration, date, area fished, and catch by species. Headboats typically accommodate 15 or more anglers on half- or full-day for-hire trips. Headboat encounters (landings plus releases) were summarized by species, year, month, and area fished. Reporting of area fished has improved through time, with resolution ranging from state level to 0.17° by 0.17° grids. Area fished is self-reported, and vessels fishing in multiple areas during a trip were constrained by the data form design to report only one area fished for the trip. As such, the spatial reliability of headboat data, especially for rarely encountered species, is questionable. Depth fished was not reported.

In July 2006, NMFS-SEFSC began a voluntary reef fish observer program (RFOP) to characterize fishery landings and bycatch in the Atlantic Ocean off the southeastern U.S. This program is limited in geographic scope, but provides accurate set-level geographic location and discard information for fish encountered using bottom longline, electric (bandit) reel, and hand lines. Depth fished was reported for each set.

Between 1972–1979, scientists from NMFS-SEFSC's Fisheries Research Group (FRG) collected fish from offshore waters between Cape Lookout and Cape Fear, North Carolina. Numbers, size, and collection location were recorded by species at three primary sites in Onslow Bay, NC [8].

Since 1990, the Reef Environmental Education Foundation (REEF) survey has collected standardized information from volunteer divers and snorkelers on marine fish populations. Using a roving diver technique, volunteers recorded the geographic location and approximate abundance of species sited (www.reef.org). Only two REEF speckled hind records, one by a novice and one by an expert, were incorporated into the analysis; both observed two speckled hind on the USS Wilkes Barre wreck.

From 1979 to 2012, NMFS-SEFSC and the University of Miami, in conjunction with various federal, state and academic partners, have conducted a reef fish visual census (RVC) in the Florida Keys and Dry Tortugas [12]. In this two-stage sampling design, trained divers conduct a stationary point count of all reef fish stocks within a given distance of the sampling site, and record species, abundance, and various size metrics.

In 1985 and 2002, Dr. George Sedberry (South Carolina Department of Natural Resources) participated in research submarine dives off the southeastern U.S. (Sedberry-Sub). Speckled hind and warsaw grouper were observed during some of these dives and the locations of the observations were recorded.

Since 1884, various U.S. museums have maintained collections of speckled hind and warsaw grouper, including the Florida Museum of Natural History, Gainesville (www.flmnh.ufl.edu/scripts/dbs/fish_pub.asp), the North Carolina State Museum of Natural Sciences, Raleigh (www.naturalsciences.org), and Smithsonian National Museum of Natural History, Washington, DC (www.mnh.si.edu). Geographic coordinates for capture locations were either downloaded directly from online catalogs or specifically requested (W. Laney, United States Fish and Wildlife Service, pers. comm.).

Through public comment and a series of expert workshops, several recreational and commercial fishermen contributed catch location information to SAFMC staff. An additional warsaw grouper site was identified from two complementary sources [13,14]. Historical photographs and underwater videos were used to groundtruth several anecdotal sites.

Spatial classification model

To classify shelf-edge habitats in the Atlantic Ocean off the southeastern U.S. with regards to their utility to speckled hind and warsaw grouper, a simple spatial classification model was developed as follows. Offshore habitats between 25–100 fathoms

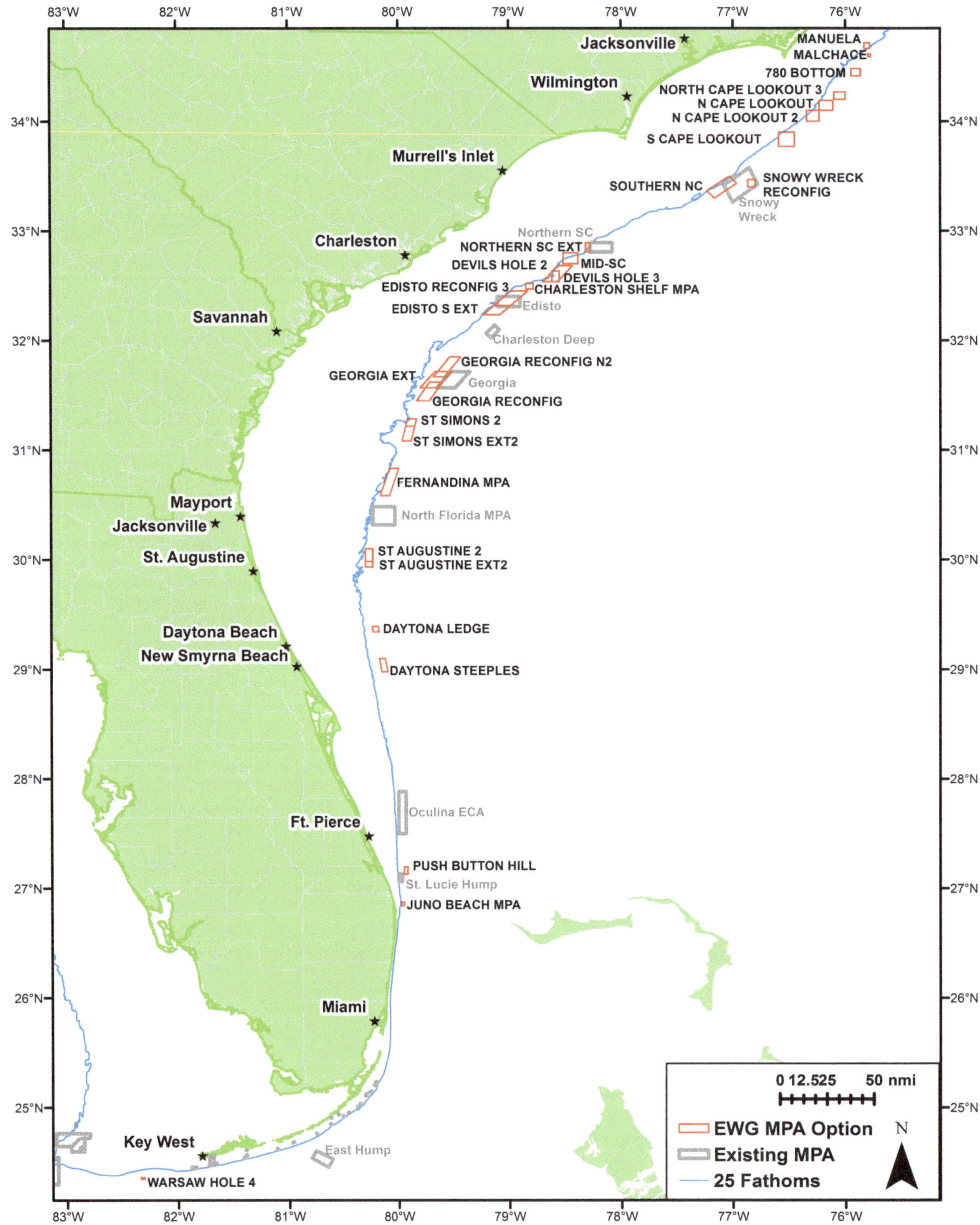

Figure 1. Existing and proposed protected areas. Existing (gray) marine protected areas (MPAs) and MPA options (red) developed by the South Atlantic Fisheries Management Council's MPA Expert Workgroup (SAFMC 2013).

Table 1. Point data sources evaluated in meta-analysis, indicating whether data was included in spatial classification model (SCM) or geographic distribution model (GDM).

Data Source	Years	Resolution	Discards?	Depth?	SCM	GDM
Headboat Log	1973–2011	Some 0.17°×0.17°	2004-present	No	X	X
Reef Fish Observer	2006–2011	Lat/Long	Yes	Yes	X	X
MARMAP	1977–2011	Lat/Long	Yes	Yes	X	X
DW ROV Survey	2004–2011	Lat/Long	Yes	Not provided	X	X
Fisher Reports	1960s–2011	Loran and Lat/Long	Yes	Some	X	
SEFIS	2010–2011	Lat/Long	Yes	Yes	X	X
REEF	1980s–2011	Lat/Long	Yes	No	X	
Oculina ROV	2003–2005	Lat/Long	Yes	Yes	X	X
Manooch Fisheries Research Group (FRG)	1972–1977	Lat/Long	Yes	Yes	X	X
Sedberry Sub	1985, 2002	Lat/Long	Yes	Yes	X	
Rudershausen et al.	2007	Lat/Long	Yes	Yes	X	
Museum Collections	1884–1991	Lat/Long	Yes	Yes	X	

(45.7–182.9 meters) from North Carolina to the Florida Keys were gridded following the Southeast Area Monitoring and Assessment Program (SEAMAP) 1199 grid [15]. Each grid cell was one-minute latitude by one-minute longitude. The grid extended from the shoreline to approximately five nautical miles beyond the 200 m depth contour (roughly the continental shelf break). Each grid cell within the one-minute grid was coded to a bottom type of Hard Bottom (HB), Possible Hard Bottom (PH), or Not Hard Bottom (NH), based on the categorization by FWC (2001) of the SEAMAP Bottom Mapping data that intersected the grid cell. If a cell had any HB data in it, it was coded to HB regardless of any NH data in the cell. If a cell had NH and no other type of data, it was coded to NH. If a cell was not sampled, it was coded as Unknown (UN).

A variety of supplemental bathymetric layers were assimilated from the National Oceanographic and Atmospheric Administration (NOAA), SEFIS, USGS, US Navy, and NCCOS (A. David and G. Şedberry, NOAA, pers. comms.; NCCOS data available from: http://ccma.nos.noaa.gov/ecosystems/sanctuaries/south_atlantic/data/). Data were merged into a layer, clipped by the SEAMAP grid, and evaluated using surface statistics for maximum percent slope. Because the average maximum percent slope across SEAMAP cells categorized as HB by FWC (2001) was 1.45, SEAMAP cells categorized as UN were recategorized as PH if their max slope from the supplemental bathymetric sources was greater than 1.45.

Using the Coastal Relief Model (www.ngdc.noaa.gov/mgg/coastal/startcrm.htm), the habitat categorization grids were clipped by 5-fathom bins, creating depth-grids. Information was projected as UTM NAD83 Zone 17N and areas (km^2) were assigned to clipped depth-grids using Hawth's Tools [16]. Point data were plotted at the set level for observations of speckled hind, warsaw grouper, and all sets. These data were counted within depth-grids. Analyses focused on the 30,275 depth-grids within the 25–100 fathom depth range; this encompassed the majority of observations of mature fish and was the primary area of concern with regards to barometric trauma and associated high release mortality.

Each SEAMAP depth-grid was classified as follows: 'Known' – A speckled hind was observed by a data source other than HBS; 'Not suitable' – Habitat type was 'NH' if no speckled hind were observed and more than 5 samples were taken in a depth-grid; 'Probable' – A HBS observation fell within the depth-grid or the habitat type was 'HB' or 'PH'; 'Unknown' – Fewer than 5 negative samples and no identified habitat within the depth-grid. Observations for headboat (HBS) were treated differently due to concerns about the reliability of headboat spatial reporting. The percentage of area falling into the various habitat classifications was computed, and the proportion of these habitat classifications already contained within currently established and proposed MPAs was determined. The same process was followed for warsaw grouper.

Geographic distribution model

Logistic regression analysis was used to develop a quantitative model predicting the spatial distribution of speckled hind and warsaw grouper. The logistic regression modeled the probability of detecting an individual within a given depth-grid as a function of gear type, depth, latitude, and habitat (Table 2). Gear type and habitat were treated as factors, and depth and latitude effects were tested in the model as continuous variables, squared terms, and factors of varying bin sizes. Because the recreational headboat logbook records contain self-reported spatial locations constrained to a $1/6° \times 1/6°$ grid, and are not necessarily reliable, we reran our models with and without this data type included to assess the effect of the headboat data on the results. The headboat logbook records make up 70% of the total observations, and thus have the greatest influence on the analysis; we also tested the exclusion of other gear types which made up more than 5% of observations (MMAP, RFOP) to ensure that our results were robust. Logistic regression analysis using a logit link was implemented using R version 2.13.2 [17]. The same modeling process was followed for both speckled hind and warsaw grouper separately. Models treated data in binary form because multiple observations were rare, and because 'catchability' differed among the various survey methods used. The expected grouper abundance for each habitat, depth and latitude were computed using the logistic model parameter estimates, controlling for the sampling effects of the gear.

We evaluated potential models of grouper probability of occurrence including all combinations of factors. Model selection applied a cross-validation procedure in which the data were split

Table 2. Input variables considered in logistic geographic distribution model.

Variable	Description
lat_cont	WGS Latitude; treated as continuous variable
lat_sq	Quadratic function of WGS Latitude
lat	WGS Latitude, treated as categorical, binned at a 1-degree resolution
dep	10-fathom depth bins; treated as categorical
dep_fine	5-fathom depth bins; treated as categorical
dep_sq	Quadratic function of 5-fathom depth bins
hab	Habitat classification: 'HB'- hardbottom or possible hardbottom, 'NH'- not hardbottom, 'UN'- unknown
gear	Gear classification*: MARMAP, HBS, FRG, SEFIS Trap, SEFIS Video, Oculina ROV, DW-ROV, RFOP, REEF

*See text for description of data sources.

up into training and testing sets, and a model was fit to the training set and subsequently tested on the "unseen" testing set. We employed 10-fold cross validation [18] such that the data set was randomly split into 10 groups, each group serving once as a testing set for a model trained on the other 9 groups of data. For each of the 10 folds of cross-validation, all possible models including the different combinations of factors were trained on the training set. For each model, we created a receiver operating characteristic (ROC) curve, which expresses the performance of a binary classification method such as the logistic regression used here (R 'pROC' library; [19]). Using the ROC curve for each model, we calculated the threshold at which the proportion of correctly classified positive observations plus the proportion of correctly classified negative observations are maximized. Using the parameters defined by each model, as well as the threshold defined by the ROC curve for each model using the training set, we then made predictions for the testing set. Model performance was calculated by creating a contingency table, which specified the rate of false positive predictions and false negative predictions for the testing set. Better performing models were those with lower false positive rates (FPR) and false negative rates (FNR). Model performance was also measured according to the reduction in Akaike's Information Criterion (AIC; [20]).

The confidence intervals around the FNR and FPR rates were quite high, due to the nature of subsetting an already sparse data set (i.e., occurrence rates of 2–7%) for the 10 fold cross-validation procedure. To further test the robustness of our model selection procedure, we repeated the cross-validation at least 10 times for each species, by re-randomizing the testing and training data sets each time and repeating the procedure. We also carried out 5-fold cross-validation, to determine whether the number of folds had any bearing on the results. Throughout these procedures, a single model for each species stood out as the model with the highest prediction power as exhibited by the lowest FNR and FPR rates. Thus, we felt justified in using the cross-validation technique to select a single best model. In the case of speckled hind, reserve protection predicted by the best predictive model and the model with the lowest AIC were presented as a range to quantify inter-model uncertainty.

MPA Protections

To evaluate the impacts of existing and proposed spatial closures, closures were overlaid on speckled hind and warsaw grouper probability of occurrence maps. For the spatial classification model, the total area of each habitat classification contained within each MPA was summed for each stock. This was subsequently expressed as a percentage of the total area of that habitat classification within the entire SAFMC shelf-edge (25–100 fathoms) jurisdiction:

$$\sum area_{MPA}^{\text{known \& probable}} \Big/ \sum area_{SAFMC}^{\text{known \& probable}}$$

For the geographic distribution model, the probability of detection weighted by area within each depth-grid within each MPA was tallied. This was subsequently expressed as a percentage of the total area-weighted probability of detection within the entire SAFMC shelf-edge (25–100 fathoms) jurisdiction:

$$\sum (p(\text{detect})_{MPA} \times area_{MPA}) \Big/ \sum (p(\text{detect})_{SAFMC} \times area_{SAFMC})$$

Reserve protection was also computed per unit area, allowing for comparison of tradeoffs between conservation of stock versus area closed to fishing.

Results

Data Sources

Plots of point observations of speckled hind and warsaw grouper indicated that the stocks were predominantly distributed on the shelf edge between 25–100 fathoms (45.7–182.9 meters), with concentrations in certain locations in 30–45 fathoms (54.9–82.9 m; Fig. 2). The spatial distribution of headboat observations suggested positioning inaccuracies when compared with other, more reliable, point data sources (Fig. 2). Observations were heavily concentrated in heavily-sampled areas such as hardbottom habitat features within and adjacent to the existing Northern South Carolina MPA, Edisto MPA, North Florida MPA, and Oculina Experimental Closed Area. Concentrations of observations visually corresponded to areas with hardbottom; this trend was most obvious in areas with high-resolution habitat mapping (Fig. 3).

Spatial classification model

The spatial classification modeling approach identified known and probable speckled hind habitats following a consistent hardbottom ridge that moved between depth contours of 25–100 fathoms from North Carolina to southeast Florida (Fig. 4a). Warsaw grouper were more rarely encountered, but their range also appeared to encompass more of the southern end of the SAFMC's jurisdiction (Fig. 4b). The spatial classification model indicated that of the 23,592 km^2 of habitat between 25–

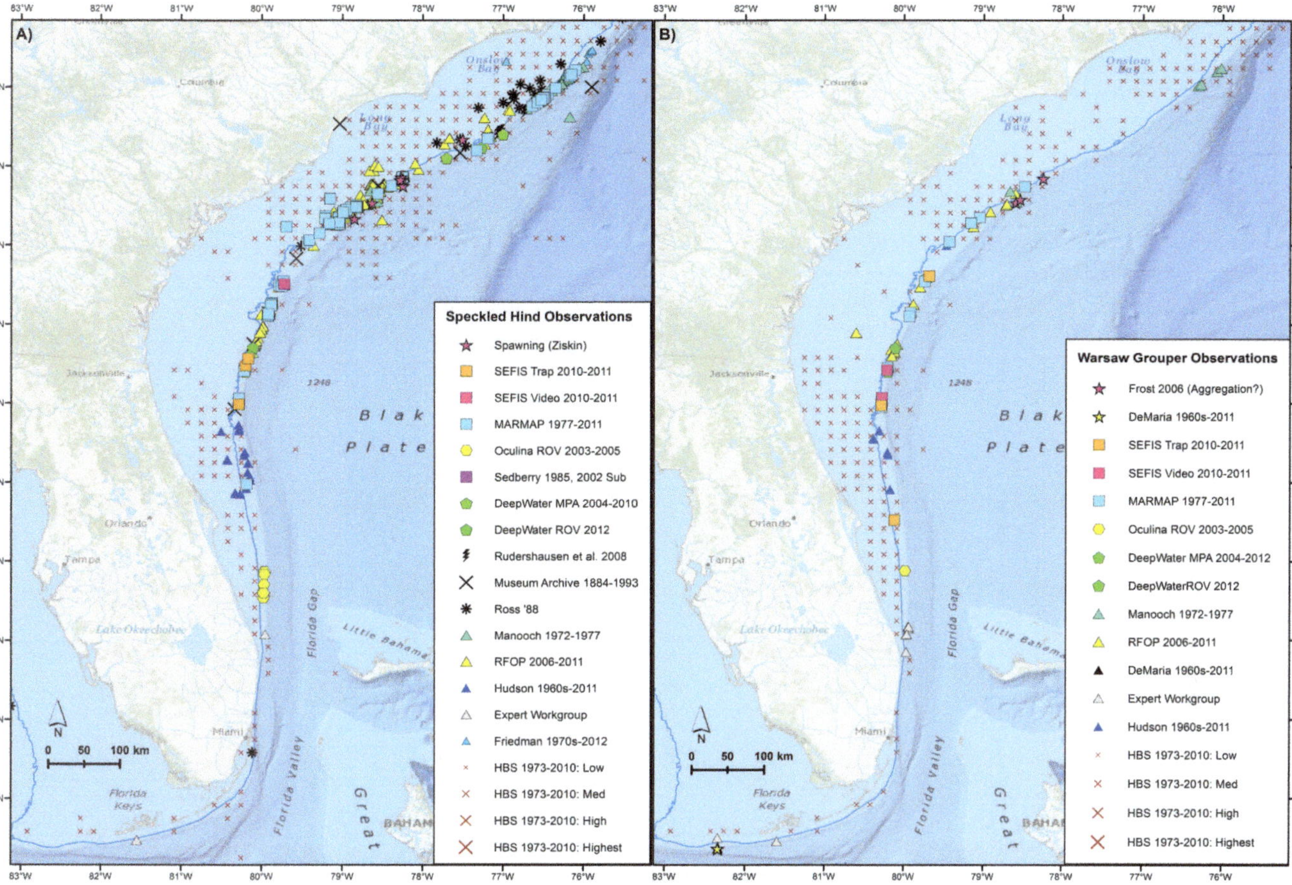

Figure 2. Point observations. A) Speckled hind and **B)** warsaw grouper encounters reported by various data sources. 25 fathom (45.7 m) bathymetric line in blue. Basemap courtesy of Esri Ocean Basemap and its partners.

100 fathoms in the SAFMC's jurisdiction, very little had been positively identified as 'Known' habitat (speckled hind: 329 km^2, warsaw grouper: 76 km^2). By contrast, the spatial classification model identifed a substantial quantity of 'Probable' habitat (speckled hind: 6984 km^2, warsaw grouper: 7090 km^2). Approximately 28% of the habitat between 25–100 fathoms was identified as unsuitable for speckled hind or warsaw grouper, and an additional 41% was unidentified.

Geographic distribution model

Logistic regression models for probability of detection for speckled hind and warsaw grouper found latitude, habitat type, and sampling gear to be important predictors of the probability of a positive observation (Table 3). The model with the lowest AIC was not necessarily the best predictive model, per 10-fold cross-validation (Table S1). For speckled hind, the model with the lowest AIC included the gear effect, latitude as a categorical variable, depth as categorical variable, and habitat. The false positive rate (FPR $\pm$ 1 S.D.) for this model was 0.248$\pm$0.166 and the false negative rate (FNR) was 0.269$\pm$0.181. The speckled hind model with the highest predictive power as assessed by cross-validation (i.e., the lowest summed FPR and FNR) included only gear, latitude as a continuous variable, and depth as a squared term. For this model, the FPR was 0.246$\pm$0.162 and the FNR was 0.234$\pm$0.148, and the deviance explained was 36.8 percent. When excluding the headboat logbook data for the speckled hind model, results in terms of model selection via both AIC and cross-validation were exactly the same, but FPR and FNR ratios for the best predictive model were increased to 0.255 and 0.357, respectively. This indicates that the information from the headboat sector in regards to distribution of specked hind was in agreement with information from other gear types, and that inclusion of headboat data improved model performance. Exclusion of other gear types also yielded similar results in terms of model selection, and therefore headboat data and all other gear types were retained for the final results. For warsaw grouper, probability of occurrence was only modeled for latitudes greater than 28 degrees north, because observations south of this point were extremely scarce (only 4 positive observations out of 11,146 data points were available). For warsaw grouper, the same model produced both the lowest AIC value and the highest predictive power (FPR = 0.282$\pm$0.227, FNR = 0.330$\pm$0.281). This model included gear effect, latitude as a categorical variable, and a squared depth term, and explained 10.8 percent of the variability in the probability of detection.

Maps of probability of occurrence across space for the two species were produced based on the best model as defined by the highest prediction power. Generally, speckled hind probability of occurrence increased with latitude, whereas occurrence of warsaw grouper decreased with latitude (Fig. 5). Speckled hind distributions were more shallow (e.g., 25–50 fathoms) relative to warsaw grouper, which were more evenly distributed across the 25–100 fathom range. Warsaw grouper probability of occurrence was highest off the coast of Georgia, and also between 28 and 29

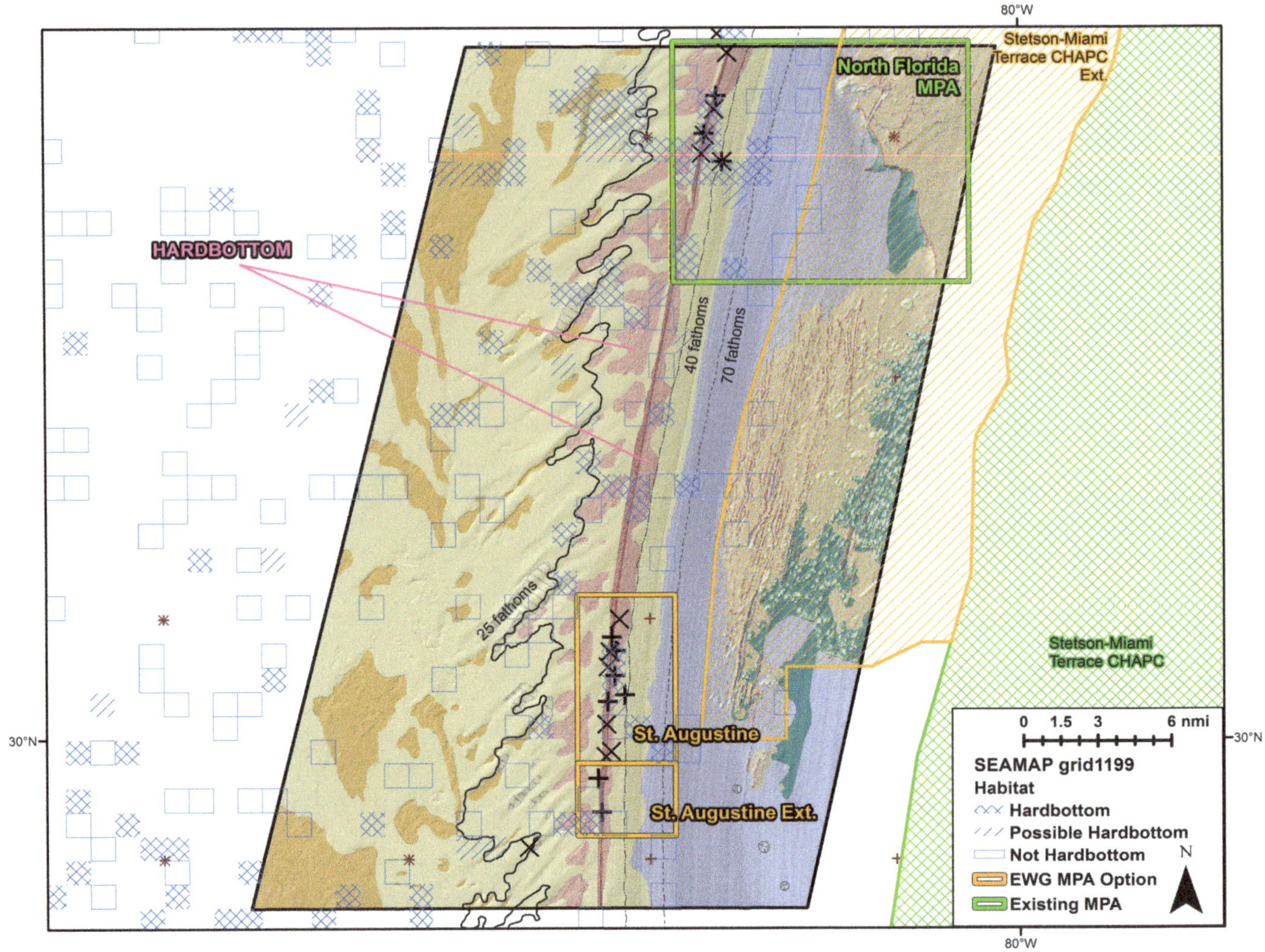

Figure 3. Point observations relative to habitat. Speckled hind (X) and warsaw grouper (+) encounters reported by various data sources versus habitat off Northeast Florida. Habitat data courtesy U.S. Navy, NOAA, and USGS (Andy David, NOAA, pers. comm.).

degrees N, off the coast of Florida. Anecdotal information [21] suggests relatively high encounter rates with warsaw grouper south of Cape Canaveral, but we did not have enough data coverage to assess occurrence in this region.

MPA Protections

Proposed and existing MPAs varied in the estimated level of protection they provided to speckled hind and warsaw grouper habitats and the percent of grouper estimated to be contained within their boundaries (Table 4). Dynamic hardbottom habitats appeared to yield the highest conservation benefit per unit area and some also contained observed spawning condition fish (Fig. 6). The highest percentage of known habitat for speckled hind was reflective of the concentration of scientific sampling around the proposed Edisto Reconfig 3, Edisto S Ext, and existing Edisto MPA. Likewise, for warsaw grouper, the highest percentage of known habitat was contained in the proposed Georgia Reconfig, Edisto S Extension, Fernandina MPA, and the existing North Florida MPA. The highest estimated percentage of known and probable habitat for speckled hind and warsaw grouper was contained within the proposed Oculina Coral Habitat of Particular Concern (CHAPC) Extension, the existing Oculina Experimental Closed Area (ECA), and the proposed Edisto Reconfig 3 and Fernandina MPA sites.

The best predictive model for speckled hind (per cross-validation) suggested the highest estimated percentage of the stock was contained in the proposed Edisto Reconfig 3 and three proposed extensions/reconfigurations of the Georgia MPA (Table 4). The best-fitting model for speckled hind (per AIC) indicated the highest estimated percentage of the stock was contained within the proposed Southern NC, Edisto Reconfig 3, and South Cape Lookout MPAs. The highest estimated percentage of warsaw grouper stock was contained within the proposed Georgia Extension, the existing North Florida MPA, and the proposed Fernandina closed area.

The most efficient reserves for the two stocks together, based on spatial classification model predictions, were the proposed Malchace Wreck and Devil's Hole 3 MPAs (Fig. 6), reconfigurations and extensions of the Edisto MPA, Push Button Hill, and the existing Oculina ECA (Fig. 7A). The most efficient reserves for both stocks together, based on geographic distribution model predictions, were the Georgia Extension, the Georgia Reconfig N2, the St. Simons 2, the Georgia Reconfig, Charleston Shelf MPA, and Edisto Reconfig 3 (Fig. 7B). The most efficient reserves

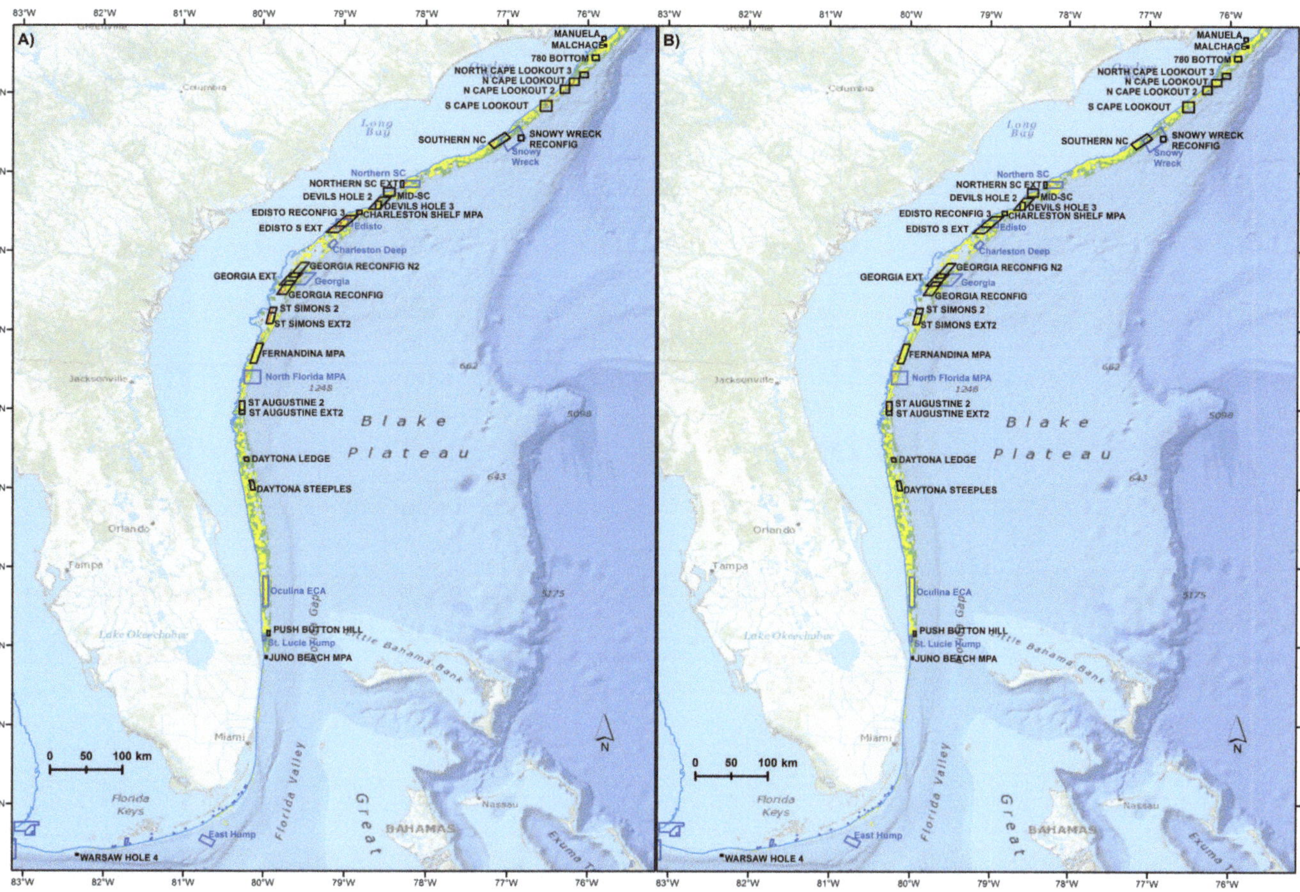

Figure 4. Habitat classification model output. Habitat classifications (red: 'known', yellow: 'probable', green: 'not suitable') for **A**) speckled hind and **B**) warsaw grouper relative to existing (blue) and proposed (black) marine protected areas. 25 fathom (45.7 m) bathymetric line in blue. Basemap courtesy of Esri Ocean Basemap and its partners.

for speckled hind, based on the spatial classification model, were the Malchace Wreck, Charleston Shelf, Edisto Reconfig 3, and 780 Bottom (Fig. 7A, black fill). The most efficient reserves for speckled hind, based on the geographic distribution model, were the Charleston Shelf, Edisto Reconfig 3, and Georgia Reconfig N2 (Fig. 7B, black fill). The most efficient reserves for warsaw grouper, based on geographic distribution model predictions, were the Malchace Wreck, Devil's Hole 3, Charleston Shelf, and 780 Bottom (Fig. 7A, gray fill). The most efficient reserves for warsaw grouper, based on geographic distribution model predictions, were the St. Augustine 2, Fernandina, and St. Simons Ext2 (Fig. 7B, gray fill).

Overall, our models estimated the 2,352 km^2 of existing deep-water MPAs and CHAPCs covered 19% of the 'Known' habitat and 10% of the 'Known & Probable' habitat for speckled hind, and protected between 3–8% of speckled hind between 25–100 fathoms. For warsaw grouper, 12% of 'Known' habitat and 10% of 'Known & Probable' habitats were estimated to be within existing MPAs, and 8% of warsaw grouper between 25–100 fathoms (north of 28° latitude) were estimated to be protected. The 3,093 km^2 of non-overlapping (e.g., excluding Devil's Hole 2 and Georgia Ext.) proposed MPA and CHAPC options were estimated to cover 72% of 'Known' and 24% of 'Known & Probable' speckled hind habitats, containing between 15–25% of speckled hind between 25–100 fathoms. The proposed closed area options were estimated to cover 68% of 'Known' and 23% of 'Known & Probable' warsaw grouper habitats, containing approximately 18% of warsaw grouper between 25–100 fathoms (north of 28° latitude).

Discussion

Conclusions regarding the status, distribution, and impacts of spatial protection for speckled hind and warsaw grouper remain uncertain. This uncertainty stems from a lack of available habitat data on scales relevant to the habitat usage of these species, as well as the rarity of the stocks. Records of warsaw grouper and speckled hind in commonly used fishery-dependent and fishery-independent data sources were limited. When data were available, catch location was often unavailable or very coarse in resolution, and thus linking these point observations to specific habitat features would be challenging even if improved habitat data were available. Prior to the early 1990s, speckled hind and warsaw grouper were not identified to species in the commercial logbooks, and a harvest prohibition began in 1994. As such, conclusions that might be drawn about the distribution of the stock from post-1994 data suffer from biases for under-representation due to the disincentive to retain the fish, and incentives to misidentify the fish if kept and sold. Depth was unavailable for most datasets. For data sources with depth, samples were most frequent from depths beyond 160 ft (48.7 m), but sampling/fishing pressure were much higher at shallower depths.

Table 3. Logistic regression model maximum likelihood parameter estimates for speckled hind and warsaw grouper probability of detection, with deviance explained (i.e. percent variability explained by inclusion of additional variable).

Parameter	Estimate	Std. Error	z value	Pr(>|z|)	Deviance explained
Speckled hind (model 5)					
Parameter	**Estimate**	**Std. Error**	**z value**	**Pr(>|z|)**	**Deviance explained**
Intercept	−20.00	0.56	−35.94	<0.001	-
gear FRG	0.53	0.37	1.44	0.15	2.8%
gear HBS	0.67	0.27	2.52	0.01	-
gear MMAP	−1.81	0.28	−6.58	<0.001	-
gear Oculina	2.36	0.44	5.36	<0.001	-
gear Reef	0.61	0.65	0.95	0.34	-
gear RFOP	−0.62	0.28	−2.23	0.03	-
gear SEFIS-trap	−2.55	0.64	−4.00	<0.001	-
gear SEFIS-video	−2.64	0.76	−3.48	<0.001	-
lat_cont	0.59	0.02	38.81	<0.001	33.1%
dep_sq	0.00	0.00	−7.21	<0.001	0.9%
					36.8%
Warsaw grouper (model 14)					
Parameter	**Estimate**	**Std. Error**	**z value**	**Pr(>|z|)**	**Deviance explained**
Intercept	−2.92	0.66	−4.44	<0.001	-
gear FRG	1.01	0.85	1.18	0.24	8.4%
gear HBS	0.29	0.62	0.46	0.65	-
gear MMAP	−2.33	0.68	−3.43	<0.001	-
gear Oculina	−12.70	334.00	−0.04	0.97	-
gear RFOP	−1.17	0.66	−1.79	0.07	-
gear SEFIS-trap	−1.87	0.84	−2.24	0.03	-
gear SEFIS-video	−1.35	0.79	−1.71	0.09	-
lat [29,30]	−0.74	0.79	−0.93	0.35	2.5%
lat [30,31]	0.39	0.43	0.90	0.37	-
lat [31,32]	−0.16	0.39	−0.40	0.69	-
lat [32,33]	−0.93	0.25	−3.70	<0.001	-
lat [33,34]	−1.77	0.47	−3.79	<0.001	-
lat [34,35]	−0.52	0.41	−1.26	0.21	-
dep_sq	0.00	0.00	0.47	0.64	0.02%
					10.92%

To control for all these confounding factors, and attempt to overcome the challenges associated with the lack of fine-scale habitat data, we consolidated a broad variety of fishery-independent, fishery-dependent, and anecdotal data sources. All data sources appeared to tell a consistent story regarding the habitats, depths, and latitudinal distribution of speckled hind and warsaw grouper. Both stocks were heavily associated with the shelf-edge between 25–100 fathoms (45.7–182.9 meters) on hardbottom habitats. Neither species was found with any frequency in the mostly mud-bottom habitats on the shelf-edge north of Cape Hatteras, North Carolina. Speckled hind were most commonly observed south of Cape Hatteras, North Carolina to northeast Florida. Warsaw grouper were most commonly observed from South Carolina to northeast Florida.

Modeling approaches

Using a suite of qualitative and quantitative approaches, we were able to generate reasonable estimates for the occurrence of speckled hind and warsaw grouper across space, and were therefore able to estimate the conservation benefits of existing and proposed marine protected areas. The spatial classification model provided a comprehensive semi-quantitative method for assimilating all available observation and habitat data. The spatial classification model appeared to provide useful predictions; following model development, additional point data were obtained from [22] and the NOAA Deepwater ROV 2012 survey. These new point observations were located within model-identified 'Known' and 'Probable' habitats. The spatial classification model provides information for warsaw grouper south of 28°N latitude, where data was too scarce for useful geographic distribution model predictions. Anecdotal information suggests the southeast Florida shelf-edge may have been historically important warsaw grouper habitat in the western North Atlantic Ocean, and this impression is supported by the spatial classification model. The spatial classification model, due to its simplistic nature, was able to provide broader predictive coverage than the more rigorous

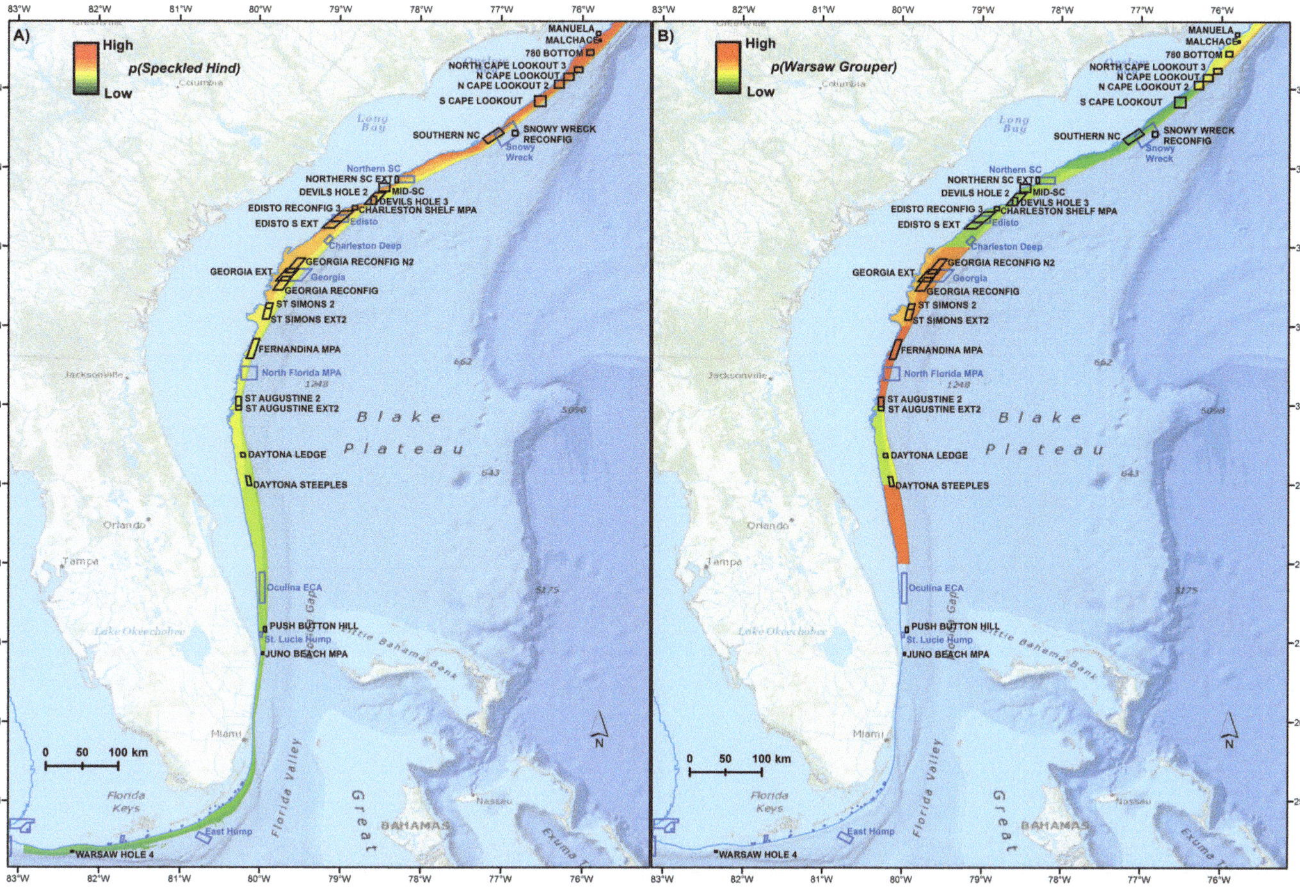

Figure 5. Geographic distribution model predictions. Probability of encounter with **A**) Speckled hind and **B**) warsaw grouper relative to existing (blue) and proposed (black) marine protected areas. 25 fathom (45.72 m) bathymetric line in blue. Basemap courtesy of Esri Ocean Basemap and its partners.

distribution model. A weakness of this simplified modeling approach was the coarse designation of habitat within each SEAMAP sampling grid (see Fig. 3). Any hardbottom or possible hardbottom within one of these one arc-minute squares would result in the whole grid ($\sim$3 km^2) being categorized as suitable habitat. This approach fails to consider factors that might make certain hardbottom habitats more 'suitable' than others – for example, speckled hind and warsaw grouper may prefer high-relief rocky reef near reef slopes, but the available data was too coarse to distinguish this preference. Additionally, the coarse categorization of the habitat data might overestimate the total 'Known' and 'Probable' habitat; however, this bias was systematic, and suitable habitat within reserves was expressed as a proportion of the total suitable habitat within the entire SAFMC jurisdiction. By expressing suitable habitat as a proportion, inter-reserve comparisons should be valid unless habitats in certain areas were less effectively categorized than others, or if certain areas featured higher concentrations of a type of hardbottom more preferred by the stock.

A geographic distribution model based on several predictor variables – gear type, latitude, and depth – was able to explain approximately 10% of the variability in warsaw grouper probability of detection and over one-third of the variability in speckled hind probability of detection. A comparable modeling attempt with a data-rich species, using a comprehensive fishery-independent survey, yielded similar results in terms of total variability explained by depth and geographical bins [23]. In this study, the presence of red snapper was modeled as a function of depth and longitude bin, among other sampling factors such as gear type. Despite the relatively high occurrence rate of red snapper (fish present at 507 of 3102 sites), and the reliability of the synoptic fishery-independent survey, depth and longitude explained only 8.5% and 6.6% of the variability in presence of red snapper after the removal of sampling artifacts. In our speckled hind model, latitude and depth explained over twice as much deviance as the red snapper model, after the removal of gear effects. This comparison suggests the overall low explanatory power of our models is due to the lack of appropriate explanatory variables, rather than a lack of reliability in the data sets used. For warsaw grouper, the limited model deviance explained by latitude and depth was likely due to the very low occurrence rates observed for this species.

While only a modest amount of the variability in detections was explained, the geographic distribution models still performed fairly well in terms of their predictive ability. Rates of false positive identifications and false negative identifications were below 25% for the best speckled hind model, and for warsaw grouper, false identification rates were only slightly higher (28–33% for the best model). Given that false positive and false negative identification rates would be about 50% for a completely random model, based purely on chance, our quantitative models reduced the rate of incorrectly predicted observations by about one-half.

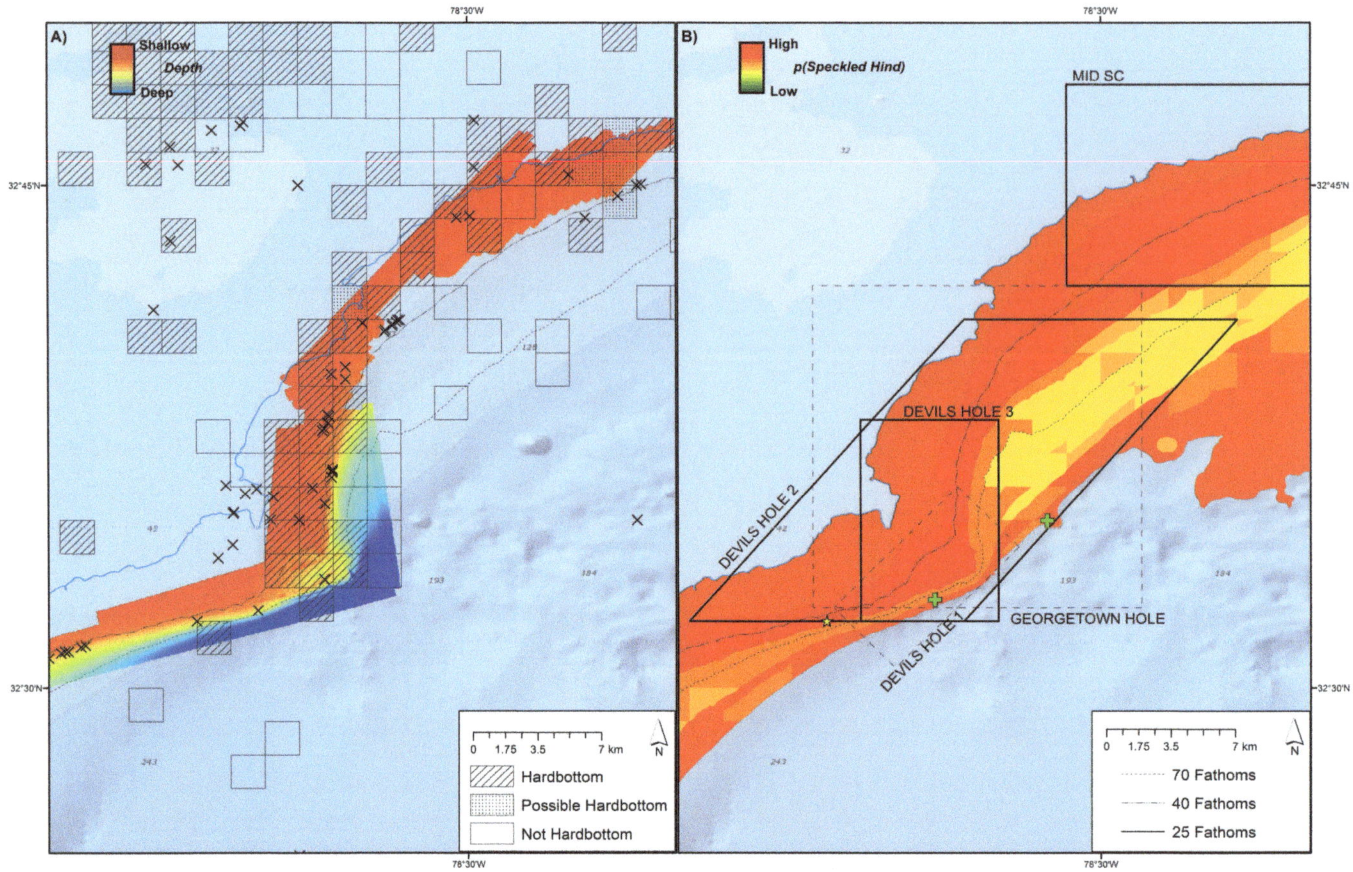

Figure 6. Point and spawning observations. A) Point observations of speckled hind (X) and warsaw grouper (+) relative to bathymetry and **B**) anecdotal spawning or aggregation observations of speckled hind (yellow star) and warsaw grouper (green crosses) relative to speckled hind geographic distribution model output and rejected (dashed lines) and proposed (solid lines) marine protected areas east of Murrell's Inlet, SC. Basemap courtesy of Esri Ocean Basemap and its partners.

The detailed cross-validation procedure carried out in this study yielded some interesting results in terms of model performance and selection. For both species, gear effects and latitude were the primary drivers for probability of detection. Once these two factors had been accounted for, additional factors had minimal impact on the predictive efficiency. Confidence intervals around the false identification rates suggested that all models including at least gear and latitude as factors were statistically equivalent in terms of prediction power. The suite of models including both gear and latitude effects did, however, differ widely in their AIC values, and in the predictions that resulted from these models. In other words, while many of the candidate models had similar predictive capabilities as tested by cross-validation, the ultimate outputs in terms of the percentage of grouper protected by each MPA were quite sensitive to the model chosen (see Table 4). Because the procedure for selecting the "best" model (e.g., AIC versus cross-validation) had an important influence on the final results for speckled hind, both models were presented, to more effectively capture the uncertainty associated with model predictions.

While the modeling approaches used here give us some confidence in identifying potential areas for improved protection of the study species, ultimately our ability to definitively distinguish the benefits of these area closures and to map these species with increased confidence will require additional data. It was surprising to find that including a habitat factor in the geographic distribution model did not improve model performance, particularly given that both speckled hind and warsaw grouper are found almost exclusively on hard bottom habitats. We believe the best explanation for the apparent inutility of the habitat data is that important habitat features are usually present at scales well below the resolution of the habitat data. For example, speckled hind and warsaw groupers are well-known to inhabit wrecks, but these small features will often be located within large areas of no hard bottom habitat. Thus, additional high-resolution habitat mapping may be necessary to enhance the predictive utility of this variable. As previously discussed, simple 'hardbottom' habitat may be an inadequate classification scheme for these stocks, which may require particular hardbottom features such as ledges, pinnacles, or other fine-scale features beyond the resolution of currently available data. As such, estimates of reserve protection should be validated with empirical observations demonstrating that suitable habitat is present within the proposed closure area. Our findings emphasize the need for detailed geomorphological maps using multibeam or sidescan technologies backed with groundtruthing along the entire southeastern U.S. shelf edge. Around 41% of the shelf-edge remains uncategorized with regards to habitat type, and the resolution of the habitat categorizations in the areas that have been studied is insufficient for many management needs. Additionally, other variables, such as physical oceanographic metrics, or location in reference to features such as channels may also be useful in predicting occurrence of grouper species [24].

Table 4. Evaluation of existing (lower case) and proposed (italicized, all caps) no-take marine reserves for speckled hind and warsaw grouper relative to coverage of viable habitats and percent of grouper protected, as predicted by geographic distribution models.

		SPECKLED HIND HABITAT SUITABILITY				WARSAW GROUPER HABITAT SUITABILITY		
Name	**Area (km²)**	**Known**	**Known & Probable**	**% Stock [X-val]**	**% Stock [AIC]**	**Known**	**Known & Probable**	**% Stock**
NORTH CAROLINA								
780 BOTTOM	56.9	0.00%	0.74%	0.99%	0.36%	0.00%	0.75%	0.46%
MANUELA WRECK	25.5	0.00%	0.00%	0.01%	0.%	0.00%	0.00%	0.00%
MALCHACE WRECK	6.4	0.00%	0.14%	0.14%	0.05%	0.00%	0.14%	0.05%
N CAPE LOOKOUT 2	114.8	2.63%	0.83%	1.1%	0.38%	2.44%	0.83%	0.62%
N CAPE LOOKOUT NC	110.9	2.33%	1.09%	1.13%	0.39%	0.00%	1.09%	0.62%
NORTH CAPE LOOKOUT 3	68.4	0.72%	0.20%	0.57%	0.17%	1.76%	0.20%	0.34%
S CAPE LOOKOUT NC	187.5	1.80%	1.07%	1.4%	1.4%	0.00%	1.02%	0.26%
SOUTHERN NC	229.9	0.74%	1.69%	1.53%	1.68%	0.00%	1.73%	0.34%
SOUTH CAROLINA								
Charleston Deep	66	0.00%	0.09%	0.28%	0.63%	0.00%	0.09%	0.31%
CHARLESTON SHELF MPA	34.8	3.58%	0.55%	0.29%	0.68%	0.00%	0.44%	0.16%
DEVILS HOLE 2	208.3	6.47%	1.72%	1.12%	2.11%	1.81%	1.72%	0.78%
Edisto	191.4	9.16%	1.85%	1.15%	2.88%	1.24%	1.65%	0.79%
EDISTO RECONFIG 3	208.7	20.09%	2.96%	1.52%	4.03%	2.39%	2.45%	0.93%
EDISTO S EXT	130.6	10.08%	1.10%	0.8%	2.07%	10.14%	0.88%	0.51%
DEVILS HOLE 3	69.4	2.60%	0.99%	0.41%	0.88%	0.70%	0.99%	0.26%
MID SC MPA	138.7	2.69%	0.69%	0.52%	1.1%	0.15%	0.63%	0.30%
Northern SC	173.2	3.81%	1.46%	1.02%	1.99%	2.52%	1.36%	0.76%
NORTHERN SC EXT	32.5	3.35%	0.22%	0.17%	0.28%	0.00%	0.10%	0.08%
GEORGIA								
Georgia	262.9	0.00%	0.76%	0.39%	1.17%	0.00%	0.78%	1.34%
GEORGIA MPA RECONFIG	204.7	4.89%	2.38%	0.97%	3.41%	11.20%	2.32%	1.87%
GEORGIA EXT	236.6	0.00%	2.01%	1.38%	4.01%	8.20%	2.14%	2.52%
GEORGIA RECONFIG N2	192.5	0.00%	1.57%	1.11%	3.31%	0.00%	1.60%	1.95%
ST SIMONS 2	58.6	3.27%	0.19%	0.32%	0.89%	2.23%	0.06%	0.66%
ST SIMONS EXT2	117.4	5.26%	1.09%	0.42%	1.45%	5.21%	0.98%	0.96%
NORTHEAST FLORIDA								
FERNANDINA MPA	221.1	1.11%	2.46%	0.48%	1.13%	7.02%	2.57%	2.80%
North Florida	354.9	2.46%	1.86%	0.36%	0.96%	4.53%	1.88%	2.56%
Oculina Bank CHAPC (excluding ECA)	753.7	0.00%	1.03%	0.11%	0.08%	0.00%	1.05%	2.05%
OCULINA BANK CHAPC EXTENSION (excluding DAYTONA STEEPLES and DAYTONA LEDGE)	627.7	0.46%	2.29%	0.34%	0.32%	0.00%	2.33%	2.33%
Oculina ECA	279.2	3.79%	3.19%	0.11%	0.28%	2.61%	3.26%	*Not eval.*

Table 4. Cont.

Name		SPECKLED HIND HABITAT SUITABILITY				WARSAW GROUPER HABITAT SUITABILITY		
	Area (km^2)	Known	Known & Probable	% Stock [X-val]	% Stock [AIC]	Known	Known & Probable	% Stock
DAYTONA STEEPLES	68.9	1.51%	0.76%	0.1%	0.12%	0.00%	0.70%	0.55%
DAYTONA LEDGE	28.4	1.79%	0.31%	0.05%	0.08%	7.76%	0.31%	0.22%
ST AUGUSTINE 2	83.1	1.92%	0.66%	0.15%	0.45%	9.17%	0.62%	1.05%
ST AUGUSTINE EXT2	35.6	0.87%	0.21%	0.07%	0.05%	1.04%	0.18%	0.20%
SOUTHEAST FLORIDA								
FKNMS SPAs & Ers	246.7	*Not evaluated*				*Not evaluated*		
JUNO BEACH MPA	9.2	0.00%	0.00%	0.%	0.%	0.54%	0.01%	*Not eval.*
PUSH BUTTON HILL	24.4	0.00%	0.27%	0.01%	0.01%	3.61%	0.30%	*Not eval.*
St. Lucie Hump	24.4	0.00%	0.19%	0.01%	0.04%	1.01%	0.20%	*Not eval.*
WARSAW HOLE 4	6.2	0.00%	0.00%	0.%	0.%	3.08%	0.03%	*Not eval.*

Output range for cross-validation best predictor (X-val) and best-fitting (AIC) models for speckled hind provided to characterize uncertainty. Note that geographic distribution model was unable to resolve probabilities south of 28' latitude for warsaw grouper.
Note: Assumes CHAPC no-anchoring provision results in 50% efficiency at eliminating bycatch mortality. Warsaw grouper percent stock estimates not generated south of 28° latitude. Oculina Bank CHAPC evaluation excludes Experimental Closed Area (ECA). Oculina Bank CHAPC Extension evaluation excludes Daytona Steeples and Daytona Ledge.

MPA Recommendations

MPAs have been endorsed as fisheries management tools that, when used in conjunction with traditional management, may help ensure sustainability of intensely exploited regional fisheries resources [25]. Theory suggests that buildup of fish biomass, density, and average size in no-take MPAs due to reduced exploitation (e.g., [26–28]) will result in density-dependent emigration of adult fish across MPA boundaries [29]. Additionally, larval production should be amplified by the larger, older population within the MPA due to its increased spawning stock biomass [30,31]. The advection of these eggs and larvae by ocean currents may enhance recruitment in fishable areas [29].

Over the past two decades, there has been much scientific discussion regarding the percent of the stock that should be protected by a MPA to provide benefits such as reduced risk of overexploitation, restoration of natural community dynamics, increased spawning stock biomass, and maximization of yield through spillover of adult biomass and larval recruits. A meta-analysis of percent closure recommendations indicated a consensus that between 20–40% of the stock should be protected unless it is heavily exploited outside the MPA system ([32]; Fig. S1). The exact amount of area or stock that should be protected will depend on the specific objectives of the MPA, and will balance the biology and status of the stocks in need of protection with the regulations that exist outside of the MPA [33]. As such, there is no 'one size fits all' answer for the appropriate size, scale, or number of MPAs needed [33].

For the specific case of speckled hind and warsaw grouper protections, a primary goal for spatial protection would be to supplement the existing prohibition of harvest with spatial closures to reduce bycatch mortality [10,21]. As such, MPAs would be most effective if located at sites where bycatch mortality is highest. Those sites would be in deep water, at the intersection of relatively high stock concentrations and high fishing pressure for associated species. Our analysis assumes these MPAs would eliminate bycatch mortality for all but the CHAPCs, which are assumed to reduce bycatch mortality by 50%. Poaching or fishing activities which violated the assumption of no bycatch mortality of speckled hind or warsaw grouper would invalidate the conclusions presented in this manuscript. MPAs would be most effective if scaled to the natural movements of the fish [34,35], with a sufficient buffer to prevent the redistribution of fishing pressure on the edges of the reserve from offsetting the benefits of protection at the core. As these species do not exist in isolation, it is important that reserves designed for stock recovery also consider ecosystem processes that may be critical to their life history, including critical habitats and the scales of movement of their prey species. Designation of large shelf-edge MPAs would protect spawning aggregations of many species, allow ecosystem recovery, and minimize perimeter-to-area ratio so that loss of fish to the outside that might dilute the benefits of the MPA [25,35].

Fish stock spatial dynamics—including preferential habitat utilization, movements and migratory behaviors—play a critical role in determining how fishing pressure will impact the stock, and result in fish stocks being heterogeneously distributed throughout the oceans [36–41]. Our meta-analysis of available fishery-dependent, fishery-independent, and anecdotal data told a consistent story with regards to the hardbottom obligate habitat preferences of speckled hind and warsaw grouper. Coupling that information with available habitat mapping and depth-grid specific computations of probability of encounter, we have provided some guidance regarding areas of higher concentration for these stocks. Use of point observations alone to guide reserve selection could lead to overly optimistic conclusions regarding the

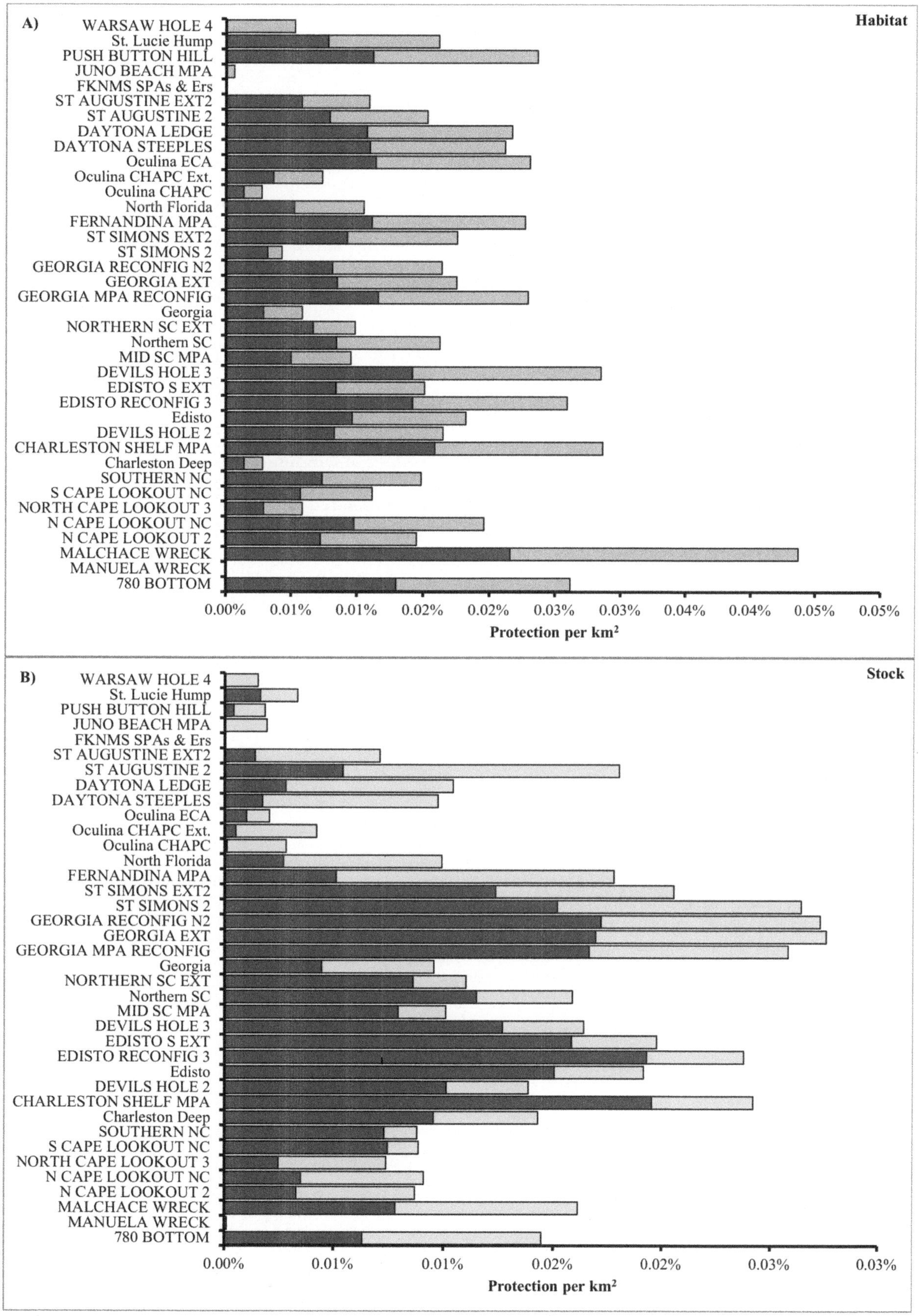

Figure 7. No-take marine reserve protection per unit area. Percent of **A**) known and probable habitat and **B**) speckled hind (black) and warsaw grouper (light gray) stock per square kilometer.

level of protection the stock is receiving. We have attempted to control for this bias using the spatial classification and grouper distribution models described above. The spatial classification model provides broad geographic coverage and incorporates information from all spatial data sources. The geographic distribution model controls for sampling biases and provides predictive utility for the percent stock occurring within various spatial closure alternatives. Future analyses should attempt to evaluate hydrographic linkages between MPA sites in the context of larval connectivity [24,42], and to identify biogeomorphic features that may serve as spawning aggregation locations [43–45]. The fishery benefits of an MPA will be most fully realized if the MPA contains spawning habitats, especially where these habitats serve as a source for other suitable habitats. If spawning aggregation sites are outside of the reserves and are subject to bycatch mortality, many of the potential benefits of spatial protection will be undermined and objectives of population recovery will not be achieved.

Selection of MPAs containing known and probable habitats for both speckled hind and warsaw grouper would be a reasonable approach towards enhancing the protection of these stocks from bycatch mortality. Our analysis suggests that the most efficient closures would be those of reasonable size (>10–20 km^2) that are sited in areas with high concentrations of quality habitat and high probabilities of encounter for each stock. Within the effective domain of the logistic model (speckled hind: 34°N to 26°S, warsaw grouper: 34°N to 28°S; 45.7–182.9 meters depth for both stocks), the probability of detection with gear effects removed are theoretically proportional to abundances. Thus, the sum of depth-grid cell probabilities within a given MPA divided by the sum of all SAFMC depth-grid probabilities may provide a reasonable estimate of the proportion of the grouper contained within the MPA, keeping in mind the uncertainties described above.

Less overall area would need to be closed to achieve the same level of estimated protection if the spatial protections are preferentially selected based on their predicted protection per unit area. There will likely be tradeoffs between distributing the socioeconomic impacts of spatial protection among fishermen from various coastal states; however, the greatest reductions in bycatch mortality will be realized by closing where fishing pressure for associated stocks is highest, unless this causes redistribution of fishing pressure onto adjacent areas where concentrations of warsaw grouper and speckled hind are even higher. In general, larger MPAs or MPAs closer to population centers are predicted to have the greatest economic impacts and lowest compliance rates; however, these MPAs could also provide the greatest proportional reduction in bycatch mortality. Given that all exploited stocks in the SAFMC are managed by annual catch limits, effort shifting may allow fishermen to compensate for spatial closures, and potential reductions in harvest may be offset unless core harvest locations are within the implemented MPA.

Overall recommendation for management

Implementation of spatial closures for speckled hind and warsaw grouper should apply adaptive management principles when possible [46]. Adaptive management modifies management practices and policies to be more successful when new science, socioeconomic information or lessons learned from previous management actions indicate that practices could be made more efficient. For spatial closures such as those discussed in this study, monitoring and evaluating, testing assumptions, and generating learning opportunities are important aspects of adaptive management. Any MPAs implemented will not exist in a vacuum, and research should be conducted to understand the level of protection afforded to the stocks by the reserves and to better describe stock status. As further information emerges regarding ecosystem conditions, fishing operations, community structures, or other social, ecological, or governance factors, MPAs could be modified, added, or removed to best address management needs. Dynamic MPA management would benefit most from improved resolution on hardbottom identification and increased fishery-independent sampling over a broader geographic range using appropriate gears. A special emphasis on building a long-term robust time series of population abundance data for both stocks to allow for an updated stock assessment is also recommended.

Supporting Information

Table S1 List of candidate models to predict probability of occurrence of specked hind and warsaw grouper, with their associated AIC values and false positive (FPR) and false negative (FNR) identification rates. Bold values denote the best model in terms of either AIC or combined FPR and FNR. Standard deviations quantify the variance around the FPR and FNR resulting from the 10 different test data sets in the cross-validation procedure.

Acknowledgments

The authors would like to thank the various individuals and agencies who contributed data and invaluable advice throughout this process. Their contributions greatly enhanced this analysis. Special thanks to the recreational and commercial fishermen who contributed their catch locations and on-the-water anecdotal observations. Special thanks to the MPA Expert Workgroup and the SAFMC Scientific and Statistical Committee for their extensive discussion of innovative analytical methods, socioeconomic considerations, proposed MPA alternatives, and valuable commentary regarding earlier drafts of this work. Special thanks to G. Sedberry, C. Grimes, C. Koenig, R.G. Gilmore, M. Barnette, D. DeMaria, R. Hudson and W. Heyman for valuable comments and discussion. Additional thanks to A. Strelcheck for multiple reviews of this manuscript.

Disclaimer:

The scientific results and conclusions, as well as any views or opinions expressed herein, are those of the authors and do not necessarily reflect the views of NOAA or the Department of Commerce.

Author Contributions

Conceived and designed the experiments: NF MK. Performed the experiments: NF MK. Analyzed the data: NF MK. Contributed reagents/materials/analysis tools: NF MK. Wrote the paper: NF MK.

References

1. National Marine Fisheries Service (2012) Status of stocks: Annual report to Congress on the status of U.S. fisheries. NOAA Fisheries Service.
2. Grimes CB, Manooch CS, Huntsman GR (1982) Reef and rock outcropping fishes of the outer continental shelf of North Carolina and South Carolina, and ecological notes on the red porgy and vermilion snapper. Bulletin of Marine Science 32: 277–289.
3. Tester PA, Wolfe CA, Dixon RL, Huntman GR (1983) Reef fish distributions off North Carolina and South Carolina as revealed by headboat catches.

4. Staff of the Beaufort Laboratory Southeast Fisheries Science Center (1991) South Atlantic snapper grouper assessment 1991. One Southpark Circle, Suite 306, Charleston, SC 29407. 21 p.
5. Huntsman G, Potts J, Mays R, Dixon R, Willis P, et al. (1992) A stock assessment of the snapper-grouper complex in the US South Atlantic based on fish caught in 1990. Report to the South Atlantic Fishery Management Council 1.
6. Potts JC, Burton ML, Manooch CS III (1998) Trends in catch data and estimated static SPR values for fifteen species of reef fish landed along the southeastern United States. One Southpark Circle, Suite 306, Charleston, SC 29407: SAFMC.
7. Potts JC, Brennan K (2001) Trends in catch data and estimated static SPR values for fifteen species of reef fish landed along the southeastern United States. Methods 2: 4.
8. Rudershausen PJ, Williams EH, Buckel JA, Potts JC, Manooch CS III (2008) Comparison of reef fish catch per unit effort and total mortality between the 1970s and 2005–2006 in Onslow Bay, North Carolina. Transactions of the American Fisheries Society 137: 1389–1405.
9. Ziskin GL (2008) Age, growth, and reproduction of speckled hind, Epinephelus drummondhayi, off the Atlantic coast of the southeast United States. CharlestonSC: College of Charleston. 130 p.
10. Ziskin GL, Harris PJ, Wyanski DM, Reichert MJ (2011) Indications of Continued Overexploitation of Speckled Hind Along the Atlantic Coast of the Southeastern United States. Transactions of the American Fisheries Society 140: 384–398.
11. Southeast Data Assessment and Review (SEDAR) (2004) Stock Assessment of the Deepwater Snapper Grouper Complex in the South Atlantic. One Southpark Circle #306CharlestonSC: SEDAR/SAFMC. 594 p.
12. Brandt ME, Zurcher N, Acosta A, Ault J, Bohnsack J, et al. (2009) A cooperative multi-agency reef fish monitoring protocol for the Florida Keys coral reef ecosystem: US Department of the Interior, National Park Service, Natural Resource Program Center.
13. Frost J (2006) Spiritual slavery to spiritual sonship: Destiny Image Publishers. 226 p.
14. Maps Unique (2012) Offshore fishing maps.
15. FWC: Florida Fish and Wildlife Conservation Commission (2001) SEAMAP Bottom Mapping Project (April 2001) Distribution of bottom habitats on the continental shelf from North Carolina through the Florida KeysD.C. Washington, D.C.: Atlantic States Marine Fisheries Commission. pp. 166.
16. Beyer HL (2004) Hawth's Analysis Tools for ArcGIS. Available: http://www.spatialecology.com/htools.
17. R Development Core Team (2009) R: a language and environment for statistical computing. Vienna, Austria:R Foundation for Statistical Computing.
18. Kohavi RA (1995) study of cross-validation and bootstrap for accuracy estimation and model selection. pp. 1137–1145.
19. Robin X, Turck N, Hainard A, Tiberti N, Lisacek F, et al. (2011) pROC: an open-source package for R and S+ to analyze and compare ROC curves. BMC bioinformatics 12: 77.
20. Akaike H (1974) A new look at the statistical model identification. Automatic Control, IEEE Transactions on 19: 716–723.
21. South Atlantic Fishery Management Council MPA Expert Workgroup (2013) Report of the Second MPA Expert Workgroup. Crowne Plaza Hotel, North Charleston, SC.
22. Rudershausen P, Mitchell W, Buckel J, Williams E, Hazen E (2010) Developing a two-step fishery-independent design to estimate the relative abundance of deepwater reef fish: Application to a marine protected area off the southeastern United States coast. Fisheries Research 105: 254–260.
23. Karnauskas M, Walter JF, Paris CB (2013) Use of the Connectivity Modeling System to estimate movements of red snapper (Lutjanus campechanus) recruits in the northern Gulf of Mexico. North CharlestonSC: SEDAR. 20 p.
24. Karnauskas M, Cherubin LM, Huntington BE, Babcock EA, Thoney DA (2012) Physical forces influence the trophic structure of reef fish communities on a remote atoll. Limnology and oceanography 57: 1403–1414.
25. Bohnsack JA, Ault JS, Causey B (2004) Why have no-take marine protected areas?. pp. 185–193.
26. Ault JS, Smith SG, Bohnsack JA, Luo J, Harper DE, et al. (2006) Building sustainable fisheries in Florida's coral reef ecosystem: Positive signs in the Dry Tortugas. Bulletin of Marine Science 78: 633–654.
27. Ault JS, Smith SG, Bohnsack JA, Luo J, Farmer NA, et al. (2007) Fishery-independent monitoring of coral reef fishes, coral reefs, and macro-invertebrates in the Dry Tortugas. Final Report to the National Park Service.
28. Bartholomew A, Bohnsack JA, Smith SG, Ault JS, Harper DE, et al. (2008) Influence of marine reserve size and boundary length on the initial response of exploited reef fishes in the Florida Keys National Marine Sanctuary, USA. Landscape Ecology 23: 55–65.
29. Crowder LB, Lyman S, Figueira W, Priddy J (2000) Source-sink population dynamics and the problem of siting marine reserves. Bulletin of Marine Science 66: 799–820.
30. Botsford LW (2001) Dependence of sustainability on the configuration of marine reserves and larval dispersal distance. Ecology Letters 4: 144–150.
31. Lubchenco J, Palumbi SR, Gaines SD, Andelman S (2003) Plugging a hole in the ocean: the emerging science of marine reserves 1. Ecological applications 13: 3–7.
32. National Research Council (NRC) Ocean Studies Board (2001) Marine Protected Areas: Tools for Sustaining Ocean Ecosystem: National Academies Press.
33. FAO (2011) Fisheries Management 4: Marine Protected Areas and Fisheries; FAO, editor. RomeItaly: FAO. 198 p.
34. Botsford LW, Brumbaugh DR, Grimes C, Kellner JB, Largier J, et al. (2009) Connectivity, sustainability, and yield: bridging the gap between conventional fisheries management and marine protected areas. Reviews in Fish Biology and Fisheries 19: 69–95.
35. Farmer NA, Ault JS (2011) Grouper and snapper movements and habitat use in Dry Tortugas, Florida. Marine Ecology Progress Series 433: 169–184.
36. Rothschild BJ (1986) Dynamics of marine fish populations: Harvard University Press.
37. Longhurst AR, Pauly D (1987) Ecology of tropical oceans: Academic Press San Diego.
38. MacCall AD (1990) Dynamic geography of marine fish populations: Washington Sea Grant Program Seattle, Washington.
39. Mann KH, Lazier J (1996) Dynamics of marine ecosystems: Blackwell Science Cambridge, Massachusetts.
40. Bakun A (1996) Patterns in the Ocean: Ocean Processes and Marine Population Dynamics. San Diego, California: University of California Sea Grant, in cooperation with Centro de Investigaciones Biologicas de Noroeste, La Paz, Baja California Sur, Mexico.
41. Humston R, Ault JS, Lutcavage M, Olson DB (2000) Schooling and migration of large pelagic fishes relative to environmental cues. Fisheries Oceanography 9: 136–146.
42. Huff DD, Lindley ST, Wells BK, Chai F (2012) Green Sturgeon Distribution in the Pacific Ocean Estimated from Modeled Oceanographic Features and Migration Behavior. PloS one 7: e45852.
43. Paz M, Sedberry GR (2008) Identifying black grouper (Mycteroperca bonaci) spawning aggregations off Belize: conservation and management. pp. 577–584.
44. Heyman WD (2011) Elements for building a participatory, ecosystem-based marine reserve network. The Professional Geographer 63: 475–488.
45. Heyman WD, Wright DJ (2011) Marine geomorphology in the design of marine reserve networks. The Professional Geographer 63: 429–442.
46. Stankey GH, Clark RN, Bormann BT (2005) Adaptive management of natural resources: Theory, concepts, and management institutions.

Redefining Thermal Regimes to Design Reserves for Coral Reefs in the Face of Climate Change

Iliana Chollett[1,2]*, Susana Enríquez[3], Peter J. Mumby[1,2]

1 Marine Spatial Ecology Lab, College of Life and Environmental Sciences, University of Exeter, Exeter, United Kingdom, **2** Marine Spatial Ecology Lab, School of Biological Sciences, Goddard Building, University of Queensland, Brisbane, Australia, **3** Instituto de Ciencias del Mar y Limnología, Universidad Nacional Autónoma de México, Puerto Morelos, México

Abstract

Reef managers cannot fight global warming through mitigation at local scale, but they can use information on thermal patterns to plan for reserve networks that maximize the probability of persistence of their reef system. Here we assess previous methods for the design of reserves for climate change and present a new approach to prioritize areas for conservation that leverages the most desirable properties of previous approaches. The new method moves the science of reserve design for climate change a step forwards by: (1) recognizing the role of seasonal acclimation in increasing the limits of environmental tolerance of corals and ameliorating the bleaching response; (2) using the best proxy for acclimatization currently available; (3) including information from several bleaching events, which frequency is likely to increase in the future; (4) assessing relevant variability at country scales, where most management plans are carried out. We demonstrate the method in Honduras, where a reassessment of the marine spatial plan is in progress.

Editor: Sebastian C. A. Ferse, Leibniz Center for Tropical Marine Ecology, Germany

Funding: The research leading to these results has received funding from the European Union 7th Framework programme (P7/2007–2013) under grant agreement no. 244161, Pew (grant 2008-000330-010) and ARC Laureate Fellowships (FL0992179) to PJM. The funders had no role in study design, data collection and analysis, decision to publish, or preparation of the manuscript.

Competing Interests: The authors have declared that no competing interests exist.

* Email: iliana.chollett@gmail.com

Introduction

Global warming and ocean acidification represent an increasing threat to reef ecosystems [1]. Many studies have assessed likely responses of coral reef organisms to thermal stress and the expected changes in aragonite saturation state associated with ocean acidification (reviewed in [2,3,4]). However, research is still ongoing to understand the physiological mechanisms that explain thermally induced coral bleaching and responses to acidification, the magnitude of their impact on coral calcification, and the ecosystem-level consequences of these disturbances.

Stabilizing carbon emissions is the most critical action needed to address global climate change impacts, but this task is beyond the scope of on-the ground conservation managers. The environmental effects of global impacts can, however, be taken into account during the design of management interventions, such as marine reserve networks, by considering the differential vulnerability of reefs to the stressor(s) and selecting sites for protection that increase the probability of persistence of the whole network under a changing climate [5,6,7,8].

Both bleaching and ocean acidification are associated with changes in oceanographic conditions: while mass bleaching is linked to acute increases in sea temperatures [9], severe reductions in coral calcification rates are expected from decreases in aragonite saturation state related to ocean acidification [10]. These variables, sea temperature and aragonite saturation state, can be mapped [11,12], depicting a heterogeneous seascape which translates into spatially variable risks to environmental changes. To date, planning for ocean acidification is in its infancy [13], the response of organisms (e.g. decreased growth and calcification) is not as unequivocal as the impact of thermal stress and coral bleaching [13,14], and the mapping of this threat is based on complex models that are still under development [11,13,15].

Spatially explicit information on patterns of sea surface temperature has been used to design reserve networks that consider the threat of bleaching (e.g. [5,7]). Broadly, the methods differ in three ways: the type of data used as input for the analyses, their temporal coverage and the type of areas targeted for protection (Figure 1). Below we assess these three aspects and identify the strongest elements of all methods, which we then include when crafting an optimal approach to plan for marine reserve design in a climate change context.

So far reserve design methods have used satellite temperature data [7,16], global climate models (GCM, [5,17]) or a combination of both [18]. GCMs have the advantage of offering insight into future patterns, but have a coarse spatial resolution (hundreds of km), do not take into account processes at the reef scale, and are therefore too uncertain to be a foundation for conservation planning [19]. Satellite data, conversely, is better at capturing reef-scale variability, but is based on historical data, and therefore assumes that future incidence of bleaching events will occur under similar conditions as past incidences. Given that GCMs are unable to predict future bleaching events at a spatial scale relevant for management, the establishment of a method based on high-

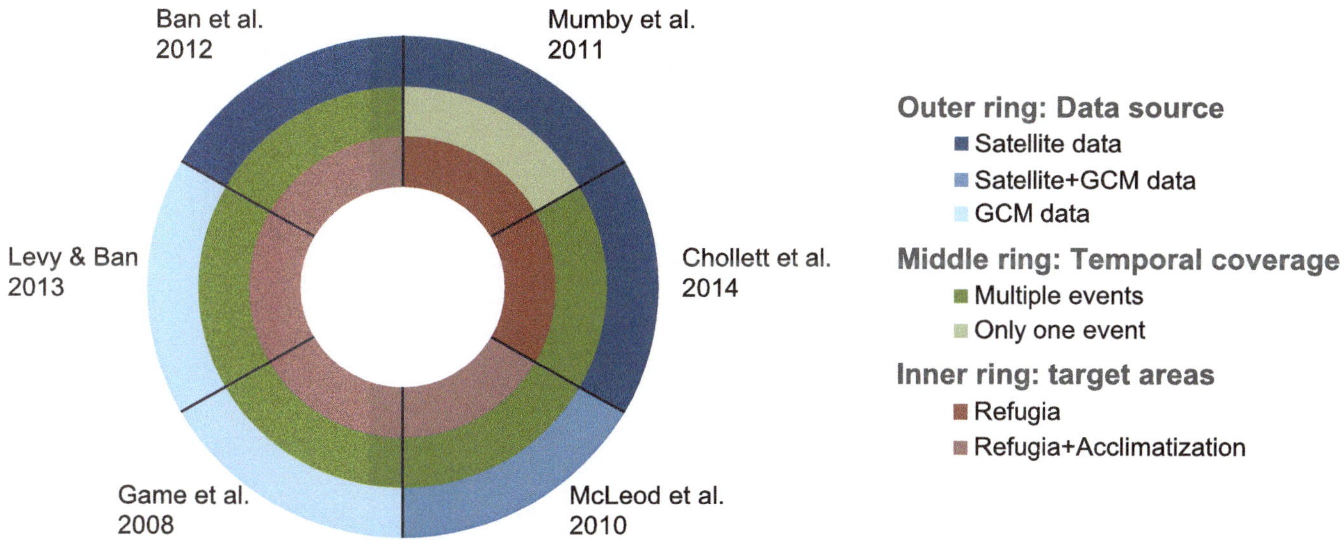

Figure 1. Comparison of methods available to prioritize areas for conservation in a climate change context. Contrasts in terms of the source of the data used (blue ring), the temporal coverage of the analyses (green ring) and the type of areas targeted for protection (red ring).

resolution historical data, which could be updated as new data becomes available, provides the most appropriate present-day option for managing reefs for climate change.

Currently planning approaches include data from a single or several bleaching events (Figure 1). Since the level of thermal stress can vary widely from one bleaching event to another and the spatial distribution of bleaching varies widely between events [16,20], a robust approach to design reserve networks should include input data from multiple years and bleaching events.

Finally, existent methods also differ regarding the type of areas to prioritize. Most efforts target areas naturally least affected by thermal anomalies, assuming that climate change needs to be avoided for reefs to survive. Mumby et al. [7] presented the only framework that also considers the scope for acclimation and adaptation to increase the survival of corals under climate change. Although the magnitude of these processes is still uncertain, their relevance in an unstoppable climate change context could be considerable [21,22], especially given the uncertain role of refugia in a changing climate [23] and therefore we suggest their inclusion in management plans.

Several environmental variables have been associated with higher rates of adaptation and/or acclimation to thermal stress, such as higher summer temperatures [7,24] or higher temperature variability [25,26], under the premise that organisms exposed to 'stressful' temperatures can develop ecological or physiological responses that will allow them to withstand acute thermal stress better. However, it has recently been recognized that one of the most relevant variables modulating bleaching occurrence is the rate of seasonal warming from spring to summer, because this variable can explain how well prepared the coral is at this critical time of the year [27,28]. Corals that are subjected to rapid seasonal increases in temperature will experience faster losses in coral pigmentation, which makes corals more sensitive to light stress (a key component of bleaching), and therefore aggravates their response in the face of acute stress conditions [27,28]. Despite its relevance for predicting the likelihood of bleaching, this new knowledge has not yet been incorporated in methods for prioritising reef areas for conservation.

Taking advantage of recent developments in the field, we propose an improved method for prioritizing areas for conservation that leverages the most desirable properties of previous methods and also incorporates the best proxies for acclimatization currently available. Specifically, the method combines freely available high-resolution satellite imagery, multiple bleaching events and target both areas naturally resistant to bleaching and potential areas of fast acclimation. We demonstrate the method in Honduras, western Caribbean, where a reassessment of the marine spatial plan is in progress.

Methods

Data set

Temperature was measured using satellite data provided by the Advanced Very High Resolution Radiometer (AVHRR) Pathfinder project v 5.2, freely available (http://pathfinder.nodc.noaa.gov). The 5.2 dataset is an updated version of the Pathfinder Version 5.0 and 5.1 collections described in Casey et al. [29].

The temperature dataset covers the period 1982–2010 and has a spatial resolution of 4×4 km. Only night data of quality 4 or higher was used [30]. Weekly averages were calculated from the daily data and data gaps were filled using temporal and spatial interpolation. First, any data gaps were identified and substituted by the average of the previous and following week for that particular pixel. Any remaining gaps were filled through spatial interpolation for the specific week using a spring metaphor, as described in Chollett and Mumby [23].

Processing

From this weekly, gap-filled dataset, two metrics were derived: total amount of acute thermal stress per year, and average rate of increase in temperature from spring to summer (hereafter only 'rate of increase in temperature', Figure 2). Acute thermal stress was quantified as the total amount of Degree Heating Weeks (DHW, [31]) experienced in a particular year (Figure 2a). DHWs are a measure of accumulated thermal stress by summing temperatures 1°C above the usual summertime maximum over a 12 week period. Temperature increases above normal summer ambient have shown to induce bleaching [32,33], and DHWs have been commonly related to the occurrence of bleaching and associated mortality [34,35]. Finally, to calculate the rate of

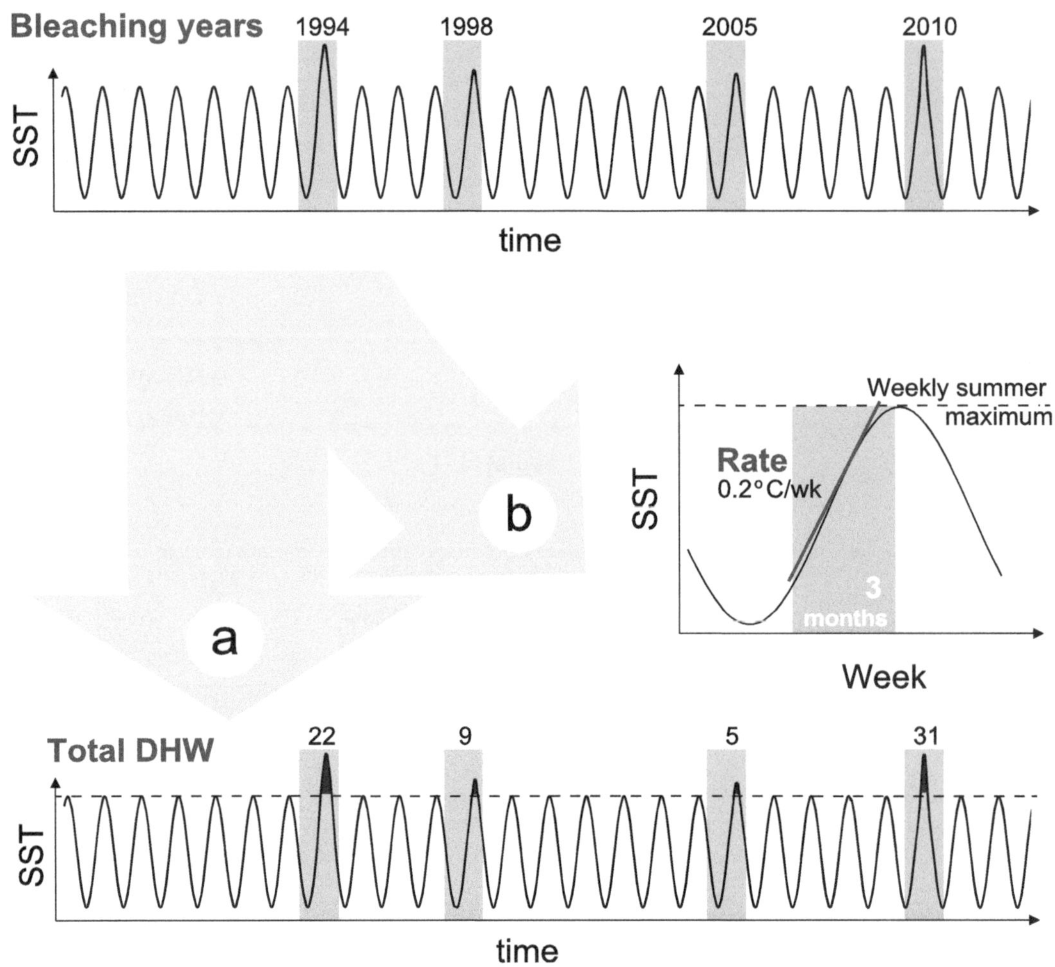

Figure 2. Temperature metrics extracted from each weekly time-series of sea surface temperature. (a) Total DHWs per bleaching year; (b) rate of increase in temperature.

increase in temperature weekly climatological means using all data were computed, and the maximum of those selected as the summer maximum. Then the slope of the line of the three-month period prior to that date was calculated (Figure 2b).

Temperature metrics were extracted for reef areas within the Exclusive Economic Zone of Honduras. EEZ's were extracted from the VLIZ Maritime Boundaries Geodatabase (http://www.marineregions.org). Data on location of coral reefs were collated by the World Resources Institute and includes information from the Millennium Coral Reef Mapping Project [36], the UNEP-WCMC coral reef map and other sources [37]. Data on temperature metrics for reef areas in the entire Caribbean are available at PANGAEA data repository (http://dx.doi.org/10.1594/PANGAEA.836153).

Mapping thermal regimes using acute stress and rates of increase in temperature

Information on the acute thermal regime and rates of increase in temperature can be used to inform spatial variability in vulnerability to bleaching and prioritize areas for conservation anywhere. These two factors affect both the long and short-term acclimation of corals to thermal stress, and incorporate the best science available.

This new method builds on information of all previous mass bleaching events (e.g. in the Caribbean 1995, 1998, 2005 and 2010: [35,38,39,40]) and occurs in a two-step fashion, first incorporating data on how intense and frequent acute thermal stress events have been, and then adding information on the rate of increase in temperature in each area, which can modulate the expected outcome (Figure 3).

Acute stress. Acute temperature stress, the total amount of DHWs, is first quantified for each location and bleaching year (Figure 3b), the values are tabulated (Figure 3c) and transformed into percentiles so for each location we have 4 different percentile values, one for each mass bleaching event (Figure 3d). Raw DHWs are transformed into percentiles to rank locations and allow easier interpretation and comparisons among them. Subsequently the average and standard deviation of stress is calculated for each location. Note that areas with low standard deviation have either rarely experienced a thermal event or always experienced one. For example, site 3 in Figure 3 has been affected by thermal stress in every single event with high intensity (Figure 3c), which translate into high percentiles (88th–100th percentile), and therefore high average (95) and low variability (6).

Average and variability measures are then divided independently into three groups (terciles) and locations at the extremes of each stress measure (i.e. upper and lower thirds) are used to

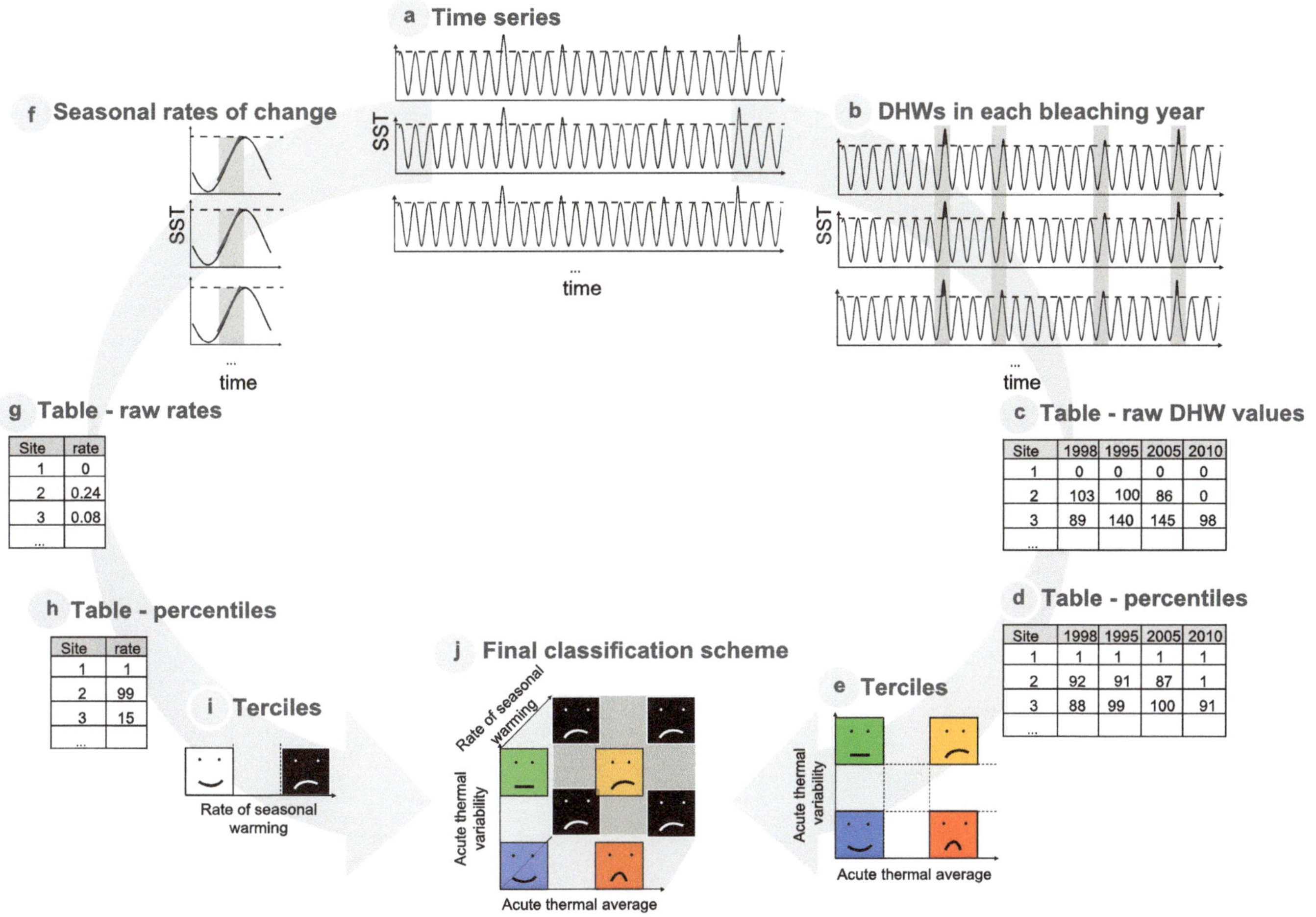

Figure 3. Diagram showing the overall methods used. (a) In each location, sea surface temperature time series are extracted and (b) the total amount of DHWs per bleaching event is calculated in each site. These values are (c) tabulated, (d) transformed into percentiles, and then (e) the average and standard deviation is calculated and plotted. This way the sites can be separated according to the intensity of acute thermal stress (weak or severe) and its variability (recurrent or not). Additionally, in each location, (f) the rate of change in temperature is quantified from the time series, (g) tabulated, (h) transformed into percentiles and (i) divided in terciles. (j) These two classifications are then incorporated into one final classification scheme.

generate four contrasting temperature stress regimes (Figure 3e). Using this information, reefs can be classified into four categories based on their exposure to high temperature: Blue, Green, Yellow and Red. (Blue) areas that have received weak acute thermal stress during all bleaching events and constitute a refugia against thermal disturbances; (Green) areas that have been subjected to generally weak, albeit variable, acute stress, providing some potential for adaptation; (Yellow) areas that have suffered severe but variable acute stress, should do worse than (Green), and still have potential for adaptation but with differential fitness among species and therefore lower diversity; and (Red) areas that have suffered severe acute stress in a constant manner, which will likely experience higher reductions in coral fitness and fare worst by climate change.

Rates of increase in temperature. The other element to take into consideration is the average rate of change in temperature between spring and summer for each location. For all years, weekly average temperatures are calculated, and the maximum of those selected as summer maximum. Then the slope of the line of the three month period prior to that date is calculated (Figure 3f). Rates are tabulated (Figure 3 g), transformed into percentiles (Figure 3 h) and then divided into terciles (Figure 3i), allowing contrasting conditions to be identified. This information can be integrated with data on acute thermal stress events (Figure 3j) such that high rates of change in temperatures aggravate the prognosis of the reef and its likely fate after bleaching. The physiological rationale of including rates of increase in temperature as a proxy for short-term acclimation can be found in the file S1.

Results

Honduran waters are largely heterogeneous in terms of acute thermal stress and rate of increase in temperature (Figure 4). Acute stress is more intense in the mid and southern Miskito Cays (Figure 4a), and more variable in the south-west Miskito Cays and the western side of the Bay Islands (Figure 4b). In contrast, higher rates of seasonal warming are found in the eastern portion of the Bay Islands and areas along the coast south of the Islands. As a result, distinct thermal regimes are well distributed across Honduras (Figure 5), with areas of low overall acute thermal stress (Blue) mostly concentrated in the north of the region along the Bay and Swan Islands and northern Miskito Cays, and areas of high acute thermal stress (regimes Yellow and Red) in the central and southern section of the Miskito Cays.

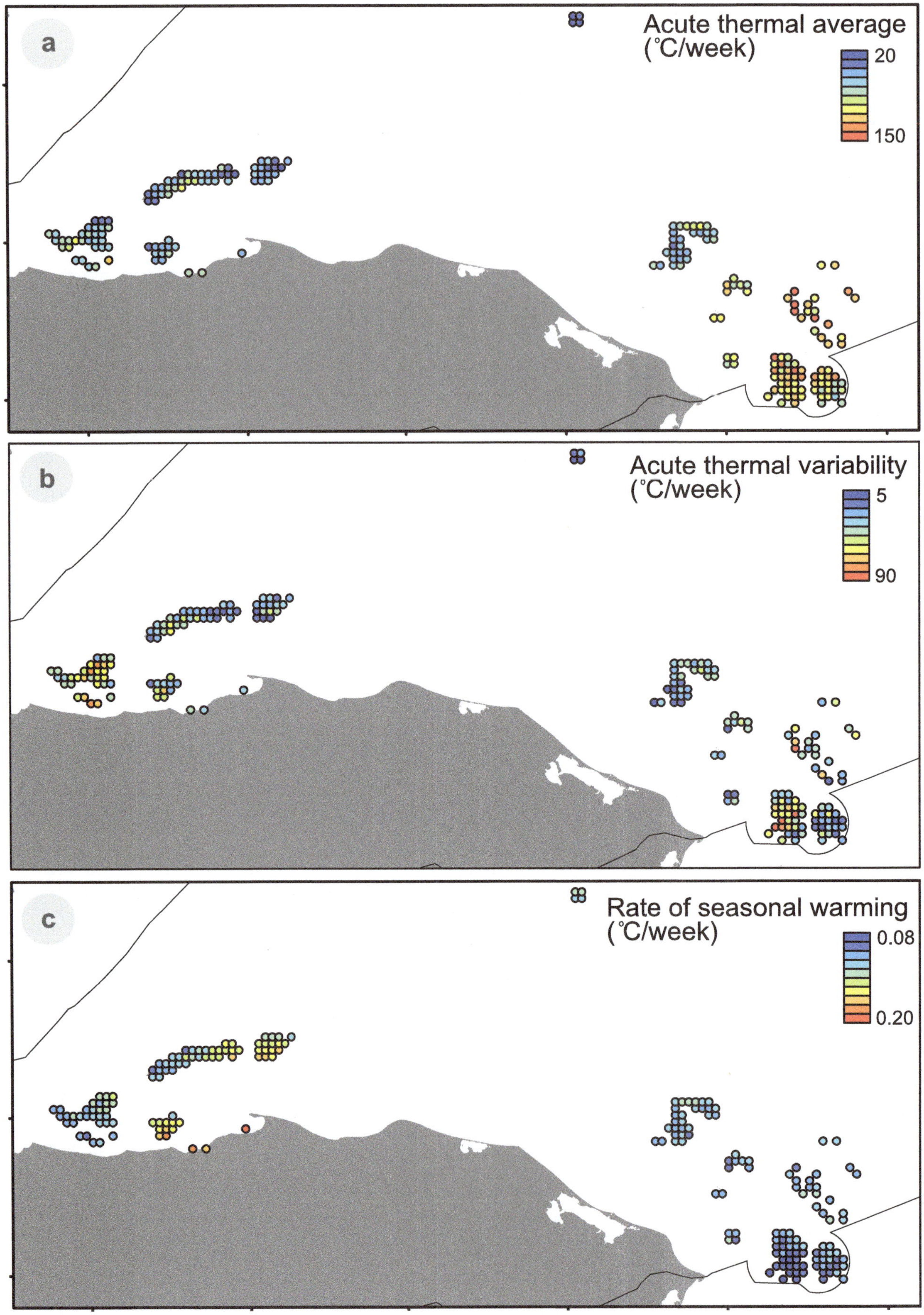
a
Acute thermal average
(°C/week)
20
150
b
Acute thermal variability
(°C/week)
5
90
c
Rate of seasonal warming
(°C/week)
0.08
0.20

Figure 4. Temperature patterns in reef areas inside the Honduran EEZ. (a) Average acute thermal stress; (b) variability in acute thermal stress; and (c) rate of warming.

Interestingly, areas of fast rates of summer warming mostly (88% of the time) occur in Blue refugia regimes, rarely in Yellow (8%) or Green (4%) regimes, and never in Red regimes. The results for Honduras suggest that a network of Marine Protected Areas that accounts for climate change impacts and scope for adaptation should include representation in the southern Miskito Cays and the Bay Islands (particularly in the north of the middle island –Roatan- and eastern island -Guanaja), probably also incorporating reefs in the northern Miskito Cays and the Swan Islands to work as stepping stones if relevant for connectivity within the area.

Discussion

Freely available sea surface temperature data can be used to plan for global warming by integrating information from multiple bleaching events and considering the capability of corals to adapt to the changes while using the best science available. The method introduced here has two main advantages over alternative approaches for reserve design in a climate change context. First, unlike most methods (Figure 1), it takes into account the ability of corals to adjust to climatic changes by using the latest advances in bleaching science available. Secondly, it incorporates the impacts of multiple bleaching events (lacking in [7]), and therefore includes natural temporal (as well as spatial) variability in the stress associated with bleaching.

Our temperature regimes coincide with well-defined oceanographic regions. Refuge areas (Blue regime) encompass environmental conditions that afford natural resistance to bleaching, and as a consequence are relatively unaffected by thermal anomalies [41]. Refugia include areas of moderate depths, regions exposed to vigorous circulation or high wave exposure, upwelling and high-latitude areas [41]. The protection afforded by areas naturally resistant to bleaching has long been identified as an important component of MPA networks in a changing climate [5,6,16,42]. If a conventional approach to reserve design has to be taken, then reserves should be located in these areas. However, we recommend following a more unconventional approach, and also include areas that have experienced stress in the past, which might prepare corals better for future temperature anomalies. This approach recognises the potential of corals to adapt and acclimatise, and also takes into account the fact that many areas allegedly resistant to bleaching are incapable of avoiding thermal stress in a significant and consistent manner through time, at least in the Caribbean [23].

Reefs where thermal stress is intense and recurrent (Red regime), will experience high coral mortality and less opportunity to recover, acclimatise or adapt [33]. Longer time periods between disturbances, and therefore high standard deviations in acute stress, are more likely to lead to higher recovery and reef persistence over time. These regimes (Yellow and Green) are key if long-term acclimation of corals is to be considered in marine reserve design. A large number of empirical studies suggest that corals that have been previously exposed to moderate levels of thermal stress have greater resistance to future events [43,44,45]. These changes might occur due to community shifts towards dominance of coral-tolerant taxa [44], or acclimation due to changes in symbiont community composition through switching or shuffling [46] or changes in the host [47,48]. Under intense, albeit variable bleaching (Yellow regime) the potential for acclimation/adaptation will depend both on the survival of corals after bleaching and the time frames required for acquiring genetic adaptation and phenotypic acclimation, which are uncertain.

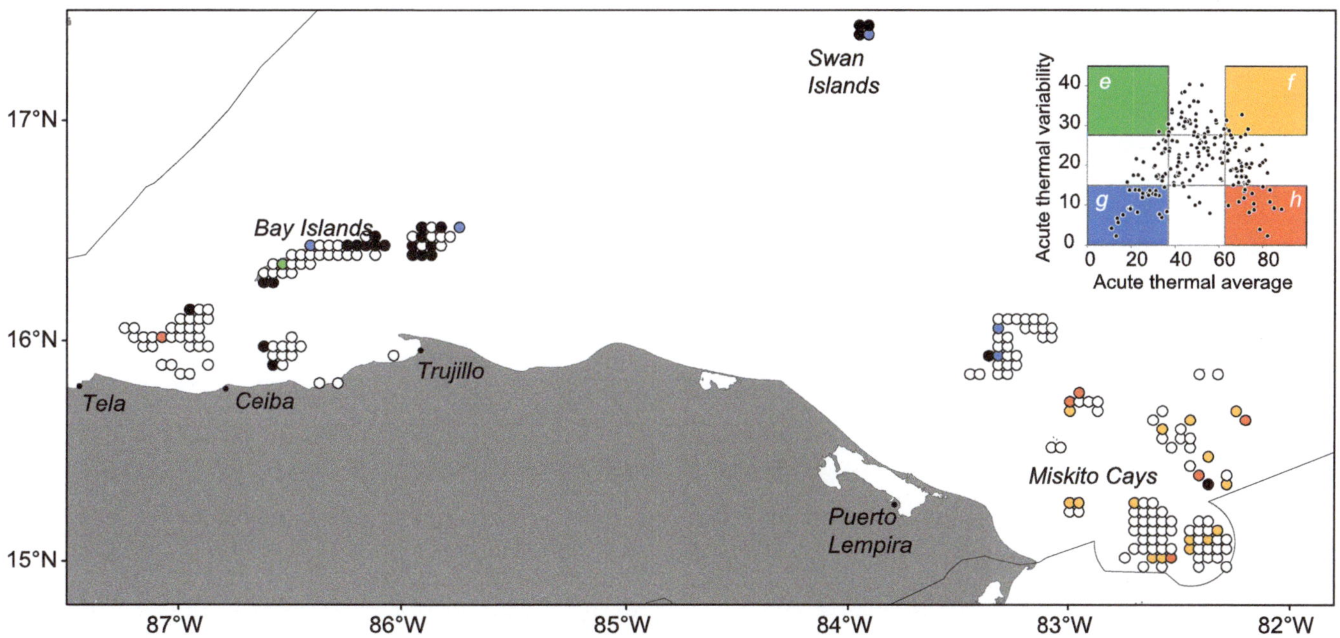

Figure 5. Spatial arrangement of thermal regimes in Honduras. Classification of extreme categories following the colour scheme in Figure 3j with areas with high rate of warming are depicted in black. Unclassified areas are depicted in white. Inset shows average vs. standard deviation percentiles of thermal anomalies in four Caribbean-wide bleaching events (1995, 1998, 2005 and 2008) in reef habitats in Honduras.

Selecting a variety of temperature regimes might increase the likelihood that some reefs will survive future bleaching [6]. To maximize chances of success, managers might focus protection on areas that are predicted to cope better with climate change, which have been less impacted by thermal stress at all times (Blue regime). If feasible, including areas that offer the greatest potential for acclimation (areas with high variability in incidence of thermal stress such as the Green or Yellow regimes) could serve as a bet-hedging strategy. Areas subjected to severe, constant stress (Red regime) should be avoided, as well as areas with rapid increases of temperature from spring to summer. This information should be factored in with other relevant layers (e.g. connectivity, habitat maps, maps of current uses and threats) to prioritize conservation actions. See Magris et al. [49] for a recent review on how to incorporate thermal regimes and connectivity patterns into marine spatial planning, or Zanakis et al. [50] for methods to incorporate multiple attributes into decision making and spatial planning.

Satellites measure the first few millimeters of the ocean surface, literally the "skin" of the ocean. The water column is dynamic and satellite temperature data do not mirrors entirely the temperature patterns on the seabed, where corals are located, missing some small-scale variability and extremes [20,51]. In spite of this limitation, satellites are unique at allowing the assessment of large areas simultaneously, and a key tool for marine spatial planning. Additionally, another constraint of the method described here is that is based on historical data, which has some limitations for use in planning for climate change when assessing rare (in a climatological context), extreme disturbances such as bleaching events, given that implicitly assumes that future incidence of bleaching events will be similar than past incidence. The method, however, could be refined by adding new information as it becomes available (i.e., as additional bleaching events occur) and be incorporated into an adaptive planning process. As we mentioned before, Global Circulation Model outputs, given their coarse resolution, are not appropriate for site selection of MPAs [19]. Techniques are available to obtain local weather patterns from GCM results, based either on statistical downscaling (by using historical information: [52]) or dynamical downscaling (by integrating global and regional models). Statistical downscaling assumes that the relationship established between the modeled data and the historical data for the present day climate will hold in the future, and that there will be no systematic changes in regional forcing at scales not considered by the global model [53], which makes them unrealistic for the assessment of future changes in coastal areas. Dynamical downscaling would be the best alternative to plan for climate change at a spatial resolution relevant for reefs; however, this method is computationally intensive and as a consequence it has never been used for marine spatial prioritization to date.

Finally, we anticipate that further research when planning for coral bleaching under a changing climate will move away from looking merely at sea temperature data to look into new synoptic satellite products that incorporate both light and temperature stress, which are currently under development [54]. While mass-bleaching events have been associated with anomalous increases in sea surface temperature, other environmental factors can also contribute to modulate the bleaching response. Among them, light plays a key role in the determination of the severity of the temperature impact, as it exacerbates its effect increasing the levels of oxidative stress accumulated in the coral tissue [27,28,55], and determining the levels of pigmentation in the coral and therefore the response to thermal stress [56,57]. These new synoptic satellite products that incorporate both light and temperature constitute the next frontier when planning for coral bleaching under different global warming scenarios.

Acknowledgments

We are very grateful to two anonymous reviewers and the handling PLOS ONE editor for their constructive comments on the manuscript.

Author Contributions

Conceived and designed the experiments: IC SE PJM. Performed the experiments: IC. Analyzed the data: IC. Contributed to the writing of the manuscript: IC SE PJM.

References

1. Hoegh-Guldberg O, Mumby PJ, Hooten AJ, Steneck RS, Greenfield P, et al. (2007) Coral reefs under rapid climate change and ocean acidification. Science 318: 1737–1742.
2. Baker AC, Glynn PW, Riegl B (2008) Climate change and coral reef bleaching: an ecological assessment of long-term impacts, recovery trends and future outlook. Estuarine, Coastal and Shelf Science 80: 435–471.
3. Munday PL, Jones GP, Pratchett MS, Williams AJ (2008) Climate change and the future for coral reef fishes. Fish and Fisheries 9: 261–285.
4. Andersson AJ, Gledhill D (2013) Ocean acidification and coral reefs: effects on breakdown, dissolution, and net ecosystem calcification. Annual Review of Marine Science 5: 321–348.
5. Game ET, Watts ME, Wooldridge S, Possingham HP (2008) Planning for the resistance in marine reserves: a question of catastrophic importance. Ecological Applications 18: 670–680.
6. McLeod E, Salm R, Green A, Almany J (2009) Designing marine protected area networks to address the impacts of climate change. Frontiers in Ecology and the Environment 7: 362–370.
7. Mumby PJ, Elliott IA, Eakin CM, Skirving W, Paris CB, et al. (2011) Reserve design for uncertain responses of coral reefs to climate change. Ecology Letters 14: 132–140.
8. Groves C, Game E, Anderson M, Cross M, Enquist C, et al. (2012) Incorporating climate change into systematic conservation planning. Biodiversity and Conservation 21: 1651–1671.
9. Hoegh-Guldberg O (1999) Climate change, coral bleaching and the future of the world's coral reefs. Marine and Freshwater Research 50: 839–866.
10. Kleypas JA, Buddemeier RW, Archer D, Gattuso, Langdon C, et al. (1999) Geochemical consequences of increased atmospheric carbon dioxide on coral reefs. Science 284: 118–120.
11. Gledhill DK, Wanninkhof R, Millero FJ, Eakin M (2008) Ocean acidification of the Greater Caribbean region 1996–2006. Journal of Geophysical Research - C Oceans 113: C10031.
12. Strong AE, Kearns EJ, Gjovig KK (2000) Sea surface temperature signals from satellites - An update. Geophysical Research Letters 27: 1667–1670.
13. McLeod E, Anthony KRN, Andersson A, Beeden R, Golbuu Y, et al. (2013) Preparing to manage coral reefs for ocean acidification: lessons from coral bleaching. Frontiers in Ecology and the Environment 11: 20–27.
14. Doney SC, Fabry VJ, Feely RA, Kleypas JA (2009) Ocean acidification: the other CO_2 problem. Annual Review of Marine Science 1: 169–192.
15. Eakin CM, Nim CJ, Brainard RE, Aubrecht C, Elvidge C, et al. (2010) Monitoring coral reefs from space. Oceanography 23: 118–133.
16. Ban NC, Pressey RL, Weeks S (2012) Conservation objectives and sea-surface temperature anomalies in the Great Barrier Reef. Conservation Biology 26: 799–809.
17. Levy JS, Ban NC (2013) A method for incorporating climate change modelling into marine conservation planning: An Indo-west Pacific example. Marine Policy 38: 16–24.
18. McLeod E, Moffitt R, Timmermann A, Salm R, Menviel L, et al. (2010) Warming seas in the coral triangle: coral reef vulnerability and management implications. Coastal Management 38: 518–539.
19. Kwiatkowski L, Halloran P, Mumby P, Stephenson D (2013) What spatial scales are believable for climate model projections of sea surface temperature? Climate Dynamics: 1–14.

20. Selig ER, Casey KS, Bruno JF (2010) New insights into global patterns of ocean temperature anomalies: implications for coral reef health and management. Global Ecology and Biogeography 19: 397–411.
21. Rowan R (2004) Coral bleaching: thermal adaptation in reef coral symbionts. Nature 430: 742–742.
22. Howells EJ, Beltran VH, Larsen NW, Bay LK, Willis BL, et al. (2012) Coral thermal tolerance shaped by local adaptation of photosymbionts. Nature Clim Change 2: 116–120.
23. Chollett I, Mumby PJ (2013) Reefs of last resort: Locating and assessing thermal refugia in the wider Caribbean. Biological Conservation 167: 179–186.
24. Castillo K, Helmuth B (2005) Influence of thermal history on the response of *Montastraea annularis* to short-term temperature exposure. Marine Biology 148: 261–270.
25. McClanahan TR, Ateweberhan M, Muhando CA, Maina J, Mohammed MS (2007) Effects of climate and seawater temperature variation on coral bleaching and mortality. Ecological Monographs 77: 503–525.
26. Oliver T, Palumbi S (2011) Do fluctuating temperature environments elevate coral thermal tolerance? Coral Reefs 30: 429–440.
27. Enríquez S, Mendez ER, Iglesias-Prieto R (2005) Multiple scattering on coral skeletons enhances light absorption by symbiotic algae. Limnology and Oceanography 50: 1025–1032.
28. Terán E, Méndez ER, Enríquez S, Iglesias-Prieto R (2010) Multiple light scattering and absorption in reef-building corals. Applied Optics 49: 5032–5042.
29. Casey KS, Brandon TB, Cornillon P, Evans R (2010) The past, present and future of the AVHRR pathfinder SST program. In: Barale V, Gower J, Albertotanza L, editors. Oceanography from space, revisited. New York: Springer. 273–288.
30. Kilpatrick KA, Podestá GP, Evans R (2001) Overview of the NOAA/NASA advanced very high resolution radiometer Pathfinder algorithm for sea surface temperature and associated matchup database. Journal of Geophysical Research C Oceans 106: 9179–9197.
31. Liu G, Strong AE, Skirving W, Arzayus LF (2006) Overview of NOAA coral reef watch program's near-real time satellite global coral bleaching monitoring activities. Proceedings of the 10th International Coral Reef Symposium. Okinawa, Japan. 1783–1793.
32. Glynn PW, D'Croz L (1990) Experimental evidence for high temperature stress as the cause of El Niño-coincident coral mortality. Coral Reefs 8: 181–191.
33. Jokiel PL, Coles SL (1990) Response of Hawaiian and other Indo-Pacific reef corals to elevated temperature. Coral Reefs 8: 155–162.
34. Liu G, Strong AE, Skirving W (2003) Remote sensing of sea surface temperature during 2002 Barrier Reef coral bleaching. EOS, Transactions, American Geophysical Union 84: 137–144.
35. Eakin CM, Morgan JA, Heron SF, Smith TB, Liu G, et al. (2010) Caribbean corals in crisis: record thermal stress, bleaching, and mortality in 2005. PLoS ONE 5: e13969.
36. Andréfouët S, Muller-Karger F, Robinson J, Kranenburg C, Torres-Pulliza D, et al. (2006) Global assessment of modern coral reef extent and diversity for regional science and management applications: a view from space. Proceedings of the 10th International Coral Reef Symposium. Okinawa, Japan. 1732–1745.
37. Burke L, Reytar K, Spalding M, Perry A (2011) Reefs at risk revisited. Washington: World Resources Institute. 114 p.
38. Goreau T, McClanahan T, Hayes R, Strong A (2000) Conservation of coral reefs after the 1998 global bleaching event. Conservation Biology 14: 5–15.
39. CARICOMP (1997) Studies on coral bleaching, 1995–96. Proceedings of the 8th International Coral Reef Symposium. Panama City, Panama. 673–678.
40. Kintisch E (2010) Caribbean coral die-off could be worst ever. Science now 14 10 2010.
41. Glynn PW (1996) Coral reef bleaching: facts, hypotheses and implications. Global Change Biology 2: 495–509.
42. West JM, Salm RV (2003) Resistance and resilience to coral bleaching: implications for coral reef conservation and management. Conservation Biology 17: 956–967.
43. Thompson DM, van Woesik R (2009) Corals escape bleaching in regions that recently and historically experienced frequent thermal stress. Proceedings of the Royal Society B: Biological Sciences 276: 2893–2901.
44. Maynard JA, Anthony KRN, Marshall PA, Masiri I (2008) Major bleaching events can lead to increased thermal tolerance in corals. Marine Biology 155: 173–182.
45. Carilli J, Donner SD, Hartmann AC (2012) Historical temperature variability affects coral response to heat stress. PLoS ONE 7: e34418.
46. Jones AM, Berkelmans R, van Oppen MJH, Mieog JC, Sinclair W (2008) A community change in the algal endosymbionts of a scleractinian coral following a natural bleaching event: field evidence of acclimatization. Proceedings of the Royal Society B: Biological Sciences 275: 1359–1365.
47. Bellantuono AJ, Hoegh-Guldberg O, Rodriguez-Lanetty M (2012) Resistance to thermal stress in corals without changes in symbiont composition. Proceedings of the Royal Society B: Biological Sciences 279: 1100–1107.
48. Bellantuono AJ, Granados-Cifuentes C, Miller DJ, Hoegh-Guldberg O, Rodriguez-Lanetty M (2012) Coral thermal tolerance: tuning gene expression to resist thermal stress. PLoS ONE 7: e50685.
49. Magris RA, Pressey RL, Weeks R, Ban NC (2014) Integrating connectivity and climate change into marine conservation planning. Biological Conservation 170: 207–221.
50. Zanakis SH, Solomon A, Wishart N, Dublish S (1998) Multi-attribute decision making: a simulation comparison of select methods. European Journal of Operational Research 107: 507–529.
51. Smale DA, Wernberg T (2009) Satellite-derived SST data as a proxy for water temperature in nearshore benthic ecology. Marine Ecology Progress Series 387: 27–37.
52. Pielke RA, Wilby RL (2012) Regional climate downscaling: What's the point? Eos, Transactions American Geophysical Union 93: 52–53.
53. IPCC (2013) Climate Change 2013: The Physical Science Basis. Contribution of Working Group I to the Fifth Assessment Report of the Intergovernmental Panel on Climate Change. Cambridge University Press, Cambridge, United Kingdom and New York, NY, USA, 1535 p.
54. Liu G, Eakin CM, Rauenzahn JL, Christensen TRL, Heron SF, et al. (2012) NOAA Coral Reef Watch's decision support system for coral reef management. Proceedings of the 12th International Coral Reef Symposium. Cairns, Australia. 5A.
55. Marcelino LA, Westneat MW, Stoyneva V, Henss J, Rogers JD, et al. (2013) Modulation of light-enhancement to symbiotic algae by light-scattering in corals and evolutionary trends in bleaching. PLoS ONE 8: e61492.
56. Takahashi S, Nakamura T, Sakamizu M, Woesik Rv, Yamasaki H (2004) Repair machinery of symbiotic photosynthesis as the primary target of heat stress for reef-building corals. Plant and Cell Physiology 45: 251–255.
57. Lesser MP (1996) Elevated temperatures and ultraviolet radiation cause oxidative stress and inhibit photosynthesis in symbiotic dinoflagellates. Limnology and Oceanography 41: 271–283.

6

Satellite Tracking of Sympatric Marine Megafauna Can Inform the Biological Basis for Species Co-Management

Christian Gredzens[1,2]*, Helene Marsh[1,3], Mariana M. P. B. Fuentes[4], Colin J. Limpus[5], Takahiro Shimada[1], Mark Hamann[1,3]

1 School of Earth and Environmental Sciences, James Cook University, Townsville, Queensland, Australia, 2 School of Marine and Tropical Biology, James Cook University, Townsville, Queensland, Australia, 3 Centre for Tropical Water and Aquatic Ecosystem Research, James Cook University, Townsville, Queensland, Australia, 4 ARC Centre of Excellence for Coral Reef Studies, James Cook University, Townsville, Queensland, Australia, 5 Aquatic Threatened Species Unit, Queensland Department of Environment and Heritage Protection, Brisbane, Queensland, Australia

Abstract

Context: Systematic conservation planning is increasingly used to identify priority areas for protection in marine systems. However, ecosystem-based approaches typically use density estimates as surrogates for animal presence and spatial modeling to identify areas for protection and may not take into account daily or seasonal movements of animals. Additionally, sympatric and inter-related species are often managed separately, which may not be cost-effective. This study aims to demonstrate an evidence-based method to inform the biological basis for co-management of two sympatric species, dugongs and green sea turtles. This approach can then be used in conservation planning to delineate areas to maximize species protection.

Methodology/Results: Fast-acquisition satellite telemetry was used to track eleven dugongs and ten green turtles at two geographically distinct foraging locations in Queensland, Australia to evaluate the inter- and intra-species spatial relationships and assess the efficacy of existing protection zones. Home-range analysis and bathymetric modeling were used to determine spatial use and compared with existing protection areas using GIS. Dugong and green turtle home-ranges significantly overlapped in both locations. However, both species used different core areas and differences existed between regions in depth zone use and home-range size, especially for dugongs. Both species used existing protection areas in Shoalwater Bay, but only a single tracked dugong used the existing protection area in Torres Strait.

Conclusions/Significance:: Fast-acquisition satellite telemetry can provide evidence-based information on individual animal movements to delineate relationships between dugongs and green turtles in regions where they co-occur. This information can be used to increase the efficacy of conservation planning and complement more broadly based survey information. These species also use similar habitats, making complimentary co-management possible, but important differences exist between locations making it essential to customize management. This methodology could be applied on a broader scale to include other sympatric and inter-related species.

Editor: Graeme Hays, Deakin University, Australia

Funding: The Torres Strait field work was funded by an Australian National Environment Research Program scholarship. The project was funded by the Australian Government's Australian Marine Mammal Centre, Marine and Tropical Science Research Facility and National Environment Research Program, the Torres Strait Regional Authority, the Great Barrier Reef Marine Park Authority, the Queensland Department of Environment and Heritage Protection, and an anonymous donor with a passion for dugongs. The identity of the donor is not considered relevant to the editors or reviewers' assessment of the validity of the work. None of the research was funded, in whole or in part, by a tobacco company. The funders had no role in study design, data collection and analysis, decision to publish, or preparation of the manuscript.

Competing Interests: The authors have declared that no competing interests exist.

* E-mail: Christian.Gredzens@my.jcu.edu.au

Introduction

Understanding how animals use space can be fundamental to identifying the components necessary for the conservation of a species [1]. These components may include key resources, critical habitats, movement patterns, and distributions; all of which can be used to determine the spatial dynamics of a population [2–4]. Such knowledge has the potential to play a key role in the designation of marine protected areas (MPAs), guiding management decisions, predicting the effects of environmental change, and aiding in the calculation of population estimates [5–7]. Knowledge of these components is especially important as the human population continues to grow and anthropogenic influences become more widespread throughout the natural environment [8].

MPAs are being increasingly used to protect marine species and environments [9,10], especially as the number of marine species listed as threatened increases [11]. However, single species approaches to MPA design are the subject of increased scrutiny with ecosystem-based approaches being increasingly favored [12–15]. Contemporary ecosystem-based approaches depend on using acceptable surrogates for the presence of species or species groups in conjunction with sighting records, density estimates, and appropriate modeling techniques [9,16]. All of these factors rely on assumptions that are vulnerable to errors of commission and omission as well as sampling bias which can affect MPA efficacy [16,17]. These concerns are exacerbated when designing MPAs to conserve species that are mobile and/or dependent on benthic habitats of variable quality.

The use of MPAs to protect mobile marine species, such as marine megavertebrates (e.g., sea turtles, marine mammals, and elasmobranchs), can be difficult as their ranges may encompass significant portions of coastlines and even entire ocean basins [9,18–20] (but see [21]). Additionally, a diverse range of cultural, economic, political, and legal obstacles can impede the establishment and enforcement of large reserves, especially transnational reserves [22,23]. However, many mobile marine megavertebrates display strong site fidelity to specific regions on a seasonal or yearly basis allowing the protection of key areas using permanent or seasonal spatial closures [9,20,24]. This protection can be particularly important for marine megavertebrates as many species display life-history characteristics (e.g., long lived, low fecundity, high adult survivorship) [25–27] that increase their vulnerability to human-induced mortality and destruction of critical habitat.

Using a systematic conservation planning approach to designate high conservation value areas to conserve marine megavertebrates has increased over the past decade [10,15,20,28]. This planning methodology was a key driver in the implementation of conservation areas for dugongs (*Dugong dugon*) and green sea turtles (*Chelonia mydas*) in the rezoning of the Great Barrier Reef Marine Park (GBRMP), Australia in 2003 and has been suggested for use in implementing reserves in Torres Strait, Australia [10,29]. These regions both contain globally significant populations of dugongs and green sea turtles, with Torres Strait being home to the World's largest dugong population (>12,000 individuals) [26] while the Great Barrier Reef (GBR) and Torres Strait region contain some of the most important nesting grounds for the largest population of green sea turtles [30]. The rezoning of the GBR in 2003 increased the level of protection of high conservation value foraging grounds in no-take zones for dugongs from 17% to 42% and green turtles from 0.03% to 13% of the total area of occupancy identified for each species [10,31,32]. The areas for protection built on the pre-2003 management arrangements including the Dugong Protection Areas (DPAs) [33] and were predominantly identified using expert opinion and data from aerial surveys, seagrass mapping, and sighting records [31,32,34]. Such approaches may not satisfy stakeholders [35] who can better identify with information on individual animals.

Dugongs and green turtles are often found together, sometimes at high densities [26], and thus may be managed together (e.g., "Go-slow" areas in Moreton Bay near Brisbane, www.nprsr.qld.gov.au/parks/moreton-bay/about.html). However, a comprehensive understanding of the comparative ecology of dugongs and green turtles is required to optimize management because a comparative study on the stomach contents of dugongs and green turtles in Torres Strait found that dugongs fed exclusively on seagrasses whereas turtles consumed both seagrasses and macroalgae [36]. Thus a detailed and concurrent study of the movements of both species is required.

We used fast-acquisition GPS satellite telemetry to quantify the fine-scale distributions of sympatric dugongs and green turtles to enhance their management by tracking them at two locations of high conservation value in northeastern Australia (Shoalwater Bay and Torres Strait). Home-range analysis and bathymetric modeling were then used to compare spatial use within and between species and within and between geographic areas. Our objectives were to: (1) determine the spatial relationships between dugongs and green turtles in regions where they co-occur and between different locations; (2) assess the biological appropriateness of developing techniques to co-manage these two species; (3) evaluate the efficacy of the existing protection zones at our study locations. We also discuss how the management of dugongs and green sea turtles and other sympatric, mobile marine megafauna can be enhanced by detailed studies of the movements of individual animals.

Materials and Methods

Ethics Statement

All necessary permits were obtained for the described study. The animal use protocol for this research was approved by the James Cook University Animal Ethics Committee and Queensland Parks and Wildlife. All protected species were handled in strict accordance with local and international regulations. The dugong research was conducted under James Cook University Animal Ethics Committee Permits A1498 and A1683 and Queensland Parks and Wildlife Scientific Purpose Permit WISP006774410. The turtle research was conducted under James Cook University Animal Ethics Committee Permits A1229 and A1683 and Queensland Parks and Wildlife Scientific Purpose Permits WISP02742310 and WISP09469711.

Satellite Transmitter Attachment and Tracking

Raw, unfiltered tracking data were collected using fast-acquisition GPS satellite transmitters attached to six dugongs (three females and three males) and four adult female green sea turtles near Mabuiag Island, Torres Strait, Australia (9° 57'S, 142° 10'E) in July 2009 and September 2010, and five dugongs (four females and one male) and six female green sea turtles (five adults and one prepubescent) in Shoalwater Bay, Australia (22° 25'S, 150° 25'E) in June/July 2012 (Table 1).

The dugongs were captured using the dermal holdfast technique [39] in Torres Strait and the standard rodeo technique in Shoalwater Bay [40]. At both locations, the dugongs were fitted with Telonics Gen 4 GPS/ARGOS marine units attached to a 3 m tether linked to a padded tailstock harness (for details see [41]).

The green turtles were captured using the standard rodeo technique [42], brought to Mabuiag Island (Torres Strait) or MacDonald Point (Shoalwater Bay), and fitted with one of four types of satellite transmitters (Sirtrack F4G 291A, Wildlife Computers SPLASH10 BF-273A and Splash10 BF-273C, or SMRU SRDL 9000x). Each transmitter was attached to the carapace using the methods described in Shimada et al. (2012) [43]. Each turtle was released from shore the day after capture.

Dugong units were programmed to collect a GPS position hourly; turtle units every 30 minutes. All units were programmed with a five minute repeat in case a signal was not received when the animal surfaced.

Data Processing

All raw data were transmitted via the ARGOS network, downloaded from ARGOS via the transmitter manufacturer supplied software, converted to the Universal Transverse Mercator (UTM) coordinate system, filtered, and processed.

Dugong location data were used from the time of release until a transmitter stopped transmitting or detached. Post-release locations were not removed from the dataset as previous studies indicated no behavioral changes after capture and handling [44]. Transmitter detachments were identified by: (1) acquisition of successive succeeded GPS location classes at consistent time

Table 1. Information on the dugongs and green sea turtles tagged in Shoalwater Bay, Australia in 2012 and Torres Strait, Australia in 2009 and 2010.

Individual	Date tagged	Sex	Maturity	Length (cm)*	Days+	Tag type
Shoalwater Bay						
Dugongs						
652631A	01 July 2012	M	Unknown	262	141	Telonics
652636A	30 June 2012	F	Adult	288	51	Telonics
652640A	01 July 2012	F	Unknown	252	6	Telonics
652642A	01 July 2012	F	Immature	179	34	Telonics
652643A	30 June 2012	F	Immature	229	54	Telonics
Turtles						
96777	01 July 2012	F	Adult	96.1	145	Sirtrack
96780	03 July 2012	F	Adult	97.3	121	Sirtrack
108469	29 June 2012	F	Adult	104.5	147	SMRU
108472	29 June 2012	F	Adult	100.5	148	SMRU
120640	30 June 2012	F	Prepubescent	102.1	137	Wildlife Computers
120641	01 July 2012	F	Adult	95.5	131	Wildlife Computers
Torres Strait						
Dugongs						
641060A	18 September 2010	F	Adult	250	69	Telonics
641058A	13 September 2010	F	Likely Adult	224	33	Telonics
641052A	22 September 2010	M	Large Juvenile	182	79	Telonics
641054A	22 September 2010	M	Likely Adult	224	30	Telonics
641057A	14 September 2010	F	Adult	335	22	Telonics
641055A	14 September 2010	M	Likely Adult	223	7	Telonics
Turtles						
70455	22 September 2010	F	Adult	118	134	Sirtrack
95889	25 July 2009	F	Adult	98	27	Sirtrack
95891	25 July 2009	F	Adult	102.1	28	Sirtrack
95892	21 September 2010	F	Adult	105.6	47	Sirtrack

*Dugong length was measured as straight total length [37], turtle length was measured as curved carapace length (CCL) [38].
+Total days tracked while foraging.

intervals (e.g., on the hour, every hour), (2) prolonged straight line movement, or (3) signals from one location for an extended period of time (e.g., days).

The location data for turtles were not used in our analysis until the target animal returned to its capture location or exhibited foraging behavior. Foraging behavior was identified by non-directed movement (i.e., tortuous, short distance movements) for a minimum of three consecutive days after an initial directed movement (i.e., straight-line movement covering a significant distance over a short period of time) [45]. The tortuosity of individual movements was evaluated through visual inspection of connecting lines between consecutive location points using GIS. Straight-line movement was defined by the trajectory between two locations being less than 45° off to either side from the extension of the previous two locations [45].

Different filtering techniques were required for each type of transmitter because of differences between transmitters, locational accuracy, and species life-history characteristics. The Telonics units used for dugongs employed a Quick Fix Pseudoranging (QFP) alternative when a GPS location could not be acquired. QFP locations were categorized by locational accuracy into three categories: resolved QFP, resolved QFP (uncertain), and unresolved QFP. Telonics (2012) [46] states that 98.4% of resolved QFP positions are within 30 m of the actual position, resolved QFP (uncertain) positions are generally within 75 m, and unresolved QFP positions are within several hundred meters. Because the use of QFP technology in animal tracking is relatively new, objective filtering techniques were developed to maintain accuracy of acquired locations. For fine-scale analysis, data were initially filtered by location class, using only succeeded GPS, resolved QFP, and resolved QFP (uncertain) location classes to maintain accuracy ≤75 m. After initial filtering, over-speed errors were removed; these were identified by the distance and time between successive fixes necessitating speeds beyond the maximum sustained swimming speed for dugongs of 10 km hr^{-1} [47]. Locations more than 30 m inland were also removed as this is the error estimate for the most precise QFP locations.

For turtles, all units used fast-acquisition GPS technology and were filtered in two ways. First, following the manufacturer's instructions, initial filtering used the residual error and number of satellite uplinks used for each location. Locations with a residual error greater than 30 or with fewer than four satellite uplinks were excluded [48]. Second, data were filtered using the data-driven method described by Shimada et al. (2012) [43], which removes locations for which speeds between successive locations exceed the maximum linear speed (V_{max}) of an animal or if all of the following apply: (a) the number of source satellites is four, (b) the inner angle is acute, and (c) the speed either from or to a subsequent location exceeds a maximum "loop-trip" speed (V_{lp}) calculated for an animal [43]. V_{max} for green sea turtles was determined to be 9.9 km hr^{-1} and V_{lp} was 2.0 km hr^{-1} (for details on how V_{max} and V_{lp} were calculated see [43]). Our calculated V_{max} for green turtles is consistent with maximum speeds calculated by Heithaus et al. (2002) [49] (median: 11 km hr^{-1}; inter-quartile range: 10–12 km hr^{-1}). Finally, locations were compared with Landsat imagery and digital elevation models to determine if locations found on land were accessible to a basking turtle and implausible locations were removed.

After filtering, the location data from each dugong or turtle were standardized by dividing the remaining location points into 3 hour duty-cycles and selecting the most accurate location within each duty-cycle. This duration was chosen to retain as many location points as possible while minimizing differences in the number of location points per day per animal. In addition, duty-cycles were used in an effort to reduce the effects of autocorrelation and effects resulting from differences in transmitter performance. These measures were necessary as sample size has been shown to significantly affect home-range estimates [50,51]. The best location was chosen using two criteria: (1) the location with the greatest number of satellite uplinks, and (2) if multiple locations had the same number of uplinks, the final choice was the location with the lowest residual error (Sirtrack, Wildlife Computers, and SMRU units) or the lowest positional dilution of precision error (Telonics units). In the event multiple locations had the same error value, the location closest to the median time point within each duty-cycle was chosen.

Home-range Estimation and Spatial Use

Home-ranges were calculated using fixed kernel density estimation with bandwidths selected by likelihood cross-validation (CVh) [52]. The fixed kernel method was selected over the adaptive kernel method because it has been shown to produce more accurate and precise home-range estimates that are less sensitive to autocorrelation [51,53]. While least-squares cross-validation (LSCV) has been suggested as the best bandwidth selector by many studies [51,54], we found that it produced elongated ranges that were skewed along a single axis; a by-product of the layout of the locational data of animals within this study and most likely a misrepresentation of each animal's actual use of space. After testing and interpreting the ecological/behavioral relevance of several other bandwidth selectors, CVh was found to produce ranges which were not skewed and the most ecologically appropriate. Hemson et al. (2005) [55] and Horne & Garton (2006) [56] also found that this method sometimes performs better than LSCV.

Kernel densities and bandwidths were calculated using the Geospatial Modelling Environment (GME), an extension to ArcGIS [57,58], with a resolution of 50 m. This resolution was selected because the mean accuracy of filtered fast-acquisition GPS locations is within 50 m of the true location [43] and the majority of filtered Telonics locations in this study were classified as resolved QFP or higher (mean = 83.5, SE = 5.3%, n = 11). GME was also used to generate 95% and 50% contour polygons from the calculated kernel densities in vector format. 95% home-range and 50% core range areas were calculated in ArcMap 10 [59] from the 95% and 50% contour polygons. Regions of contour polygons that overlapped land were removed before final areas were calculated. A core area was defined as the area in which an individual is predicted to be 50% of the time.

Home-ranges were calculated for each animal using data from the entire period in which they were tracked. We plotted days tracked versus range size to determine if there was a relationship and found no significant correlation between days tracked and range size with the majority of animals showing range stabilization at or near 20 days of tracking (Figure S1; Table S1). Diurnal comparisons were made by calculating two home-ranges for each animal using all filtered data, recorded during the day (0600–1800 hours) or night (1800–0600 hours). Differences between day and night ranges were determined in two ways: (1) visual determination of differences in location of animal movements (e.g., distance from geographic features) and (2) differences in estimated range size. Combined home-ranges were calculated for each species by merging the 95% and 50% ranges of all individuals to determine the total area covered.

Depth Zone Distribution

Bathymetry layers of each region were used to determine the depth zones used by each tracked animal. The Torres Strait layer was developed in 2008 at a resolution of 110 m [60] and the Shoalwater Bay layer was developed in 2010 at 100 m resolution [61]. Depths were measured at mean sea level and depth profiles were calculated by interpolation of data points from a variety of survey methods [60,61]. Each layer was stored in raster format, reclassified into 5 m depth zones and converted to vector format to allow overlay operations with home-range vector polygons. Total individual ranges and the diurnal ranges of each individual (both 95% home-ranges and 50% core areas) were then overlaid on the reclassified bathymetry layer and the total area over each depth zone was calculated for each individual.

Results

Tracking Duration and Location Points

The dugongs were tracked for between 6 and 141 days (mean = 47.8 days, SE = 11.7 days, n = 11) and foraging female turtles were tracked between 27 and 148 days (mean = 106.5 days, SE = 16.1 days, n = 10) (Table 1). Unfiltered data points ranged between 100 and 2846 locations (median: 465 locations) for dugongs and 196 and 2471 locations (median: 521 locations) for turtles. After data filtering, dugong transmitters provided a per-dugong median of 193 locations (range: 41–965 locations) and turtle transmitters provided a per-turtle median of 210 locations (range: 118–732 locations) over the total tracking time for each individual. All transmitters provided roughly an equal number of locations during the day and night after data filtering (Table S2). One dugong from Shoalwater Bay (652642A – female) became stranded during tracking with the transmitter unit still attached. Location fixes and time-depth recorder (TDR) data suggested the dugong started drifting with the current after 33 tracking days. All location points after the approximate time of death were excluded from the analysis. Inspection of the dugong carcass and photos of the stranded animal did not reveal any apparent cause of death.

Home-ranges and Core Areas - Shoalwater Bay

Four of the five dugongs used relatively small ranges with 95% home-range areas ranging from 15.9 km^2 to 72.8 km^2 (median: 49.5 km^2) and 50% core areas ranging between 2.6 km^2 and 21.3 km^2 (median: 4.2 km^2), encompassing a total area of 123.7 km^2 (95%) and 28.5 km^2 (50%) (Figure 1; Table S2). The fifth individual (652636A – female) had a significantly larger range

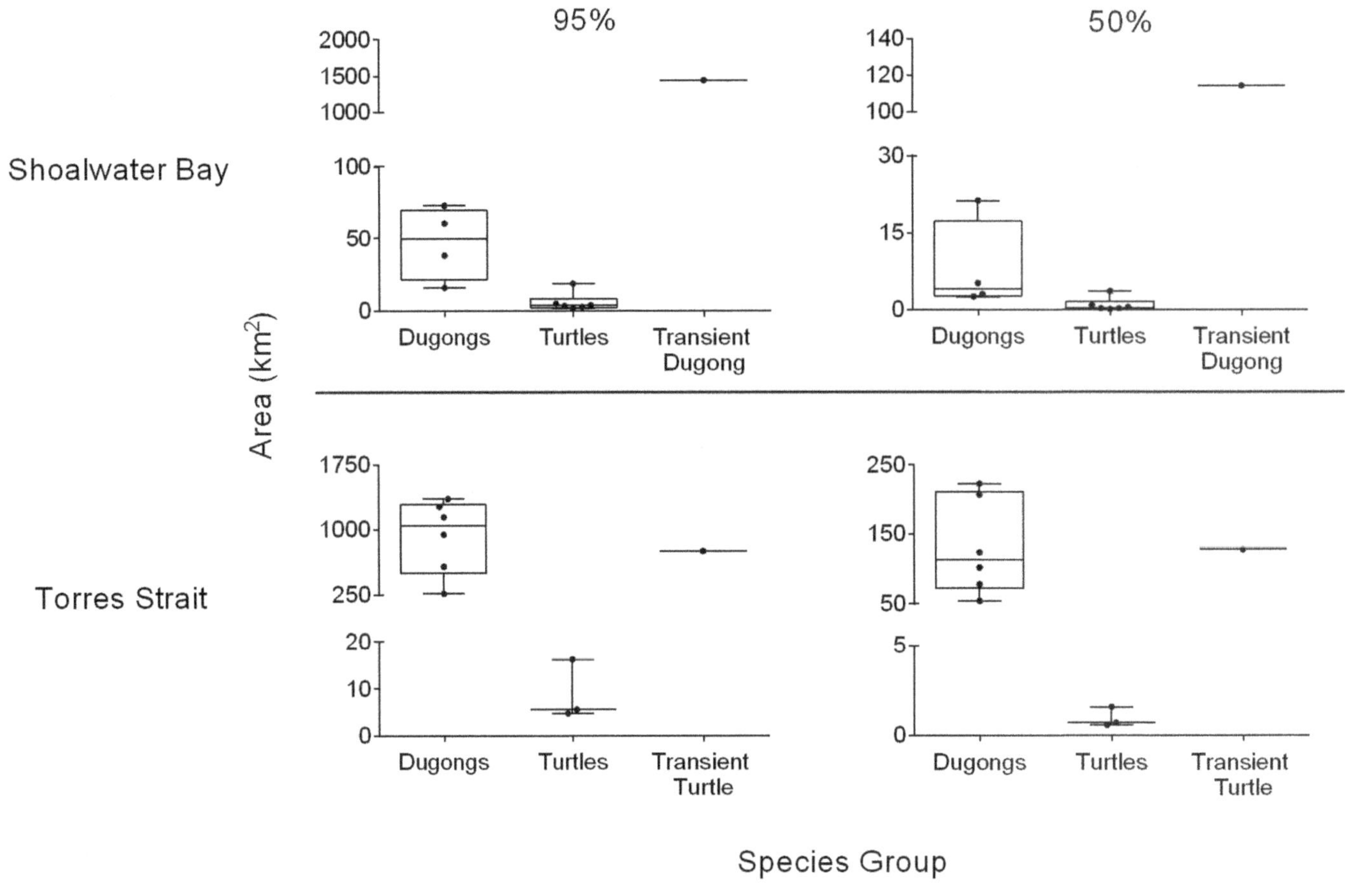

Figure 1. Home-range and core area size of tracked dugongs and green turtles. Comparison of 95% home-ranges (left) and 50% core areas (right) between Shoalwater Bay, Australia dugongs (n = 5), green sea turtles (n = 6) and a transient dugong (n = 1) (top) with Torres Strait, Australia dugongs (n = 6), reef associated green sea turtles (n = 3), and a transient green sea turtle (n = 1) (bottom) over the total tracking time of each animal. Lines within boxes represent the median, boxes represent interquartile range, whiskers represent minimum and maximum values, and dots indicate values for each individual. Note differences in scale on y axes.

with a 95% home-range of 1444.6 km^2 and 50% core area of 114.4 km^2 and exhibited behavior consistent with a transient animal.

All five dugongs were individualistic in their movements. The four dugongs with small 95% home-ranges (<75 km^2) were resident within the northwest portion of Shoalwater Bay between Clara Island and MacDonald Point for most of the tracking period (Figure 2B,C). One of these dugongs (652631A – male, 141 days) relocated approximately 90 km to the west near Clairview on 26 September 2012 (after two days at large) and spent the remainder of its tracking time at this location.

Four dugongs used the existing Dugong Protection Area (DPA [restrictions of gill netting; [33]]) and Great Barrier Reef Marine Park (GBRMP) marine national park zone (green zone [all fishing banned]) within Shoalwater Bay. Two of these animals (652643A – female and 652642A- female) spent their entire tracked time within the DPA and green zone, one relocated (652631A – male) from the Shoalwater Bay DPA (some restrictions of gill netting [33]) and green zone directly to the Clairview Bluff – Carmilla Creek DPA and GBRMP conservation park zone (yellow zone), thus all of its presumed foraging range was within the boundaries of these zones. The fourth dugong (652636A – female) made several large-scale movements around the region, moving throughout Shoalwater Bay and making an exploratory loop-trip 90 km to the west to Carmilla and then back into Shoalwater Bay over three days (Figure 2A). This animal spent two days in Port Clinton, 75 km away along the eastern side of the Warginburra Peninsula, before returning to Shoalwater Bay. Despite making this larger scale movement, this dugong had a preferred core area located at the northwest edge of Shoalwater Bay near Stanage Point and most (60%) of its 95% home-range and core areas fell outside DPA and GBRMP protection zones.

Five of the six green turtles tracked in Shoalwater Bay had very small coastal ranges that were closer to shore than those of the dugongs (Figure 2D,E,F). The sixth turtle (120641) was much more mobile than the others and had a significantly larger range (95%: 18.6 km^2, 50%: 3.6 km^2). This animal showed space use similar to that of dugongs within the area (Figure 2F). Combined turtle 95% home-range areas ranged from 1.4 km^2 to 18.6 km^2 (median: 3.5 km^2) and 50% core areas ranged between 0.1 km^2 and 3.6 km^2 (median: 0.4 km^2), encompassing a total area of 25.5 km^2 (95%) and 5.1 km^2 (50%) (Figure 1; Table S2). All six individuals remained near the coast in the northwest portion of Shoalwater Bay between MacDonald Point and Clara Island, close to their points of capture. Each turtle displayed strong site fidelity and had its own distinct core areas with very little overlap between other individuals' core areas. Four individuals (96777, 96780, 108469, and 108472) used mangrove areas and travelled up coastal streams with all but one of these individuals (108469) having core areas within these habitats. Location points along the

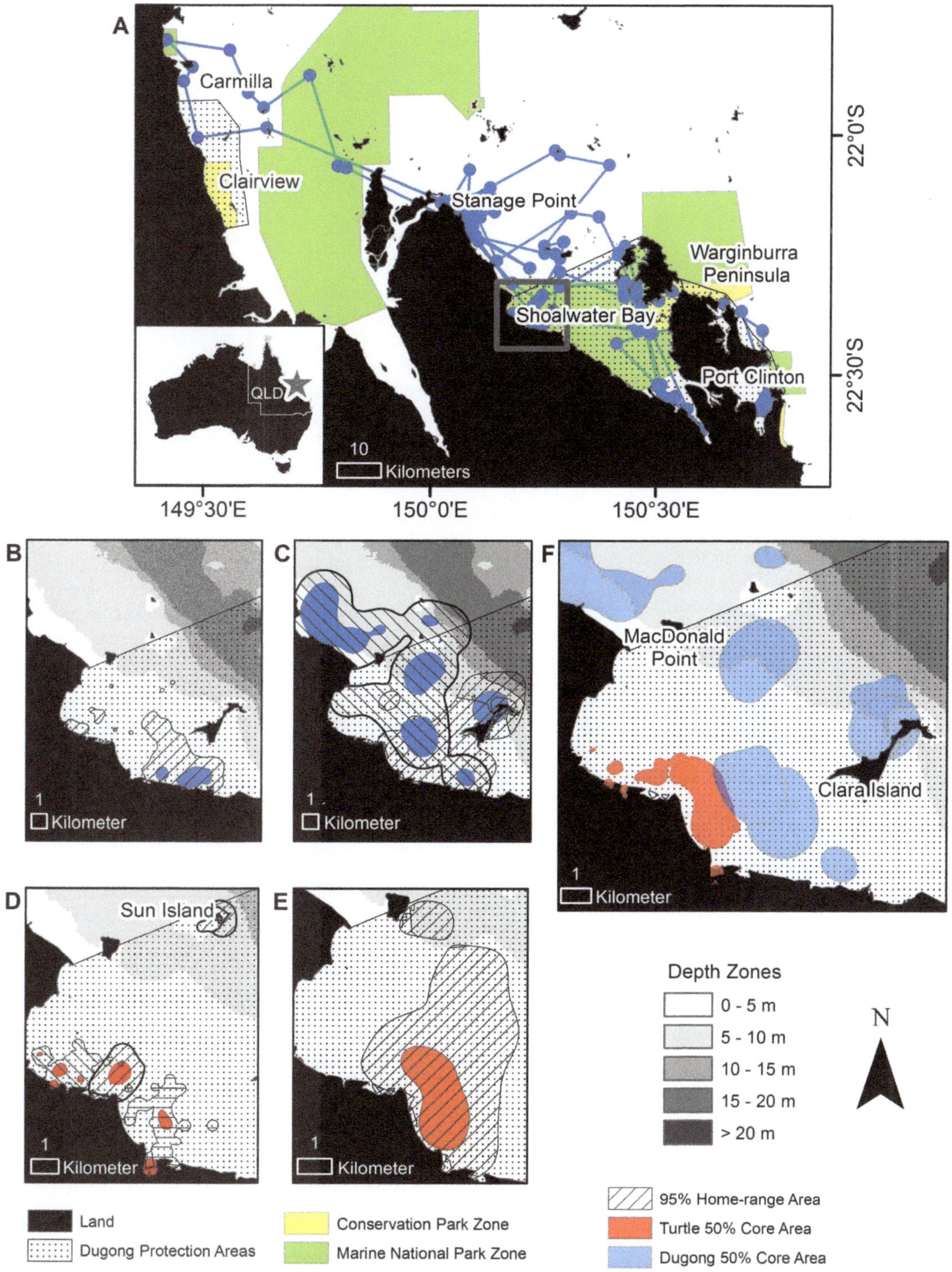

Figure 2. Shoalwater Bay dugong and green turtle spatial use. Dugong and green sea turtle tracking and foraging home-ranges within Shoalwater Bay, Australia using the total tracking duration of each animal. Blue features indicate dugongs and red features indicate green sea turtles. (**A**) Tracking locations and migratory path of the transient dugong (652636A). (**B** and **C**) Example home-ranges of dugongs (**B**: 652642A, **C**: 652640A (left) and 652643A (right)). (**D** and **E**) Example home-ranges of green sea turtles (**D**: 96780 (left), 108469 (middle), and 120640 (right); **E**: 120641). (**F**) Comparison of dugong and green sea turtle 50% core areas in northwestern Shoalwater Bay. Gray box indicates region displayed in maps **B-F**. For maps with multiple individuals plotted, each individual's 95% home-range is delineated by hatching at a different angle. Note differences in scale for each map.

shoreline suggested that all the turtles basked on land. One individual (108469) spent 2.5 weeks near Sun Island before relocating to the coast south of MacDonald Point. All turtles stayed within the Shoalwater Bay DPA and green zone and only a single turtle (108469) had a range that crossed outside of the DPA boundary, but this only accounted for <5% of this individuals

95% home-range and was still within the GBRMP green zone boundary. For both the dugongs and the green turtles, spatial use did not significantly differ in relation to geographic areas or between species on a monthly basis (Figure S2).

Home-ranges and Core Areas – Torres Strait

The six dugongs tracked in Torres Strait used 95% home-ranges ranging from 264.3 km^2 to 1269.2 km^2 (median: 1042.9 km^2) and 50% core areas ranging between 54.3 km^2 and 222.8 km^2 (median: 112.6 km^2), encompassing a total area of 3861.9 km^2 (95%) and 640.9 km^2 (50%) (Figure 1; Table S2). These ranges were significantly larger than four of the Shoalwater Bay dugongs, but not the transient dugong (652636A – female).

As in Shoalwater Bay, the movements of the Torres Strait dugongs were individualistic with multiple core areas. Most animals spent a proportion of their time at a foraging site between Mabuiag and Buru Islands known as the Yarral Gumi Maza region, which is generally over 5 meters deep (Figure 3A,D). Two animals (641052A – male and 611057A – female) made movements towards the southeast boundary of the Torres Strait Protected Zone Joint Authority (PZJA) region, one individual (641054A – male) spent the majority of its time around the Orman Reefs, another animal (641055A – male) moved north to the coast of Papua New Guinea (Figure 3E), and the remaining two dugongs (641060A – female and 641058A – female) remained within the Yarral Gumi Maza region. Only one dugong (641054A – male) ventured over the reef flat and lagoon of the Orman Reefs, but this animal spent the majority of its time away from reefs in deeper-water areas and relocated off the east coast of Moa Island on 16 October 2010 where it remained for the remainder of the time it was tracked (Figure 3D). A single dugong (641058A – female) crossed into the existing Dugong Sanctuary; however, 85% of its range was outside of the Sanctuary. One individual (641052A – male) made an exploratory loop-trip of 80 km south to the Cape York Peninsula and northern GBR over a period of two days. This individual did not enter the GBRMP.

Three Torres Strait turtles used space very similarly to the Shoalwater Bay turtles with 95% ranges between 4.8 km^2 and 16.2 km^2 (median: 5.6 km^2) and 50% core areas between 0.6 and 1.6 km^2 (median: 0.7 km^2), encompassing a total area of 25.0 km^2 (95%) and 2.9 km^2 (50%) (Figure 1; Table S2). Similar to the Shoalwater Bay turtles, the Torres Strait turtles showed strong site fidelity; each individual had its own distinct core areas. However, unlike turtles in Shoalwater Bay, which preferred coastal areas, Torres Strait turtles were reef associated with three of four turtles inhabiting the Orman Reefs and two of these individuals (70455 and 95892) remained on the northern side of Gariar Reef over almost the entire tracking period. These turtles only made short excursions (<1 km) over the reef flat and into the lagoon, spending most of their time near the reef crest and reef slope (Figure 3C). A single turtle (95889) was classified as transient, having a 95% range of 749.1 km^2 and 50% core area of 127.4 km^2, similar to the ranges of the Torres Strait dugongs (Figure 1). The transient turtle (95889) also spent a significant portion of time within the Yarral Gumi Maza region and northwest portion of Torres Strait towards Papua New Guinea (Figure 3B). No turtles crossed into the Dugong Sanctuary. For both the dugongs and the green turtles, spatial use did not significantly differ in relation to geographic areas or between species on a monthly basis (Figure S3).

Diurnal Comparisons

There was no overall trend in differences in the size of dugong 95% home-ranges and 50% core areas between day and night. Conversely, green turtles in Shoalwater Bay and reef associated turtles in Torres Strait had more restricted ranges during the night, with significant decreases of up to 76% in nighttime home-ranges and core areas (Table S2). Dugongs and turtles in both regions did not show major differences in the geographic locations of their home-ranges and core areas between day and night.

Depth Zone Distributions - Shoalwater Bay

Within Shoalwater Bay, the tracked dugongs and turtles frequented shallow water areas between 0–5 m (Figure 4). Turtles were found almost exclusively in intertidal areas with all core areas found in depth zones less than 5 m and over tidal flats, within mangrove zones, or up tidal creeks (Figure 2D,F). Only three turtles (108469, 108472, and 120641) had 95% home-ranges extending over regions >5 m in depth and these deeper areas accounted for only a small percentage of their ranges (mean = 8.3%, SE = 4.6%). No turtles were found in waters >10 m deep. Resident dugongs were found predominantly in subtidal zones with around 80% of their 95% and 50% ranges found in depth zones between 0–5 m (Figure 4). The remainder of the dugong ranges were in the 5–10 m zone; some individuals occasionally ventured over deeper areas up to a maximum of 20 m in depth. The transient dugong (652636A – female) was found over a range of depth zones between 0–20 m, with roughly equal portions of its range found over each 5 m depth zone (Figure 4). None of the dugongs had core areas over the tidal flats preferred by turtles (Figure 2F). There were no diurnal differences in the depth preferences of dugongs or turtles in Shoalwater Bay.

Depth Zone Distributions - Torres Strait

Torres Strait dugongs and the transient turtle had similar distributions over depth zones. Animals were found mostly over deeper-water areas with depths of 5-15 m. Core areas were centered around the 5-10 m depth zone with 75% of each individual's range lying within these depths. Over half of the 95% home-range of two dugongs (641057A – female and 641052A – male) were over depths >10 m, with a maximum depth of 25 m. Reef associated turtles showed preference for regions with depths between 0–10 m, with >90% of their 95% ranges over these depths (Figure 4). All Torres Strait animals displayed significantly different depth zone preferences to animals in Shoalwater Bay with dugongs and turtles found over deeper depth zones in Torres Strait. Diurnal comparisons showed no differences in the depth preferences of dugongs, reef associated turtles, or the transient turtle.

Discussion

This study is the first to simultaneously track dugongs and green sea turtles within their foraging grounds to compare the habitat use of these two sympatric megaherbivores. Fast-acquisition GPS tracking of dugongs and green turtles confirmed that they use similar areas in two separate regions where they co-occur with a significant amount of overlap in their ranges, despite most of the turtles using much smaller areas than the dugongs. However, at both locations some turtles showed patterns of space use similar to the dugongs. These results suggest that spatial management could be systematically designed to co-manage these species using fine-scale satellite tracking to identify key areas of habitat use to supplement survey data. Satellite tracking also showed that management plans may need to be designed to accommodate specific differences in range sizes and usage of different depth zones specific to each region.

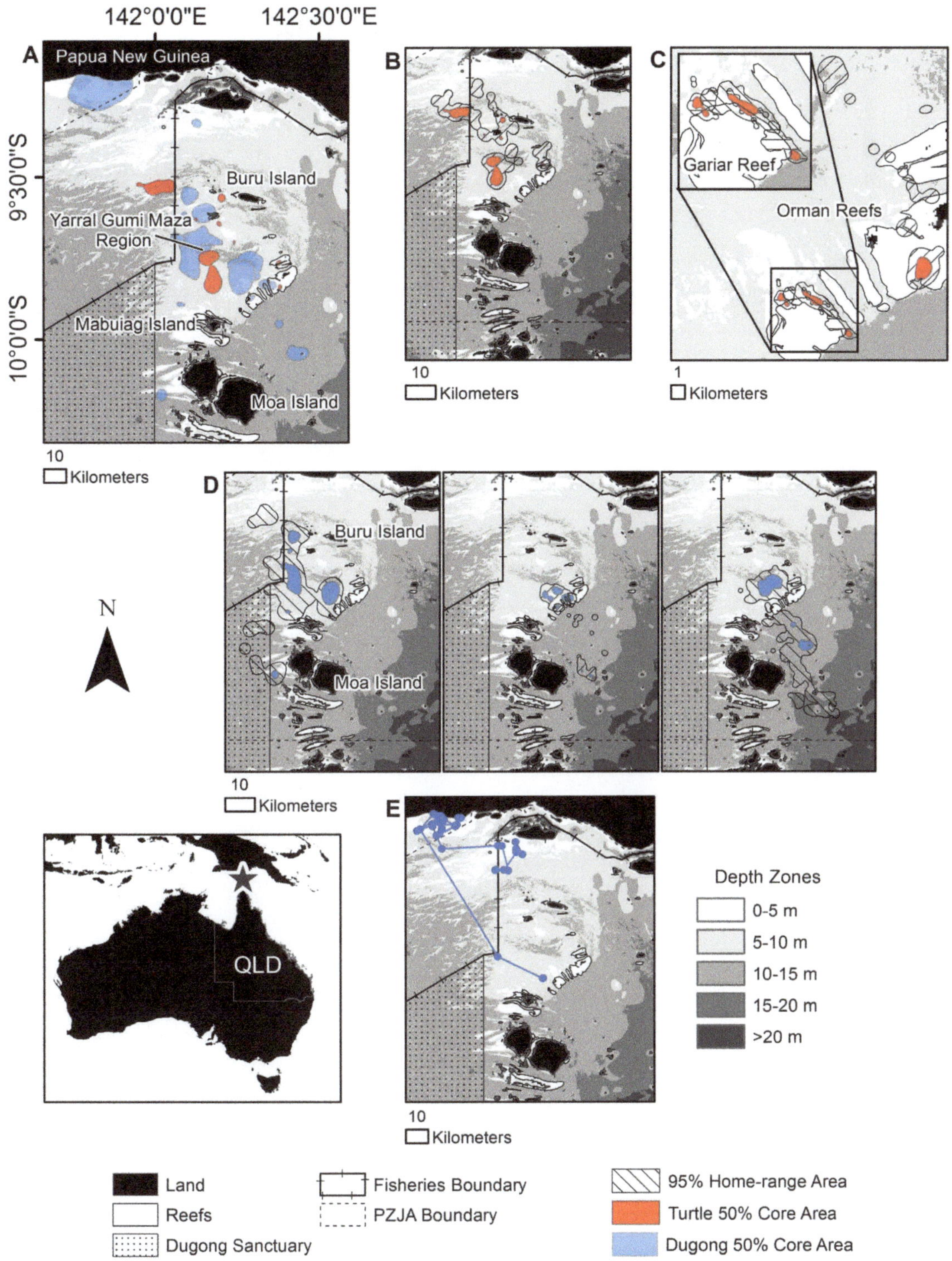

Figure 3. Torres Strait dugong and green turtle spatial use. Dugong and green sea turtle tracking and foraging home-ranges within Torres Strait, between the northern tip of Australia and Papua New Guinea using the total tracking duration of each animal. Blue features indicate dugongs and red features indicate green sea turtles. (**A**) Comparison of dugong and green sea turtle 50% core areas in Torres Strait. (**B**) Home-range of the transient turtle (95889). (**C**) Home-ranges of reef associated turtles (inset is a close up of Gariar Reef where two turtles were resident (70455 and 95892). (**D**) Example home-ranges of dugongs (left: 641058A, middle: 641054A, right: 641052A). (**E**) Tracking locations and migratory path of the dugong that traveled to Papua New Guinea (641055A). PZJA = Protected Zone Joint Authority. For maps with multiple individuals plotted, each individual's 95% home-range is delineated by hatching at a different angle. Note differences in scale for each map.

Spatial Relationships and Regional Comparisons

Dugong range sizes and depth zone preferences differed between localities with dugong ranges in Torres Strait being much larger than those in Shoalwater Bay and other regions where they have been tracked (e.g., Hervey Bay and Cleveland Bay, Australia; Lease Islands, Indonesia), by several orders of

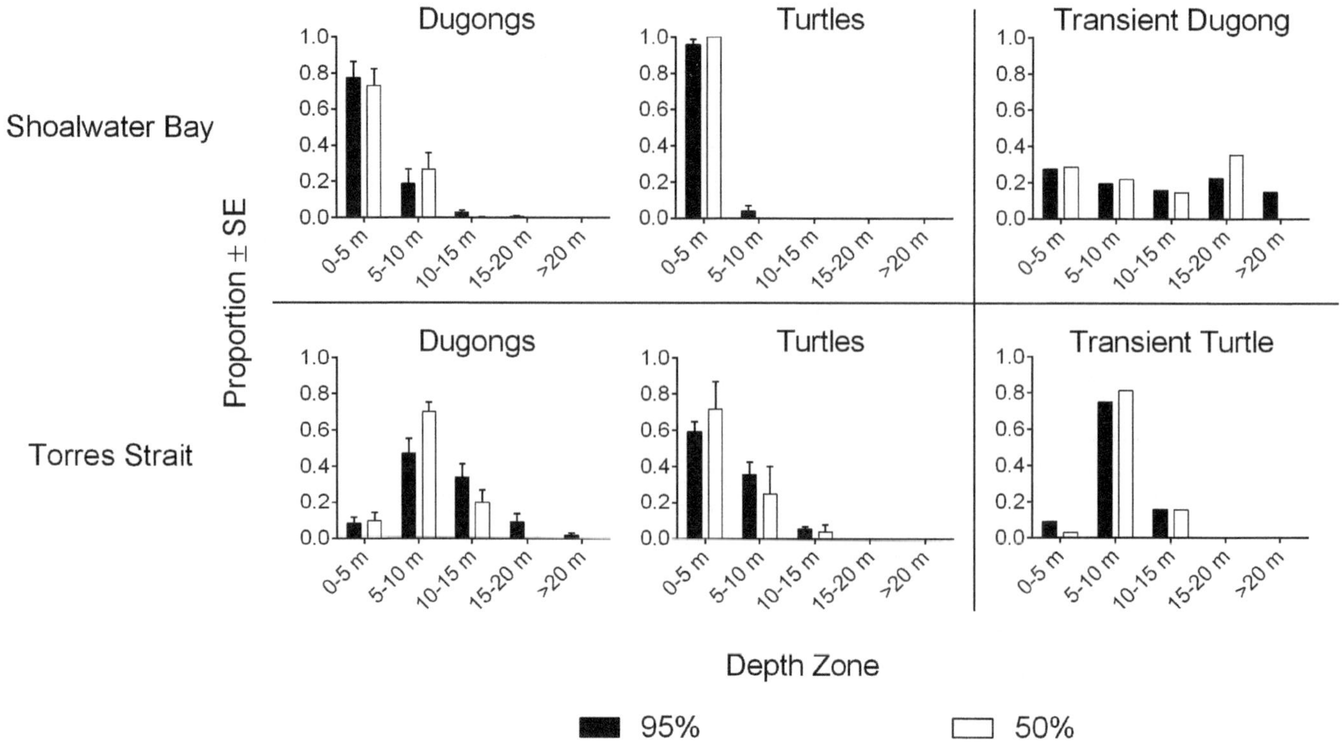

Figure 4. Depth zone use of tracked dugongs and green turtles. Depth comparison of 95% home-ranges and 50% core areas between Shoalwater Bay, Australia dugongs (n = 5), green sea turtles (n = 6), and a transient dugong (n = 1) (top) with Torres Strait, Australia dugongs (n = 6), reef associated green sea turtles (n = 3), and a transient green sea turtle (n = 1) (bottom).

magnitude (average range from other regions: 24.0–63.6 km^2; this study: Torres Strait; median = 942.6 km^2, Shoalwater Bay; median = 60.6 km^2) [41,62,63]. Dugongs in Torres Strait also preferred deeper-water habitats (>5 m), rather than the tidal and shallow sub-tidal habitats observed in Shoalwater Bay (this study) and other areas (e.g., Moreton Bay, Shark Bay, and Hervey Bay, Australia) [64,65].

In contrast, most of the green sea turtles (n = 8) we tracked used small, restricted ranges in relatively shallow water, with high site fidelity, and limited nighttime ranges; a result similar to green turtles in other regions [66–68]. However, the presence of green turtles displaying behavior similar to dugongs (n = 2) in both regions suggests that some green turtles may be influenced by environmental factors and geographic features in a similar way to dugongs. These differences and similarities between species and regions may be associated with the relative availability of deeper-water and shallow-water habitat, the presence of reefs, the location and size of available seagrass meadows, and the regional dugong and turtle population sizes.

Torres Strait provides considerable areas of deeper-water habitat between 5 and 15 m in depth with shallow-water areas accounting for only a small percentage of available area which is primarily restricted to reef-top zones [60]. Additionally, most seagrass in Torres Strait is found within these deeper-water zones, which support an estimated 13,425 km^2 to 17,500 km^2 of seagrass habitat, the largest continuous seagrass area in Australia [69–71]. Reef-top areas also support productive seagrass pastures, but at a significantly smaller scale (e.g., ~95 km^2 at the Orman Reefs) [72,73]. Dugong access to reef-top seagrass meadows is limited by tides [74–76] whereas the deeper-water seagrass is continuously available.

In contrast, along the coastline of eastern Australia south of Cooktown, which includes Shoalwater Bay, most seagrass meadows are much smaller (range: 55–1843 km^2), primarily restricted to coastal tidal and sub-tidal flats [77] in shallow, sheltered bays, and thus fragmented into multiple, disjunct pastures, creating a much smaller area of available habitat for dugongs and turtles. Shoalwater Bay supports the largest area of seagrass (130 km^2) in the southern GBR [77,78]. This pattern of seagrass distribution may explain the smaller range sizes of the dugongs tracked in Shoalwater Bay and in other studies [41] along with the shallower depths used by both species.

Spatial Zoning and Regional Management

The regional differences in the spatial patterns of dugongs and green sea turtles within Torres Strait and Shoalwater Bay have significant implications for the management and conservation of these two species. In Torres Strait, dugongs and green sea turtles cross jurisdictional boundaries, and they cross these boundaries in all directions. This situation requires management to be co-developed with neighboring regions to enable protection throughout each species' range. This approach should be used to extend the community based management planning, currently developed only at the scale of the sea-country of individual communities [29]. In contrast, dugongs and green turtles along the GBR have much more restricted ranges and do not cross international boundaries while on their feeding grounds. Thus, coordinated management decisions at state and federal levels are appropriate for most of the life cycles of both species. However, because some GBR green turtles cross international boundaries during breeding migrations [79] and dugongs occasionally make individualistic large-scale movements to foraging grounds outside the GBR [41], management plans also need to incorporate the likelihood of cross-jurisdictional movements.

The overlap in the home-ranges of dugongs and green turtles shows that protection measures, such as MPAs, no-take reserves, and seasonal spatial closures, could be developed for either species and, by default, spatial protection would encompass at least a portion of the range of the other. However, focusing conservation efforts on the space use of both species in a coordinated management scheme may be a more cost-effective approach as dugongs and green turtles appear to have different preferences for core areas and depth zones which need to be taken into account when designating protection zones. Also, co-management of dugongs and green turtles may act as an umbrella for many other species that share these habitats within the GBR/Queensland coast and Torres Strait.

Under current Australian national legislation, green turtles are listed as threatened species, while dugongs are not. Because of this, the triggers for significant impact are easier to meet for green turtles than they are for dugongs. This difference enables dugong management to be triggered by green turtle management. Alternatively, the broad-scale spatial distribution of dugongs is much better known than for green turtles [80] as dugongs being larger are more reliably detected using aerial-surveys than turtles, which are also difficult to identify species from the air. Thus, marine spatial management in Australia is generally better developed for dugongs than it is for green turtles. Through complimentary co-management of both species aided by satellite tracking of individual animals, dugong spatial management could be used to designate regions for protection that are important to both species, while green turtles may act as a catalyst for setting other conservation and environmental management plans into action.

Efficacy of Existing Protection Zones

The efficacy of current protection zones also appears to differ among regions, although our conclusion is tentative because of our small sample sizes of tracked animals. The existing Torres Strait Dugong Sanctuary was only used by a single tracked dugong and accounted for less than 15% of this individual's total range. This lack of spatial use of the Dugong Sanctuary by our tracked turtles and dugongs _ENREF_57supports conclusions based on extensive aerial surveys that the Sanctuary should be extended [29]. Spatial closures for dugongs and green turtles in the Torres Strait region would be more effective if the current Dugong Sanctuary also protected green turtles and sanctuary areas were extended; negotiations are currently progressing to this end.

In contrast, existing protection areas in Shoalwater Bay appear to be adequate for the protection of dugongs and turtles within the region, with all tracked turtles and the majority of dugongs found within protection zone boundaries. However, the effectiveness of these reserves is unknown if the geographic distribution of food resources changes due to environmental and anthropogenic factors or if local dugong and turtle populations increase. Adaptive management of the region in addition to combined management of terrestrial and marine systems will be required to maintain the effectiveness of these protected areas, especially as the human population increases in Australia. Additionally, identification of migratory corridors, dugong breeding grounds, and source/sink habitats will further increase the efficacy of protection.

More broadly, these results show that fine-scale tracking of individual animals can provide important information on where animals travel and where their preferred habitats lie. This information can then aid in the delineation of MPA boundaries and assess the efficacy of existing MPAs that were created using other approaches such as density surveys, habitat mapping, and expert opinion.

Conclusions

We have demonstrated the potential of fine-scale satellite tracking for improving species co-management, increasing the efficacy of existing MPAs and other protected zones, and identifying new areas for protection. Our results are preliminary and cannot be applied at the population level as sample sizes for both species were low, only two turtles classified as transient were tracked, and animals were only tracked during a single season in each location. Distribution patterns may be markedly different between seasons and/or years, especially in relation to changes in seagrass and algae distributions. Continued monitoring and larger sample sizes are required to further inform this evidence-based approach. It should also be noted that range sizes for individuals with tracking durations less than 20 days may be slightly larger or smaller than what we have shown due to fluctuations in range size correlated with tracking durations. However, the general areas they inhabit can be captured with short tracking periods providing useful information for comparisons between species. The findings of the present study have broad implications for the management of other sympatric species, mobile species dependent on benthic environments of variable quality, and predator-prey relationships. Examples of other co-occurring species that may benefit from this approach include manatees and green turtles; tiger sharks, dugongs, and sea turtles; and dolphin species which form mixed-species assemblages.

Acknowledgments

This research would not have been possible without the help of the Traditional Owners of Mabuiag Island who allowed access to their lands, the help and support of the community on Mabuiag Island in locating and capturing animals, and the diverse group of researchers from various local and regional organizations who collected the data at both locations. We would also like to thank Dr. James Daniell for providing the high resolution bathymetry map of Torres Strait, Dr. Mark Read from the GBRMPA for assistance with the Shoalwater Bay field trip, Terrence Whap, David Amber, and Charlie Hankin from the Prescribed Body Corporate of Torres Strait for help with the Torres Strait field trip, and Daniel Zeh for his preliminary work on the Shoalwater Bay data.

Author Contributions

Conceived and designed the experiments: CG HM MH. Analyzed the data: CG. Contributed reagents/materials/analysis tools: TS. Wrote the paper: CG HM MMPBF MH CJL TS. Organized the field work: MH HM CJL. Conducted the field work: MMPBF MH CJL TS.

References

1. Hazen EL, Maxwell SM, Bailey H, Bograd SJ, Hamann M, et al. (2012) Ontogeny in marine tagging and tracking science: technologies and data gaps. Mar Ecol Prog Ser 457: 221–240.
2. Bailey H, Foley DG, Block BA, Piedra R, Hitipeuw C, et al. (2012) Identification of distinct movement patterns in Pacific leatherback turtle populations influenced by ocean conditions. Ecol Appl 22: 735–747.
3. Cuevas E, Liceaga-Correa MDA, Garduno-Andrade M (2007) Spatial characterization of a foraging area for immature hawksbill turtles (*Eretmochelys imbricata*) in Yucatan, Mexico. Amphib-reptil 28: 337–346.
4. Lee KA, Huveneers C, Macdonald T, Harcourt RG (2014) Size isn't everything: movements, home range, and habitat preferences of eastern blue gropers (Achoerodus viridis) demonstrate the efficacy of a small marine reserve. Aquat Conserv: n/a-n/a.
5. Bowler DE, Benton TG (2005) Causes and consequences of animal dispersal strategies: relating individual behaviour to spatial dynamics. Biol Rev 80: 205–225.
6. Parnell PE, Dayton PK, Lennert-Cody CE, Rasmussen LL, Leichter JJ (2006) Marine reserve design: optimal size, habitats, species affinities, diversity, and ocean microclimate. Ecol Appl 16: 945–962.
7. Schofield G, Dimadi A, Fossette S, Katselidis KA, Koutsoubas D, et al. (2013) Satellite tracking large numbers of individuals to infer population level dispersal and core areas for the protection of an endangered species. Divers Distrib 19: 834–844.
8. Halpern BS, Fox HE, Fujita R, Heinemann D, Lenihan HS, et al. (2008) A global map of human impact on marine ecosystems. Science 319: 948–952.
9. Hooker SK, Gerber LR (2004) Marine reserves as a tool for ecosystem-based management: The potential importance of megafauna. Bioscience 54: 27–39.
10. Fernandes L, Lowe D, Innes J, Tanzer J, Chadwick V, et al. (2005) Establishing representative no-take areas in the Great Barrier Reef: Large-scale implementation of theory on marine protected areas. Conserv Biol 19: 1733–1744.
11. IUCN (2012) IUCN Red List of Threatened Species. Available: www.iucnredlist.org. Accessed 20 December 2013.
12. Roberts CM, Mason L, Hawkins JP (2006) Roadmap to recovery: A global network of marine reserves. York, United Kingdom: University of York. 60 p.
13. Gerber LR, Botsford LW, Hastings A, Possingham HP, Gaines SD, et al. (2003) Population models for marine reserve design: A retrospective and prospective synthesis. Ecol Appl 13: S47–S64.
14. Dryden J, Grech A, Moloney J, Hamann M (2008) Rezoning of the Great Barrier Reef World Heritage Area: does it afford greater protection for marine turtles? Wildl Res 35: 477–485.
15. Lynch TP, Harcourt R, Edgar G, Barrett N (2013) Conservation of the Critically Endangered Eastern Australian Population of the Grey Nurse Shark (Carcharias taurus) Through Cross-Jurisdictional Management of a Network of Marine-Protected Areas. Environmental Management 52: 1341–1354.
16. Rondinini C, Wilson KA, Boitani L, Grantham H, Possingham HP (2006) Tradeoffs of different types of species occurrence data for use in systematic conservation planning. Ecol Lett 9: 1136–1145.
17. Pressey RL (2004) Conservation planning and biodiversity: Assembling the best data for the job. Conserv Biol 18: 1677–1681.
18. Boersma PD, Parrish JK (1999) Limiting abuse: marine protected areas, a limited solution. Ecol Econ 31: 287–304.
19. Block BA, Jonsen ID, Jorgensen SJ, Winship AJ, Shaffer SA, et al. (2011) Tracking apex marine predator movements in a dynamic ocean. Nature 475: 86–90.
20. Pendoley KL, Schofield G, Whittock PA, Lerodiaconou D, Hays GC (2014) Protected species use of a coastal marine turtle migratory corridor connecting Australian MPAs. Mar Biol.
21. Scott R, Hodgson DJ, Witt MJ, Coyne MS, Adnyana W, et al. (2012) Global analysis of satellite tracking data shows that adult green turtles are significantly aggregated in Marine Protected Areas. Glob Ecol Biogeogr 21: 1053–1061.
22. Grech A, Marsh H, Coles R (2008) A spatial assessment of the risk to a mobile marine mammal from bycatch. Aquat Conserv 18: 1127–1139.
23. Christie P, McCay BJ, Miller ML, Lowe C, White AT, et al. (2003) Towards developing a complete understanding: social science research agenda for marine protected areas. Fisheries 28: 22–25.
24. Roberts CM, Halpern BS, Palumbi SR, Warner RR (2001) Designing marine reserve networks: Why small, isolated protected areas are not enough. Conservat Pract 2: 10–17.
25. Lutz PL, Musick JA (1997) The Biology of Sea Turtles. Boca Raton: CRC Press. 432 p.
26. Marsh H, O'Shea TJ, Reynolds III JE (2011) Ecology and Conservation of the Sirenia: Dugongs and Manatees. Cambridge: Cambridge University Press. 536 p.
27. Hoelzel AR (2009) Marine Mammal Biology: An Evolutionary Approach. Hoboken: John Wiley & Sons, Ltd. 448 p.
28. Maiorano L, Bartolino V, Colloca F, Abella A, Belluscio A, et al. (2009) Systematic conservation planning in the Mediterranean: a flexible tool for the identification of no-take marine protected areas. ICES J Mar Sci 66: 137–146.
29. Marsh H, Grech A, Hagihara R (2011) Aerial survey of Torres Strait to evaluate the efficacy of an enforced and possibly extended Dugong Sanctuary as one of the tools for managing the dugong fishery. Townsville, Australia: James Cook Univeristy. 33 p.
30. Limpus CJ, Miller JD, Parmenter CJ, Limpus DJ (2003) The green turtle, *Chelonia mydas*, population of Raine Island and the northern Great Barrier Reef: 1843–2001. Mem Queensl Mus 49: 349–440.
31. Dobbs K, Fernandes L, Slegers S, Jago B, Thompson L, et al. (2008) Incorporating dugong habitats into the marine protected area design for the Great Barrier Reef Marine Park, Queensland, Australia. Ocean & Coastal Management 51: 368–375.
32. Dobbs K, MacDonald F, Limpus CJ, Fernandes L, Slegers S, et al. (2007) Incorporating marine turtle habitats into the marine protected area design for the Great Barrier Reef Marine Park, Queensland, Australia. Pac Conserv Biol 13: 293–302.
33. Marsh H (2000) Evaluating management initiatives aimed at reducing the mortality of dugongs in gill and mesh nets in the Great Barrier Reef World Heritage area. Mar Mamm Sci 16: 684–694.
34. Dobbs K (2007) Marine turtle and dugong habitats in the Great Barrier Reef Marine Park used to implement biophysical operational principles for the Represenlative Areas Programme. Townsville, Australia: Great Barrier Reef Marine Park Authority. 35 p.
35. Thompson PM, Wilson B, Grellier K, Hammond PS (2000) Combining power analysis and population viability analysis to compare traditional and precautionary approaches to conservation of coastal cetaceans. Conserv Biol 14: 1253–1263.
36. André J, Gyuris E, Lawler IR (2005) Comparison of the diets of sympatric dugongs and green turtles on the Orman Reefs, Torres Strait, Australia. Wildl Res 32: 53–62.
37. Heinsohn GE (1981) Methods of taking measurements, other data and specimen material from dugong carcassess. In: H Marsh, editors. The dugong Proceedings of Workshop 8–13 May 1979. Townsville, Australia: James Cook Univeristy. pp. 130–135.
38. Bolten AB (1999) Techniques for measuring sea turtles. In: K. L Eckert, K. A Bjorndal, F. A Abreu-Grobois and M Donnelly, editors. Research and Management Techniques for the Conservation of Sea Turtles. Pennsylvania, USA: IUCN/SSC Marine Specialist Group. pp. 110–114.
39. Fuentes MMPB, Cleguer C, Liebsch N, Bedford G, Amber D, et al. (2013) Adapting dugong catching techniques to different cultural and environmental settings. Mar Mamm Sci 29: 159–166.
40. Lanyon JM, Slade RW, Sneath HL, Broderick D, Kirkwood JM, et al. (2006) A method for capturing dugongs (*Dugong dugon*) in open water. Aquat Mamm 32: 196–201.
41. Sheppard JK, Preen AR, Marsh H, Lawler IR, Whiting SD, et al. (2006) Movement heterogeneity of dugongs, *Dugong dugon* (Muller), over large spatial scales. J Exp Mar Bio Ecol 334: 64–83.
42. Limpus CJ (1978) The Reef. In: H. J Lavery, editors. Exploration north: Australia's wildlife from desert to reef. Richmond, Victoria, Australia: Richmond Hill Press. pp. 187–222.
43. Shimada T, Jones RE, Limpus C, Hamann M (2012) Improving data retention and home range estimates by data-driven screening. Mar Ecol Prog Ser 457: 171–180.
44. Hagihara R, Jones RE, Grech A, Lanyon JM, Sheppard JK, et al. (2013) Improving population estimates by quantifying diving and surfacing patterns: A dugong example. Mar Mamm Sci 30: 348–366.
45. Zbinden JA, Aebischer A, Margaritoulis D, Arlettaz R (2008) Important areas at sea for adult loggerhead sea turtles in the Mediterranean Sea: satellite tracking corroborates findings from potentially biased sources. Mar Biol 153: 899–906.
46. Telonics (2012) Gen4 GPS Systems Manual. Mesa, Arizona: Telonics. 105 p.
47. Marsh H, Gardner BR, Heinsohn GE (1981) Present-day hunting and distribution of dugongs in the Wellesley Islands (Queensland) - Implications for conservation. Biol Conserv 19: 255–267.
48. Sirtrack (2010) Fastloc^{T} User Guide version: 1.3. Havelock North, New Zealand: Sirtrack.
49. Heithaus MR, Frid A, Dill LM (2002) Shark-inflicted injury frequencies, escape ability, and habitat use of green and loggerhead turtles. Mar Biol 140: 229–236.
50. Boyle SA, Lourenço WC, da Silva LR, Smith AT (2009) Home range estimates vary with sample size and methods. Folia Primatol 80: 33–42.

51. Seaman DE, Millspaugh JJ, Kernohan BJ, Brundige GC, Raedeke KJ, et al. (1999) Effects of sample size on kernel home range estimates. J Wildl Manage 63: 739–747.
52. Silverman BW (1986) Density estimation for statistics and data analysis. New York: Chapman and Hall. 176 p.
53. Seaman DE, Powell RA (1996) An evaluation of the accuracy of kernel density estimators for home range analysis. Ecology 77: 2075–2085.
54. Gitzen RA, Millspaugh JJ, Kernohan BJ (2006) Bandwidth selection for fixed-kernel analysis of animal utilization distributions. J Wildl Manage 70: 1334–1344.
55. Hemson G, Johnson P, South A, Kenward R, Ripley R, et al. (2005) Are kernels the mustard? Data from global positioning system (GPS) collars suggests problems for kernel home-range analyses with least-squares cross-validation. Journal of Animal Ecology 74: 455–463.
56. Horne JS, Garton EO (2006) Likelihood cross-validation versus least squares cross-validation for choosing the smoothing parameter in kernel home-range analysis. J Wildl Manage 70: 641–648.
57. Beyer HL (2012) Geospatial Modelling Environment (Version 0.7.2.1). Spatial Ecology, LLC.
58. R Core Team (2012) R: A language and environment for statistical computing. R Foundation for Statistical Computing. Vienna, Austria.
59. ESRI (2011) *ArcGIS Desktop v 10.0.* Environmental Systems Research Institute, Inc. Redlands, California, USA.
60. Daniell JJ (2008) Development of a bathymetric grid for the Gulf of Papua and adjacent areas: A note describing its development. J Geophys Res -Earth Surf 113.
61. Beaman RJ (2010) Project 3DGBR: A high-resolution depth model for the Great Barrier Reef and Coral Sea. Cairns, Australia: Marine and Tropical Sciences Research Facility. 13 p.
62. de Iongh HH, Langeveld P, Van der Wal M (1998) Movement and home ranges of dugongs around the Lease Islands, East Indonesia. Mar Ecol-P S Z N I 19: 179–193.
63. Marsh H, Rathbun GB (1990) Development and application of conventional and satellite radio tracking techniques for studying dugong movements and habitat use. Aust Wildlife Res 17: 83–100.
64. Chilvers BL, Delean S, Gales NJ, Holley DK, Lawler IR, et al. (2004) Diving behaviour of dugongs, *Dugong dugon*. J Exp Mar Bio Ecol 304: 203–224.
65. Preen AR (1992) Interactions between dugongs and seagrasses in a subtropical environment. PhD Thesis, School of Earth and Environmental Sciences. James Cook Univeristy, Townsville, Australia.
66. Seminoff JA, Resendiz A, Nichols WJ (2002) Home range of green turtles *Chelonia mydas* at a coastal foraging area in the Gulf of California, Mexico. Mar Ecol Prog Ser 242: 253–265.
67. Hazel J, Hamann M, Lawler IR (2013) Home range of immature green turtles tracked at an offshore tropical reef using automated passive acoustic technology. Mar Biol 160: 617.
68. Hays GC, Glen F, Broderick AC, Godley BJ, Metcalfe JD (2002) Behavioural plasticity in a large marine herbivore: contrasting patterns of depth utilisation between two green turtle (*Chelonia mydas*) populations. Mar Biol 141: 985–990.
69. Coles RG, McKenzie LJ, Campbell SJ (2003) The seagrasses of Eastern Australia. In: E. P Green, F. T Short and M. D Spalding, editors. The World Atlas of Seagrasses. Berkeley: University of California Press. pp. 119–133.
70. Taylor HA, Rasheed MA (2010) Torres Straits Dugong Sanctuary Seagrass Baseline Survey, March 2010. Cairns, Australia: DEEDI. 26 p.
71. Poiner IR, Peterkin C (1996) Seagrasses. Townsville, Australia: Great Barrier Reef Marine Park Authority.
72. Long BG, Skewes TD, Taranto TJ, Jacobs DR, Dennis DM (1997) Torres Strait reef resource inventory and reef habitat mapping. Final report on CSIRO Research 1993–1996. Cleveland, Australia: CSIRO. 23 p.
73. Rasheed MA, Dew KR, McKenzie LJ, Coles RG, Kerville SP, et al. (2008) Productivity, carbon assimilation and intra-annual change in tropical reef platform seagrass communities of the Torres Strait, north-eastern Australia. Cont Shelf Res 28: 2292–2303.
74. Sheppard JK, Jones RE, Marsh H, Lawler IR (2009) Effects of tidal and diel cycles on dugong habitat use. J Wildl Manage 73: 45–59.
75. Nietschmann B, Nietschmann J (1981) Good dugong, bad dugong: Bad turtle, good turtle. Nat Hist 90: 55–63.
76. Sheppard JK, Marsh H, Jones RE, Lawler IR (2010) Dugong habitat use in relation to seagrass nutrients, tides, and diel cycles. Mar Mamm Sci 26: 855–879.
77. McKenzie LJ, Yoshida R, Grech A, Coles R (2010) Queensland seagrasses: Status 2010 - Torres Strait and East Coast. Cairns, Australia: Fisheries Queensland (DEEDI). 6 p.
78. O'Neill P (2009) Chapter 7 Seagrass, Mangroves and Saltmarshes. In: W. Lee Long, R Melzer and J Robertson, editors. State of the Environment Report for Shoalwater Bay Training Area 2008. Australia: Australian Department of Defence. pp. 180–208.
79. Limpus CJ, Miller JD, Paramenter CJ, Reimer D, McLachlan N, et al. (1992) Migration of green *(Chelonia mydas)* and loggerhead (*Caretta caretta*) turtles to and from eastern Australian rookeries. Wildl Res 19: 347–357.
80. Grech A, Sheppard JK, Marsh H (2011) Informing species conservation at multiple scales using data collected for marine mammal stock assessments. PLoS One 6: e17993.

7

Protection Enhances Community and Habitat Stability: Evidence from a Mediterranean Marine Protected Area

Simonetta Fraschetti[1]*, Giuseppe Guarnieri[1], Stanislao Bevilacqua[1], Antonio Terlizzi[1], Ferdinando Boero[1,2]

1 Department of Biological and Environmental Sciences and Technologies, University of Salento, Consorzio Nazionale Interuniversitario per le Scienze del Mare, Lecce, Italy, **2** Institute of Marine Sciences of National Research Council, Genova, Italy

Abstract

Rare evidences support that Marine Protected Areas (MPAs) enhance the stability of marine habitats and assemblages. Based on nine years of observation (2001–2009) inside and outside a well managed MPA, we assessed the potential of conservation and management actions to modify patterns of spatial and/or temporal variability of *Posidonia oceanica* meadows, the lower midlittoral and the shallow infralittoral rock assemblages. Significant differences in both temporal variations and spatial patterns were observed between protected and unprotected locations. A lower temporal variability in the protected vs. unprotected assemblages was found in the shallow infralittoral, demonstrating that, at least at local scale, protection can enhance community stability. Macrobenthos with long-lived and relatively slow-growing invertebrates and structurally complex algal forms were homogeneously distributed in space and went through little fluctuations in time. In contrast, a mosaic of disturbed patches featured unprotected locations, with small-scale shifts from macroalgal stands to barrens, and harsh temporal variations between the two states. Opposite patterns of spatial and temporal variability were found for the midlittoral assemblages. Despite an overall clear pattern of seagrass regression through time, protected meadows showed a significantly higher shoot density than unprotected ones, suggesting a higher resistance to local human activities. Our results support the assumption that the exclusion/management of human activities within MPAs enhance the stability of the structural components of protected marine systems, reverting or arresting threat-induced trajectories of change.

Editor: Philippe Archambault, Université du Québec à Rimouski, Canada

Funding: The research leading to these results also received funding from the Italian Ministry of the Environment (Afrodite project), the European Community's 7th Framework Programmes (FP7/2007–2013) for the project COCONET (Grant agreement No. 287844, http://www.coconet-fp7.eu/) and PERSEUS (Grant agreement No. 287600, http://www.perseus-net.eu/site/content.php). The support by the European Union from VECTORS (http://www.marine-vectors.eu/) and from the Italian Ministry of the Research PRIN TETRIS is also acknowledged. The funders had no role in study design, data collection and analysis, decision to publish, or preparation of the manuscript.

Competing Interests: The authors declare that no competing interests exist since there are no financial, personal, or professional interests that could be construed to have influenced the paper.

* E-mail: simona.fraschetti@unisalento.it

Introduction

Over the past decades, both marine and terrestrial biodiversity experienced rapid global erosion [1]. Ensuing concern about the consequences of biodiversity loss fuelled theoretical and empirical studies aimed at unravelling the role of biodiversity in maintaining ecosystem properties and the goods and services they provide to humans [2]. Most research indicates that highly diverse assemblages increase the efficiency of ecosystem processes, being also less variable in space and time and more resistant to invasion and disturbance than low-diversity assemblages [1–3]. However, results are far from being unequivocal and the potential link between biodiversity and several ecosystem processes is still debated [4,5]. Understanding if and how conservation and management actions, while restoring biodiversity, could have a role in maintaining the functional properties of marine and terrestrial ecosystems is therefore overriding [6].

Rising temporal and spatial variability is often a subtle outcome of human disturbance on ecological systems [7–9], undermining community structure, leading to decreased resilience and to increased potential for regime shifts [10]. Regime shifts are largely unpredictable due to the inherent complexity of ecological systems [11]. The spatial and temporal heterogeneity of a given system might shed light on determinants of its vulnerability and guide actions to mitigate the risk of critical transitions towards degraded states [12]. Changes in spatial and temporal variability may inform about the ongoing effects of natural or anthropogenic disturbance [13,14], or serving as warning signals of approximating transitions [15].

Tropical coral reefs [16–18], temperate and boreal intertidal and subtidal rocky reefs [19,20–22], and temperate coastal pelagic systems [23,24] are all experiencing dramatic changes in populations, species, or entire functional groups leading to regime shifts that could be long-lasting or even irreversible [25]. The returning to previous undisturbed conditions is difficult, unless the major drivers of change, such as terrestrial runoffs, nutrient loading, pollution or fishing pressure and/or their destabilizing effects are reduced [26]. Also, while reverting from altered states, damaged marine ecosystems often follow different trajectories of recovery from that observed during decline [26].

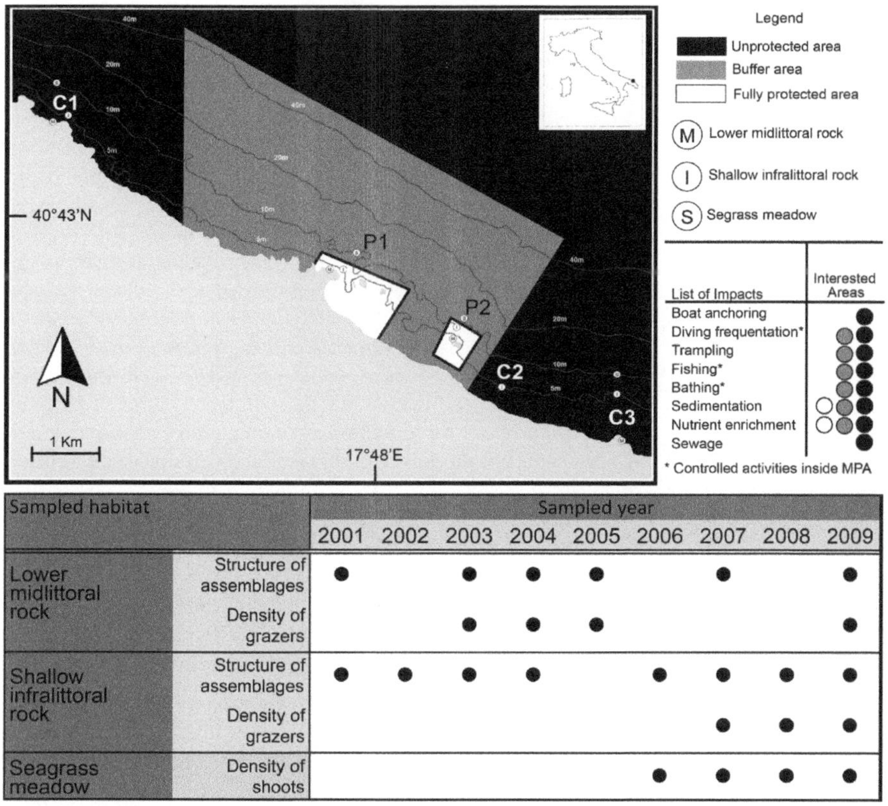

Figure 1. Framework of the study area and sampling timesheet. Map of the study area. No-take, no access areas are given in white, the buffer area of the MPA in grey, in black the unprotected area. P1, P2 = protected locations; C1, C2, C3 = unprotected locations. The main sources of human disturbance acting within and outside the MPA, along with the years in which sampling was carried out in each habitat are also provided (details in the legend).

Marine Protected Areas (MPAs) cannot be considered a global solution to biodiversity loss [27]. However, if properly managed, they can play a major role in reducing cumulative impacts [22,28–30]. By excluding (or regulating) human activities, MPAs can enhance fisheries yields outside their boundaries through spill over [31,32], promote biodiversity recovery with greater richness and abundance (or biomass) of species within protected conditions [33,34], empower local communities and indirectly provide additional income from tourism [3].

These effects should also enhance the stability of single species and whole communities (i.e. the persistence in abundance and species composition through time) within MPAs, increasing their resistance (i.e. the ability to remain unperturbed despite the occurrence of a given disturbance) and/or resilience (i.e. the ability to absorb recurrent natural and human perturbations without slowly degrading or unexpectedly flipping into alternate states) and providing an insurance against the consequences of large-scale human threats. Variations in functional traits, identity, evenness and spatial distribution of species structuring communities might trigger different responses to disturbances in protected systems [35]. An effective protection, thus, may reduce the variability of ecological responses to natural and/or anthropogenic disturbance [30,36], generating a 'buffer effect' that could minimize variations in density, biomass, and diversity of protected populations over time [37]. However, evidences of the role of MPAs in sustaining the stability of communities and ecosystems are scant [3,38–40]. Theoretical work on stability has outpaced experimental work stressing the need for long-term experiments to assess temporal stability, as well as recovery from a variety of disturbances [4]. Decadal-scale observations of MPAs documented increased resilience [41], suggesting that the limited availability of long-

Table 1. Summary of PERMANOVAs testing for the effect of protection on the lower midlittoral and on the shallow infralittoral rocky assemblages.

	Lower midlittoral			Shallow infralittoral		
Source of variation	**MS**	***F***	***P***	**MS**	***F***	***P***
Time = T	42693.0	2.71	**0.003**	38853.0		
Protection = P	17534.0	0.30	0.999	50270.0		
Location(P) = L(P)	91908.0	4.09	**0.000**	25098.0		
Site(L(P)) = S(L(P))	8200.5	1.34	0.101	5704.7		
T × P	17235.0	1.09	0.375	18169.0	2.31	**0.000**
T × L(P)	15734.0	2.58	**0.000**	7969.8	2.32	**0.000**
T × S(L(P))	6098.2	5.44	**0.000**	3460.6	2.80	**0.000**
Residuals	1121.5			1235.8		

Analyses were based on Bray-Curtis dissimilarities and each test was performed using 4999 permutations of appropriate units. Only tests for the terms relevant to hypothesis have been reported. Significant *P*-values are given in bold (see text for details).

term monitoring programs can be one of the reasons behind the general paucity of evidences on the stability of the effects of protection [42]. In addition, most studies focused on populations of single target species [40,42] with few exceptions involving whole assemblages [43]. Attempts to document and compare changes in spatial and temporal variability of habitats and assemblages in response to protection regimes are extremely scant [44].

Here, we use a nine-year data set to examine the potential of full protection regime of a Mediterranean MPA to affect the spatial and temporal variability of the assemblages inhabiting the rocky lower midlittoral (–0.1 to 0.1 m across mean-low-level water), the assemblages of the shallow infralittoral rocks (5–7 meter depth), and the *Posidonia oceanica* meadows (approximately 8–10 meter depth). Since its institution, outfall discharge and trawling are excluded from the whole MPA. Artisanal and recreational fishery, anchoring, trampling, diving frequentation, and maritime traffic are also severely regulated within the buffer area of the MPA and completely excluded from the two fully protected areas (Fig. 1). All these activities have been documented to directly and indirectly affect benthic habitats and assemblages [45]. Macroalgal canopies and seagrasses, in particular, can largely benefit from the exclusion of human activities. For example, predatory interactions are re-established when protection from exploitation is effective, causing the decrease of grazing pressure on rocky reefs [46]. The regulation of direct disturbances such as trawling, artisanal fishery, anchoring can halt the decrease in the shoot density and the ultimate regression of *P. oceanica* meadows [47]. In addition, the local control of the water quality can limit the shift to species with lower structural complexity such as turf forming, filamentous or other ephemeral seaweeds [21]. This study tests the hypothesis that the exclusion of human activities occurring within MPAs can be effective in reverting or halting threat-induced trajectories of change of *P. oceanica* meadows, of the rocky midlittoral and the shallow infralittoral assemblages.

Materials and Methods

Study Area

The study was carried out in the MPA of Torre Guaceto (40°42′N; 17°48′E), SE Italy (Adriatic Sea) (Fig. 1), covering a total surface of about 2.207 ha, consisting in two fully protected areas (i.e. no-take, no-access), and a buffer area, where few human activities are permitted and strictly regulated (Fig. 1). The MPA, instituted in 1991, owns an adequate enforcement [48] combined with a full involvement of fishermen in conservation and sustainable practices [49] that determined a significant recovery of target fish populations (mainly sparids and labrids) [34,46]. A full description of habitats and assemblages present in the MPA is available in Fraschetti et al. [50], which also includes a bathymetric map of the whole MPA with the georeferred spatial information about the distribution and extent of habitats. The list of taxa found in the lower midlittoral and the shallow infralittoral rocks are reported in the Supporting Information S1.

All necessary permits were obtained from the Italian Ministry of the Environment and Protection of Land and Sea, and from the Direction of the MPA.

Sampling Design

Benthic assemblages of the lower midlittoral rock. This is an algal-dominated assemblage mostly characterized by erect (e.g. *Corallina officinalis*, *Jania rubens*) and encrusting coralline algae (e.g. *Lithophyllum* spp.), *Laurencia* spp. and filamentous algae (e.g. Ceramiales) [50]. Sessile invertebrates, boring sponges (i.e. *Cliona* spp.) and anthozoans (e.g. *Actinia equina*) can be occasionally present. Sampling was carried out in six dates from 2001 to 2009 separated by at least one year at a depth comprised between - 0.1 to 0.1 m across mean-low-level water. Rocky substrates account for the 50% of the midlittoral of the MPA [50]. All sampling activities were carried out in the late spring-early summer to avoid possible effects of seasonality in the data. Samplings were

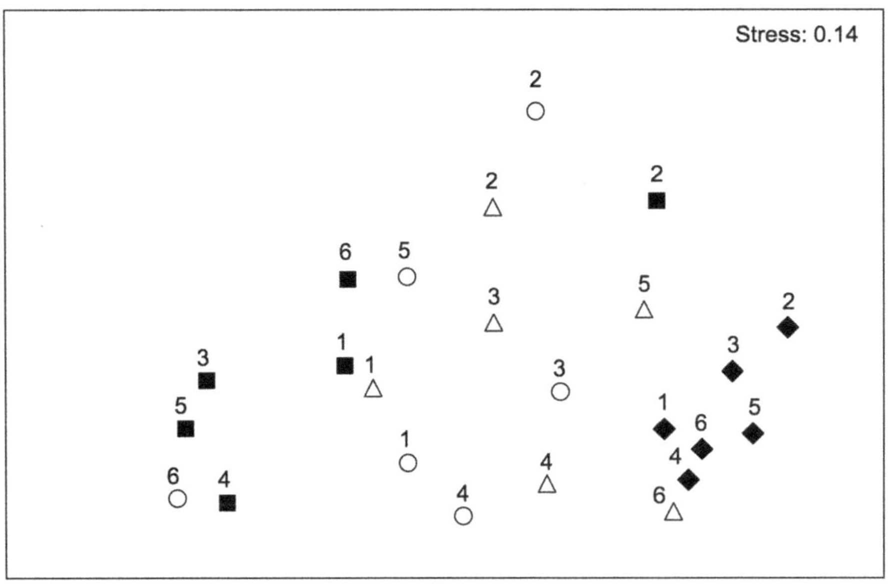

Figure 2. Differences in the multivariate structure of midlittoral assemblages. Non-metric multidimensional scaling ordinations (nMDS) of T × L(P) centroids based on Bray-Curtis dissimilarity measures for the midlittoral assemblages (○ = P1; △ = P2; ■ = C1; ◆ = C2; P = protected locations; C = unprotected locations). Numbers (from 1 to 6) indicate the sampling times.

Table 2. Summary of ANOVAs testing for the effect of protection on multivariate estimates of spatial variability for the lower midlittoral and the shallow infralittoral rocky assemblages.

	Lower midlittoral			Shallow infralittoral		
Source of Variation	**MS**	***F***	***P***	**MS**	***F***	***P***
Time = T	381348.4	6.81		369939.5	6.10	
Protection = P	254713.6	4.55		834647.2	13.76	**0.000**
Scale = Sc	705000.6	12.58		7345335.6	121.11	
T × P	180268.2	3.22		108618.0	1.79	0.111
T × Sc	539617.6	9.63		136143.5	2.24	**0.019**
P × Sc	1620263.0	28.92		126932.8	2.09	0.134
T × P × Sc	131370.3	2.34	**0.030**	77770.6	1.28	0.253
Residuals	56022.6				60651.4	

SNK Lower midlittoral

	Time					
Scale	**1**	**2**	**3**	**4**	**5**	**6**
Location	P<Cs	P<Cs	P<Cs	P<Cs	P<Cs	P = Cs
Site	P>Cs	P = Cs	P>Cs	P = Cs	P = Cs	P>Cs
Replicate	P = Cs	P = Cs	P = Cs	P = Cs	P = Cs	P>Cs

P = Protected locations; Cs = Unprotected locations. Terms already involved in significant higher-order interactions were not reported. Significant *P*-values are given in bold. SNK pair-wise tests for significant interaction terms involving the factor Protection (i.e. only T × P × Sc, in the lower midlittoral rocks) are given below in the table.

undertaken at two locations within the no-take zones (P1 and P2) and at two unprotected locations (C1 and C3) outside the boundaries of the MPA (Fig. 1). Locations were positioned few kilometres from each other and were characterized by similar features in terms of substrate slope and wave exposure. Three sites (approximately 100–300 m apart) were randomly sampled at each location. At each site, ten 20×20 cm random quadrats were sampled to estimate *in situ* the abundance of sessile organisms. The division of each quadrat into 25 sub-quadrats eased visual estimates of the presence of taxa [51]. Final values were expressed as percentages of cover. Organisms that were not easily identifiable at species level were lumped into higher taxonomic

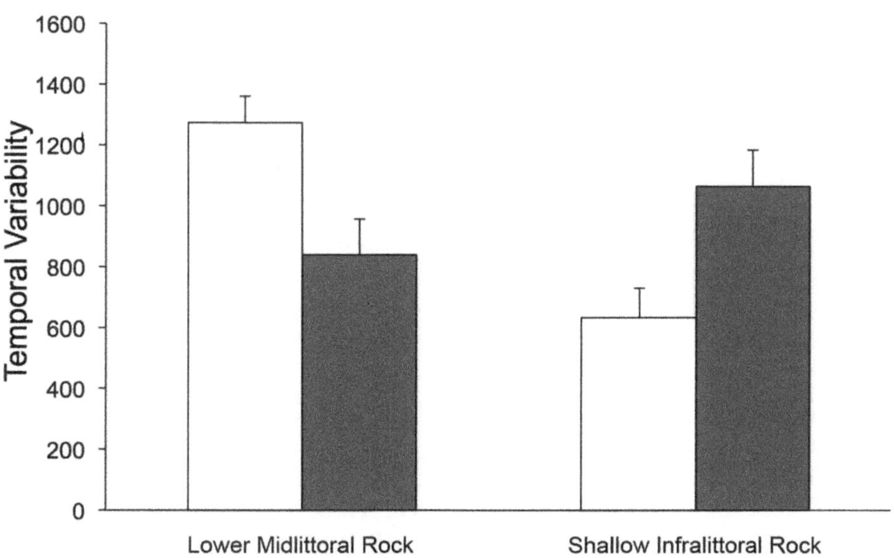

Figure 3. Temporal variability of rocky bottom assemblages. Mean temporal variability (i.e. estimates of variance associated to factor Time, see methods for further details) ± SE ($n = 6$) of the lower midlittoral and the shallow infralittoral rocky assemblages. White bars = protected assemblages; grey bars = unprotected assemblages.

Table 3. Summary of ANOVAs testing for the effect of protection on the total abundance of midlittoral grazers in each sampling time.

	2003			2004			2005			2009		
Source of variation	**MS**	*F*	*P*	**MS**	*F*	*P*	**MS**	*F*	*P*	**MS**	*F*	*P*
Protection = P	8.5	0.02	0.896	246.5	0.48	0.561	468.1	2.37	0.263	6063.4	6.03	0.133
Location(P) = L(P)	392.4	2.20	0.173	517.3	8.46	**0.011**	197.5	3.68	0.074	1005.7	14.91	**0.002**
Site(L(P)) = S(L(P))	178.1	4.84	**0.000**	61.2	2.02	0.050	53.7	0.47	0.472	67.4	1.68	0.112
Residuals	36.8			30.2			56.0			40.2		

Significant *P*-values are given in bold.

groups or into morphological groups [52] (details are reported in the Supporting Information S1).

Due to the potential role of herbivores in structuring macroalgal assemblages [53], at each site, ten additional 20×20 cm random quadrats were used to count *in situ* the densities of dominant grazers (i.e. Polyplacophora, the gastropods of the genera *Patella* spp. and *Phorcus* spp.) in four sampling times, from 2003 to 2009, in the same locations and sites where sessile assemblages were also sampled.

Benthic community of the shallow infralittoral rock. This is an algal-dominated community at 5–7 m depth, where species such as *Flabellia petiolata*, *Halimeda tuna*, *Padina pavonica*, *Laurencia* spp. and the order of Dictyotales, dark filamentous algae (*Ectocarpus* spp. and *Sphacelaria* spp.), Cladophorales, encrusting coralline red algae (such as *Lithophyllum frondosum*, *L. incrustans*, *Mesophyllum alternans*) can be dominant. Sponges such as *Chondrilla nucula*, *Aplysina aerophoba*, *Ircinia variabilis*, *Cliona* spp. and *Phorbas* spp.) can be also present. In the infralittoral, rocky substrates account for the 10% of bottom surface within the MPA [50]. Sampling was conducted in eight random occasions from 2001 to 2009 (same season, i.e. late spring-early summer, as for midlittoral assemblages) at two protected (P1, P2) and two unprotected locations (C1, C3), with three sites in each location. A further unprotected location (C2) was also sampled from 2003 to 2006 (Fig. 1). The variable number of unprotected locations sampled during the study depends on the fact that the sampling program combines data coming from different projects funded for different periods. Assemblages were sampled photographically using a Nikonos V underwater camera, 28 mm focal length, close-up macro-system and two SB 105-Nikon electronic strobes. Thirteen random photographic sample of 16×23 cm were photographed at each site and 10 of them were randomly selected and analysed. This prevented the risk of having blurred, unclear photographic samples. In each photographic sample, the cover of sessile organisms was estimated under magnification by superimposing a transparent grid of 24 sub-quadrats on the entire photographed surface, and final values were expressed as a percentage. Destructive samples were collected for later identification of organisms present in the slides. Organisms not identified at species level were lumped into higher taxonomic groups or into morphological groups. Full taxonomic details are reported in the Supporting Information S1.

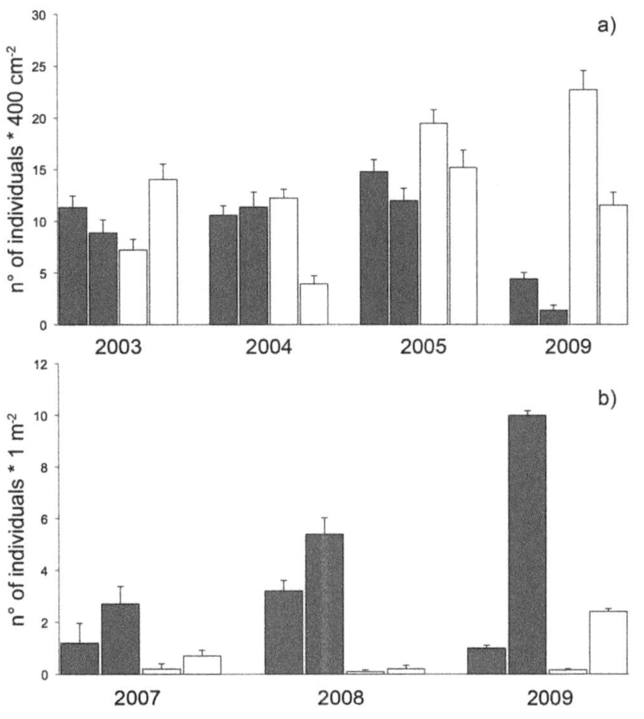

Figure 4. Relative amount of grazers on both rocky habitats. Mean density (± SE) of grazers found in the (a) lower midlittoral (gastropods) and in the (b) shallow infralittoral (sea urchins) rocky assemblages in each time of sampling. White bars = protected locations; grey bars = unprotected locations.

The sea urchins *Paracentrotus lividus* and *Arbacia lixula* are both present in the algal-dominated community of the infralittoral rocks of the MPA. Density of both species was estimated in three sampling dates (once per year from 2007 to 2009) in two protected locations (P1, P2) and two unprotected locations (C1, C3) where the sessile assemblages were also sampled (Fig. 1), with two sites for each location. At each site, underwater identification and counts of sea urchins within 20 replicate 1 m^2 quadrats were performed in the infralittoral rocks. Counts were made at approximately 3–8 m depth during the daylight. Care was taken to search for urchins in crevices.

Seagrass Meadows

Posidonia oceanica accounts for about the 20% of the infralittoral of the MPA [50]. Sampling was carried out in July, once per year, from 2006 to 2009. Due to the lack of seagrass beds within the no-take zones, the density of shoots was sampled in close proximity of P1 and P2 but in the buffer zone, where the seagrass forms extensive meadows [50] but human activities potentially affecting this habitat (i.e. anchoring, trawling) are also banned (see the Introduction and Fig. 1). Also in this case, two unprotected

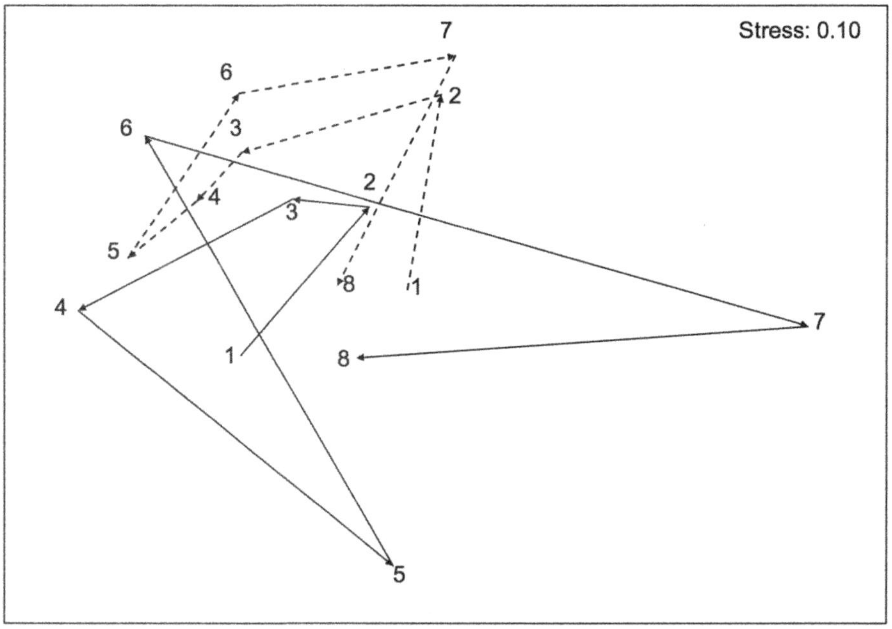

Figure 5. Differences in the multivariate structure of the shallow infralittoral rocky assemblages. Non-metric multidimensional scaling ordinations (nMDS) of T × P centroids based on Bray-Curtis dissimilarity measures. Dotted trajectories = protected assemblages; solid trajectories = unprotected assemblages. Numbers (from 1 to 8) indicate the sampling times.

locations outside the MPA (C1 and C3) were sampled (Fig. 1). At each location, two patches (100–300 m apart) were randomly chosen within beds at 8–10 m depth. In each patch, the density of shoots was estimated *in situ* within five 1 m^2 random quadrats. The status of *P. oceanica* beds was evaluated on the basis of the number of shoots per square meter following Pergent et al. [54].

Statistical Analyses

Multivariate analyses. A distance-based permutational multivariate analysis of variance (PERMANOVA, [55]) was performed separately on the data sets relative to the lower midlittoral and shallow infralittoral rocky assemblages (sessile benthos only, grazers not included) to test for the effect of protection on their structural features (species composition and abundance). The analyses were based on Bray-Curtis dissimilarities calculated on untransformed data and each term was tested using 4999 random permutations [56].

For the analysis of the lower midlittoral rocky assemblages, the experimental design consisted of four factors: Time (T, 6 levels, random), Protection (P, 2 level, fixed), Location (L(P), 2 levels, random, nested in P) and Site (S(L(P)), 3 levels, random, nested in L(P)), with $n = 10$. For the shallow infralittoral rock community, the design and the factor labelling were the same but the number of levels differed for the factor T (8) and the number of sites and/or unprotected locations actually varied at different sampling times (see 'Sampling design' section for details). However, formal tests were still possible since PERMANOVA allows the handling of complex unbalanced designs [56]).

As the analysis of the lower midlittoral rocky assemblages showed significant temporal variations of differences among locations (see Results), a non-metric multidimensional scaling ordination (nMDS) of T × L(P) centroids was plotted to visualize patterns of variation among location through time. In the case of the shallow infralittoral, nMDS ordination of T × P centroids was plotted to visualize multivariate patterns of differences between protected and unprotected assemblages through time, since PERMANOVA revealed a significant effect of protection on temporal trajectories of assemblages (see Results). Centroids were obtained calculating principal coordinates (PCO) on the basis of the Bray–Curtis dissimilarity matrices among all pairs of units.

For the shallow infralittoral rocky assemblages, a canonical analysis of principal coordinates (CAP, [57]) was also performed for the T × P interaction term, calculating the distance matrix among sites in protected and unprotected locations in each time of sampling. Distinctness among T × P groups was assessed using leave one-out allocation success [58]. Individual taxa that might be responsible for any group differences seen in the CAP plot were investigated by calculating product–moment correlations of original variables (taxa) with canonical axes [57]. These correlations of individual variables with the two canonical axes (r^1 and r^2) were then represented as lines in the CAP plot. Taxa were included in the plot only if exceeding an arbitrarily chosen value of correlation (i.e. $\sqrt{r_1^2 + r_2^2}) \geq 0.2$ [57].

Univariate analyses. ANOVA on multivariate estimates of spatial variability was employed to test for differences between protected and unprotected assemblages (sessile benthos only, grazers not included) at all investigated spatial scales (i.e. replicates, sites, locations), separately for the lower midlittoral and the shallow infralittoral rocky assemblages.

Components of variation from hierarchical analyses of variance allow separating sampling error from estimates of true variability associated to each spatial scale of observation [59]. However, a formal test requires multiple estimates of variance components, whereas, starting from a single set of data, only a single estimate is available for each source of variation. To deal with this issue, a single data set could be split into subsets allowing calculation of replicated independent estimates of variance components [60]. Each of the two data sets (i.e., lower midlittoral and shallow infralittoral rocky assemblages) was thus split into two halves, randomly selecting five replicates out of ten for each site in each

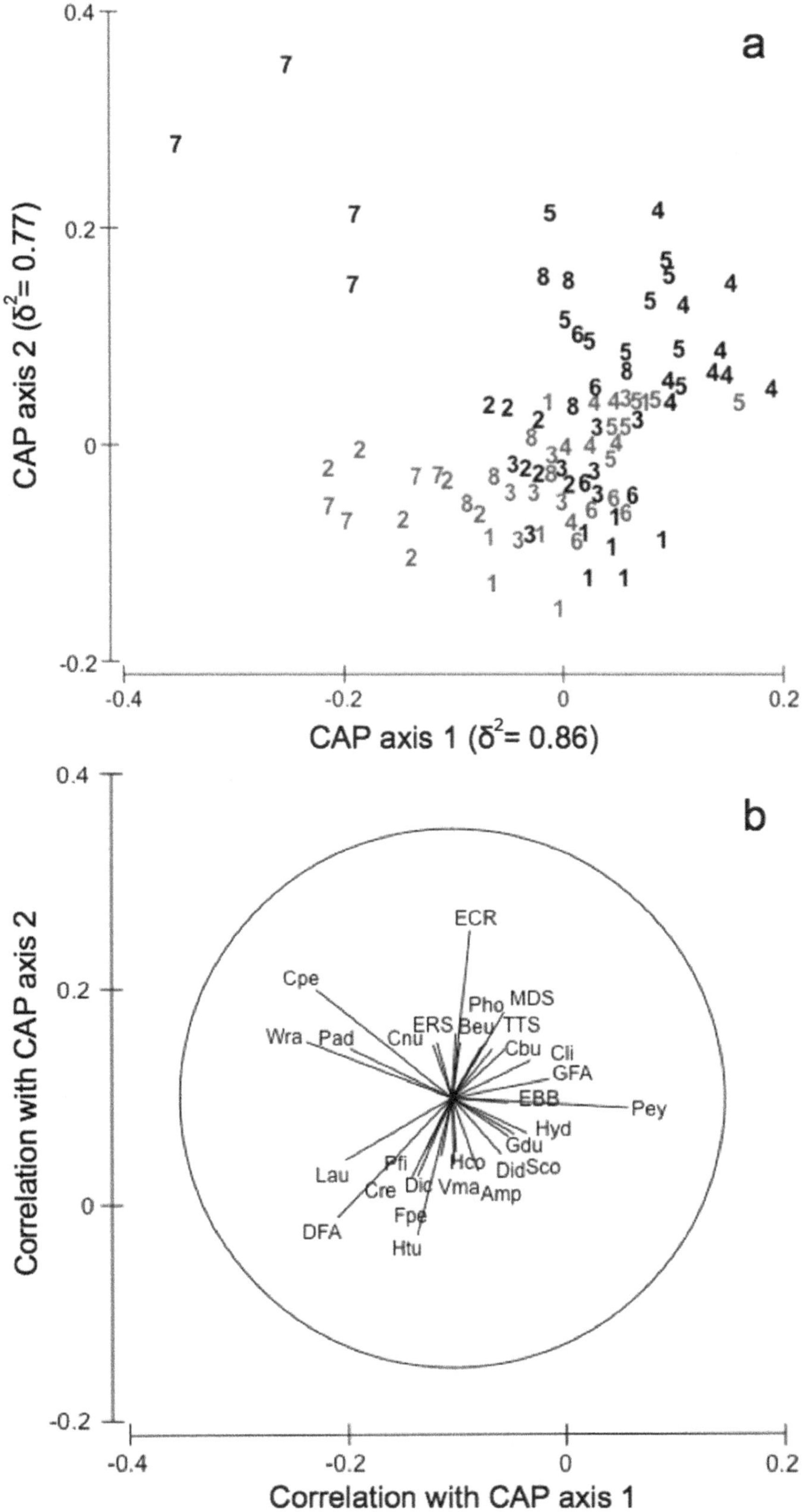

Figure 6. Canonical analysis of principal coordinates for protected *vs.* unprotected shallow infralittoral rocky assemblages and discriminating taxa. CAP for factor T × P based on the distance matrix of sites of the shallow infralittoral rocky assemblages (a). Numbers (from 1 to 8) indicate progressive times of sampling. Grey numbers = protected assemblages; black numbers = unprotected assemblages. Individual taxa highly correlated with canonical axes were also shown (b) (see Supporting Information S1 for taxa abbreviations).

Table 4. Summary of ANOVAs testing for the effect of protection on the total abundance of sea urchins of the shallow infralittoral in each time of sampling.

	2007			2008			2009		
Source of variation	**MS**	*F*	*P*	**MS**	*F*	*P*	**MS**	*F*	*P*
Protection = P	2.8	0.69	0.494	688.9	15.48	0.059	57.6	0.96	0.431
Location(P) = L(P)	4.0	1.68	0.295	44.5	0.47	0.655	60.1	6.08	0.061
Site(L(P)) = S(L(P))	2.4	6.02	**0.002**	94.5	4.95	**0.001**	9.9	4.39	**0.002**
Residuals	0.1			19.1			2.2		

Significant *P*-values are given in bold.

time of sampling. The two subsets were then analysed running separate PERMANOVAs for each Time and Protection level, following the corresponding full hierarchical design. This allowed obtaining two independent components of pseudo-variance for each spatial scale (i.e. location, site, replicate) in each time of sampling for both protected and unprotected assemblages thus allowing to perform an ANOVA considering Time, Protection, and Scale as factors. For both communities, the design consisted of three factors: Time (T, 6 levels for the lower midlittoral and 8 levels for shallow infralittoral, random), Protection (P, 2 levels, fixed), Scale (Sc, 3 levels, fixed), with $n=2$. For significant interaction terms involving the factor Protection, the Student Newman-Keuls (SNK) test was used for *post-hoc* pair-wise comparisons of spatial variability between protected and unprotected assemblages.

ANOVA on multivariate estimates of temporal variability was done to test the hypothesis that protection could increase the stability of sessile benthic assemblages. For the shallow infralittoral rocky assemblages, location C2 was excluded from the analysis because not available in all sampling occasions. For each data set, data were analysed separately for each of the twelve sites (i.e. the six protected sites and the six unprotected sites) using PERMANOVA, obtaining pseudo-variance components associated to factor Time. This allowed calculating six multivariate estimates of temporal variability for protected and unprotected assemblages. The design for the analysis consisted of only one factor, Protection (2 levels, fixed) with $n=6$.

ANOVA was employed to test for differences in shoot density of *P. oceanica* beds between protected and unprotected locations. The analysis was done separately for each time of sampling. The design for the analyses consisted of three factors: Protection (P, 2 level, fixed), Bed (B(P), 2 levels, random, nested in P), Patch (Pa(B(P)), 2 levels, random, nested in B(P)), with $n=5$.

ANOVA was also used to test for differences in density of midlittoral and shallow infralittoral grazers (i.e. gastropods and sea urchins respectively) between protected and unprotected locations. The analyses were done separately for each time of sampling. The three-factorial design for analyses consisted of factor Protection (P, 2 level, fixed), Location (L(P), 2 levels, random, nested in P), Site (S(L(P)), 3 levels for the lower midlittoral and 2 levels for the infralittoral grazers, random, nested in L(P)), with $n=10$ for the lower midlittoral and $n=20$ for the shallow infralittoral.

Prior to all ANOVAs the homogeneity of variance was examined using the Cochran's *C* test. All analyses were performed using the GMAV 5 software (University of Sydney, Australia).

Results

Benthic Assemblages of the Lower Midlittoral

Multivariate analyses showed no effects of protection on the structure of assemblages (Table 1). After more than 20 years from its institution, inside and outside the MPA, benthic assemblages were largely dominated by erect and encrusting coralline algae, few boring sponges and anthozoans [50]. Results of PERMANOVA showed significant T × L(P) and T × S(L(P)) interactions, indicating a significant assemblage variability among locations, and among sites within locations, that varied in time (Table 1). In other words, despite similarities in taxa composition were found between assemblages from protected and unprotected locations, their patterns of distribution differed in space and time. Temporal variations among locations were portrayed in the nMDS ordination of T × L(P) centroids (Fig. 2), which also showed a higher scattering of centroids of protected locations with respect to unprotected locations, suggesting higher temporal variations in protected assemblages than in unprotected ones. An exception to the observed patterns was detected especially in Time 2, where

Table 5. Summary of ANOVAs testing for the effect of protection on the shoot density of *P. oceanica* beds in each sampling time.

	2006			2007			2008			2009		
Source of variation	**MS**	*F*	*P*	**MS**	*F*	*P*	**MS**	*F*	*P*	**MS**	*F*	*P*
Protection = P	513022.5	11.82	0.075	361000.0	2.60	0.248	253446.4	27.35	**0.035**	292410.0	13.02	0.070
Beds(P) = B(P)	43397.0	3.03	0.375	138707.2	1.99	0.251	9267.2	6.21	0.059	22454.8	1.17	0.397
Patch(B(P)) = Pa(B(P))	14337.1	2.49	0.063	69535.2	19.30	**0.000**	1492.8	0.35	0.839	19122.0	5.00	**0.003**
Residuals	5765.6			3602.2			4215.8			3825.0		

Significant *P*-values are given in bold.

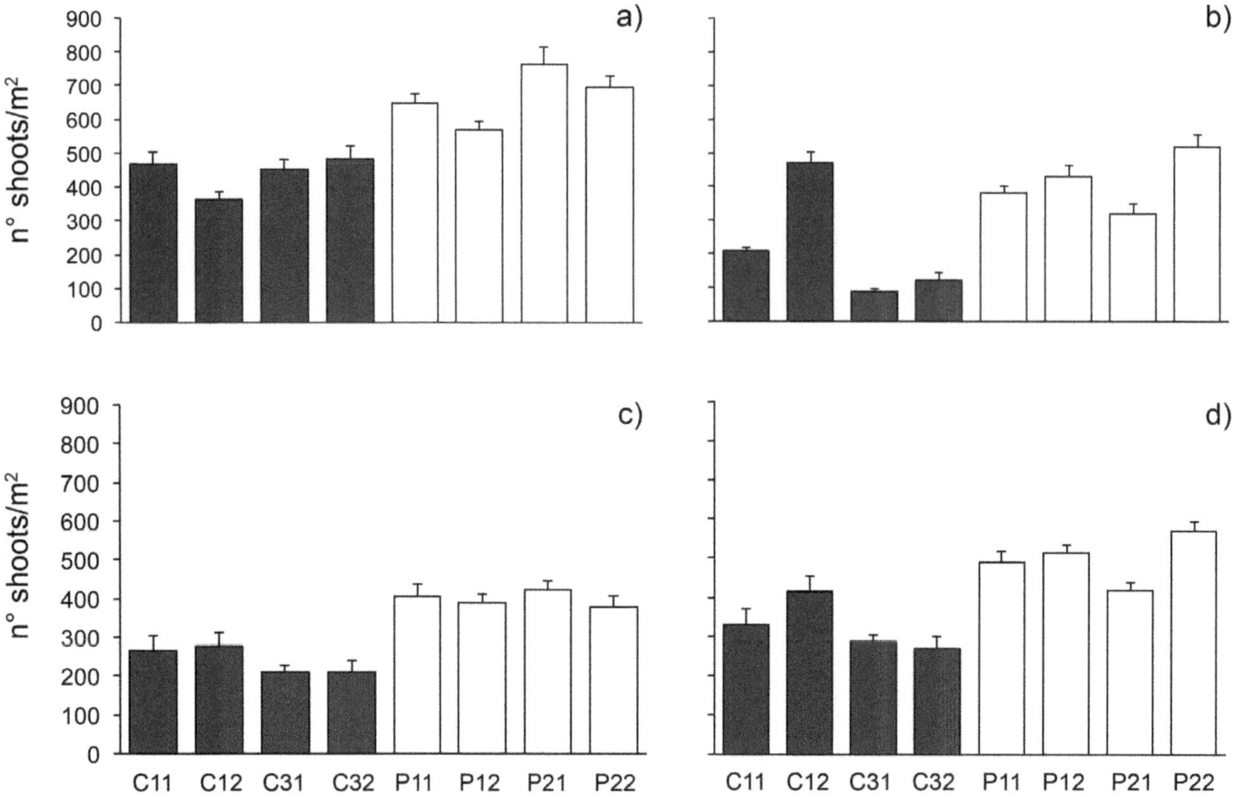

Figure 7. Variations of seagrass shoot density in the different locations. Mean (± SE, $n = 5$) n° of shoots/m^2 of *P. oceanica* in each patch in (a) 2006, (b) 2007, (c) 2008 and (d) 2009. White bars = protected beds; grey bars = unprotected beds.

one of the unprotected locations was featured by particularly high abundances of articulated corallines, such as *Corallina* and *Jania.*

Results of ANOVA on multivariate estimates of spatial variability of assemblages showed a significant T × P × Sc interaction, indicating an effect of protection on spatial heterogeneity of assemblages, varying in time and with scale (Table 2). *Post-hoc* pair-wise comparisons indicated an increase through time of spatial heterogeneity in protected assemblages at the scale of replicates (10 s of centimetres) and locations (kilometres), but not at the scale of sites (100 s of metres) (Table 2). ANOVA on estimates of multivariate temporal variability detected a significant effect of protection ($F = 8.75$, $P < 0.05$), with a higher temporal variability of protected assemblages with respect to unprotected ones (Fig. 3). In other words, protected assemblages were more spatially heterogeneous and with more pronounced temporal fluctuations than the assemblages outside the MPA.

ANOVA on density of grazers (i.e. *Patella* spp. and *Phorcus* spp.) did not highlight a significant effect of protection through time (Table 3). However, inspection of the graph in Fig. 4a suggested a decrease of grazers in the unprotected locations and, on average, an increase in protected locations in time. More particularly, in 2009 (Time 4), the density of these grazers was five times higher in protected locations than in unprotected ones, although such differences were not significant probably due to the high variability at the scale of sites.

Table 6. Classification of the status of *P. oceanica* beds based on shoot density following Pergent et al. [54].

Location	Patch	2006	2007	2008	2009
P1	1	undisturbed	disturbed	Undisturbed	undisturbed
P1	2	undisturbed	undisturbed	Undisturbed	undisturbed
P2	1	undisturbed	disturbed	Undisturbed	undisturbed
P2	2	undisturbed	undisturbed	Undisturbed	undisturbed
C1	1	undisturbed	very disturbed	very disturbed	disturbed
C1	2	undisturbed	very disturbed	very disturbed	undisturbed
C3	1	disturbed	undisturbed	Disturbed	Disturbed
C3	2	undisturbed	very disturbed	very disturbed	Disturbed

The state of the seagrass beds has been reported for each investigated patch in each time (see text for details). P1, P2 = Protected locations;
C1, C3 = Unprotected locations.

Benthic Assemblages of the Shallow Infralittoral Rock

A significant effect of protection in modifying temporal trajectories of sessile assemblages was detected by PERMANOVA, as indicated by the significant T × P interaction (Table 1). These differences in temporal trajectories between protected *vs.* unprotected assemblages were clearly portrayed in the nMDS ordination plot of T × P centroids, which also underlined more severe temporal changes of unprotected *vs.* protected assemblages (Fig. 5). In Time 1, protected and unprotected assemblages were quite close to each other and showed similar patterns of temporal variation until Time 3, whereas, starting from Time 4, their trajectories diverged following different directions that led to a clear separation of centroids in Time 5 and 7, coming very close again at Time 8.

The canonical analysis of principal coordinates for the term T × P achieved the highest allocation success (52.75%) using $m = 9$ principal coordinate (PCO) axes, which explained 90.1% of variation in the original dissimilarity matrix. The two canonical axes had very high canonical correlations with the multivariate assemblages (δ^2, Fig. 6a, $P = 0.001$). The CAP analysis revealed that harsh temporal changes outside the MPA corresponded to alternate states of assemblage structure whereas, within the MPA, assemblage structure remained relatively homogeneous through time (Fig. 6). Starting from Time 1, unprotected assemblages move counter clockwise from the down right corner of the graph, reaching the up left corner (Time 7) and coming back on the up right side (Time 8) (Fig. 6a), shifting from assemblages characterized by erect-canopy [*Sphaerococcus coronopifolius* (Sco)], turf-forming [*Amphiroa* spp. (Amp), *Valonia macrophysa* (Vma)] algae, and several invertebrates [*Hemimycale columella* (Hco), Didemnidae (Did), Hydroids (Hyd), *Gastrochaena dubia* (Gdu)] (Time 1, 2, 3), to assemblages mostly characterized by encrusting [Encrusting Calcified Rhodophytes (ECR), *Peyssonnelia* spp. (Pey)], ephemeral [Green Filamentous Algae (GFA)] algae, encrusting invertebrates [e.g., Encrusting Brown Bryozoans (EBB), *Balanophyllia europaea* (Beu), Encrusting Red Sponges (ERS), *Chondrilla nucula* (Cnu)] and boring sponges [Clionidae (Cli)] (Time 4, 5, 6, 8), passing by assemblages dominated by turf-forming algae [*Wrangelia penicillata* (Wra), *Padina pavonica* (Pad)] and the anthozoan *Cereus pedunculatus* (Cpe) (Time 7) (Fig. 6b). In contrast, protected assemblages in all times clustered together on the down side of the graph (Fig. 6a) being distinguished by taxonomic groups of erect-canopy [*Dictyota* spp. (Dic), *Laurencia* complex (Lau), *Sphaerococcus coronopifolius* (Sco)], turf-forming [*Flabellia petiolata* (Fpe), *Amphiroa* spp. (Amp), *Halimeda tuna* (Htu), Dark Filamentous Algae (DFA)] algae, sponges [*Petrosia ficiformis* (Pfi), *Chondrosia reniformis* (Cre), *Hemimycale columella* (Hco),] and other invertebrates [Didemnidae (Did), Hydroids (Hyd), *Gastrochaena dubia* (Gdu)] (Fig. 6b).

Results of ANOVA on multivariate estimates of spatial variability of assemblages detected significant differences between protected and unprotected locations that were consistent in time and across scales (Table 2), indicating a significant effect of protection in decreasing spatial heterogeneity of assemblages at the scale of tens of centimetres, among replicate units, up to kilometres (results not showed). In other words, long-lived and relatively slow-growing invertebrates and structurally complex algal forms were homogeneously distributed in space and were featured by less fluctuations in time compared to unprotected assemblages.

The analysis on estimates of multivariate temporal variability also detected a significant effect of protection ($F = 4.99$, $P < 0.05$) with a lower value in protected than in unprotected assemblages (Fig. 3), indicating an effect of protection in smoothing out the temporal variability of assemblages.

Inspection of graph in Figure 4b suggested an increase of grazers (i.e. sea urchins) in unprotected locations through time. However, ANOVA did not highlight a significant effect of protection on their density. Also, a significant variability in their pattern of distribution was documented at the scale of sites in all sampling times (Table 4).

Seagrass Meadows

Results of ANOVA on shoot density (Table 5) showed an overall lack of significant differences between protected and unprotected *P. oceanica* seagrass beds (Fig. 7). Inspection of graphs in Figure 7 suggested a general reduction of shoot density through time that was lower in protected than in unprotected beds. Even if the pattern is quite clear, in one of unprotected areas, the number of shoots increased in 2007, reaching densities comparable to protected areas. The classification of the status of *P. oceanica* beds following Pergent et al. [54] revealed that, in contrast to beds within the MPA, which were characterized by general undisturbed conditions in all times, unprotected beds showed a progressive shift from undisturbed towards disturbed or very disturbed conditions (Table 6).

Discussion

Our case study supports the recent findings on the efficacy of MPAs, documenting further ecological consequences of reserve implementation. These results show that the effective protection of this area, while restoring biodiversity, can enhance assemblage stability, by increasing the spatial homogeneity of protected assemblages, decreasing their temporal variability and increasing their resistance to large-scale disturbance. After more than 20 years from the institution of the MPA, the presence of highly performing species with the re-establishment of continuous and persistent macroalgal canopies, and the maintenance of more dense and healthy meadows are an important outcome of a successful regulation of human activities. Our study contributes to demonstrate that reducing the effects of local sources of disturbance by excluding direct human activities within MPAs may represent an effective strategy in mitigating also the effects of large-scale threats [40]. Even though the MPA is embedded within a landscape featured by a low population density and coastal development, the whole area has been recently ranked as under the pressure of moderate to high human impact [61]: fifty-five fishermen are active in the area, outside the MPA, an industrial/commercial harbour with a cargo tonnage of 10 million tons and a sewage treatment for 130,000 inhabitants are at a distance of about 15 kilometres from the MPA. Results showing an effective mitigation from threats are clear for seagrass meadows and for assemblages from the shallow infralittoral, while different patterns have been found for midlittoral assemblages.

In the shallow infralittoral rocks, at fully protected locations, the exclusion of fishery, anchoring and diving frequentation had important consequences at assemblage level. Within the MPA, macrobenthic assemblages with long-lived and relatively slow-growing invertebrates (e.g. sponges, madreporarians) and structurally complex algal forms (i.e. erect-canopy forming algae like Dictyotales), were homogeneously distributed in space and went through little fluctuations in time. In contrast, a mosaic of disturbed patches featured unprotected locations, with small-scale shifts from macroalgal stands to barrens, and harsh temporal variations between the two states and an increase of sea urchins through time. Such differences in the structure of assemblages determined the different patterns of spatio-temporal variability between protected and unprotected areas (Fig. 3). Canopy-forming species create a complex three-dimensional habitat and can increase community stability by reducing the variability of environmental factors [62,63], even though this effect could be context dependent [64,65]. The effects of mechanical disturbance such as anchoring on erect algae have been largely demonstrated [66]. There are also strong evidences that the exclusion of fishery can re-establish lost predatory interactions, decreasing the density of sea urchins and enhancing top-down mechanisms able to drive the recovery of macroalgae in MPAs through cascade effects [34,39,46,67]. Within an MPA, moreover, predators might dampen high-recruitment episodes of sea urchins, stabilizing their populations [42]. Such indirect effects may lead to an increased resilience of protected benthic assemblages by accelerating the processes of colonization after disturbance [43]. As also documented in previous studies [46], during the last decade, our results show that sea urchins were highly variable in abundance at unprotected locations and more stable and less abundant at protected locations. These patterns suggest that, in the absence of the selecting sea urchin removal by fish predators under protection regimes, fluctuating levels of biological disturbance on macroalgae determined by sea urchin grazing may be also responsible of the higher spatial and temporal variability in unprotected areas. Such findings reinforce the assumption that increased variability in space and time can be considered an indicator of (natural and/or anthropogenic) stress in benthic assemblages [14,68–70] and support the existence of positive feedback related to dense macroalgal canopies on assemblage resistance potential.

Opposite patterns of spatial and temporal variability were found for the lower midlittoral rocky assemblages, with no effect of protection detected on their multivariate structure, largely dominated by articulate corallines, *Laurencia* and other Ceramiales inside and outside the MPA. Higher temporal variations and a general increase through time of spatial heterogeneity of assemblages within the MPA were also found. Interestingly, in protected locations (where the collection of organisms is forbidden), sea urchin populations of the shallow infralittoral rock decreased, whereas an increase in gastropod grazers was observed in the midlittoral. Idiosyncratic responses to protection among habitats and ecological compartments have been recently found also in others MPAs [69,71]. While in the infralittoral the re-establishment of predatory interactions led to a control of grazing pressure, in the lower midlittoral, the exclusion of human collection of grazers such as *Patella* spp. and *Phorcus* spp. within no-take areas had a clear effect on their abundances and was possibly responsible for the general increase through time of the spatial heterogeneity in protected assemblages at all investigated spatial scales [72–74].

Direct effects of protection were observed also on the seagrass *P. oceanica*. Recently, a combination of available aerial photographs and results from the report of the European project EUROSION (http://www.eurosion.org) showed that the recent evolution of the littoral (ratio between the length of the stretches of coast in recession and the total coastal length) is towards a clear regression, probably causing the high sediment load observed in the whole area [75], which, in turn, can result in increased stress levels for seagrass meadows [76,77]. Although an overall clear pattern of regression was observed in the whole region, a significantly higher shoot density characterized protected meadows with respect to unprotected ones. Such a pattern was consistent through time and suggests a higher resistance to coastal pressures of the protected meadows. In this case, the local management of fishing and anchoring inside the MPA seems to provide some insurance against large-scale pressures, such as sedimentation, which are impractical to manage directly. Montefalcone et al. [78] conclude that MPAs alone are not sufficient to guarantee the protection of *P. oceanica* meadows and that management tools other than MPAs are needed to enhance the large-scale persistence of this habitat [79,80]. Activities that reduce coastal pollution and eutrophication by the establishment of management plans for water resources, for instance, are also key elements for the recovery of habitats such as *Cystoseira* spp. fringes [81–83]. Clear indications of thresholds and variables involved in these observed recovery trajectories have also been documented [84–86] and, at least for *P. oceanica*, it is apparent that the management of direct disturbances, such as trawling, anchoring, dredging and pipeline refilling, can help its restoration, even though over long time scales.

Natural fluctuations in abundance of species in space and time are extremely common in the marine environment [87] and deriving the conclusion that the protection of this area is the only process driving the observed changes is not realistic. Multiple processes acting simultaneously have the potential to interact causing changes that are difficult to interpret. The exceptions observed in our results (see Fig. 2 for the lower midlittoral rocks and the high shoot density in *P. oceanica* in one unprotected location) represent clear evidence that detailed information on the environmental context might have largely improved our potential to understand the variability in the observed outcomes. Even with these limits, our results suggest that the reduction of human activities, and especially overexploitation, can have clear positive consequences at habitat scale [88], but also that the ensuing direct effects on target populations can affect indirectly non-target species. A major limit to the understanding of interacting outcomes of protection and natural or anthropogenic disturbance on marine ecosystems relies on the scarcity of long-term studies. Having long-term funding to support studies as this one is still a challenge and this analysis represents an effort to ensure data comparability despite the lack of funding continuity. A long-term perspective in assessing the effects of protection is critical since different ecological components may respond differently and over varying temporal scales [41]. In addition, a deeper insight on causal processes underlying the dynamic responses of communities could not disregard their spatial and temporal heterogeneity.

The present study represents one of the first attempts in this direction. Even though we recognize that other processes might be involved in explaining the observed changes, by combining decadal time series of data on different habitats and assemblages, we reinforced the view that MPAs might provide an insurance against the consequences of local and large-scale human disturbances promoting the persistence of desirable ecological conditions and enhancing the stability of marine communities. Under future scenarios of frequent and/or persistent disturbance, conservation strategies based on MPA networks aimed at enhancing resilience may be the most effective tool to limit the negative impacts of the complex suites of threats on marine ecosystems.

Acknowledgments

Thanks are due to the Marine Protected Area of Torre Guaceto for the technical support.

Author Contributions

Conceived and designed the experiments: SF GG SB AT. Performed the experiments: SF GG SB AT. Analyzed the data: SF GG SB AT. Contributed reagents/materials/analysis tools: SF AT FB. Wrote the paper: SF GG SB AT FB.

References

1. Hooper DU, Adair EC, Cardinale BJ, Byrnes JEK, Hungate BA, et al. (2012) A global synthesis reveals biodiversity loss as a major driver of ecosystem change. Nature 486: 105–U129.
2. Cardinale BJ, Duffy JE, Gonzalez A, Hooper DU, Perrings C, et al. (2012) Biodiversity loss and its impact on humanity. Nature 486: 59–67.
3. Worm B, Barbier EB, Beaumont N, Duffy JE, Folke C, et al. (2006) Impacts of biodiversity loss on ocean ecosystem services. Science 314: 787–790.

4. Hooper DU, Chapin FS, Ewel JJ, Hector A, Inchausti P, et al. (2005) Effects of biodiversity on ecosystem functioning: A consensus of current knowledge. Ecol Monogr 75: 3–35.
5. Srivastava DS, Vellend M (2005) Biodiversity-ecosystem function research: Is it relevant to conservation? Annu Rev Ecol Evol Syst. 267–294.
6. Schwartz MW, Brigham CA, Hoeksema JD, Lyons KG, Mills MH, et al. (2000) Linking biodiversity to ecosystem function: implications for conservation ecology. Oecologia 122: 297–305.
7. Warwick RM, Clarke KR (1993) Increased variability as a symptom of stress in marine communities. J Exp Mar Biol Ecol 172: 215–226.
8. Hsieh CH, Reiss CS, Hunter JR, Beddington JR, May RM, et al. (2006) Fishing elevates variability in the abundance of exploited species. Nature 443: 859–862.
9. deYoung B, Barange M, Beaugrand G, Harris R, Perry RI, et al. (2008) Regime shifts in marine ecosystems: detection, prediction and management. Trends Ecol Evol 23: 402–409.
10. Hewitt JE, Thrush SF (2009) Do Species' abundances become more spatially variable with stress? Open Ecol J 2: 37–46.
11. Folke C, Carpenter S, Walker B, Scheffer M, Elmqvist T, et al. (2004) Regime shifts, resilience, and biodiversity in ecosystem management. Annu Rev Ecol Evol Syst 35: 557–581.
12. Scheffer M, Carpenter SR, Lenton TM, Bascompte J, Brock W, et al. (2012) Anticipating Critical Transitions. Science 338: 344–348.
13. Fraterrigo JM, Rusak JA (2008) Disturbance-driven changes in the variability of ecological patterns and processes. Ecol Lett 11: 756–770.
14. Hewitt J, Thrush S, Lohrer A, Townsend M (2010) A latent threat to biodiversity: consequences of small-scale heterogeneity loss. Biodivers Conserv 19: 1315–1323.
15. Carpenter SR, Cole JJ, Pace ML, Batt R, Brock WA, et al. (2011) Early Warnings of Regime Shifts: A Whole-Ecosystem Experiment. Science 332: 1079–1082.
16. Hughes TP, Baird AH, Bellwood DR, Card M, Connolly SR, et al. (2003) Climate change, human impacts, and the resilience of coral reefs. Science 301: 929–933.
17. Mumby PJ, Dahlgren CP, Harborne AR, Kappel CV, Micheli F, et al. (2006) Fishing, trophic cascades, and the process of grazing on coral reefs. Science 311: 98–101.
18. Hoey AS, Pratchett MS, Cvitanovic C (2011) High Macroalgal Cover and Low Coral Recruitment Undermines the Potential Resilience of the World's Southernmost Coral Reef Assemblages. Plos One 6.
19. Gorman D, Connell SD (2009) Recovering subtidal forests in human-dominated landscapes. J Appl Ecol 46: 1258–1265.
20. Petraitis PS, Methratta ET, Rhile EC, Vidargas NA, Dudgeon SR (2009) Experimental confirmation of multiple community states in a marine ecosystem. Oecologia 161: 139–148.
21. Perkol-Finkel S, Airoldi L (2010) Loss and Recovery Potential of Marine Habitats: An Experimental Study of Factors Maintaining Resilience in Subtidal Algal Forests at the Adriatic Sea. Plos One 5.
22. Sala E, Kizilkaya Z, Yildirim D, Ballesteros E (2011) Alien Marine Fishes Deplete Algal Biomass in the Eastern Mediterranean. Plos One 6.
23. Boero F, Bonsdorff E (2007) A conceptual framework for marine biodiversity and ecosystem functioning. Mar Ecol-Evol Persp 28: 134–145.
24. Fauchald P (2010) Predator-prey reversal: A possible mechanism for ecosystem hysteresis in the North Sea? Ecology 91: 2191–2197.
25. Bellwood DR, Hughes TP, Folke C, Nystrom M (2004) Confronting the coral reef crisis. Nature 429: 827–833.
26. Hughes TP, Bellwood DR, Folke C, Steneck RS, Wilson J (2005) New paradigms for supporting the resilience of marine ecosystems. Trends Ecol Evol 20: 380–386.
27. Allison GW, Lubchenco J, Carr MH (1998) Marine reserves are necessary but not sufficient for marine conservation. Ecol Appl 8: S79–S92.
28. Halpern BS, Lester SE, McLeod KL (2010) Placing marine protected areas onto the ecosystem-based management seascape. P Natl Acad Sci USA 107: 18312–18317.
29. Fraschetti S, Claudet J, Grorud-Colvert K (2011) Transitioning from single-sector management to ecosystem-based management: What can marine protected areas offer? In Claudet J (ed.) Marine Protected Areas - A multidisciplinary approach. Cambridge University Press. 11–34.
30. Linares C, Garrabou J, Hereu B, Diaz D, Marschal C, et al. (2012) Assessing the Effectiveness of Marine Reserves on Unsustainably Harvested Long-Lived Sessile Invertebrates. Conserv Biol 26: 88–96.
31. Goñi R, Quetglas A, Reñones O (2006) Spillover of spiny lobsters *Palinurus elephas* from a marine reserve to an adjoining fishery. Mar Ecol-Prog Ser 308: 207–219.
32. Guidetti P (2006) Potential of marine reserves to cause community-wide changes beyond their boundaries. Conserv Biol 21: 540–545.
33. Lester SE, Halpern BS, Grorud-Colvert K, Lubchenco J, Ruttenberg BI, et al. (2009) Biological effects within no-take marine reserves: a global synthesis. Mar Ecol-Prog Ser 384: 33–46.
34. Sala E, Ballesteros E, Dendrinos P, Di Franco A, Ferretti F, et al. (2012) The Structure of Mediterranean Rocky Reef Ecosystems across Environmental and Human Gradients, and Conservation Implications. Plos One 7.
35. Garcia-Charton JA, Perez-Ruzafa A, Marcos C, Claudet J, Badalamenti F, et al. (2008) Effectiveness of European Atlanto-Mediterranean MPAs: Do they accomplish the expected effects on populations, communities and ecosystems? J Nat Conserv 16: 193–221.
36. Pauly D, Christensen V, Guenette S, Pitcher TJ, Sumaila UR, et al. (2002) Towards sustainability in world fisheries. Nature 418: 689–695.
37. Francour P (1994) Pluriannual analysis of the reserve effect on ichthyofauna in the Scandola Natural Reserve (Corsica, Northwestern Mediterranean). Oceanol Acta 17: 309–317.
38. Hughes TP, Graham NAJ, Jackson JBC, Mumby PJ, Steneck RS (2010) Rising to the challenge of sustaining coral reef resilience. Trends Ecol Evol 25: 633–642.
39. Mumby PJ, Harborne AR (2010) Marine Reserves Enhance the Recovery of Corals on Caribbean Reefs. Plos One 5.
40. Micheli F, Saenz-Arroyo A, Greenley A, Vazquez L, Montes JAE, et al. (2012) Evidence That Marine Reserves Enhance Resilience to Climatic Impacts. Plos One 7.
41. Babcock RC, Shears NT, Alcala AC, Barrett NS, Edgar GJ, et al. (2010) Decadal trends in marine reserves reveal differential rates of change in direct and indirect effects. P Natl Acad Sci USA 107: 18256–18261.
42. Hereu B, Linares C, Sala E, Garrabou J, Garcia-Rubies A, et al. (2012) Multiple Processes Regulate Long-Term Population Dynamics of Sea Urchins on Mediterranean Rocky Reefs. Plos One 7.
43. Bevilacqua S, Terlizzi A, Fraschetti S, Russo GF, Boero F (2006) Mitigating human disturbance: can protection influence trajectories of recovery in benthic assemblages? J Anim Ecol 75: 908–920.
44. Navarrete SA, Gelcich S, Castilla JC (2010) Long-term monitoring of coastal ecosystems at Las Cruces, Chile: Defining baselines to build ecological literacy in a world of change. Rev Chil Hist Nat 83: 143–157.
45. Claudet J, Fraschetti S (2010) Human-driven impacts on marine habitats: A regional meta-analysis in the Mediterranean Sea. Biological Conservation 143: 2195–2206.
46. Guidetti P (2006) Marine reserves reestablish lost predatory interactions and cause community changes in rocky reefs. Ecol Appl 16: 963–976.
47. Boudouresque CF, Bernard G, Pergent G, Shili A, Verlaque M (2009) Regression of Mediterranean seagrasses caused by natural processes and anthropogenic disturbances and stress: a critical review. Bot Mar 52: 395–418.
48. Guidetti P, Milazzo M, Bussotti S, Molinari A, Murenu M, et al. (2008) Italian marine reserve effectiveness: Does enforcement matter? Biol Conserv 141: 699–709.
49. Claudet J, Guidetti P (2010) Improving assessments of marine protected areas. Aquat Conserv-Mar Freshw Ecosyst 20: 239–242.
50. Fraschetti S, Terlizzi A, Bussotti S, Guarnieri G, D'Ambrosio P, et al. (2005) Conservation of Mediterranean seascapes: analyses of existing protection schemes. Mar Environ Res 59: 309–332.
51. Dethier MN, Graham ES, Cohen S, Tear LM (1993) Visual versus random-point percent cover estimations - Objective is not always better. Mar Ecol-Prog Ser 96: 93–100.
52. Steneck RS, Dethier MN (1994) A functional-group approach to the structure of algal-dominated communities. Oikos 69: 476–498.
53. Coleman RA, Underwood AJ, Benedetti-Cecchi L, Aberg P, Arenas F, et al. (2006) A continental scale evaluation of the role of limpet grazing on rocky shores. Oecologia 147: 556–564.
54. Pergent G, Pergent-Martini C, Boudouresque CF (1995) Utilisation de l'herbier à *Posidonia oceanica* comme indicateur biologique de la qualité du milieu littoral en Méditerranée: Etat des connaissances. Mésogée 54: 3–29.
55. Anderson MJ (2001) Permutation tests for univariate or multivariate analysis of variance and regression. Can J Fish Aquat Sci 58: 626–639.
56. Anderson MJ, Gorley RN, Clarke KR (2008) PERMANOVA+ for PRIMER: guide to software and statistical methods. PRIMER-E, Plymouth, UK: 214 p.
57. Anderson MJ, Willis TJ (2003) Canonical analysis of principal coordinates: A useful method of constrained ordination for ecology. Ecology 84: 511–525.
58. Anderson MJ, Robinson J (2003) Generalized discriminant analysis based on distances. Aust NZ J Stat 45: 301–318.
59. Gaston KJ, McArdle BH (1994) The temporal variability of animal abundances - measures, methods and patterns. Philos Trans R Soc Lond Ser B-Biol Sci 345: 335–358.
60. Fraschetti S, Terlizzi A, Bevilacqua S, Boero F (2006) The distribution of hydroids (Cnidaria, Hydrozoa) from micro- to macro-scale: Spatial patterns on habitat-forming algae. J Exp Mar Biol Ecol 339: 148–158.
61. Coll M, Piroddi C, Albouy C, Lasram FB, Cheung WWL, et al. (2012) The Mediterranean Sea under siege: spatial overlap between marine biodiversity, cumulative threats and marine reserves. Global Ecol Biogeogr 21: 465–480.
62. Polley HW, Wilsey BJ, Derner JD (2007) Dominant species constrain effects of species diversity on temporal variability in biomass production of tallgrass prairie. Oikos 116: 2044–2052.
63. Grman E, Lau JA, Schoolmaster DR, Gross KL (2010) Mechanisms contributing to stability in ecosystem function depend on the environmental context. Ecol Lett 13: 1400–1410.
64. Bulleri F, Alestra T, Ceccherelli G, Tamburello L, Pinna S, et al. (2011) Determinants of *Caulerpa racemosa* distribution in the north-western Mediterranean. Mar Ecol-Prog Ser 431: 55–67.
65. Valdivia N, Golléty C, Migne A, Davoult D, Molis M (2012) Stressed but Stable: Canopy Loss Decreased Species Synchrony and Metabolic Variability in an Intertidal Hard-Bottom Community. Plos One 7.

66. Milazzo M, Chemello R, Badalamenti F, Camarda R, Riggio S (2002) The Impact of Human Recreational Activities in Marine Protected Areas: What Lessons Should Be Learnt in the Mediterranean Sea? Mar Ecol 23: 280–290.
67. Hughes TP, Bellwood DR, Folke CS, McCook LJ, Pandolfi JM (2007) No-take areas, herbivory and coral reef resilience. Trends Ecol Evol 22: 1–3.
68. Bevilacqua S, Plicanti A, Sandulli R, Terlizzi A (2012) Measuring more of beta-diversity: Quantifying patterns of variation in assemblage heterogeneity. An insight from marine benthic assemblages. Ecol Indic 18: 140–148.
69. Fraschetti S, Bevilacqua S, Guarnieri G, Terlizzi A (2012) Idiosyncratic effects of protection in a remote marine reserve. Mar Ecol-Prog Ser 466: 21–U41.
70. Guarnieri G, Terlizzi A, Bevilacqua S, Fraschetti S (2012) Increasing heterogeneity of sensitive assemblages as a consequence of human impact in submarine caves. Mar Biol 159: 1155–1164.
71. O'Sullivan D, Emmerson M (2011) Marine reserve designation, trophic cascades and altered community dynamics. Mar Ecol-Prog Ser 440: 115–U128.
72. Ceccherelli G, Casu D, Pala D, Pinna S, Sechi N (2006) Evaluating the effects of protection on two benthic habitats at Tavolara-Punta Coda Cavallo MPA (North-East Sardinia, Italy). Mar Environ Res 61: 171–185.
73. Bertocci I, Dominguez R, Freitas C, Sousa-Pinto I (2012) Patterns of variation of intertidal species of commercial interest in the Parque Litoral Norte (north Portugal) MPA: Comparison with three reference shores. Mar Environ Res 77: 60–70.
74. Shears NT, Kushner DJ, Katz SL, Gaines SD (2012) Reconciling conflict between the direct and indirect effects of marine reserve protection. Environ Conserv 39: 225–236.
75. Guarnieri G, Terlizzi A, Bevilacqua S, Fraschetti S (2009) Local vs regional effects of substratum on early colonization stages of sessile assemblages. Biofouling 25: 593–604.
76. Badalamenti F, Carlo G, D'Anna G, Gristina M, Toccaceli M (2006) Effects of dredging activities on population dynamics of *Posidonia oceanica* (L.) Delile in the Mediterranean sea: The case study of Capo Feto (SW Sicily, Italy). Hydrobiologia 555: 253–261.
77. Gonzalez-Correa JM, Torquemada YF, Lizaso JLS (2008) Long-term effect of beach replenishment on natural recovery of shallow *Posidonia oceanica* meadows. Estuar Coast Shelf S 76: 834–844.
78. Montefalcone M, Albertelli G, Morri C, Parravicini V, Bianchi CN (2009) Legal protection is not enough: *Posidonia oceanica* meadows in marine protected areas are not healthier than those in unprotected areas of the northwest Mediterranean Sea. Mar Pollut Bull 58: 515–519.
79. Montefalcone M, Albertelli G, Morri C, Bianchi CN (2007) Urban seagrass: Status of *Posidonia oceanica* facing the Genoa city waterfront (Italy) and implications for management. Mar Pollut Bull 54: 206–213.
80. Dominguez M, Celdran D, Munoz-Vera A, Infantes E, Martinez-Banos P, et al. (2012) Experimental Evaluation of the Restoration Capacity of a Fish-Farm Impacted Area with *Posidonia oceanica* (L.) Delile Seedlings. Restor Ecol 20: 180–187.
81. Hereu B, Mangialajo L, Ballesteros E, Thibaut T (2008) On the occurrence, structure and distribution of deep-water *Cystoseira* (Phaeophyceae) populations in the Port-Cros National Park (north-western Mediterranean). Eur J Phycol 43: 263–273.
82. Hereu B, Zabala M, Sala E (2008) Multiple controls of community structure and dynamics in a sublittoral marine environment. Ecology 89: 3423–3435.
83. Sales M, Cebrian E, Tomas F, Ballesteros E (2011) Pollution impacts and recovery potential in three species of the genus *Cystoseira* (Fucales, Heterokontophyta). Estuar Coast Shelf S 92: 347–357.
84. Bellan G, Bourcier M, Salen-Picard C, Arnoux A, Casserley S (1999) Benthic ecosystem changes associated with wastewater treatment at Marseille: Implications for the protection and restoration of the Mediterranean coastal shelf ecosystems. Water Environ Res 71: 483–493.
85. Aguado-Gimenez F, Piedecausa MA, Gutierrez JM, Garcia-Charton JA, Belmonte A, et al. (2012) Benthic recovery after fish farming cessation: A "beyond-BACI" approach. Mar Pollut Bull 64: 729–738.
86. Bustamante M, Bevilacqua S, Tajadura J, Terlizzi A, Saiz-Salinas JI (2012) Detecting human mitigation intervention: Effects of sewage treatment upgrade on rocky macrofaunal assemblages. Mar Environ Res 80: 27–37.
87. Boero F (1994) Fluctuations and variations in coastal marine environments. Mar Ecol 15: 3–25.
88. Halpern BS, White C, Lester SE, Costello C, Gaines SD (2011) Using portfolio theory to assess tradeoffs between return from natural capital and social equity across space. Biol Conserv 144: 1499–1507.

8

Does a No-Take Marine Protected Area Benefit Seahorses?

David Harasti[1]*, Keith Martin-Smith[2], William Gladstone[3]

1 Fisheries Research, Marine Ecosystems, NSW Department of Primary Industries, Nelson Bay, New South Wales, Australia, **2** School of Zoology, University of Tasmania, Hobart, Tasmania, **3** School of the Environment, University of Technology, Sydney, New South Wales, Australia

Abstract

Seahorses are iconic charismatic species that are often used to 'champion' marine conservation causes around the world. As they are threatened in many countries by over-exploitation and habitat loss, marine protected areas (MPAs) could help with their protection and recovery. MPAs may conserve seahorses through protecting essential habitats and removing fishing pressures. Populations of White's seahorse, *Hippocampus whitei*, a species endemic to New South Wales, Australia, were monitored monthly from 2006 to 2009 using diver surveys at two sites within a no-take marine protected areas established in 1983, and at two control sites outside the no-take MPA sites. Predators of *H. whitei* were also identified and monitored. *Hippocampus whitei* were more abundant at the control sites. Seahorse predators (3 species of fish and 2 species of octopus) were more abundant within the no-take MPA sites. Seahorse and predator abundances were negatively correlated. Substantial variability in the seahorse population at one of the control sites reinforced the importance of long-term monitoring and use of multiple control sites to assess the outcomes of MPAs for seahorses. MPAs should be used cautiously to conserve seahorse populations as there is the risk of a negative impact through increased predator abundance.

Editor: Pauline Ross, University of Western Sydney, Australia

Funding: The authors have no support or funding to report.

Competing Interests: The authors have declared that no competing interests exist.

* Email: david.harasti@dpi.nsw.gov.au

Introduction

Human uses of the marine environment have caused declines in species worldwide [1]. Over-fishing, pollution, introduction of invasive species, climate change and habitat loss continue to threaten marine species [2]. It has been estimated that the global abundance of marine fishes has declined ~38% between 1970 and 2007 [3] and the IUCN Red List has approximately 800 marine fish species listed as threatened". One group of fishes, the seahorses (*Hippocampus* spp.) of the family Syngnathidae, have 11 species assessed as threatened on the IUCN Red List. In several countries they have been over-harvested for traditional medicines, curios and the aquarium trade and several species face population declines as a result of loss of essential habitats and over-fishing [5,6]. Concerns over the unsustainable trade in seahorses led to them being listed on Appendix II of the Convention on International Trade in Endangered Species (CITES) [6]. Appendix II still allows trade in *Hippocampus* spp.; however, exporting countries must be able to certify that export of seahorses is not causing a decline or damage to wild populations.

Various management options have been proposed or implemented to protect *Hippocampus* spp. in the wild including the application of minimum size limits [7], implementation of temporary fishing closures during recruitment periods [8], the protection of essential habitats [6], providing seahorses with a conservation status prohibiting collection [9], and the implementation of no-take marine protected areas (MPAs) [8,10–12].

The benefits of MPAs for conserving marine biodiversity are well documented [13–16]; however, the potential benefit of MPAs for conserving seahorse populations is relatively unknown. It has been suggested that *Hippocampus* spp. with small over-lapping home ranges would benefit from the creation of small scale no-take MPAs [17] by protecting critical spawning biomasses [18]. The creation of no-take MPAs would also contribute towards conserving seahorse habitats by removing damaging processes, such as destructive fishing practises including dynamite fishing [11] and demersal seine netting [19].

As seahorses are charismatic species that garner considerable public support, it has been suggested they could be used as flagship species to assist with the protection of marine biodiversity around the world [6]. It has been shown that selecting MPAs for estuarine seagrass habitats, based on the density and assemblage variations of syngnathids, would benefit other fish species [20]. Seahorses have been used as a flagship marine species to help establish MPAs in the Philippines; however, the MPAs had no significant effect on seahorse densities and little effect on seahorse size [21]. In this example, the removal of fishing from the MPA did not increase densities of seahorses. This may have been because of poor habitat quality within the MPA, the biology of seahorses, and the small population sizes of seahorses outside the MPA to supply the MPA [21]. Calls for MPAs to be used generally for syngnathid conservation should be treated cautiously. The biological attributes of syngnathids, such as limited movement and strong site fidelity [22], small home range [17], early reproduction [23], and (for some species) lack of a dispersive pelagic larval phase [24], suggest that local populations are likely to respond positively to an MPA. However, there are other reasons why MPAs may not be effective for syngnathids, including specific habitat preferences of all life stages of syngnathids not being met within an MPA [21], habitat changes that follow MPA establishment leading to a decline in the availability of preferred habitat [25–26], larval dispersal by some species limiting opportunities for local recruitment and population replenishment [27], and the build-up of predators within an MPA causing a decline in prey species [25], potentially including syngnathids. In addition, the effectiveness of an MPA for syngnathids might be compromised by activities occurring outside the boundaries that affect habitats within the MPA, such as pollution [21]. To date, apart from Yasué *et al.* (2012), there have been no studies that have specifically tested the effects of an MPA for syngnathids.

The aim of this study was to assess the benefits of no-take MPAs on seahorses. This was done by quantifying the relative abundance of the White's seahorse *Hippocampus whitei* within multiple no-take MPAs and multiple control sites, by identifying and quantifying predators of *H. whitei*, and testing for correlations between the abundance of predators and *H. whitei*. *Hippocampus whitei* is a medium-sized seahorse (maximum length (L_T) of 162 mm) that is considered endemic to several estuaries along the New South Wales (NSW) coast [23] and is protected under NSW fisheries legislation ensuring it cannot be taken from the wild [9]. The species exhibits initial rapid growth, reaches sexually maturity at approximately 6 mo and has a lifespan in the wild of 5–6 yr [23]. It occurs in a range of habitats including artificial structures [29], sponge gardens [24] and seagrass habitats [17].

Materials and Methods

Study Sites

This study was undertaken at four sites near Nelson Bay in the Port Stephens-Great Lakes Marine Park in Port Stephens, NSW, Australia (32°43′04.63″S, 152°08′29.27″E) (Figure 1). Each site was approximately 6000 m^2 and ranged in depth from 2–13 m with a variety of habitat types, such as *Dendronephthya australis* soft coral, *Posidonia australis* seagrass and sponge gardens, located at each of the sites. Two of the sites (Fly Point and Little Beach) are located within the Fly Point Sanctuary Zone, a no-take zone that has been protected since 1983 with all forms of fishing excluded. The other two sites (Pipeline and Seahorse Gardens) are located in a Habitat Protection Zone, which has restrictions on commercial fishing activities such as no trawling whilst fishing and anchoring are permitted, and both are popular fishing locations (personal observations). Habitats across the four sites consisted of sponge, soft coral and seagrass habitats and it was found that there was no significant difference in habitat availability amongst three of the sites (Pipeline, Seahorse Gardens and Little Beach) [30]. Fly Point was found to contain significantly more available habitat for seahorses, as this site had the most extensive sponge garden habitat and the least amount of sand (Harasti *unpublished data*). The research undertaken in this project was done in accordance with NSW DPI Animal Care and Ethics Committee (ACEC) permit 01/05 and University of Newcastle ACEC permit 9610708.

Relative abundance of *H. whitei*

The hypothesis that seahorse abundance would differ between the sanctuary and non-sanctuary sites was tested with data gathered during monthly surveys of each site between January 2006 and December 2009 (n = 48 monthly surveys). Seahorse abundance in each site was assessed with a 60 min random roving diver search [31], which involved the observer (DH) haphazardly swimming over the site searching for seahorses amongst the various habitats while swimming at a constant speed. To minimise problems associated with non-independence, the start and end point varied from survey to survey. When a seahorse was encountered, it was classified as male, female or juvenile. Adult males were determined by the presence of a brood pouch whilst females lacked a brood pouch and were greater than L_T 75 mm. Juveniles were considered less than L_T 75 mm as ~75 mm was found to be the mean size for sexual maturity for *H. whitei* in Port Stephens [23].

Pilot study

To determine if time of day affected the observability of *H. whitei* at the sites, a pilot study was done to test the null hypothesis that *H. whitei* abundance would not differ between day and night, as has been found for another similar sized seahorse *H. comes* that was considered to be easier to detect at night [32]. The pilot study involved conducting a 60 min diver search (as described above) at the site during daylight hours (0700-1700) then followed up by a repeat survey during the night (1800-0600); both dives were done on the high tide approximately 12 hr apart. Surveys were conducted at one of the no-take sanctuary sites (Fly Point) and one of the non-sanctuary sites (Pipeline) with each site being surveyed on six occasions between October and December 2005. Sites were both sampled within 48 hr of each other. The hypothesis that seahorse abundance would not vary between day and night was tested with a 2-factor analysis of variance (ANOVA) with the factor time treated as fixed with two levels (day, night) and the factor site treated as random and orthogonal with two levels. There was no significant difference in the mean abundance of *H. whitei* between night and day surveys ($F_{1,24} = 0.45$, $P > 0.5$), and the time x site interaction was also non-significant ($F_{1,24} = 0.02$, $P > 0.5$). Therefore, time of surveying was considered irrelevant and all sampling occurred between 0600 and 2200.

Predator abundance

During 2006, as part of the 60-min monthly surveys and additional dives at the four locations (N = ~100 dives across four sites), predation events on *H. whitei* were observed and recorded. Species that were classified as predators of *H. whitei* were observed to attack or feed at *H. whitei*. From 2007–2009, during the 60-min monthly abundance surveys (n = 36 monthly surveys), the numbers of predators observed at each site were identified and recorded.

Data analysis

The hypothesis that mean seahorse abundance would differ between the sanctuary zone and non-sanctuary zone sites was tested by 3-factor permutational multivariate analysis of variance (PERMANOVA) using PERMANOVA+1.0.5 within PRIMER-E 6 (Plymouth Routines in Multivariate Ecological Research http://www.primer-e.com/) [33]. The factor Status was analysed as fixed with 2 levels (sanctuary, non-sanctuary), the factor Site was analysed as random with 2 levels and nested in Status, and the factor Year was analysed as random with 4 levels (2006, 2007, 2008, 2009). Each monthly survey was treated as a replicate (n = 12) for each year. The analysis was done on the Euclidean distance similarity matrix with significance determined from n = 9999 permutations. The same 3-factor PERMANOVA design was applied to test the hypothesis that predator abundance would differ between sanctuary and non-sanctuary sites with the factor Year having 3 levels (2007, 2008, 2009). Post-hoc evaluations of significant results were done using pair-wise t-tests. The hypothesis that there would be a relationship between the abundance of seahorses and abundance of predators was tested, using the combined data for all sites from all monthly surveys between 2007 and 2009, by Pearson product-moment correlation coefficient with SPSS 20.

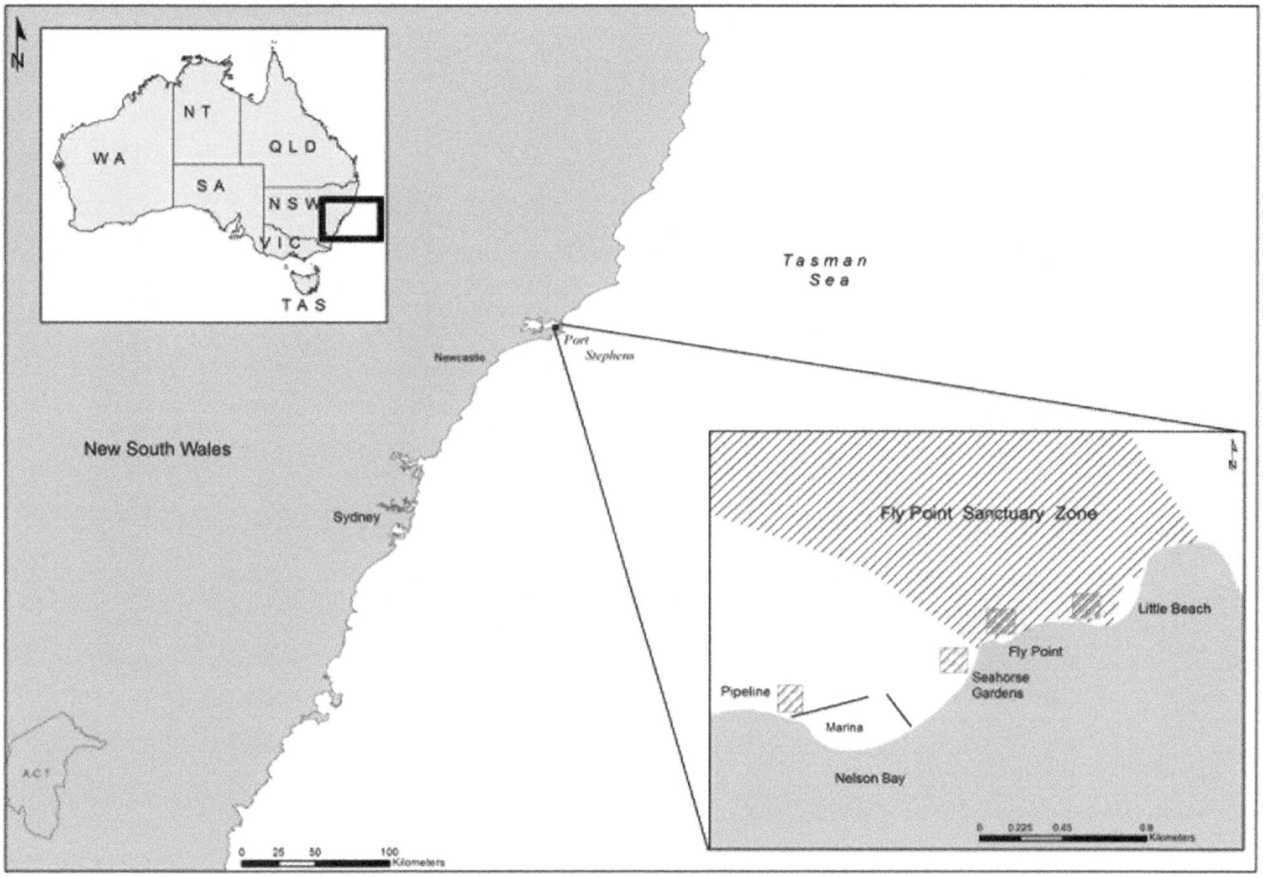

Figure 1. Location of study sites, Port Stephens, New South Wales - Australia.

Results

Relative abundance of *H. whitei*

A grand total of 2,104 *H. whitei* (1953 adult and 151 juvenile) were observed in the monthly surveys from 2006–2009, with 1802 observed in the non-sanctuary zone (control) sites and 302 observed in the sanctuary zone sites. Mean monthly abundance of *H. whitei* in the sanctuary zone (mean 3.1±0.3 S.E.) was significantly less than the non-sanctuary zone (18.8±0.9) (Table 1, Figure 2) therefore the hypothesis that seahorse abundance differed between the sanctuary and non-sanctuary sites was supported. The significant year x site(MPA) interaction occurred because mean seahorse abundance differed between the two non-sanctuary zone sites in some years but not all years and did not differ between the two sanctuary zone sites in any year (Table 1(a)).

Numbers of *H. whitei* varied greatly at the non-sanctuary sites with a large decline in the *H. whitei* population at the Seahorse Gardens in 2007. The decline commenced in October 2006 and continued until March 2007 (Figure 3), during which the monthly mean abundance of *H. whitei* was 4.8±1.8 compared to the mean monthly abundance of 17.2±1.4 for the site across all years. From January to February 2007, 0 adult *H. whitei* and only 1 small juvenile were observed at the Seahorse Gardens. This was the only time across all four sites and all years when no adult seahorses were observed.

Predator abundance

Five different species preyed on *H. whitei* across the four sites. Three species of fish (dusky flathead *Platycephalus fuscus*, eastern red scorpionfish *Scorpaena jacksoniensis*, and striped anglerfish *Antennarius striatus*) and two species of octopus (Sydney octopus *Octopus tetricus* and blue-lined octopus *Hapalochlaena fasciata*), were recorded either attacking or feeding on *H. whitei*. A total of 13 predation events were recorded from 2006 to 2009 (9 in the sanctuary sites and 4 in the non-sanctuary), with the most frequently observed predation events involving *S. jacksoniensis* ($n=5$) and *O. tetricus* ($n=4$). These five species were surveyed monthly from 2007 to 2009 and it was found that the mean number of predators in the sanctuary zone sites (11.4±0.4 S.E.) was significantly greater than the mean number of predators in the non-sanctuary zone sites (3.5±0.3) (Table 1 (b), Figure 4). Therefore, the hypothesis that predator abundance differed between the sanctuary and non-sanctuary sites was supported. The significant Site(MPA) effect occurred because mean predator abundance differed between sites in the non-sanctuary zone but not between the two sites in the sanctuary zone. The most abundant predators in the sanctuary zone sites were *S. jacksoniensis, O. tetricus* and *P. fuscus* (Figure 5). There was a significant, negative correlation between monthly seahorse abundance and predator abundance ($r=-0.69$, $n=144$, $P<0.001$; Figure 6), with high abundance of predators associated with lower abundance in seahorses.

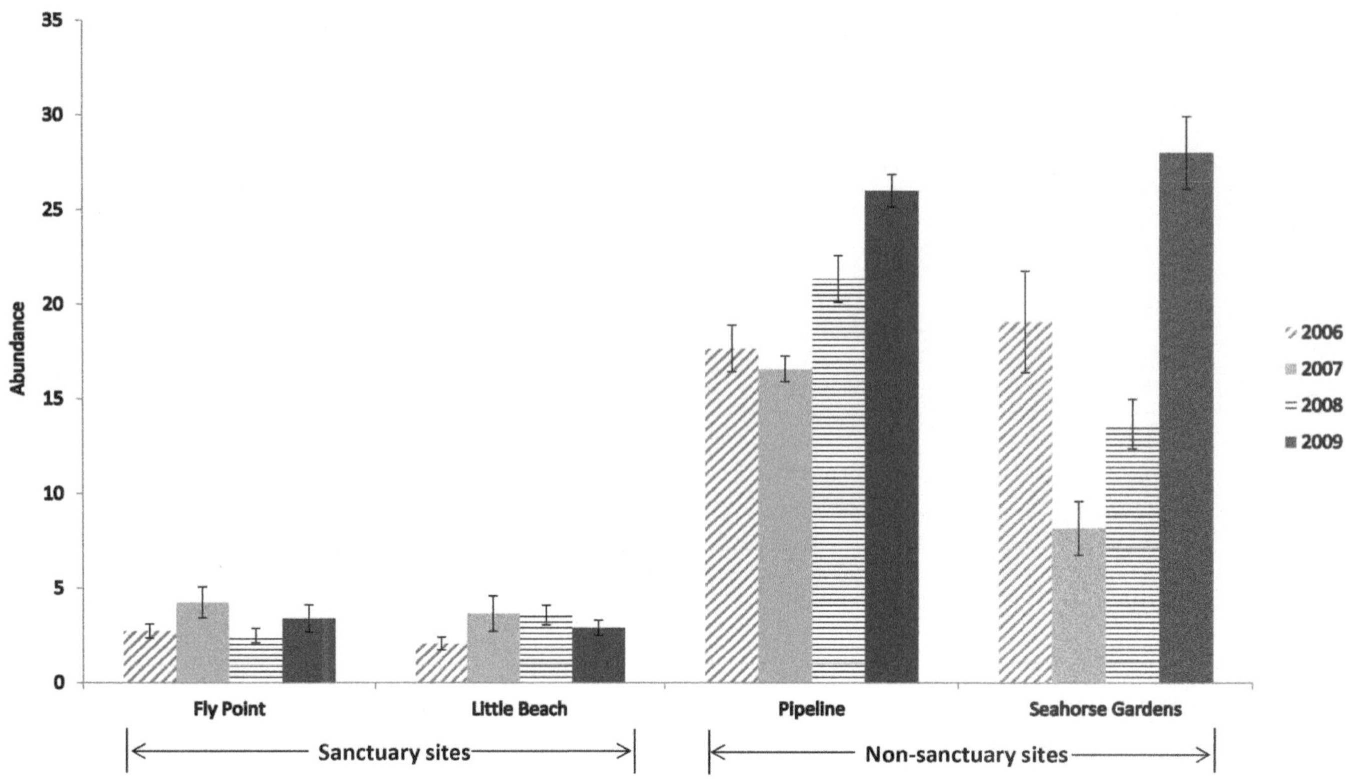

Figure 2. Mean monthly abundance of *H. whitei* (± S.E.) at four sites within Port Stephens for 2006–2009.

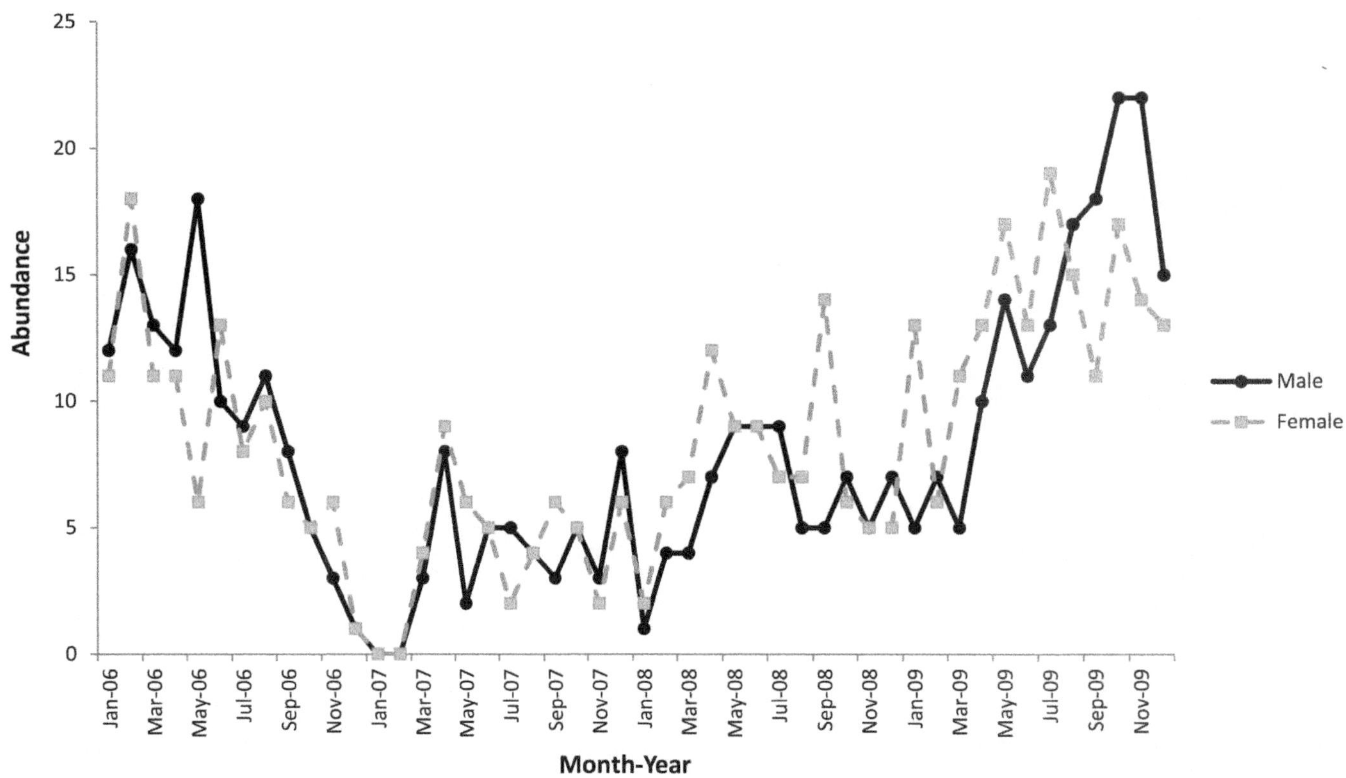

Figure 3. Monthly abundance of adult *H. whitei* recorded in 60 min dive surveys from 2006–2009 at the non-sanctuary site Seahorse Gardens, Port Stephens.

Table 1. Summary of hypotheses tested in the long term monitoring of *H. whitei* and predator abundance within and outside a marine protected area (MPA) with details of statistical analysis performed and PERMANOVA results.

Hypotheses	Source	df	MS	Pseudo-F	P(perm)
(a) Seahorse abundance would vary between sanctuary zone sites and non-sanctuary zone sites and years.	MPA site (fixed) Year x Site(MPA)	1	11781	19.67	0.012
	Year (random)	3	409.54	4.19	0.067
	Site(MPA) (random)	2	120.67	1.24	0.358
	MPA x Year	3	483.21	4.95	0.052
	Year x Site(MPA)	6	97.61	6.01	0.001
	Residual	191			
(b) Predator abundance would vary between sanctuary zone sites and non-sanctuary zone sites and years.	MPA site (fixed)	1	2272.11	36.24	0.006
	Year (random)	2	13.03	1.99	0.24
	Site(MPA) (random)	2	61.18	9.33	0.04
	MPA x Year	2	1.69	0.26	0.79
	Year x Site(MPA)	4	6.55	2.17	0.09
	Residual	143			

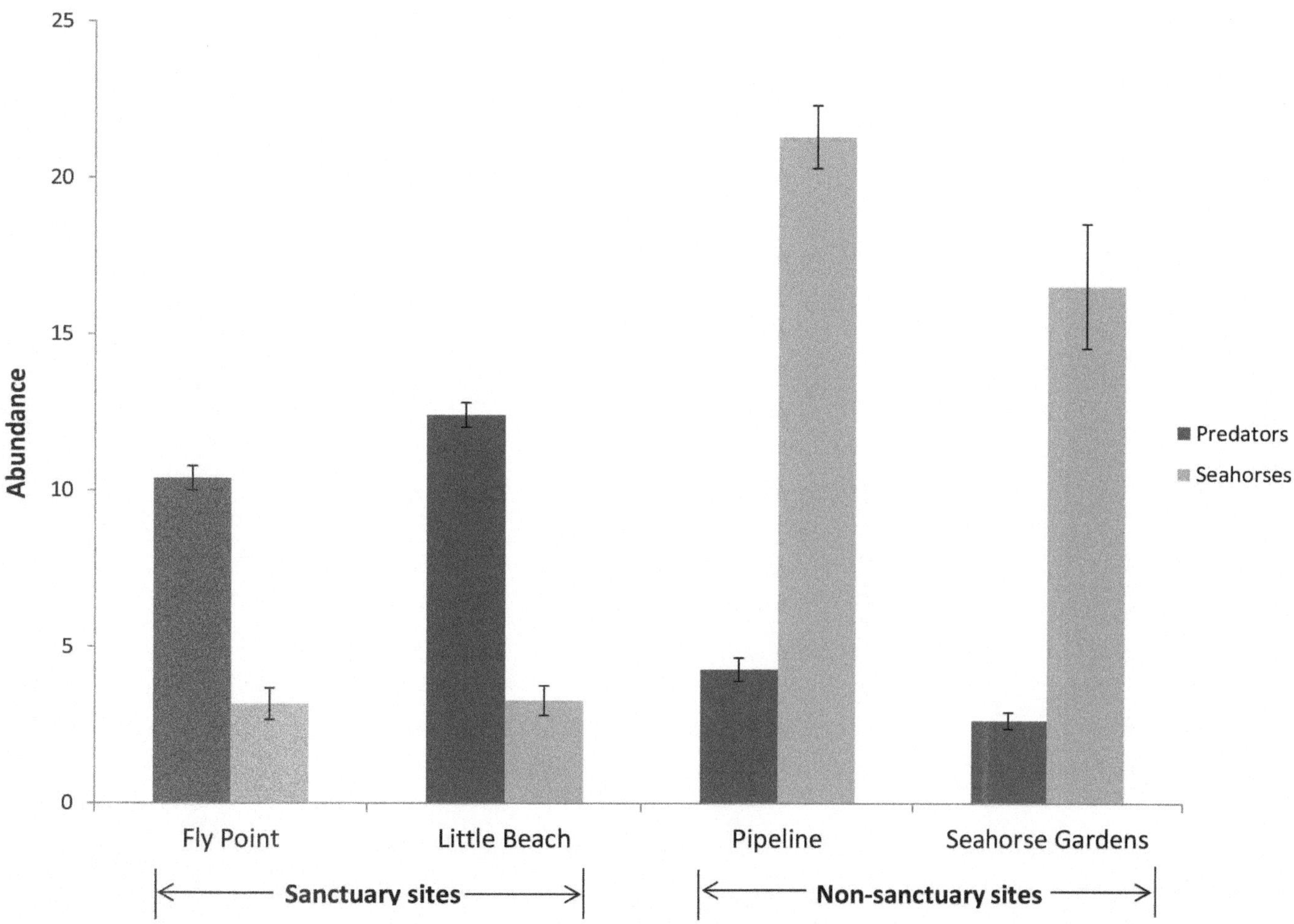

Figure 4. Monthly mean abundance (± S.E.) for *H. whitei* and predators (fish and octopus) for each site from 2007–2009.

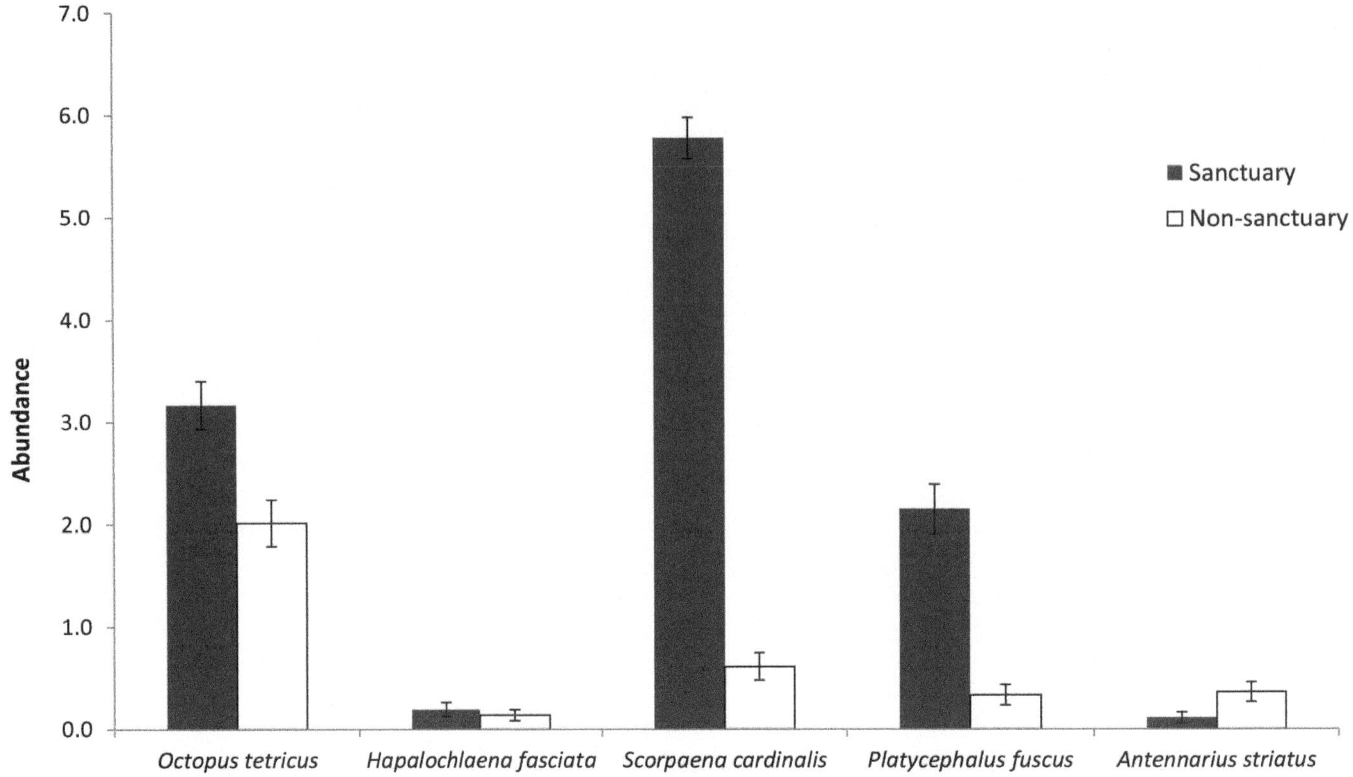

Figure 5. Monthly mean abundance (± S.E.) in 2007–2009 of seahorse predators at two sites within the Fly Point Sanctuary Zone and at two sites outside the Sanctuary Zone.

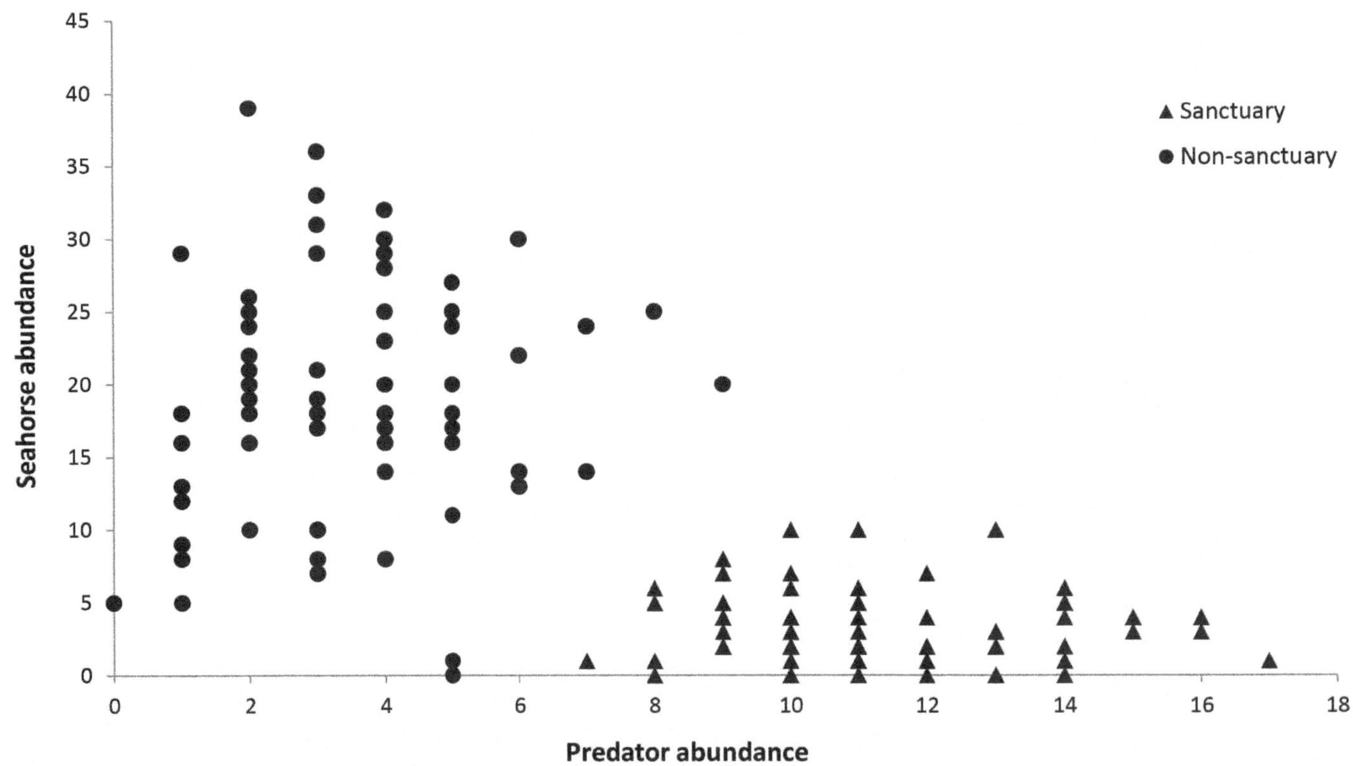

Figure 6. Relationship between monthly seahorse abundance and predator abundance from 2007–2009 at each site.

Discussion

Seahorse abundance

The main finding of this study that the abundance of seahorses was significantly lower within a no-take MPA, compared with sites open to fishing, was unexpected as it has been suggested that seahorse abundance would benefit from small-scale no-take MPAs [10–12,17]. The most likely cause of this was the greater risk of predation in the no-take sanctuary zone sites, as suggested by the greater abundance of seahorse predators in these sites and the negative correlation between predator abundance and seahorse abundance. In a study on marine reserves in the Philippines, it was found that syngnathid abundance was lower within the no-take MPA compared to the fished sites outside; however, the difference was not significant and possibly confounded by differences in habitat [34]. In this example, the authors acknowledged that it was difficult to determine if observed effects were real responses to changed increases within the protected MPA (e.g. increased predator abundance) or other factors such as increased species visibility in the non-protected MPA sites as result of less structurally-complex habitats [34].

It is unlikely that the significant difference in *H. whitei* abundance between the no-take sites and control sites was related to habitat differences. Habitat types did not differ between sites and Fly Point, one of the sanctuary zone sites, had the greatest coverage of sponge habitat; a known habitat for *H. whitei* [24]. The observed differences between the no-take and control sites are also unlikely to have been confounded by differences in the detectability of seahorses. Although seahorse detectability may differ among habitat types, there is no evidence that the occurrence of cryptic behaviour among seahorses differed amongst the sites (Harasti *unpublished data*).

One of the often-stated goals of MPAs is the preservation of areas with species and assemblages occurring in an undisturbed state, at least from the exclusion of fishing pressure, for the benefit of scientific research, education and public awareness [35,36]. Seahorses and other syngnathids are charismatic species that attract support for marine conservation [6,37]. The findings of this study suggest that seahorses might not benefit from the use of MPAs for marine conservation; however, the finding of this study linking decreased seahorse abundance with increased predator abundance is based on correlative evidence. There are no data available on the abundances of seahorse and their predators prior to the establishment of the no-take MPA. To test and validate this finding, field experiments are needed to determine the actual rates of seahorse predation between sites closed and open to fishing, and to determine other predator species [38,39].

Predators of Hippocampus whitei

As seahorses are a slow-moving species, they rely on crypsis through colour changes and algal-like filaments that mimic their habitat, to avoid predation [24,40]. Eighty-two predators of syngnathids are known, including fishes, turtles, sea birds, invertebrates and marine mammals [41]. None of the predators recorded in this study were included in the [41] review, with the only recorded predator of *H. whitei* in the literature being the little penguin *Eudyptula minor* whilst [24] observed the striated frogfish *Antennarius striatus* predating on *H. abdominalis*, a seahorse known to occur in the same region as *H. whitei* [29].

The cephalopod *Octopus tetricus* and the scorpionfish *Scorpaena jacksoniensis* are believed to the most frequent predators of *H. whitei* as they were responsible for the majority of observed predation events and were the two most abundant predators. However, given the large diversity and size of fishes found within the Fly Point no-take MPA site [42], there are potentially other predators of *H. whitei* that were not detected. During monthly surveys from 2008–2009, both snapper *Pagrus auratus* and leatherjacket *Nelusetta ayraudi* were observed to attack *H. whitei* following their release after being handled underwater (for measuring or tagging as part of other studies). This occurred if the seahorse swam away from the holdfast it was placed on after handling. However, there were no observations of either species attacking *H. whitei* that had not been 'disturbed'. Another cephalopod species, the mourning cuttlefish *Sepia plangon*, was observed to prey on juvenile *H. whitei* on two separate occasions within the sanctuary zone; however, it was not included in the monthly predator surveys as the observations occurred in 2008 and 2009, prior to the predator study reported here.

Predator abundance

Seahorse predators were more abundant within the no-take sanctuary zone sites, which is similar to findings of other studies from around the world that have reported greater abundance and/or biomass of predator fishes in areas protected from fishing [34,43–45]. The three most abundant predator species (*Platycephalus fuscus*, *Octopus tetricus* and *Scorpaena jacksoniensis*) were more abundant within the sanctuary zone and are considered to be important recreational and commercial species that are targeted by fishers in NSW [46], therefore these species are likely to benefit from the exclusion of fishing. Additionally, data collected from baited underwater remote video systems has found that the Fly Point sanctuary zone has greater diversity and larger fish species than the non-sanctuary zone sites (NSW DPI *unpublished data*). This is also supported by the findings of Edgar *et al.* (2009) that demonstrated that Fly Point was high in fish biomass and in density of larger fish species. The increased numbers of predators within the sanctuary zone sites is not surprising, as the sanctuary zone has been protected for 30 years (since 1983) with no fishing allowed, and numerous studies have shown that fish biomass and density increased over time within MPA's [13,15,42–45]. With the implementation of MPAs, there will be 'winners and losers', with some species benefiting from protection by increases in size and abundance [47,48]. Other species showed no change in abundance or abundance decreased as a result of increased predation and interspecific competition [28,49,50], particularly the smaller cryptic fishes [34,51,52]. Protected areas have been shown to help promote recovery of predatory species [43,53], which potentially can have indirect negative effects on prey species in the protected areas [28].

Whilst this study suggests that *H. whitei* has been negatively impacted by a no-take protected area most likely through increased predation, other species of seahorse and other syngnathids might be affected in different ways by MPAs. Species' responses to MPAs will depend on a range of factors including the availability of preferred habitats, potential predators in the area and factors occurring outside an MPA.

Decline in *Hippocampus whitei* abundance

Population estimates and monthly relative abundance data show that *Hippocampus whitei* populations across the four sites in Port Stephens were stable with the exception of the Seahorse Gardens, which experienced a large population decline in late 2006. As the species is protected and not exploited by fishing, such an abrupt decline is unusual and the cause of the decline is unknown. Population declines in *Hippocampus sp.* in the absence of fishing pressure have been recorded elsewhere, with *H. abdominalis* populations declining 79–98% over 3 years [54] and populations of *H. guttulatus* and *H. hippocampus* declining by

94% and 73%, respectively over a seven year period [55]. Given that the decline of *H. whitei* in this study occurred only at the Seahorse Gardens site, it is unlikely that the decline can be attributed to ecosystem-wide stressors such as disease or environmental variables as populations at the nearby three sites should also have been affected. Throughout the study, there was no noticeable change in currents or water temperature at the Seahorse Gardens. These two variables were, however considered to influence changes in seahorse abundance in Ria Formosa lagoon [55]. There were also no recorded incidents of illegal collecting, nor was the site affected by trawling, netting or dredging which are prohibited in the area. Seahorse predator abundance at the site did not increase during the study, nor were there observations of increases in other species that may prey on *H. whitei*.

A potential hypothesis for the decline is that the seahorses may have moved off the site into deeper water; however, *H. whitei* is known to have small home ranges and displays site fidelity [17,22]. Support for the movement hypothesis is that several seahorses tagged at the site in 2006, disappeared during the decline period, but started to be resighted again from late 2007 until 2010 [23]. Numerous exploratory dives were undertaken in the deeper water (12–18 m) surrounding the Seahorse Gardens site from 2006–2009; however, no *H. whitei* were encountered deeper than 12 m (maximum depth of study site), so the location to which seahorses might have migrated is unknown. With such a population decline there is concern that reproduction would be reduced as a result of Allee effects [56], especially with the high level of monogamy displayed by *H. whitei* [23], as mature animals could find it difficult to find a mate. Although there was a rapid decrease in population abundance, the actual recovery of the population to almost pre-decline levels occurred within three years with the highest number of juveniles at the site occurring in 2009. As *H. whitei* is considered an R-selected species with rapid growth, early age at maturity and sexually mature at approximately six months [23], the species has the potential to repopulate a site if sufficient breeding adults return to the site, or recruitment from adjacent sites is successful.

Conclusions

This study illustrates the importance of long-term monitoring of seahorse populations as it was shown that seahorse numbers varied considerably over a 12-month period. Long-term monitoring of multiple sites is necessary for a good understanding of seahorse population changes in the wild and allows for better assessment on the status of seahorse populations. The study indicates that caution should be used when investigating the use of MPAs to conserve seahorse populations as there is potential for negative impacts on seahorse abundance through potentially increased predator abundance. Other management interventions may be more suitable such as entire protection of the seahorse species, removal of destructive fishing practises that damage essential habitats, restoration of natural habitats or creation of artificial habitats. A range of management measures are needed to conserve threatened populations of seahorses and the declaration of a marine protected area may not be the ideal solution.

Acknowledgments

Pam and Chris Norman, formerly of Pro Dive Nelson Bay, provided great support with numerous airfills. Thanks to Chris Gallen for developing Fig. 1 and Dr Tim Glasby for advice on statistical analysis. We are grateful to the editor and two anonymous reviewers who provide feedback on the initial draft of this manuscript.

Author Contributions

Conceived and designed the experiments: DH KMS WG. Performed the experiments: DH KMS WG. Analyzed the data: DH WG. Contributed reagents/materials/analysis tools: DH KMS WG. Wrote the paper: DH.

References

1. Butchart SHM, Walpole M, Collen B, van Strien A, Scharlemann PW Jr, et al. (2010) Global biodiversity: indicators of recent declines. Science 328: 1164–1168.
2. McClenachan L, Cooper AB, Carpenter KE, Dulvy NK (2010) Extinction risk and bottlenecks in the conservation of charismatic marine species. Conservation Letters 5: 73–80.
3. Hutchings JA, Minto Ci, Ricard D, Baum JK, Jensen OP (2010) Trends in the abundance of marine fishes. Canadian Journal of Fisheries and Aquatic Sciences 67: 1205–1210.
4. IUCN (2013) 2013 IUCN Red List of Threatened Species. Available at: http://www.iucnredlist.org/. Accessed 15 January 2014.
5. Foster SJ, Vincent ACJ (2004) Life history and ecology of seahorses: implications for conservation and management. Journal of Fish Biology 65: 1–61.
6. Vincent ACJ, Foster SJ, Koldewey HJ (2011) Conservation and management of seahorses and other Syngnathidae. Journal of Fish Biology 78: 1681–1724.
7. Foster SJ, Vincent ACJ (2005) Enhancing sustainability of the international trade in seahorses with a single minimum size limit. Conservation Biology 19: 1044–1050.
8. Vincent ACJ, Meeuwig JJ, Pajaro MG, Perante NC (2007) Characterizing a small-scale, data-poor, artisanal fishery: Seahorses in the central Philippines. Fisheries Research 86: 207–215.
9. DPI N (2005) Seahorses and their relatives. Available: http://www.dpi.nsw.gov.au/fisheries/species-protection/protected-species/marine-or-estuarine-species/syngnathiformes. Accessed 15 January 2014.
10. Martin-Smith KM, Samoilys MA, Meeuwig JJ, Vincent ACJ (2004) Collaborative development of management options for an artisanal fishery for seahorses in the central Philippines. Ocean & Coastal Management 47: 165–193.
11. Marcus JE, Samoilys MA, Meeuwig JJ, Villongco ZAD, Vincent ACJ (2007) Benthic status of near-shore fishing grounds in the central Philippines and associated seahorse densities. Marine pollution bulletin 54: 1483–1494.
12. Morgan S, Vincent ACJ (2013) Life-history reference points for management of an exploited tropical seahorse. Marine and Freshwater Research 64: 185–200.
13. Halpern BS (2003) The impact of marine reserves: do reserves work and does reserve size matter? Ecological applications 13: 117–137.
14. Roberts CM, Bohnsack JA, Gell F, Hawkins JP, Goodridge R (2001) Effects of marine reserves on adjacent fisheries. Science 294: 1920–1923.
15. Babcock RC, Shears NT, Alcala AC, Barrett NS, Edgar GJ, et al. (2010) Decadal trends in marine reserves reveal differential rates of change in direct and indirect effects. Proceedings of the National Academy of Sciences 107: 18256–18261.
16. Lester SE, Halpern BS, Grorud-Colvert K, Lubchenco J, Ruttenberg BI, et al. (2009) Biological effects within no-take marine reserves: a global synthesis. Marine Ecology Progress Series 384: 33–46.
17. Vincent ACJ, Evans KL, Marsden AD (2005) Home range behaviour of the monogamous Australian seahorse, *Hippocampus whitei*. Environmental Biology of Fishes 72: 1–12.
18. Curtis JMR, Vincent ACJ (2006) Life history of an unusual marine fish: survival, growth and movement patterns of *Hippocampus guttulatus* Cuvier 1829. Journal of Fish Biology 68: 707–733.
19. Curtis JMR, Ribeiro J, Erzini K, Vincent ACJ (2007) A conservation trade-off? Interspecific differences in seahorse responses to experimental changes in fishing effort. Aquatic Conservation: Marine and Freshwater Ecosystems 17: 468–484.
20. Shokri MR, Gladstone W, Jelbart J (2009) The effectiveness of seahorses and pipefish (Pisces: Syngnathidae) as a flagship group to evaluate the conservation value of estuarine seagrass beds. Aquatic Conservation: Marine and Freshwater Ecosystems 19: 588–595.
21. Yasué M, Nellas A, Vincent ACJ (2012) Seahorses helped drive creation of marine protected areas, so what did these protected areas do for the seahorses? Environmental Conservation 39: 183–193.
22. Harasti D, Gladstone W (2013) Does underwater flash photography affect the behaviour, movement and site persistence of seahorses? Journal of Fish Biology 83: 1344–1353.
23. Harasti D, Martin-Smith K, Gladstone W (2012) Population dynamics and life history of a geographically restricted seahorse, *Hippocampus whitei*. Journal of Fish Biology 81: 1297–1314.
24. Kuiter RH (2009) Seahorses and their relatives. Seaford: Aquatic Photographics. 334 p.

25. Babcock RC, Kelly S, Shears NT, Walker JW, Willis TJ (1999) Changes in community structure in temperate marine reserves. Marine Ecology Progress Series 189: 125–134.
26. Shears NT, Babcock RC, Salomon AK (2008) Context-dependent effects of fishing: variation in trophic cascades across environmental gradients. Ecological Applications 18: 1860–1873.
27. Morgan SK, Vincent ACJ (2007) The ontogeny of habitat associations in the tropical tiger tail seahorse Hippocampus comes Cantor, 1850. Journal of Fish biology 71: 701–724.
28. Graham NAJ, Evans RD, Russ GR (2003) The effects of marine reserve protection on the trophic relationships of reef fishes on the Great Barrier Reef. Environmental Conservation 30: 200–208.
29. Harasti D, Glasby TM, Martin-Smith KM (2010) Striking a balance between retaining populations of protected seahorses and maintaining swimming nets. Aquatic Conservation: Marine and Freshwater Ecosystems 20: 159–166.
30. Harasti D, Martin-Smith K, Gladstone W (in press) Ontogenetic and sex-based differences in habitat preferences and site fidelity of the White's seahorse Hippocampus whitei. Journal of Fish Biology.
31. Kingsford M, Battershill C (1998) Studying temperate marine environment-s: a handbook for ecologists. Christchurch: Canterbury University Press. 335 p.
32. Perante NC, Pajaro MG, Meeuwig JJ, Vincent ACJ (2002) Biology of a seahorse species, *Hippocampus comes* in the central Philippines. Journal of Fish Biology 60: 821–837.
33. Anderson MJ (2001) A new method for non-parametric multivariate analysis of variance. Austral Ecology 26: 32–46.
34. Samoilys MA, Martin-Smith KM, Giles BG, Cabrera B, Anticamara JA, et al. (2007) Effectiveness of five small Philippines coral reef reserves for fish populations depends on site-specific factors, particularly enforcement history. Biological conservation 136: 584–601.
35. Allison GW, Lubchenco J, Carr MH (1998) Marine reserves are necessary but not sufficient for marine conservation. Ecological applications 8: S79–S92.
36. Agardy T, Bridgewater P, Crosby MP, Day J, Dayton PK, et al. (2003) Dangerous targets? Unresolved issues and ideological clashes around marine protected areas. Aquatic Conservation: Marine and Freshwater Ecosystems 13: 353–367.
37. Scales H (2009) Poseidon's steed: the story of seahorses, from myth to reality. New York: Gotham Books. 272 p.
38. Mislan KAS, Babcock RC (2008) Survival and behaviour of juvenile red rock lobster, Jasus edwardsii, on rocky reefs with varying predation pressure and habitat complexity. Marine and Freshwater Research 59: 246–253.
39. Bassett DK, Jeffs AG, Montgomery JC (2009) Identification of predators using a novel photographic tethering device. Marine and Freshwater Research 59: 1079–1083.
40. Schmid MS, Senn DG (2002) Seahorses - Masters of adaptation. Vie Et Milieu-Life and Environment 52: 201–207.
41. Kleiber D, Blight LK, Caldwell IR, Vincent ACJ (2011) The importance of seahorses and pipefishes in the diet of marine animals. Reviews in Fish Biology and Fisheries 21: 205–223.
42. Edgar GJ, Barrett NS, Stuart-Smith RD (2009) Exploited reefs protected from fishing transform over decades into conservation features otherwise absent from seascapes. Ecological applications 19: 1967–1974.
43. Williamson DH, Russ GR, Ayling AM (2004) No-take marine reserves increase abundance and biomass of reef fish on inshore fringing reefs of the Great Barrier Reef. Environmental Conservation 31: 149–159.
44. Ashworth J, Ormond R (2005) Effects of fishing pressure and trophic group on abundance and spillover across boundaries of a no-take zone. Biological Conservation 121: 333–344.
45. Currie JC, Sink KJ, Le Noury P, Branch GM (2012) Comparing fish communities in sanctuaries, partly protected areas and open-access reefs in South-East Africa. African Journal of Marine Science 34: 269–281.
46. DPI N (2013) Species - Common recreational saltwater. Available: http://www.dpi.nsw.gov.au/fisheries/info/nsw-fish-species#Common-saltwater-recreational-species. Accessed 15 January 2014.
47. Barrett NS, Edgar GJ, Buxton CD, Haddon M (2007) Changes in fish assemblages following 10 years of protection in Tasmanian marine protected areas. Journal of Experimental Marine Biology and Ecology 345: 141–157.
48. Watson DL, Anderson MJ, Kendrick GA, Nardi K, Harvey ES (2009) Effects of protection from fishing on the lengths of targeted and non-targeted fish species at the Houtman Abrolhos Islands, Western Australia. Marine Ecology Progress Series 384: 241–249.
49. Watson D, Harvey E, Kendrick G, Nardi K, Anderson M (2007) Protection from fishing alters the species composition of fish assemblages in a temperate-tropical transition zone. Marine Biology 152: 1197–1206.
50. Götz A, Kerwath SE, Attwood CG, Sauer WH (2009) Effects of fishing on a temperate reef community in South Africa 1: ichthyofauna. African Journal of Marine Science 31: 241–251.
51. Willis TJ, Anderson MJ (2003) Structure of cryptic reef fish assemblages: relationships with habitat characteristics and predator density. Marine Ecology Progress Series 257: 209–221.
52. Edgar GJ, Stuart-Smith RD (2009) Ecological effects of marine protected areas on rocky reef communities-a continental-scale analysis. Marine Ecology Progress Series 388: 51–62.
53. Willis TJ, Millar RB, Babcock RC (2003) Protection of exploited fish in temperate regions: high density and biomass of snapper Pagrus auratus (Sparidae) in northern New Zealand marine reserves. Journal of Applied Ecology 40: 214–227.
54. Martin-Smith KM, Vincent ACJ (2005) Seahorse declines in the Derwent estuary, Tasmania in the absence of fishing pressure. Biological Conservation 123: 533–545.
55. Caldwell IR, Vincent ACJ (2012) Revisiting two sympatric European seahorse species: apparent decline in the absence of exploitation. Aquatic Conservation: Marine and Freshwater Ecosystems 22: 427–435.
56. Kramer A, Dennis B, Liebhold A, Drake J (2009) The evidence for Allee effects. Population Ecology 51: 341–354.

9

The Crowded Sea: Incorporating Multiple Marine Activities in Conservation Plans Can Significantly Alter Spatial Priorities

Tessa Mazor[1]*, Hugh P. Possingham[1,2], Dori Edelist[3], Eran Brokovich[4], Salit Kark[1]

1 ARC Centre of Excellence for Environmental Decisions, School of Biological Sciences, The University of Queensland, Brisbane, Queensland, Australia, **2** Grand Challenges in Ecosystems and the Environment, Silwood Park, Imperial College, London, United Kingdom, **3** Leon Recanati Institute for Maritime Studies, Department of Maritime Civilizations, University of Haifa, Mount Carmel, Haifa, Israel, **4** Department of Geography, The Hebrew University of Jerusalem, Mount Scopus, Jerusalem, Israel

Abstract

Successful implementation of marine conservation plans is largely inhibited by inadequate consideration of the broader social and economic context within which conservation operates. Marine waters and their biodiversity are shared by a host of stakeholders, such as commercial fishers, recreational users and offshore developers. Hence, to improve implementation success of conservation plans, we must incorporate other marine activities while explicitly examining trade-offs that may be required. In this study, we test how the inclusion of multiple marine activities can shape conservation plans. We used the entire Mediterranean territorial waters of Israel as a case study to compare four planning scenarios with increasing levels of complexity, where additional zones, threats and activities were added (e.g., commercial fisheries, hydrocarbon exploration interests, aquaculture, and shipping lanes). We applied the marine zoning decision support tool Marxan to each planning scenario and tested a) the ability of each scenario to reach biodiversity targets, b) the change in opportunity cost and c) the alteration of spatial conservation priorities. We found that by including increasing numbers of marine activities and zones in the planning process, greater compromises are required to reach conservation objectives. Complex plans with more activities incurred greater opportunity cost and did not reach biodiversity targets as easily as simplified plans with less marine activities. We discovered that including hydrocarbon data in the planning process significantly alters spatial priorities. For the territorial waters of Israel we found that in order to protect at least 10% of the range of 166 marine biodiversity features there would be a loss of ~15% of annual commercial fishery revenue and ~5% of prospective hydrocarbon revenue. This case study follows an illustrated framework for adopting a transparent systematic process to balance biodiversity goals and economic considerations within a country's territorial waters.

Editor: Elliott Lee Hazen, UC Santa Cruz Department of Ecology and Evolutionary Biology, United States of America

Funding: This research was conducted with the support of funding from the Australian Research Council Centre of Excellence for Environmental Decisions. T. M was a recipient of a University of Queensland Library OA Award, was funded by an Australian Postgraduate Award and gratefully acknowledges the financial support of the Australian-Israel Scientific Exchange Foundation (AISEF). The funders had no role in study design, data collection and analysis, decision to publish, or preparation of the manuscript.

Competing Interests: The authors have declared that no competing interests exist.

* Email: t.mazor@uq.edu.au

Introduction

Implementing marine conservation plans is a major challenge. Plans that determine priority areas for conservation are often based solely on biological and ecological information [1] One of the main factors inhibiting the uptake of marine conservation plans by decision makers is inadequate consideration of the broader social and economic context within which conservation operates [2–4]. Marine waters and their biodiversity are shared by a host of stakeholders and interest groups, such as commercial fishers, recreational users and offshore developers [5]. Inclusion of the activities of these multiple marine users within conservation plans is critical for achieving plans which are realistic and achievable in the real world, thereby moving from paper to action [2].

Conservation planners must try to explicitly consider other marine activities within conservation plans, to ensure no time is wasted over trying to conserve areas essential for other uses [6]. Competition for ocean space is becoming increasing intensified as resource extraction and developments are expanding to include the marine realm [7]. Offshore activities such as commercial fishing, aquaculture facilities, sand mining, desalination plants, offshore wind farms and offshore power plants, provide countries with substantial economic gains [5]. Currently, hydrocarbon operations are one of the largest economic stakeholders in the sea [8], and provide countries with huge potential and realized monetary benefits, and are expected to increase economic and political independence [9,10]. However, incorporation of such economic activities is often absent from marine conservation planning literature. Despite the little willingness for countries to protect marine areas that are deemed economically important

[11], excluding other marine activities in conservation planning means we may not be able to design a marine reserve network that is representative or economically viable [12].

Disregarding other marine activities in marine conservation planning may also mean that anthropogenic threats to biodiversity are being ignored. When planning marine reserves that aim to reap sustainable long-term benefits it is important to examine the threats to biodiversity of the system that could impair this goal. However, reserve planning should not be solely based upon threat data [13]. Examples of threats to biodiversity for consideration in reserve planning include: shipping lanes which pose a collision risk to marine mammals [14], trawlers and demersal longliners which are damaging to benthic environments and responsible for the majority of annual sea turtles deaths via by-catch [15], and marine energy installations which have been linked to habitat loss, noise pollution and invasive species [16]. In some cases marine users have made changes or modifications, such as altering the path of shipping lanes for cetaceans [17]. However, in cases where compromises cannot be met, conservation planners must be able to incorporate the potential threats to biodiversity into the planning process.

A common misconception is that marine zoning itself is a conservation planning tool. Marine zoning is the allocation of particular activities to specified marine areas [5,18]. This practice can help reduce user conflict by separating incompatible activities [7,18,19]. Several countries have stepped up to implement zoning strategies for their waters, the largest and perhaps most successful example of marine zoning is the Great Barrier Reef Marine Park off the coast of Queensland, Australia [20,21]. More recent zoning efforts occurring around the globe include the United Kingdom Irish Sea Pilot [22], the Belgian Exclusive Economic Zone [11], the waters of Norway [18], Australia's entire commonwealth waters [23] and the zoning of China's territorial sea [24]. However, key elements are often missing from some zoning plans to ensure biodiversity goals are met. For marine zoning to be used as an appropriate method or tool for protecting marine biodiversity it must enable an explicit consideration of the trade-off between biodiversity and socio-economic objectives [25]. Furthermore, zoning plans need to ensure that the zoning system provides protection that is representative of as many biodiversity features as possible [12,25].

The concept of including other activities within marine conservation planning is slowly emerging. Unlike marine spatial planning (MSP) which aims to plan water spaces to meet objectives of multiple marine users and stakeholders, [19,26], marine conservation planning (MCP) is centred on one primary goal - achieving biodiversity protection [18]. Recently, several systematic conservation plans in the marine realm have focused on a hybrid approach; reaching conservation objectives while also minimizing the opportunity cost to fishery stakeholders [25,27,28]. However, only some of these plans have been expanded to other social and economic contexts (e.g., [3,29]). Facilitating the inclusion of other activities into marine conservation planning is the emerging development of zoning software that enables multiple objectives to be considered (e.g., Marxan with Zones [27]). Up to now there has been little application of these new tools to address the complexity of marine conservation planning at regional scales or an entire country scale. As many countries around the globe aim to implement conservation measures by zoning their waters [18], it is important to develop an explicit zoning process which integrates the current spatial occupancy of other marine activities and where possible their economic objectives. The inclusion of other marine uses in marine conservation planning means that we need to carefully consider the trade-offs that underpin the resulting conservation plans and ensure that biodiversity goals are adequately achieved.

In this study we follow a framework (Fig. 1) using a systematic approach for zoning territorial waters to achieve the protection of marine biodiversity in the face of multiple anthropogenic threats and economic activities. Within this context, we aim to test how increased complexity (by the inclusion of zones, multiple activities and economic factors) in marine conservation planning alters: a) the ability to reach biodiversity targets, b) the opportunity cost, and c) the spatial conservation priorities. Furthermore, we aim to examine the explicit incorporation of prospective hydrocarbon extraction into marine conservation planning [8].

Methods

Here our methods follow the steps outlined in Figure 1.

Spatial setting and study area

As a case study, we examined Israel's complete Mediterranean territorial waters. Israel is located in the eastern Mediterranean Sea and has relatively small territorial waters (~4200 km^2) compared with other coastal countries around the world. Currently, it faces rapid exploitation of its marine resources and aims to expand its protection of marine biodiversity [30]. Israel's Mediterranean Sea territorial waters are defined by the National Planning Authority of Israel and are used by The Israel Nature and Parks Authority (NPA) for marine reserve planning. The territorial waters of Israel's Mediterranean Sea spreads along a coastline ~190 km long, and extends outwards for 12 nautical miles from the coast to a depth of ~1000 m, covering an area of ~4200 km^2 [30]. For our analyses, we divided this study area into 1×1 km planning units, resulting in a total of 4,205 planning units.

Compiling biodiversity features

In order to select marine areas which will fulfil a representative reserve network where all types of biodiversity are protected we compiled available distribution data of Israel's Mediterranean territorial waters of biotic and abiotic features. These included 166 biodiversity features, comprising of vertebrate marine species (153 fishes, 2 turtles, 1 cetacean), and 10 geomorphologic features (Fig. 2a; see Table S1 in File S1. for a list of species and features included in this study).

Marine species distribution data. We compiled data from currently available published studies on native cartilaginous and bony fishes whose distribution lies within Israel's Mediterranean waters [31–38]. All native (non-alien) fish species (153 species) present in these publications were included in our study. We digitized the documented depth ranges of these native fish species using ArcGIS ([39]; Fig. S1 in File S1.; Table S2 in File S1.) and sea floor bathymetry [40], following methods in Tognelli et al. [41] and Clark & Tittensor [42]. We derived the distributions via a number of sources; locations and depth ranges from the above eight studies, data from the Hebrew University of Jerusalem Fish Collection (accessed 2012), ranges as documented in Golani et al. [43], and by expert opinion (for further details see File S2).

The distribution of sea turtles within Israel's marine waters has not been well documented and their preferred feeding, foraging and mating areas are currently poorly known. Therefore, we used the locations of established nesting sites (within [44] in Israel for both the green (*Chelonia mydas*) and loggerhead (*Caretta caretta*) sea turtle species. The targeted nesting habitats for protection in this study were chosen as planning units adjacent to nesting beaches with over 20 nest counts (from 1993–2011) and a persistence of more than five years of nesting at a particular site, in

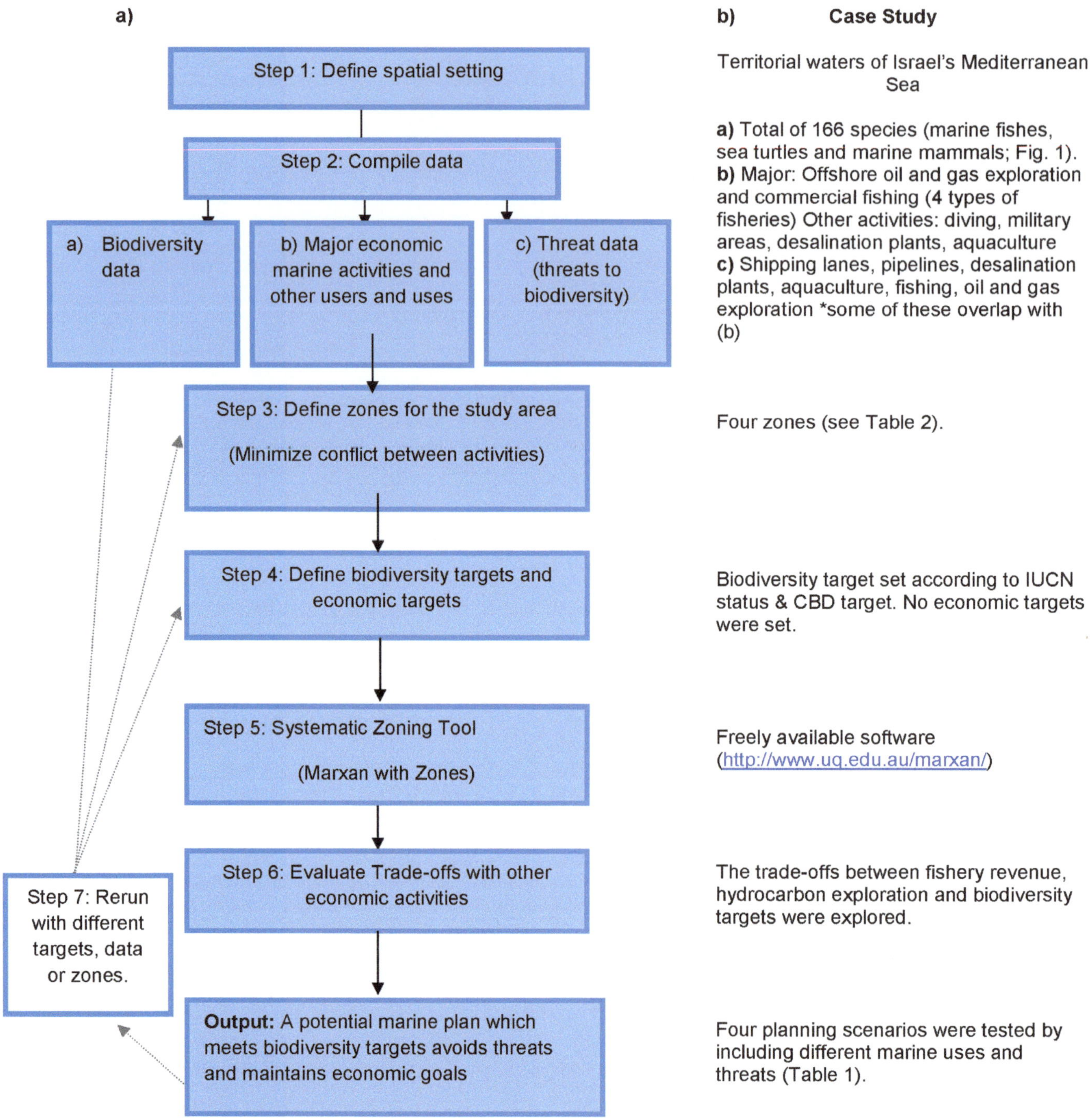

Figure 1. Proposed framework for incorporating multiple activities and threats into marine conservation planning. These show the steps followed in the case study presented in this paper that encompasses Israel's entire Mediterranean territorial waters.

accordance with expert opinion from rangers and scientists at Israel's Nature and Parks Authority and Sea Turtle Rescue Centre.

We included the distribution of the common bottlenose dolphin (*Tursiops truncatus*), the most common cetacean species in Israel's territorial waters. Other cetacean species exist in Israel's waters but not enough observational data exists to determine priority habitats for these species. The common bottlenose dolphin has been sighted throughout Israel's territorial waters, therefore to better direct our conservation efforts we have considered important habitat areas as the species distribution. Scheinin [45], identified three core areas for feeding and foraging, an area at a depth of 40–50 m near Ashkelon, an area at a depth of 30–60 m between Ashdod and Palmachim beaches and another area off the coast of Netanya at a depth of 90–120 m. These three core habitat areas cover in total 213.64 km^2 (for additional information see File S2.).

Geomorphological features. In order to represent different types of marine habitats we included geomorphologic features to serve as surrogate "biodiversity features". We used ten geomor-

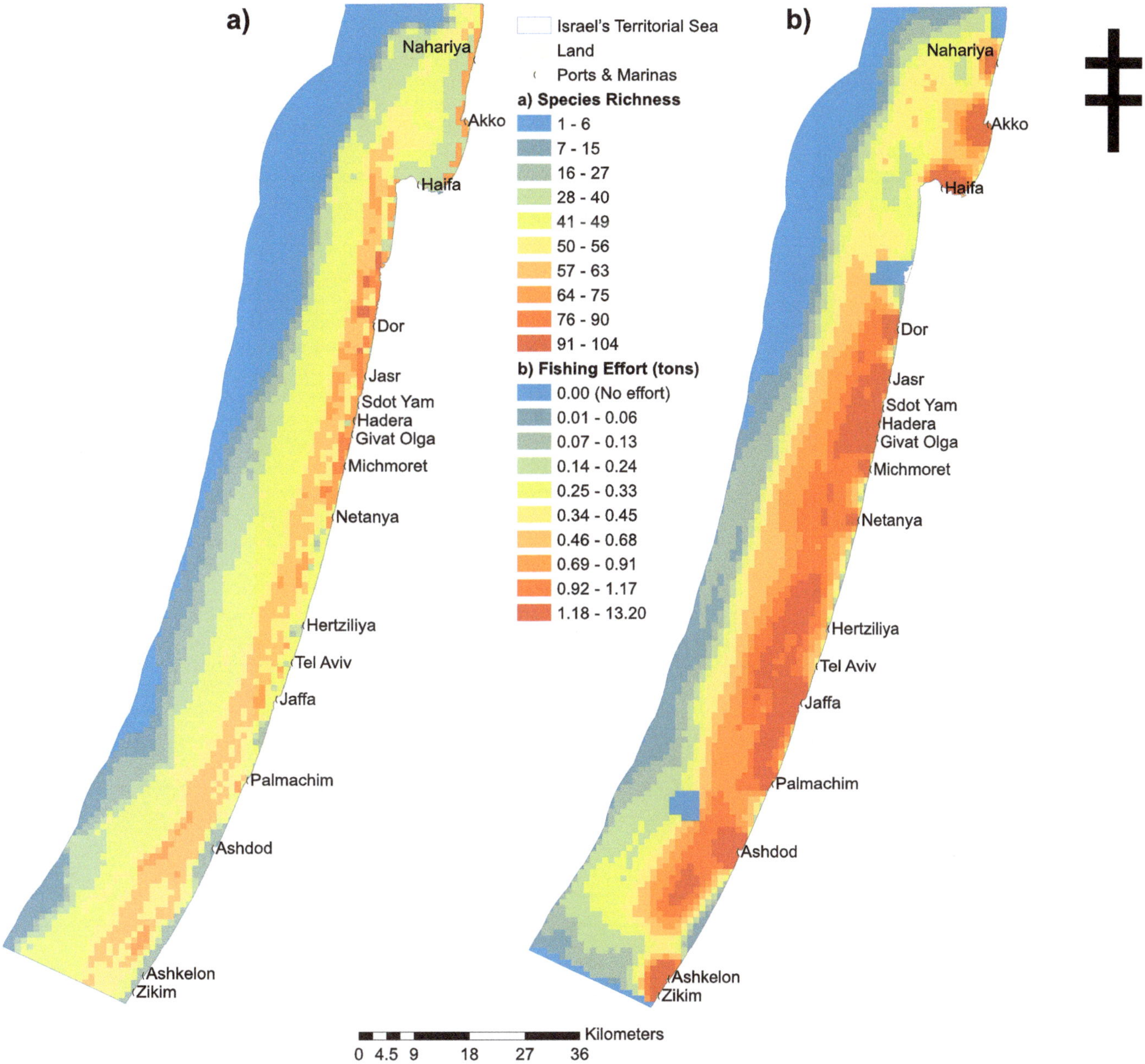

Figure 2. Biodiversity features and fishing effort in Israel's Mediterranean Sea territorial waters; a) species richness of 166 biodiversity features (species and geomorphologic features), b) combined fishing effort (entangling nets, longliners, purse seiners and trawlers), where the blue areas (no effort) are restricted fishing areas; marine reserves, military areas and aquaculture.

phologic features within Israel territorial waters that were mapped (in 2008) and provided by The Israel Nature and Parks Authority. These features include: shallow rocks, kurkar (calcareous aeolianite) ridges, kurkar bustan, deep kurkar ridges, continental shelf silt, continental shelf sand, continental ridges, large canyons, continental slope and canyons, deep sea [46].

Setting biodiversity targets. Biodiversity targets were set to protect a percentage of the species distribution according to its level of global threat based on the IUCN red list criteria ([46], Table S1 in File S1.) and current range size. We set a 10% target for species that were listed "Least Concern" by the IUCN and all other fish species that have not been evaluated by IUCN. This target was increased to 15% for species listed "Vulnerable" by the IUCN [47] and to 20% for species listed "Endangered" by the by the IUCN. Species listed "Endangered" that had a distribution of less than 1% of the study area were given a target of 50%. For the geomorphological features, we set a target to protect 5% of all features and those that are represented by an area less than 1% of Israel's territorial waters were given a 10% target. We also set a constraint that at least 5% of the distribution of all species and features must be placed within the no-take zone (Conservation Zone), meaning that the rest of the biodiversity target could be fulfilled in other zones. While our target setting approach does not consider whether the target is adequate at conserving the species or maintaining population viability, it aims to address the IUCN criteria that defines the risk of species extinction as applied in Kark et al. [48] and Lieberknecht et al [49]. To test the sensitivity of our results we also used a 10% target for each species and a 5% target for each geomorphologic feature.

Incorporating economic activities in the sea

We included the two major economic activities in the Mediterranean waters of Israel (commercial fishing and hydrocarbon operations) in the conservation planning exercise. While there are other localized marine activities and features (addressed below; Table 1) commercial fishing and hydrocarbon operations are activities that span across Israel's territorial waters and rely on resource extraction. Thus, we focused on these activities which are likely to be the main source of opportunity cost incurred when implementing marine protected areas and zones. We translated these activities into opportunity cost layers for use within Marxan. Opportunity cost in this study was defined as the value of forgone economic activities (commercial fishing and hydrocarbon operations) when a particular area (planning unit) is made into a protected area that excludes these economic activities. As spatial opportunity cost data were unavailable for these activities, we developed surrogates to represent the annual revenue (approximation of annual opportunity cost) of each economic activity within our 1 km^2 planning units. Here we used annual values to reflect the relative opportunity cost differences across the territorial waters of Israel. We used the most current available data for Israel's territorial waters for each of these activities, specifically the year 2009 for commercial fisheries and year 2012 for hydrocarbon operations. The minimal fluctuation of Israel's annual commercial fishing catch and value over the last few years suggests that the available data of these activities is relatively comparable [50].

Opportunity cost of commercial fisheries. We developed surrogate opportunity cost layers of commercial fishing by spatially mapping fishing effort for the four major commercial fishing gears used in Israel; entangling nets, longliners, purse seiners and trawlers (see Fig. S2 in File S1; Fig. S3 in File S1; File S2. for detailed methods). We derived effort maps by equations which assume effort is proportional to the number of fishing vessels at each port for each gear type and effort decreases exponentially with distance from port (methods described in Mazor et al. [51]). For each gear type we used expert opinion (total of 25 experts) to refine our effort layers. We did this by constraining our effort layer by the maximum depth that each fishing gear is used and incorporating weightings over habitats and areas that are targeted by particular gear types. For entangling nets we constrained our effort layer by a depth of 50 m (maximum depth that entangling nets are used in Israel's as confirmed by 15 entangling net fishers in Israel). For longliners, fishing effort was weighted by both distance from port and rocky habitat (targeted fishing areas) and confined to 50 m depth (confirmed by 6 longline fishers in Israel; Fig. S3 in File S1.). For purse seiners, effort was weighted across two distinct areas in the north and south at a depth between 10–50 m as determined by expert opinion (6 purse seine fishers; Fig. S2 in File S1.). Trawling effort was based on data collected from on-board GPS devices by Edelist [38] between the years 2009–2011 and trawling data from Israel's Department of Fisheries and Aquaculture ([52]; Fig. S2 in File S1.) Using these effort maps we created surrogate opportunity cost layers by overlaying the annual revenue (year 2009) reported by Edelist et al. [50] for each fishing gear type, thereby assigning monetary values to each planning unit for each fishing gear type (Fig. 2b).

Opportunity cost of hydrocarbon operations. Spatial data identifying offshore oil and gas operations and leased and licensed marine extraction areas was provided by Israel's Ministry of Interior from the National Master Plan of Israel (Tama 34b). Areas of Israel's Mediterranean waters are licensed to several oil and gas companies (e.g., Noble Energy, Shemen, Delek) for hydrocarbon exploration for a period of seven years [53]. If economically viable resources are found within these licensed areas they can then be leased by energy companies with a fifty year

Table 1. Four zones for Israel's territorial waters that restrict and permit different activities.

Activities	Zones			
	Conservation Zone "No-Take"	Benthic Protection Zone	Exploration Zone	Economic Zone "General use"
Trawling	X	X	X	✓
Purse Seiners	X	✓	X	✓
Gillnetting	X	✓	X	✓
Long liners	X	✓	X	✓
Oil and Gas Exploration	X	X	✓	✓
Additional threats and marine activities				
Aquaculture	X	X	X	✓
Current protected areas[1]	✓	X	X	X
Desalination plants	X	X	X	✓
Diving	✓	✓	✓	✓
Military areas	✓	✓	✓	✓
Pipelines	X	X	✓	✓
Safety area[2]	X	X	X	X
Shipping lanes	X	✓	X	✓

[1]Rosh HaNikra
[2]Mari B Platform
Additional threats and marine activities (listed below) in Israel's territorial waters have been locked to particular zones as per the four scenarios.
A "✓" in the column means that this activity was permitted in this zone, where an "x" it is prohibited.

production permit. Unexplored "blank" areas will be temporarily left aside as Israel is trying to limit exploration into these new areas. The licensed areas that were not explored will be recycled if there is no exploration in them.

As no reliable data sources were available for a total estimation of the value of Israel's offshore oil and gas reserves we performed calculations using data from Israel's Ministry of Energy and Water Resources [53,54] (Table S3 in File S1.) and converted these estimated reserve quantities into monetary values. We multiplied the annual average international market price of oil (NIS per barrel = 404.52 in 2012; World Bank http://www.worldbank.org/) and natural gas (NIS per thousands of cubic meters = 399.33; International Monetary Fund http://www.imf.org/external/index.htm) with Israel's estimated reserve volumes. These calculations resulted in a static estimate (year 2012 values) of the value of Israel's oil and gas reserves (not including extraction cost), but we realize that prices will fluctuate annually and are expected to reach higher values in the future, thus our calculated values are expected to be an under estimate (unless estimated reservoirs will be smaller than predicted). We have estimated the value of Israel's offshore oil and gas reserves at US$ ~324 billion (~1,250 billion NIS; Table S3 in File S1.), with 15% of this amount retrieved from the territorial waters (US$ 50 billion). Our resulting equation gives a greater weighting to the opportunity cost of leased areas (known sources of oil and gas; $\alpha = 1$) compared to licensed areas (half weighting $\alpha = 0.5$):

$$Cost\ of\ one\ exploration\ unit(EU) =$$

$$\frac{\alpha(Area\ of\ EU)}{\sum Area\ of\ all\ EU} * Value\ of\ oil\ and\ gas(US\$),$$

$$where,\ leased\ areas\ \alpha = 1\ and\ licensed\ areas\ \alpha = 0.5,$$

$$and,\ the\ Cost\ of\ each\ planning\ unit(PU) =$$

$$\frac{Area\ PU}{Area\ EU} * Value\ of\ EU\ unit(US\$).$$

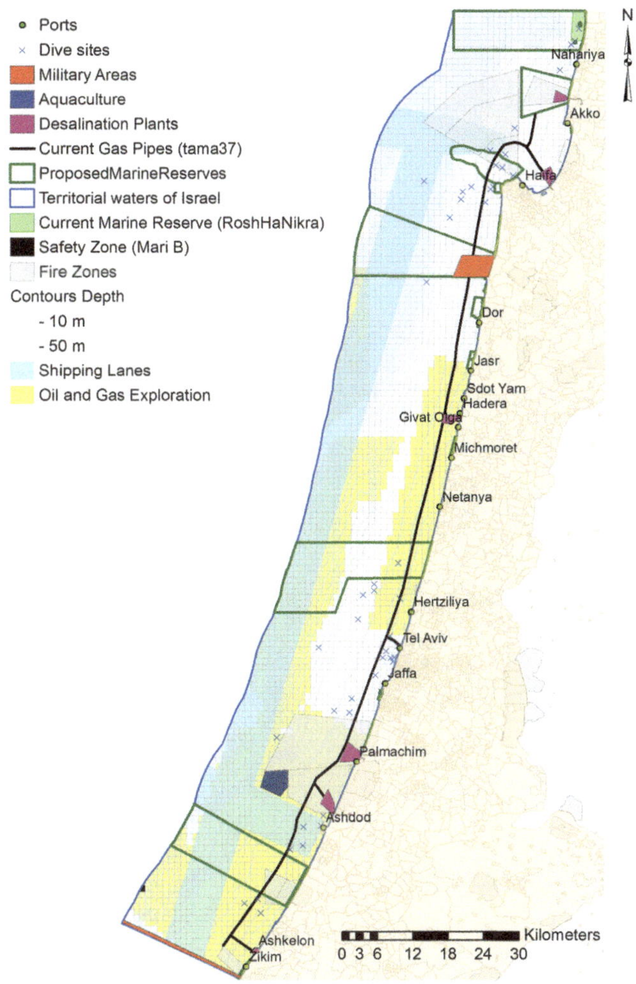

Figure 3. A map of the activities of Israel's Mediterranean territorial waters included in this study.

Considering additional marine activities in conservation planning. There are many features to consider when planning marine conservation within territorial waters. Israel has a relatively small territorial water area with a large number of marine activities (Fig. 3). In addition to the fishing and hydrocarbon operations (included as opportunity cost) we included eight additional marine activities. These include: aquaculture, desalination plants, dive sites, current protected areas, exploration safety zone (500 m buffer around hydrocarbon exploration sites), military areas (fire zones), shipping lanes and pipelines (Table 1; see File S2. for a full description of these activities and their data sources). We included these other activities by assigning their usage to specific zones (see Table 1).

Systematic planning tools and planning scenarios

Marxan with Zones is a conservation decision-support tool that enables the user to prioritize places for different zones to achieve multiple objectives [25,27]. This tool is an extension of Marxan [55], a globally used conservation planning tool for marine and terrestrial realms [27]. Marxan works by minimizing one variable (e.g., the opportunity cost of commercial fishing), creating a system that is separated into areas which are protected or non-protected [25]. In comparison, Marxan with Zones can minimize multiple variables (i.e. incorporating more than two opportunity cost layers) and enables the user to develop a more complex system of zones that provide varying degrees of protection and have zone specific actions, objectives and restrictions [27].

We applied Marxan [55] and Marxan with Zones [27] to compare four planning scenarios for Israel's Mediterranean territorial waters (see Table 2). For each planning scenario we aimed to meet the same biodiversity targets while minimizing the opportunity cost incurred by other marine activities, as described below. The four scenarios increase in complexity with the inclusion of human activities (threats) and economic objectives; Simple Planning, Basic Zoning, Intermediate Zoning and Complex Zoning (Table 2). We define the term "activities" as any other activity within Israel's marine waters that is not biodiversity protection as proposed in this study. For the first scenario, Simple Planning, we used Marxan (without zoning) and tested two sub-scenarios; Simple Planning A with six activities and commercial fishing opportunity cost, and Simple Planning B with seven activities and combined commercial fishing and hydrocarbon extraction opportunity cost. Our second scenario, Basic Zoning, used Marxan with Zones and included three zones and six other activities. The third scenario, Intermediate Zoning, used Marxan with Zones and included four zones and seven activities (three sub-scenarios A, B and C for protection effectiveness of the Exploration Zone; see File S2. for full explanation). In the fourth scenario, Complex Zoning, we used Marxan with Zones with four

Table 2. Four planning scenarios that were examined using Marxan and Marxan with Zones for Israel's territorial Mediterranean waters.

		Planning Scenarios						
		1. Simple Planning		2. Basic Zoning	3. Intermediate Zoning			4. Complex Zoning
		A	B		A	B	C	
Planning Tool:	Marxan	+	+					
	Marxan with Zones			+	+	+	+	+
Zones	Conservation Zone (No-Take)			+	+	+	+	+
	Economic Zone (General Use)			+	+	+	+	+
	Benthic Protection Zone			+	+	+	+	+
	Exploration Zone (% of effectiveness at protecting species[3])				+ (25%)	+ (50%)	+ (75%)	+ (50%)
Marine activities *(Included as opportunity cost)*[4]	Commercial Fisheries: Trawlers, Purse Seiners, Gill Nets, Long liners	+	+	+	+	+	+	+
	Hydrocarbon Operations		+		+	+	+	+
(Assigned to specific zones)[5]	Aquaculture	+	+	+	+	+	+	+
	Current Protected Areas	+	+	+	+	+	+	+
	Diving areas	+	+	+	+	+	+	+
	Military areas	+	+	+	+	+	+	+
	Safety platform	+	+	+	+	+	+	+
	Shipping lanes							+
	Desalination plants							+
	Pipelines							+
Number of Zones		**0**		**3**	**4**			**4**
Total number of marine activities included in the analysis		**6**	**7**		**7**			**10**

[3]We assigned possible percentages of biodiversity protection that may be achieved by the Exploration Zone. The Conservation Zone assumes 100% protection of .biodiversity but due to the unknown impacts of hydrocarbon exploration we tested different values (25%, 50%, and 75%). See File S2. for further details.
[4]The opportunity cost layers are the variables where are minimized in Marxan software: In Marxan this is treated as one minimized cost layer (summed together), and in Marxan with Zones these opportunity cost layers are separately minimized.
[5]See Table 1 for the zones that each marine activity is permitted or restricted within and File S2. for a detailed explanation of each activity and their data references.
The inclusion of features and data in each scenario is represented by a plus sign (+). Planning scenarios increase (from the Simple Planning to Complex Zoning scenario) in complexity by the planning tool, zones and number of activities included. For more detailed information on each of the zones see Table 1.

zones (for descriptions of each zone see Table 1) and ten other activities. For a detailed description of each scenario see File S2.

Comparing planning scenarios

Four planning scenarios (Table 2) were compared. The Simple Planning scenario (without zoning) was run using Marxan and the other three scenarios (Basic Zoning, Intermediate Zoning and Complex Zoning) used Marxan with Zones, all scenarios with 1,000 runs each. Based on the results of the 1,000 runs we calculated the average opportunity cost and number of 1 km^2 planning units within each zone that were needed to meet our biodiversity targets. We tested the ability of planning scenarios to meet all biodiversity targets. In cases where targets were unable to be reached for a particular species, we eliminated the constraint for 5% of their protection to be met in the Conservation Zone. Thus, we re-ran our results with the same altered targets for all scenarios. We then mapped the selection frequency outputs (number of time a planning unit is selected in Marxan for a particular zone) for each planning scenario and each zone. To compare between zoning configurations and scenarios we also mapped the best solution that Marxan could find. To test the similarity between the selection frequency outputs for each scenario we used the Spearman Rank Correlation (ρ). Higher values indicate a more similar spatial pattern in selection frequencies, meaning that these plans will require similar conservation actions.

Evaluating trade-offs

We evaluated the trade-off between meeting biodiversity targets and maximizing annual fishery revenue for each of the four fisheries in Israel's Mediterranean Sea, following methods described in Klein et al. [25]. These trade-offs can only be evaluated for scenarios using Marxan with Zones that enables multiple variables to be considered. We set fishery targets where we aimed to preserve an equal percentage of the total fishing revenue (from the fishing effort maps) for each of the four fishery gear types. These targets could only be met within zones that did not restrict that type of fishery (Table 1). Expanding this analysis, we tested the trade-off with areas that are leased and licensed for oil and gas (using the hydrocarbon opportunity cost layer described above). We therefore included a hydrocarbon target (preserving hydrocarbon industry revenue) as well as both biodiversity and fishery targets. Fishery targets were extracted from the previous trade-off analysis; the highest target where all biodiversity targets were met.

Results

Comparing planning scenarios for territorial waters

Here we compared our four planning scenarios (Table 2) by: a) the ability to reach biodiversity targets, b) the change in opportunity cost and c) the alteration of spatial conservation priorities.

a) Biodiversity targets. We found that meeting the same biodiversity targets became more difficult as our planning scenarios included more marine activities. The Simple Planning (without zoning and six activities) and Basic Zoning (three zones and six activities) scenarios met all biodiversity targets, the Intermediate Zoning scenario (four zones and seven activities) met 98% of targets and the Complex Zoning scenario (four zones and ten activities) met 96% of targets (Table 3). Our constraint (5% target in the Conservation Zone – no-take area) was unable to be met in the Intermediate and Complex Zoning scenarios for nine species (Table S4 in File S1.) that had restricted distribution ranges that overlapped with prospective hydrocarbon exploration areas. For each of these nine species we eliminated the constraint; however the overall biodiversity target for these nine species remained and was met within other zones. Targets were then able to be met for all planning scenarios.

b) Opportunity cost. We found that more complex planning scenarios incurred greater opportunity cost (Table 3). When comparing the two Simple Planning scenarios (Simple Planning A with six users and commercial fishing opportunity cost, and Simple Planning B with seven users and combined commercial fishing and potential hydrocarbon opportunity cost) we found that a reserve network that only included the opportunity cost of fishing had a substantially lower cost (Simple Planning A = US$2.05 million) compared to a plan that included the opportunity cost of hydrocarbon operations (Simple Planning B = US$595,132.38 million). Comparing our zoning scenarios (when targets are met 100% in each scenario) we found that the most expensive zoning scenario is the Intermediate Zoning scenario A that assumes the Exploration Zone can provide a zone effectiveness measure of twenty-five percent. This opportunity cost decreased as the Exploration Zone's ability to protect biodiversity (zone effectiveness) was increased to fifty percent (Intermediate Zoning scenario B 10.5% cost decrease) and seventy-five percent (Intermediate Zoning scenario C 14.9% cost decrease); allowing targets to be met more easily within the Exploration Zone. The Intermediate Zoning scenario increased opportunity cost by 27.8% from the Basic Zoning scenario (three zones and six activities) as we introduced the opportunity cost of prospective oil and gas reserves as well as a fourth zone (Exploration Zone). The Complex Zoning scenario also increased the opportunity cost of the Basic Zoning plan by 35.7% and Intermediate Zoning B plan by 6.2%.

c) Conservation priorities. Selection frequency outputs from our analysis indicated that spatial configurations are substantially altered by the inclusion of hydrocarbon opportunity cost (Fig. 4; Fig. 5). The scenarios Simple Planning A (without zoning and six activities) and Basic Zoning (three zones and six users), which did not include hydrocarbon opportunity cost had a high Spearman's rank correlation of $\rho = 0.84$ ($p<0.001$; Table 4).

We found that priority areas for no-take reserves (Conservation Zone) were mainly concentrated in the north and south of Israel's territorial waters. From the best solution outputs (Fig. 6) no-take areas moved from areas in the south to areas in the north with the inclusion of potential hydrocarbon extraction data. Similarly, the three scenarios that included hydrocarbon opportunity cost (Simple Planning B, Intermediate Zoning and Complex Zoning) had selection frequency outputs that were significantly correlated (Table 4). The most similar spatial outputs were between Simple Planning B and Intermediate Zoning B ($\rho = 0.86$, $p<0.001$; Table 4). In these three scenarios, we discovered that spatial priorities were much more distinct (higher spectrum of selection frequency; see Fig. 4) than the Simple Planning A and Basic Zoning scenarios. Areas with high selection frequency for placing no-take reserves were off the coast of Jaffa, in coastal waters between Dor and Haifa Bay and along northern border with Lebanon (Fig. 4).

Priority areas for each zone become more pronounced as planning scenarios became more complex and restricted by the inclusion of other marine activities. Conservation priorities for the Benthic Zone were most similar between the Intermediate Zoning B and Complex Zoning ($\rho = 0.82$, $p<0.001$; Table 4). In all scenarios we find that Benthic Protection Zone has higher selection frequency in the northern part of the Sea. In the best solution outputs we also notice how benthic protection becomes confined to the north with the inclusion of the Exploration Zone

Table 3. Results showing average opportunity cost for 1,000 Marxan runs for each planning scenario.

Planning Scenario	Opportunity cost (US$ million)	Percent of targets met	Percent of conservation zone in entire reserve design (no-take areas)
Simple Planning A	2.05	100	22
Simple Planning B	595,132.38	100	21
Marxan with Zones			
Basic Zoning	4.09	100	22
Intermediate Zoning B 50%	333,004.37	98	17
Complex Zoning	4.20	96	14
Marxan with Zones (minus nine species for the 5% Conservation Zone target)			
Basic Zoning	3.92	100	22
Intermediate Zoning A 25%	5.59	100	18
Intermediate Zoning B 50%	5.01	100	17
Intermediate Zoning C 75%	4.76	100	17
Complex Zoning	5.32	100	14

The constraint/target that 5% of the distribution of all features needs to be within the Conservation Zone (no-take zone) was unable to be reached for nine species. This constraint was removed for these species so targets could all be met. This table shows the opportunity cost of each planning scenario, the percentage of biodiversity targets met in the scenario and the percentage of "no-take area" surface coverage of the entire reserve system. For a description of planning scenarios see Table 2. Targets were set according to IUCN criteria and the size of a species distribution range (as described in the methods section).

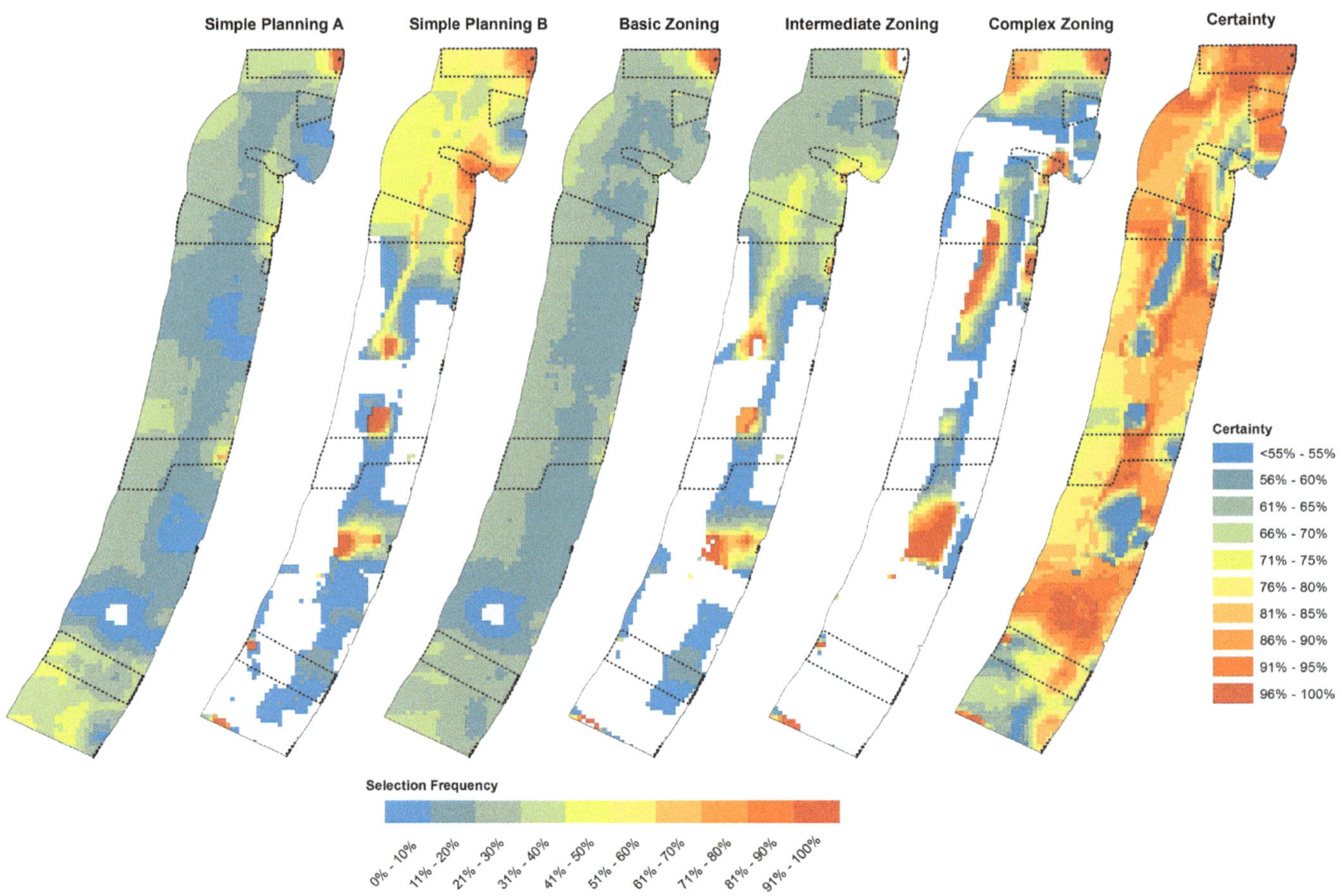

Figure 4. Selection frequency output maps (shows the percentage of times a planning unit was selected when run in Marxan 1000 times) from Marxan with Zones for each Zone and each zoning scenario. All scenarios meet biodiversity targets. The dashed black lines represent the proposed marine reserve system by Israel's Nature and Parks Authority [45]. The certainty map expresses the level of certainty/ agreement of planning units selected (either highly selected for no-take areas or low selection) across all planning scenarios. Therefore, the higher the percentage of certainty means there is more agreement between scenarios.

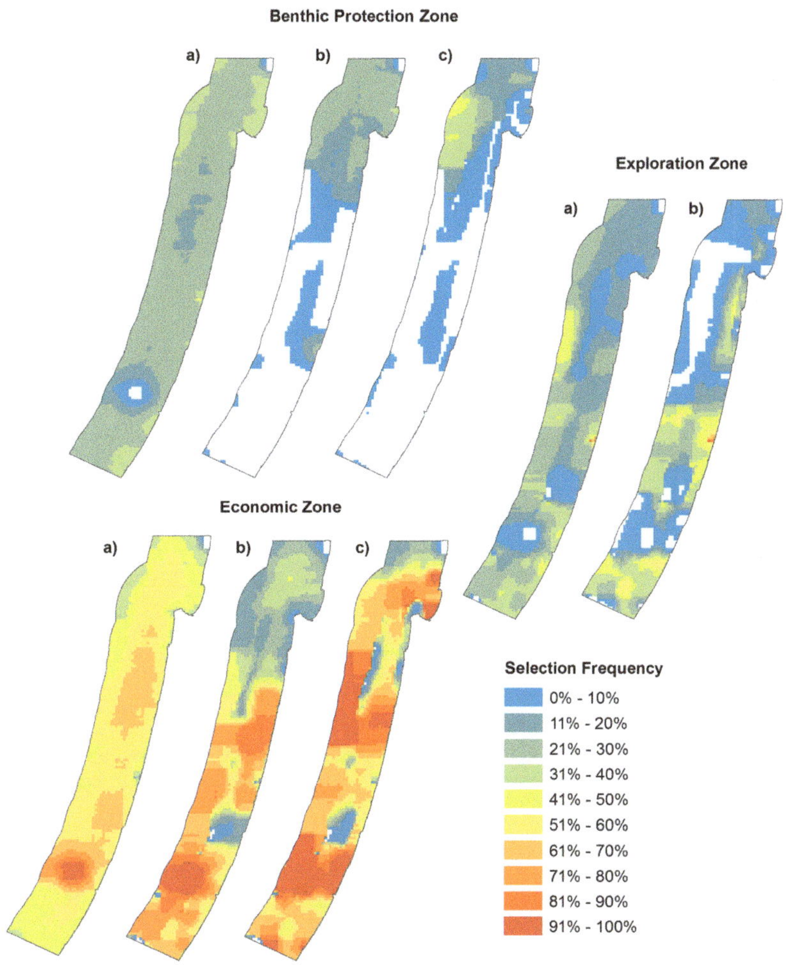

Figure 5. Selection frequency output maps (shows the percentage of times a planning unit was selected when run in Marxan 1000 times) from Marxan with Zones for each Zone and each zoning scenario. For the Benthic Protection Zone and Economic Zone the three scenarios are a) Basic Zoning, b) Intermediate Zoning, c) Complex Zoning. For the Exploration Zone the two scenarios are a) Intermediate Zoning and b) Complex Zoning.

Table 4. Spearman rank correlation (ρ) of the similarity between the selection frequency outputs of each planning scenario.

Zone	Planning Scenario	Simple Planning A	Simple Planning B	Basic Zoning	Intermediate Zoning B	Complex Zoning
Conservation Zone	Simple Planning A	-	0.69	0.84	−0.09	−0.09
	Simple Planning B	0.69	-	0.11	0.86	0.67
	Basic Zoning	0.84	0.11	-	−0.04	−0.06
	Intermediate Zoning B	−0.09	0.86	−0.04	-	0.73
	Complex Zoning	−0.09	0.67	−0.06	0.73	-
Benthic Zone	Basic Zoning	-	-	-	0.45	0.33
	Intermediate Zoning B	-	-	0.45	-	0.82
	Complex Zoning	-	-	0.33	0.82	-
Economic Zone	Basic Zoning	-	-	-	0.41	0.29
	Intermediate Zoning B	-	-	0.41	-	0.58
	Complex Zoning	-	-	0.29	0.58	-
Exploration Zone	Intermediate Zoning B	-	-	-	-	0.42
	Complex Zoning	-	-	-	0.42	-

High values (closer to 1) indicate a more similar spatial pattern in selection frequencies, meaning that these plans will require similar conservation actions. All scenarios show significant correlations ($p<0.001$).

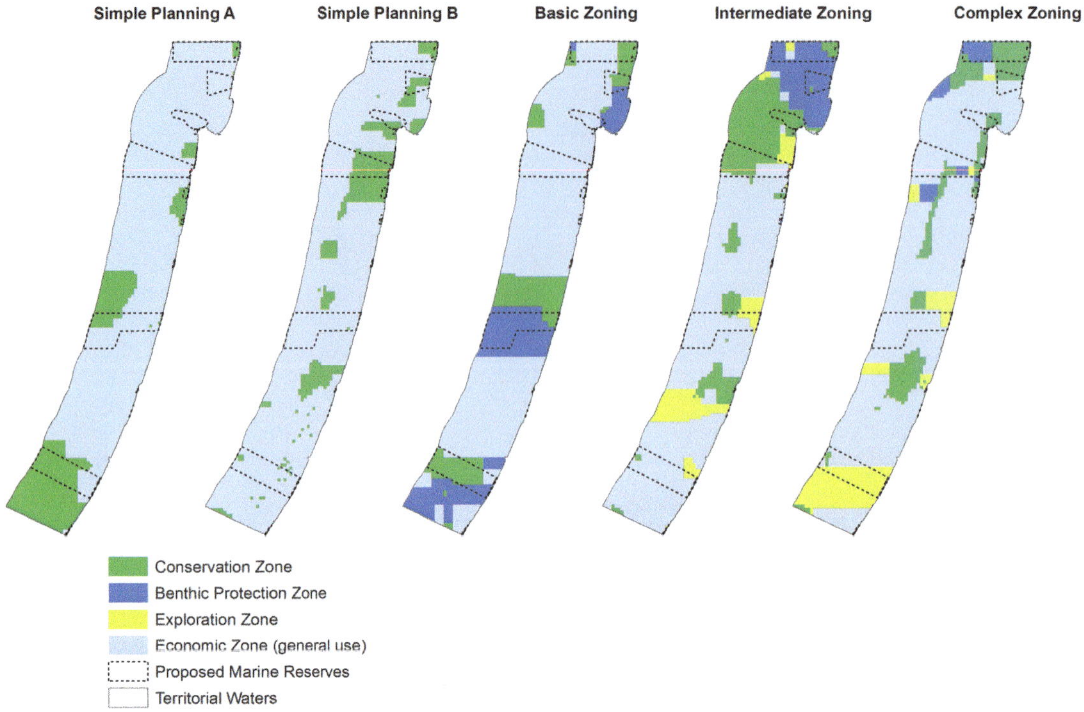

Figure 6. Marxan best solution outputs (the reserve configuration that best reduces opportunity cost and meets biodiversity targets from 1000 Marxan runs) for each planning scenario. The four colours designate the four types of zones (see Table 1).

(Fig. 6). The Economic Zone has highest selection frequency in the south for the Basic Zoning scenario where high fishing pressure is evident. In the Intermediate Zoning B scenario the high selection frequency of this zone extends over the south and central region where hydrocarbon is included. Further expansion of this zone's high selection frequency extends to the north as shipping lanes and pipelines are included in the Complex Zoning scenario. The Exploration Zone's priority areas were dissimilar between the Intermediate and Complex Planning scenarios, ($\rho = 0.42$, $p < 0.001$; Table 4). The inclusion of other marine activities affected the available area for the Exploration Zone.

Evaluating trade-offs between conservation and economic objectives

In the Basic Zoning scenario (three zones and six users) all biodiversity targets were met with a loss of 7% of commercial fishing revenue (Fig. 7a). By increasing the complexity of our planning scenarios (including more marine activities) we found that our biodiversity targets could only be met by decreasing the area of fishery grounds, consequently decreasing the revenue. Hence, the resulting fishing revenue loss was 12% for the Intermediate Zoning scenario (zoning network that includes four zones) and 15% for the Complex Zoning scenario.

In comparison, by including the economic objectives of hydrocarbon operations while meeting biodiversity and fishery targets (all four fishing gear types targeted 88% (Intermediate Zoning) and 85% (Complex Zoning) of revenue; values obtained from Fig. 7a), a small revenue loss was incurred (Fig. 7b). For the Intermediate Zoning scenario 5% of hydrocarbon revenue was lost. Similarly, the Complex Zoning scenario kept biodiversity and fishery targets with revenue losses of 6%. Therefore, for a loss of ~5% of hydrocarbon revenue, fishery and biodiversity targets could be fully met. Interestingly, we found that the drop-off rate of not meeting biodiversity and fishery targets was very minimal for the hydrocarbon industry in comparison with the rate at which biodiversity and fishery targets were traded off. Moreover, if hydrocarbon revenue was not traded-off (100% revenue was maintained), biodiversity and fishery targets could reach ~98% (Fig. 7b). However, if fishery revenue was not traded off (100% revenue was maintained), biodiversity targets could only reach between 93–84% (Fig. 7a).

Discussion

This study demonstrates how conservation objectives can be achieved while considering economic objectives where there are multiple marine activities. We found that the inclusion of many activities in marine conservation plans can significantly alter spatial priorities (Table 4; Fig. 4; Fig. 5). Economic goals are more compromised (in this case for the fisheries and hydrocarbon industries; Fig. 7) to achieve biodiversity targets when there are more marine activities in the planning process. Moreover, complex plans with more activities involved greater opportunity cost and did not reach biodiversity targets as easily as more simplified plans with less marine activities. Given that a complex plan is working with a more constrained problem, this result is expected [3,56]. Despite the increased opportunity cost and lack of spatial flexibility to achieve biodiversity goals with more complex conservation plans, planning that incorporates other activities can steer us towards areas which are feasible (greater potential for implementation success), minimize conflict with other users and reduce threats to biodiversity.

Conservation planning and zoning with multiple activities is challenging. Our case study shows that decisions made by conservation planners such as the number of zones or number of marine activities included in the planning process can substantially shape the resulting zone and reserve configuration.

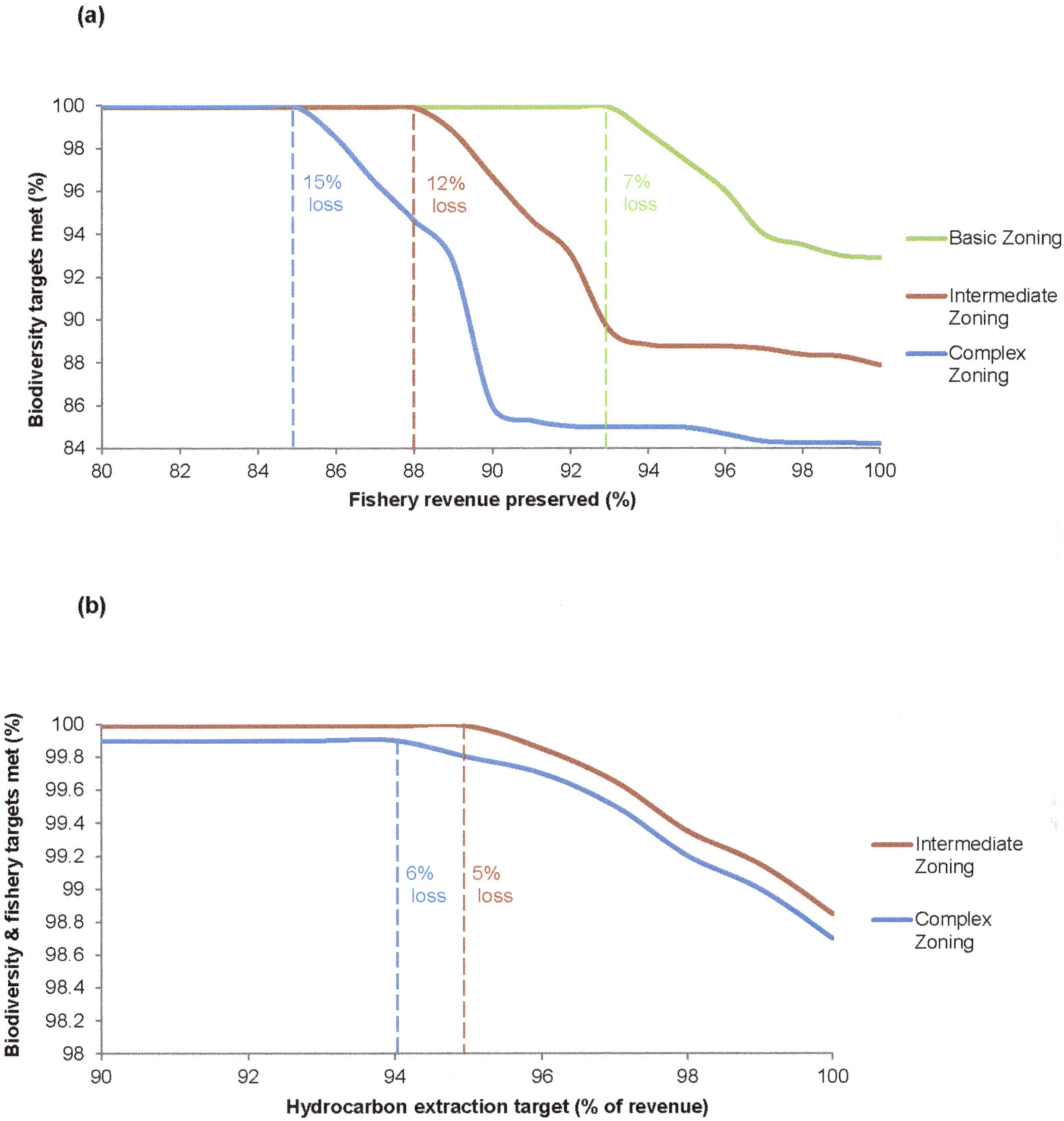

Figure 7. The trade-off between meeting biodiversity targets and maintaining economic objectives for each zoning scenario. (a) biodiversity targets are met when the fishery targets (percentage of annual fishery revenue) are less than 93% (7% revenue loss) in the Basic Zoning scenario (three zones and six activities), less than 88% (12% revenue loss) in the Intermediate Zoning B scenario (four zones and seven activities), and less than 85% (15% revenue loss) is the Complex Zoning scenario (four zones and ten activities), (b) biodiversity targets are met when hydrocarbon operations (leased and licensed expected revenue) are less than ≤95% (5% revenue loss) in the Intermediate Zoning scenario and less than 94% (6% revenue loss) in the Complex Zoning scenario.

Therefore, it is important to first identify the impact that each activity and feature has on marine biodiversity in the study system and the appropriate conservation action to take [57]. Here we follow a framework (Fig. 1) to help conservation planners address offshore activities and their potential threat in the marine realm. This framework outlines the steps needed to comprehensively zone for biodiversity protection while maintaining economic goals and can be a useful guide for countries currently striving to zone their waters [18]. One of the most important steps is testing the sensitivity of the results to user decisions (e.g., the inclusion of data, number of zones, the targets, see Step 7 Fig. 1; [58]). Other challenges that need to be accounted for when zoning include: the lack of shared information between stakeholders [59], the unknown expansion and objectives of industries [60], the unknown value of economic industries [5], and unforseen threats or disasters [18]. Given that some of these challenges can be overcome, in reality, conservation planning is largely shaped by the willingness to trade-off economic and conservation objectives. Moreover, there is no one correct solution to planning within a complex system [61].

Trade-off analysis is an important step to include in conservation planning [62,63]. It enables us to determine how much of a commercial activity must be forgone in order to achieve biodiversity targets. It also helps to address the implementation gap (the gap between conservation planning and real-world action) inherent in many conservation plans [2]. However, Hirsch et al. [62] cautions that not every problem can be solved by compromise. For example, we assume in our study that a portion of the hydrocarbon leased and licensed areas and commercial fishing grounds are available for trade-off, whereas stakeholders may disagree and reject any compromise. In marine conservation, fishing trade-offs have been the focus of several studies [3,25,28]. In this study we have incorporated economic trade-offs for both the commercial fishing and hydrocarbon industries, indicating the necessary compromises that are needed to meet our biodiversity targets in each planning scenario. Specifically, we have triaged targets for nine species that were unable to be met within no-take zones (of the Intermediate Zoning and Complex Zoning scenarios) and enabled them to be met within other zones. We suggest that future work should expand this type of analysis to examine the trade-offs with other social, economic and cultural activities where appropriate.

Marine features and activities which are confined in space may be difficult, or impossible, to trade-off. In this study we introduced a range of features into marine conservation planning in addition to the more traditionally used fishing such as pipelines, shipping lanes, desalination plants and aquaculture. In comparison to the full coverage of commercial fishing practises and the wide cover of hydrocarbon exploration across the study area, other features are restricted to a specific area (Fig. 3). Such restricted features are difficult to plan around as they often cannot be traded-off. Fishing effort for example can be redispersed to other spatial areas when an area is declared a marine reserve [64,65], whereas aquaculture farms are more difficult to relocate. We also found that linear-shaped features such as pipelines and shipping lanes influence the shape of marine zones, causing thin elongated zones (Fig. 6). Therefore, conservation planners must decide whether such features are planned around, planned with, or ignored. Performing a cost-benefit analysis of altering some of these features (e.g., rerouting shipping lanes, planning reserves over pipelines or moving planned aquaculture cages) within various planning scenarios could be a way to examine their potential flexibility within the reserve system. We suggest that future research explores the way that such features are included in conservation planning as they can have an influence on the selection of conservation priorities.

We found that the incorporation of oil and gas exploration can substantially alter spatial priorities and the opportunity cost of conservation. This is first time that offshore hydrocarbon operations are explicitly incorporated in marine conservation planning. Possible reasons for its absence in previous conservation plans are because a) the economic gains that are at stake are so large that these areas are "off-limits" to all other marine activities (e.g., Australian commonwealth zoning plan; [12]), b) uncertainty as to how to incorporate hydrocarbon information and c) the uncertain future of the industry that is dependent on new discoveries and may quickly demand large marine space (e.g., new discoveries in the Mediterranean Sea [66]). If we incorporate hydrocarbon information by assuming such areas cannot be protected, we may not be able to achieve a representative reserve network. One of the problems we encountered with including prospective hydrocarbon exploration is that sometimes biodiversity targets could not be achieved because a few species substantially overlap with hydrocarbon interests. Thus, we must carefully assess our targets and understand the compromises or actions that need to be taken in order to ensure conservation-worthy species are maintained in the face of hydrocarbon operations. We suggest that conservation plans endeavour to incorporate mining and fossil fuel data where possible to avoid costly conservation mistakes.

The ability of hydrocarbon exploration areas to provide some level of protection for biodiversity is unknown. In this study we tested different levels of protection from the "Exploration Zone" (see Intermediate Zoning Table 3; "zone effectiveness" see [67]), and found that opportunity cost is reduced if hydrocarbon areas are able to contribute to biodiversity protection. This is a novel conservation planning example that incorporates the notion of hydrocarbon areas providing some conservation benefit. The impacts of oil spills and gas leaks on marine biodiversity are severe and are well documented [68–70]. Likewise, there is some understanding of the impacts of offshore construction and extraction e.g., drilling impacts that are damaging to benthic structures [71]. However we have little understanding of the impacts posed by the ongoing maintenance of a drilling site that is dormant (leased or licensed without current activity). We suggest that future research focuses on better understanding the impacts that hydrocarbon operations pose on marine biodiversity and further develop ways to include hydrocarbon information into marine conservation plans.

Our study has interesting implications for Israel. We found that for Israel's territorial waters we can meet all our biodiversity targets (but not all no-take zone targets) for a loss of ~15% of annual commercial fishery revenue and ~5% of potential hydrocarbon revenue. A reduction of 7% of fishery revenue was needed to meet our biodiversity targets if hydrocarbon exploration is ignored. Our planning scenarios indicate that a surface area of 14–22% (Table 3) of Israel's territorial waters needs to be protected to meet biodiversity targets. The marine area reserved in Israel is currently less than 1% of the territorial waters [72], although none of these are considered no-take areas. Efforts are currently being undertaken to expand Israel's reserve network. The proposed network has been planned using species gradients and the representation of geomorphological features, without the use of Marxan or other similar tools [72]. We found that there is some overlap between the proposed marine reserves and the high priority areas found in our study (Fig. 4). The primary overlapping areas include: the proposed reserve in the north (Rosh Hanikra), the Haifa headland, the proposed reserve near Atlit and the smaller sized reserve near Dor. While different methods have been used for these two plans, some results are overlapping and we recommend that these areas that overlap should be targeted as initial reserve priorities for Israel as they are robust to the kind of process used to define priorities. However, it should be cautioned that, while overlapping priorities could be a good starting point, they will not necessarily provide a representative network that meets biodiversity targets.

Marine conservation planning often lacks good quality spatial data and must therefore rely on surrogate measures [6,59,73]. In this study, the surrogate fishing effort layers were generated with large involvement and input from experts. In comparison, our opportunity cost layer for hydrocarbon operations, although based on available government data, may less accurately reflect unpredictable shifts in future opportunity cost due to the fluctuating price of fossil fuels. Here we also set relatively low biodiversity targets because very minimal marine protection exists in this area, thus, these targets are potentially achievable (20% of Israel's Mediterranean Sea needs to be protected to meet our biodiversity targets (Table 3), corresponding with Israel's proposed target by the Israel Nature and Parks Authority [72]). These targets do not guarantee species persistence, but increasing these target may mean that other targets become unachievable, particularly within the Intermediate Zoning and Complex Zoning scenarios. We have included several novel features in our planning (e.g., aquaculture farms, desalination plants, shipping lanes and pipelines), yet there are other features that could be incorporated in future work, for example sand mining, offshore power plants, tourism and recreational fishing. Similarly, management and monitoring cost can also be included in future studies [6]. The aim of this study was to evaluate the impact of including multiple features and activities into marine conservation, however we do intend for these results to serve as useful baseline plans for the territorial waters of Israel. To improve the selection of conservation priorities in Israel's Mediterranean waters future work should attempt to build upon these scenarios, including additional species data for which data is currently limited, incorporate additional marine activities and create more robust cost layers with the availability of new data.

This case study can serve as an example for many other countries around the world, which are faced with the need to carefully balance economic considerations while protecting marine biodiversity. It is particularly relevant for countries surrounding the Mediterranean Sea that share common challenges and arising threats from developing offshore hydrocarbon exploration to biodiversity and ecosystems [66]. Our results suggest that planning with more complexity (e.g., multiple economic objectives, multiple threats and multiple zones) will be slightly more costly, have higher trade-offs with other marine activities and will require more input data. Despite these inefficiencies, a complex plan considers the objectives of more stakeholders (marine activities) and is more likely to result in successful implementation of conservation outcomes [2] and better compliance than a plan which ignores other activities. In the Mediterranean region with its many marine users, this is particularly important where compliance is often a major limiting factor in reserve design and implementation success [74]. We propose that countries aiming to protect marine biodiversity in their territorial waters should move from a single objective approach to one that links to the broader socioeconomic context incorporating multiple activities. A way forward may be the incorporation of lessons from marine spatial planning [11,19] into marine conservation planning, while aiming at maintaining biodiversity goals and examining trade-offs. Explicitly quantifying trade-offs can provide an initial starting point for discussion between stakeholders [62] and ultimately enable successful conservation outcomes which other marine users are willing to comply with.

Acknowledgments

We thank the Israel Nature and Parks Authority, particularly R. Yahel for advice and G. Weil for providing GIS data. For advice on the collation of fish species data and distribution we thank D. Golani, M. Gorem, & G. Rilov. For advice and data on sea turtles and cetaceans we thank Y. Levy and A. Scheinin. We thank the Israel Department of Fisheries and Aquaculture, especially O. Sonin and N. Mozes for providing data, advice and expert opinions and N. Mazor for Hebrew translation support. Data and advice on hydrocarbon exploration was provided by I. Nissim & M. Livnat, and Marxan software support and GIS data from N. Levin. We thank the two anonymous reviewers and the editor for their constructive comments.

Author Contributions

Conceived and designed the experiments: TM HPP SK. Performed the experiments: TM DE. Analyzed the data: TM HPP DE EB SK. Wrote the paper: TM HPP DE EB SK.

References

1. Knight AT, Cowling RM (2007) Embracing Opportunism in the Selection of Priority Conservation Areas. Conservation Biology 21: 1124–1126.
2. Knight AT, Cowling RM, Rouget M, Balmford A, Lombard AT, et al. (2008) Knowing but not doing: selecting priority conservation areas and the research-implementation gap. Conservation Biology 22: 610–617.
3. Weeks R, Russ GR, Bucol AA, Alcala AC (2010) Incorporating local tenure in the systematic design of marine protected area networks. Conservation Letters 3: 445–453.
4. Biggs D, Abel N, Knight AT, Leitch A, Langston A, et al. (2011) The implementation crisis in conservation planning: could "mental models" help? Conservation Letters 4: 169–183.
5. Douvere F (2008) The importance of marine spatial planning in advancing ecosystem-based sea use management. Marine Policy 32: 762–771.
6. Naidoo R, Balmford A, Ferraro PJ, Polasky S, Ricketts TH, et al. (2006) Integrating economic costs into conservation planning. Trends in Ecology & Evolution 21: 681–687.
7. Norse EA (2008) EBM perspective: Ocean zoning is inevitable. Washington, DC: Marine Conservation Biology Institute. Available http://mcbi.marine-conservation.org/publications/pub_pdfs/norse_MEAM_2008.pdf. Accessed 2013 October 20.
8. Butt N, Butt N, Beyer HL, Bennett JR, Biggs D (2013) Biodiversity Risks from Fossil Fuel Extraction. Science 342: 425–426.
9. Shaffer B (2011) Israel—New natural gas producer in the Mediterranean. Energy Policy 39: 5379–5387.
10. Tagliapietra S (2013) Towards a New Eastern Mediterranean Energy Corridor? Review of Environment, Energy and Economics-Re3. Available http://dx.doi.org/10.7711/feemre3.2013.02.002. Accessed 2013 August 24.
11. Douvere F, Ehler CN (2009) New perspectives on sea use management: Initial findings from European experience with marine spatial planning. Journal of Environmental Management 90: 77–88.
12. Barr LM, Possingham HP (2013) Are outcomes matching policy commitments in Australian marine conservation planning? Marine Policy 42: 39–48.
13. Pressey RL, Bottrill MC (2008) Opportunism, Threats, and the Evolution of Systematic Conservation Planning. Conservation Biology 22: 1340–1345.
14. Redfern JV, McKenna MF, Moore TJ, Calambokidis J, Deangelis ML, et al. (2013) Assessing the Risk of Ships Striking Large Whales in Marine Spatial Planning. Conservation Biology 27: 292–302.
15. Casale P (2011) Sea turtle by-catch in the Mediterranean. Fish and Fisheries 12: 299–316.
16. Inger R, Attrill MJ, Bearhop S, Broderick AC, James Grecian W, et al. (2009) Marine renewable energy: potential benefits to biodiversity? An urgent call for research. Journal of Applied Ecology 46: 1145–1153.
17. NOAA (2009) Setting an new course: shipping lane shift helps mariners steer clear of whales. SanctuaryWatch, National Oceanic and Atmospheric Administration. Washington D.C. pp.6–7. Available http://sanctuaries.noaa.gov/news/pdfs/sanctuarywatch/sw_1209.pdf Accessed 2014 January 15.
18. Agardy T (2010) Ocean Zoning: Making Marine Management More Effective: Earthscan. UK, London. pp.207.

19. Ehler C, Douvere F (2009) Marine spatial planning: a step-by-step approach toward ecosystem-based management. Intergovernmental Oceanographic Commission and Man and the Biosphere Programme. IOC Manual and Guides No. 53, ICAM Dossier No. 6. Paris: UNESCO.
20. Day JC (2002) Zoning—lessons from the Great Barrier Reef Marine Park. Ocean & Coastal Management 45: 139–156.
21. Fernandes L, Day J, Lewis A, Slegers S, Kerrigan B, et al. (2005) Establishing Representative No-Take Areas in the Great Barrier Reef: Large-Scale Implementation of Theory on Marine Protected Areas. Conservation Biology 19: 1733–1744.
22. Boyes SJ, Elliott M, Thomson SM, Atkins S, Gilliland P (2007) A proposed multiple-use zoning scheme for the Irish Sea: An interpretation of current legislation through the use of GIS-based zoning approaches and effectiveness for the protection of nature conservation interests. Marine Policy 31: 287–298.
23. DSEWPC (2013) Commonwealth marine reserves. Commonwealth Marine Reserves Branch, Parks Australia, Department of Sustainability, Environment, Water, Population and Communities Kingston, Tasmania. Available http://www.environment.gov.au/marinereserves/index.html Accessed 2014 February 22.
24. Cao W, Wong MH (2007) Current status of coastal zone issues and management in China: A review. Environment International 33: 985–992.
25. Klein CJ, Steinback C, Watts M, Scholz AJ, Possingham HP (2009) Spatial marine zoning for fisheries and conservation. Frontiers in Ecology and the Environment 8: 349–353.
26. Foley MM, Halpern BS, Micheli F, Armsby MH, Caldwell MR, et al. (2010) Guiding ecological principles for marine spatial planning. Marine Policy 34: 955–966.
27. Watts ME, Ball IR, Stewart RS, Klein CJ, Wilson K, et al. (2009) Marxan with Zones: Software for optimal conservation based land- and sea-use zoning. Environmental Modelling & Software 24: 1513–1521.
28. Grantham HS, Agostini VN, Wilson J, Mangubhai S, Hidayat N, et al. (2013) A comparison of zoning analyses to inform the planning of a marine protected area network in Raja Ampat, Indonesia. Marine Policy 38: 184–194.
29. Agostini VN, Margles SW, Schill SR, Knowles JE, Blyther RJ (2010) Marine Zoning in Saint Kitts and Nevis: A Path Towards Sustainable Management of Marine Resources. Arlington, Virginia: The Nature Conservancy. pp.39.
30. European Commission (2011) Exploring the potential of Maritime Spatial Planning in the Mediterranean Sea – Country reports: Israel. Directorate-General for Maritime Affairs and Fisheries, MARE.E.1 "Maritime Policy in the Baltic, North Sea and Landlocked Member States", European Commission, Brussels, Belgium.
31. Diamant A, Tuvia AB, Baranes A, Golani D (1986) An analysis of rocky coastal eastern Mediterranean fish assemblages and a comparison with an adjacent small artificial reef. Journal of Experimental Marine Biology and Ecology 97: 269–285.
32. Spanier E, Pisanty S, Tom M, Almog-Shtayer G (1989) The fish assemblage on a coralligenous shallow shelf off the Mediterranean coast of northern Israel. Journal of Fish Biology 35: 641–649.
33. Goren M, Galil BS (2001) Fish Biodiversity in the Vermetid Reef Of Shiqmona (Israel). Marine Ecology 22: 369–378.
34. Golani D, Reef-Motro R, Ekshtein S, Baranes A, Diamant A (2007) Ichthyofauna of the rocky coastal littoral of the Israeli Mediterranean, with reference to the paucity of Red Sea (Lessepsian) migrants in this habitat. Marine Biology Research 3: 333–341.
35. Stern N (2010) The impact of invasive species on the soft bottom fish communities in the eastern Mediterranean [PhD thesis]. Tel-Aviv University. Tel-Aviv, Israel.
36. Levit Y (2012) The influence of the depth gradient on the status of alien fish in the Mediterranean coast of Israel. [PhD thesis (in Hebrew)]. Tel-Aviv University. Tel-Aviv, Israel.
37. Lipsky G (2012) Sunken ships in the Levant: Spot of attraction for various species or invasive species paradise? [MSc thesis (in Hebrew)]. Tel-Aviv University. Tel-Aviv, Israel.
38. Edelist D (2013) Fishery management and marine invasion in Israel. [PhD thesis]. University of Haifa. Haifa, Israel.
39. ESRI (2008) ArcMap 9.3 Geographical Information System Software. ESRI Inc.
40. Amante C, Eakins BW (2009) ETOPO1 1 Arc-Minute Global Relief Model: Procedures, Data Sources and Analysis. NOAA Technical Memorandum NESDIS NGDC-24. Colorado, USA. Available http://www.ngdc.noaa.gov/mgg/global/ Accessed 2013 March 20.
41. Tognelli MF, Silva-Garcıa C, Labra FA, Marquet PA (2005) Priority areas for the conservation of coastal marine vertebrates in Chile. Biological Conservation 126: 420–428.
42. Clark MR, Tittensor DP (2010) An index to assess the risk to stony corals from bottom trawling on seamounts. Marine Ecology 31: 200–211.
43. Golani D, Öztürk B, Başusta N (2006) Fishes of the eastern Mediterranean: Turkish Marine Research Foundation.
44. Mazor T, Levin N, Possingham HP, Levy Y, Rocchini D, et al. (2013) Can satellite-based night lights be used for conservation? The case of nesting sea turtles in the Mediterranean. Biological Conservation 159: 63–72.
45. Scheinin A (2010) The Population of Bottlenose Dolphins (*Tursiops truncatus*), Bottom Trawl Catch Trends and the Interaction between the Two along the Mediterranean Continental Shelf of Israel Haifa. [PhD thesis]. Haifa University. Haifa, Israel.
46. The Israel Nature and Parks Authority (2012) Data - Ecological Geographical Information Centre, Department of Telecommunication and Information Systems. Jerusalem, Israel.
47. IUCN (2013) The IUCN Red List of Threatened Species. Version 2013.2. Available http://www.iucnredlist.org. Accessed 2013 November 27.
48. Kark S, Levin N, Grantham HS, Possingham HP (2009) Between-country collaboration and consideration of costs increase conservation planning efficiency in the Mediterranean Basin. Proceedings of the National Academy of Sciences 106: 15368–15373.
49. Lieberknecht L, Ardron JA, Wells R, Ban NC, Lötter M, et al. (2010) Addressing Ecological Objectives through the Setting of Targets. In: Ardron JA., Possingham HP, and Klein CJ (eds), Marxan Good Practices Handbook, Version 2. Pacific Marine Analysis and Research Association, Victoria, BC, Canada. pp.24–38.
50. Edelist D, Scheinin A, Sonin O, Shapiro J, Salameh P, Rilov G, (2013) Israel: Reconstructed estimates of total fisheries removals in the Mediterranean, 1950–2010. Acta Adriatica 54: 253–264.
51. Mazor T, Giakoumi S, Kark S, Possingham HP (2014) Large-scale conservation planning in a multinational marine environment: cost matters. Ecological Applications 24:1115–1130.
52. Israel Department of Fisheries (2012) Trawling routes along Israel's Mediterranean coast. 1:100000. Israel Ministry of Agriculture and Rural development - Department Of Fisheries and Aquaculture. Haifa, Israel.
53. State of Israel Ministry of Energy and Water resources (2012) Oil and Gas Section Ownership in Petroleum Rights. Haifa, Israel. Available www.energy.gov.il. Accessed 2014 December 20.
54. Varshavsky A (2012) Current Status of Offshore Oil and Gas Exploration in Israel. The Ministry of Energy and Water Resources of Israel. Haifa, Israel. Available www.energy.gov.il. Accessed 2013 September 4.
55. Ball IR, Possingham HP, Watts M (2009) Marxan and relatives: Software for spatial conservation prioritization. In: Moilanen A, Wilson KA, Possingham H, P., editors. Spatial conservation prioritization: Quantitative methods and computational tools. Oxford, UK. Oxford University Press. pp.185–195.
56. McDonald RI (2009) The promise and pitfalls of systematic conservation planning. Proceedings of the National Academy of Sciences 106: 15101–15102.
57. Pressey RL, Cabeza M, Watts ME, Cowling RM, Wilson KA (2007) Conservation planning in a changing world. Trends in Ecology & Evolution 22: 583–592.
58. Warman LD, Sinclair ARE, Scudder GGE, Klinkenberg B, Pressey RL (2004) Sensitivity of Systematic Reserve Selection to Decisions about Scale, Biological Data, and Targets: Case Study from Southern British Columbia. Conservation Biology 18: 655–666.
59. Levin N, Coll M, Fraschetti S, Gajewski J, Gal G, et al. (In Press) Review of biodiversity data requirements for systematic conservation planning in the Mediterranean Sea. Marine Ecology Progress Series.
60. Sivas DA, Caldwell MR (2008) New Vision for California Ocean Governance: Comprehensive Ecosystem-Based Marine Zoning. Stan Envtl LJ 27: 209–216.
61. Game ET, Meijaard E, Sheil D, McDonald-Madden E (2013) Conservation in a wicked complex world; challenges and solutions. Conservation Letters: 7: 271–277.
62. Hirsch PD, Adams WM, Brosius JP, Zia A, Bariola N, et al. (2011) Acknowledging Conservation Trade-Offs and Embracing Complexity. Conservation Biology 25: 259–264.
63. Halpern BS, Klein CJ, Brown CJ, Beger M, Grantham HS, et al. (2013) Achieving the triple bottom line in the face of inherent trade-offs among social equity, economic return, and conservation. Proceedings of the National Academy of Sciences 110: 6229–6234.
64. Halpern BS, Gaines SD, Warner RR (2004) Confounding effects of the export of production and the displacement of fishing effort from marine reserves. Ecological Applications 14: 1248–1256.
65. Roberts CM, Hawkins JP, Gell FR (2005) The role of marine reserves in achieving sustainable fisheries. Philosophical Transactions of the Royal Society B: Biological Sciences 360: 123–132.
66. EIA (2013) Overview of oil and natural gas in the Eastern Mediterranean region. EIA US Energy Information Administration. US Department of Energy, Washington, DC. Available http://www.eia.gov/countries/analysisbriefs/Eastern_Mediterranean/eastern-mediterranean.pdf Accessed 2013 October 3.
67. Makino A, Klein CJ, Beger M, Jupiter SD, Possingham HP (2013) Incorporating Conservation Zone Effectiveness for Protecting Biodiversity in Marine Planning. PLOS ONE 8: e78986.
68. Gomez Gesteira JL, Dauvin JC, Salvande Fraga M (2003) Taxonomic level for assessing oil spill effects on soft-bottom sublittoral benthic communities. Marine Pollution Bulletin 46: 562–572.
69. Lee L-H, Lin H-J (2013) Effects of an oil spill on benthic community production and respiration on subtropical intertidal sandflats. Marine Pollution Bulletin 73: 291–299.
70. Rooker JR, Kitchens LL, Dance MA, Wells RJD, Falterman B, et al. (2013) Spatial, Temporal, and Habitat-Related Variation in Abundance of Pelagic Fishes in the Gulf of Mexico: Potential Implications of the Deepwater Horizon Oil Spill. PLOS ONE 8: e76080.
71. Davies AJ, Roberts JM, Hall-Spencer J (2007) Preserving deep-sea natural heritage: Emerging issues in offshore conservation and management. Biological Conservation 138: 299–312.

72. Yahel R, Angert N (2012) Nature conservation policy in the Mediterranean, marine reserves as biodiversity and environmental conservation tool [in Hebrew]. Israel Nature and Parks Authority. Jerusalem, Israel.
73. Weeks R, Russ GR, Bucol AA, Alcala AC (2010) Shortcuts for marine conservation planning: The effectiveness of socioeconomic data surrogates. Biological Conservation 143: 1236–1244.
74. Fenberg PB, Caselle JE, Claudet J, Clemence M, Gaines SD, et al. (2012) The science of European marine reserves: Status, efficacy, and future needs. Marine Policy 36: 1012–1021.

Conus: First Comprehensive Conservation Red List Assessment of a Marine Gastropod Mollusc Genus

Howard Peters[1]*, Bethan C. O'Leary[1], Julie P. Hawkins[1], Kent E. Carpenter[2], Callum M. Roberts[1]

1 Environment Department, University of York, York, United Kingdom, 2 International Union for Conservation of Nature, Global Marine Species Assessment, Biological Sciences, Old Dominion University, Norfolk, Virginia, United States Of America

Abstract

Marine molluscs represent an estimated 23% of all extant marine taxa, but research into their conservation status has so far failed to reflect this importance, with minimal inclusion on the authoritative Red List of the International Union for the Conservation of Nature (IUCN). We assessed the status of all 632 valid species of the tropical marine gastropod mollusc, *Conus* (cone snails), using Red List standards and procedures to lay the groundwork for future decadal monitoring, one of the first fully comprehensive global assessments of a marine taxon. Three-quarters (75.6%) of species were not currently considered at risk of extinction owing to their wide distribution and perceived abundance. However, 6.5% were considered threatened with extinction with a further 4.1% near threatened. Data deficiency prevented 13.8% of species from being categorised although they also possess characteristics that signal concern. Where hotspots of endemism occur, most notably in the Eastern Atlantic, 42.9% of the 98 species from that biogeographical region were classified as threatened or near threatened with extinction. All 14 species included in the highest categories of Critically Endangered and Endangered are endemic to either Cape Verde or Senegal, with each of the three Critically Endangered species restricted to single islands in Cape Verde. Threats to all these species are driven by habitat loss and anthropogenic disturbance, in particular from urban pollution, tourism and coastal development. Our findings show that levels of extinction risk to which cone snails are exposed are of a similar magnitude to those seen in many fully assessed terrestrial taxa. The widely held view that marine species are less at risk is not upheld.

Editor: David L. Roberts, University of Kent, United Kingdom

Funding: The Natural Environment Research Council (NERC) and Economic and Social Research Council (ESRC) generously supported the Red List research under a studentship grant to Howard Peters http://www.nerc.ac.uk/funding/. The authors gratefully acknowledge funding from the Biodiversity Synthesis Center of the Encyclopedia of Life (EOL) at the Field Museum of Natural History, Chicago for the Conus Synthesis Workshop http://synthesis.eol.org/center/funding; also Tom Haas and the New Hampshire Charitable Foundation and the Thomas W. Haas Foundation for their support of the Global Marine Species Assessment under the IUCN Global Species Programme. The funders had no role in study design, data collection and analysis, decision to publish, or preparation of the manuscript.

Competing Interests: The authors have declared that no competing interests exist.

* E-mail: hp510@york.ac.uk

Introduction

Extinction risk of marine organisms has attracted little attention compared to that of terrestrial taxa, with a widely held view that such risk is inconsequential due to high dispersal ability and large geographic ranges [1,2] especially when taking reference from the fossil record [1,3]. These beliefs are particularly prevalent when considering marine invertebrates, where a decline in abundance of the important phylum *Mollusca* has been overshadowed by the collapse in many exploited vertebrates, especially finfish [4]. This is primarily due to their relatively minor contribution to human protein requirements and the generally held belief that molluscs possess greater resilience to extinction through their perceived wide distribution and a likelihood of hidden pockets of survivors [5]. Marine invertebrates in general are seriously under-represented within the IUCN Red List [6]. Only cuttlefish, lobsters and scleractinian corals have been fully assessed and published [6,7]. Although limited research on the impact of habitat loss and fishing pressure on marine gastropod molluscs has been undertaken on a regional scale including for shell fisheries [8,9], there have been no comprehensive assessments of trends in species abundance, commercial and environmental impacts and extinction risk to any genera with a global biogeographical distribution.

Cone snails of the genus *Conus* offer an excellent opportunity to explore global threats to marine molluscs owing to their exceptional diversity [10], wide distribution, high degree of endemism, varied depth distribution [11], and an established global market in their trade from amateur shell collectors to commercial traders [12]. In addition, cone snails are used in some communities in the Pacific as an occasional foodstuff [13] but, more importantly, they are actively targeted by international drug companies and researchers as a potential pharmacological resource [14].

Cone snails constitute the family *Conidae*, which together with the *Turridae* (turrid snails) and *Terebridae* (auger snails) comprise the superfamily *Conoidea* otherwise known as *Toxoglossa* ('poison tongue') owing to the venom apparatus they deploy for immobilising prey [15]. The *Conoidea* form part of the order *Neogastropoda* in the sub-class *Prosobranchia* of the class *Gastropoda* of the phylum *Mollusca* [11].

Cone snails live throughout the world's tropical coastal waters with a steep latitudinal diversity gradient away from the tropics, extending into cooler regions that include southern California, northern Gulf of Mexico, Florida and the Carolinas, North Africa, the Mediterranean, South Africa, Australia, southern Japan and

China [16]. Distribution varies widely with some species occurring across the entire tropical Indo-Pacific but others restricted to a single bąy or seamount [11,17].

The genus *Conus* is taxonomically challenging. Although morphological characteristics of the shell remain the initial means of species identification [11], more recently, other traits have also been employed to differentiate among species, in particular the radular teeth used in the capture of prey, whose shape and structure not only reflects the dietary preferences of the species [18] but may also be specific to a single species [19]. Separation of species through DNA sequence variations provides even greater reliability, but more recently, character-based DNA barcoding has been highly effective in distinguishing among closely-related species [20]. For this assessment we relied upon expertise from taxonomists in *Conus* to create a dataset of valid species.

The fossil record indicates that the first *Conus* appeared in a sea that covered what is now England and France during the Lower Eocene around 55 mya [21]. During subsequent radiations the genus expanded around the globe and by the Holocene had formed into four biogeographical regions: Indo-Pacific (IP), Eastern Pacific (EP), Western Atlantic (WA), and Eastern Atlantic (EA). Although widening of the Atlantic during the Cretaceous and Cenezoic has today created an impermeable barrier to *Conus* crossing the ocean, there have been some migrations in the past, as witnessed from the fossil record and more recently by *C. ermineus* extant in both the EA and WA and *C. chaldaeus, C. ebraeus* and *C. tessulatus* found in both the IP and EP [22].

The majority of the 632 species of *Conus* assessed (53.6%) occurs in the infralittoral zone of 5 m deep or less, with most of the remaining species (27.7%) at 50 m deep or less. However, there are some species such as *C. teramachii* that live in deeper parts of the continental shelf extending to 1,000 m where they may be brought to the surface as bycatch of demersal fisheries. The bathymetric ranges of individual species vary considerably with some shallow water species living within a one or two metre depth range and some deep-water species being found within a 500 m range or more [11].

Microhabitats vary by species and most often consist of sand or mud into which the cone snail may burrow, but may also include inter-tidal limestone benches (the smooth remains of reef structures from earlier geological periods when sea levels were higher [23]) with sand or algal turf, sub-tidal reef platforms with living and dead corals, or boulders with sandy layers [24]. They may also be found among coral rubble and occasionally among mangroves and seagrasses.

Cone snails are generally nocturnal in their feeding habit [25] and group-specific in their preference for worms, molluscs or fish (Fig. 1) although some species have a mixed diet [26]. The smallest groupings by diet are the obligate piscivores with around 50 species [11,27], and obligate molluscivores with approximately 80 [28]. The majority of *Conus* are vermivorous with polychaetes representing the largest dietary component, that can be the exclusive source of food for some species [11]. All cone snails use venom to immobilize their prey. The diversity of venoms employed by a particular species in the capture of prey is a reflection of the degree of specialisation in its diet [29].

From the earliest civilizations, people have prized cone shells for their exceptional beauty, with examples discovered among prehistoric artefacts used for personal adornment extending back 5,000 years [30]. Their striking patterns and wide range of colours and shades continue to attract collectors today with rare examples in perfect condition changing hands for thousands of dollars with common and abundant species traded for cents to a dollar or two each [12].

Over millions of years *Conus* has evolved a battery of peptide toxins (conopeptides/conotoxins) for immobilizing prey [31]. The venom of each species is a cocktail mixed from between 50 and 200 different peptides each of only 10 to 35 amino acids in length and is generally targeted at voltage-gated or ligand-gated ion channels [30]. These conopeptides have become a focus for biomedical research worldwide [14]. Indeed, with the probability that there are on average over 100 distinct toxins for each species [30], as a whole, the *Conidae* can probably synthesize in excess of 50,000 toxins with little, if any, replication [32].

Cone snails are therefore important to: a) biodiversity; they have evolved into one of the largest of all marine genera, b) biopharmaceutics; they offer unparalleled opportunities in the development of novel drugs, and c) economics; their shells provide income to poor fishing communities through sales to tourists, traders and a global business in the specimen shell trade.

Habitat loss is considered by many malacologists to be the primary risk factor facing tropical marine mollusc species (Bouchet pers. comm. 2011) and there is plenty of hard evidence to support this view. In Queensland, Australia, for example, abundance and species richness of mollusc assemblages have been shown to be adversely affected by removal of subtropical mangrove forests, with population declines of 83% recorded [33]. In San Diego, Southern California the endemic horn snail, *Cerithidia fuscata*, that lived along intertidal mudflats was last seen in 1935 after pollution and dredging had driven it to extinction [34]. Where coral cover has been extensively damaged or degraded through pollution, sedimentation, coastal development and destructive fishing, as witnessed throughout much of the tropics, coral-associated molluscs such as the *Conidae* are being usurped by bivalve crevice dwellers [35].

In this paper we report one of the first comprehensive extinction risk assessments of a taxonomically well-resolved marine taxon. Our research assesses the extinction risk to the global populations of *Conus*, examining each species' distribution, current and projected threats from disturbance to habitats, pollution, coastal development, and shell gathering. We have examined where possible the effects of fragmentation of populations and the likely impact of demersal fisheries on deeper water species. The assessment enables us to reappraise whether marine taxa are less extinction prone than terrestrial. In addition, our aim is to provide data in support of conservation measures for those species at the greatest risk of extinction over the short to medium term.

Methods

Red List Assessment

We used the assessment standards and procedures of the International Union for the Conservation of Nature (IUCN) Red List of Threatened Species to assess extinction risk to 632 species of *Conus*. This is the world's leading resource for describing the global conservation status of plants and animals and uses a standard methodology to classify species into one of nine categories, together with a codified set of criteria [36]. The assessment includes examination of the effects of both ecological change and commercial exploitation on the subject taxa. Data derived during the research and discovery process for each species is compiled to a standard format together with maps, images and other supporting documentation.

Following taxonomic review, we divided valid species into 12 biogeographical working sets for detailed assessment. A comprehensive assessment was not possible for those species where data was substantially absent. For example, species endemic to areas of protracted civil unrest such as the Horn of Africa may not have

Figure 1. Diet and toxicity. Left: *C. geographus* Linnaeus, 1758; piscivorous, 65–165 mm; intertidal to 20 m; significant fatality risk to humans. Centre: *C. textile* Linnaeus, 1758; molluscivorous, 40–150 mm; intertidal to 50 m; handle with extreme caution. Right: *C. betulinus* Linnaeus, 1758; vermivorous, 55–177 mm; intertidal to 20 m; minimal risk to humans; note operculum. All species Indo-Pacific.

been researched in the field for many years. Coincidentally, these regions are not generally subject to intensive coastal development, harbour works and refineries and so may offer a degree of protection to marine taxa. Similarly, species occurring in deep water, where recovery is most commonly through fisheries by-catch, often suffer a paucity of data including extent of distribution and habitat types. Furthermore, bathymetry data will often rely on the questionable estimation of fishers. Wherever possible for deep-water *Conus*, we have focussed our attention on the level of demersal fishing in the area, including destructive methods such as dredging that may seriously affect mollusc assemblages.

Most *Conus* species, however, occur in shallow water where impacts such as coastal development, pollution and habitat destruction can be more easily recorded. Such threats can give rise to population fragmentation leading to a serious decline in abundance which may be difficult to quantify until it has become extreme. However, indicators including market prices for specimen shells provide a useful guide to increasing scarcity. Knowledge voids are common for *Conus* but where they occur we have, where possible, used estimation or inference using suboptimal data permitted under Red List standards [36]. Despite this, 13.8% of *Conus* species were found to be so deficient in data we were unable to make an assessment with any degree of reliability.

Key indicators of risk

Distribution. A key indicator of potential risk to a species is the size of its geographical distribution. The Red List standard assessment uses two measures: Extent of Occurrence (EOO) and Area of Occupancy (AOO). EOO for marine species is the area within a polygon drawn around the boundary of the species' range, excluding land areas. This will include areas which may not be physically occupied by the taxon, e.g. deep water, but which could contribute to larval dispersal. AOO is the physical area within the EOO in which the taxon is known to occur. For shallow water species, this may be calculated from the perimeter of an island or length of coastline, extended by the width of habitat calculated from the known or inferred bathymetric range of the species over the area under review. However, for 'linear' habitats such as rivers and coastlines, IUCN suggests that their standard habitat width of 2 km should be used in computing AOO [36] and we have adopted this approach in the assessment for *Conus*. It should be noted that both the AOO and the EOO are only of major significance in assessing the level of threat if the species has a restricted range, and that for most wide-ranging *Conus* species that fall outside the parameters of the assessment criteria these range sizes are not calculated. However, spatial data, derived from range maps created for every species assessment, were projected using ArcGIS version 10.1 (Environmental Systems Research Institute) to generate species richness maps for a) all species and b) only those species with a range less than the median mapped range size. All data were standardised onto 1° grid cells and projected to world cylindrical equal area.

Number of Locations. It is possible that a catastrophic event could have a profound effect on the population size of some species. Although marine molluscs are resilient in being able to endure physical forces such as extreme weather events, small populations may be extirpated as a result of sudden habitat loss caused by catastrophic events such as major oil spills. The 'location' count indicates the number of areas in which a single catastrophic event could affect all individuals of the taxon present, events that may cumulatively drive a species into extinction. The value of this measure is another key factor in determining the level of risk a global population faces.

Literature search

We conducted a comprehensive search through published papers and other literature for data relating to *Conus* species' populations, depth, distribution, habitats, trade in animals and shells, use for foodstuff, pharmaceuticals, etc., together with any conservation measures in place, including indirect conservation as may be offered by marine protected areas. We sought information on current and possible future threats, including coastal development for tourism, industry or port construction, nutrient loading from agricultural run-off, pollution from domestic and industrial effluent, intensive trawling, siltation from land-based sources, dredging for shipping channels and mineral extraction. Data on activities such as these can often only be found in trade publications, contract award notifications etc.

We also examined the market in shells to determine 'collectability', pricing fluctuations, scarcity and demand. Some shells with exceptional colour and form will achieve iconic status, and if they are also rare like *C. gloriamaris* or *C. milneedwardsi*, it adds to their cachet. Species that live within a highly restricted range, within a single bay for example, are often at heightened risk from human activity. This particularly applies to shallow water species which may be gathered as curios in areas where new beach tourism projects are being developed or planned. We synthesised distribution data including observed fragmentation, location counts, marketability, population declines and threats for each species to apply one of the nine categories listed below.

Assessment categories

There are three categories of extinction risk: critically Endangered (CR), Endangered (EN) and Vulnerable (VU) that broadly define 'extremely high', 'very high, or 'high' risk of extinction respectively. In addition, there are two extinct categories, Extinct (EX) and Extinct in the Wild (EW), and three other categories: Near Threatened (NT) for species that will be elevated to a threatened category in the short term unless the potential risk is removed; Data Deficient (DD) where there is insufficient data to determine a category, and Least Concern (LC) where current and projected population levels indicate the species is not at risk. As this was a comprehensive assessment, we did not use the category Not Evaluated (NE), where the species has been recorded but no assessment has been carried out.

For *Conus*, the criteria in support of the selected category are primarily derived from a range of variables based on estimated population size and/or level of decline together with species range size and location count.

Synthesis and pre-publication checks

Following our research and assessment, the results were reviewed by a panel of fourteen international experts, each with specialist knowledge of the *Conus* species within their allotted biogeographical working sets. The review took the form of a five day synthesis workshop with teams comprising leading academics together with renowned specialists from the commercial sector with comprehensive field knowledge of species' distribution, scarcity and threats, and a facilitator experienced in Red List standards and procedures. This peer-review process confirmed or modified findings of the original assessment authors, and allowed inclusion of supplementary field-based knowledge from the participating experts. All reports were checked for consistency by the Mollusc Specialist Group of the IUCN Species Survival Commission before final approval and submission for publication through the IUCN Red List Unit.

Results

The greatest species richness for *Conus* is in the Philippines and countries to the south and east towards Papua New Guinea, the Solomon Islands, New Caledonia and Fiji (Fig. 2A). Species occurring in the Atlantic including West Africa and the Gulf of Mexico and Caribbean are fewer in number. However, when very wide-ranging species are omitted, i.e. those equal to or greater than the median mapped range size, centres of endemism are revealed in scattered locations including the Caribbean, the Marquesas, New Caledonia and the Eastern Atlantic, in particular Cape Verde, Senegal and Angola (Fig. 2B).

Global threats

Three of 632 *Conus* species assessed were considered to be Critically Endangered (CR), 11 Endangered (EN) and 27 Vulnerable (VU), which together represent 6.5% of all global species (Fig. 3), with a further 26 species (4.1%) categorised as Near Threatened (NT). Over one in ten of all *Conus* species is therefore considered at risk or may become so in the near future. 87 species (13.8%) were categorised as Data Deficient (DD) of which 75 (86.2%) occur in the Indo-Pacific.

All 14 CR and EN species occur in the waters off Cape Verde and Senegal, West Africa (Table 1). Of the 27 assessed as VU, eight are from Cape Verde and Senegal with three from Angola (Table 1), seven from the Western Atlantic (Table 2), and nine from the Indian Ocean, including two from Western Australia (Table 3). Only three threatened species occur east of longitude 60 (Oman to Mascarenes): *C. rawaiensis* from Western Thailand and *C. compressus* and *C. thevenardensis* from Western Australia – all VU. According to this assessment procedure, there are no threatened species in the Pacific (Fig. 4). Of the 26 Near-Threatened species (NT), Cape Verde and Senegal are again over-represented with 14 of the 17 species from the Eastern Atlantic. Of the remainder in this category, five are from the Western Atlantic, one from the Western Indian Ocean, and three from the Pacific (Fig. 4).

Analysis by Region

Marine molluscs that are wide-ranging are likely to be more resilient against threats than those that are range-restricted, with dispersed populations providing a reservoir for re-colonization in the event of local extirpations [2]. The Eastern Atlantic species occupy a limited length of coast with few islands when compared to the Western Atlantic and, more particularly, the Indo-Pacific. It is also intersected by large rivers draining the tropical land mass of Africa which render substantial areas of coastal water unsuitable for many marine molluscs. Conversely, islands of the tropical Indo-Pacific and Caribbean contribute substantial areas of shallow water habitat suitable for taxa such as *Conus* and do not generally suffer any significant flux of freshwater. Figure 5 shows the percentage distribution of species' range sizes within each of the four oceanic regions. This graphically illustrates that wide-ranging *Conus* species, i.e. AOO $>2{,}000$ km^2, are uncommon within the Eastern Atlantic compared to the other regions.

Eastern Atlantic. Ninety-eight species of *Conus* occur along the Eastern Atlantic seaboard from the Mediterranean and Morocco south to Namibia, with associated island archipelagos including the Canaries, Azores, and Cape Verde (plus one: *C. ermineus*, that also occurs in the Western Atlantic and was included in that region). There is one species from the island of St Helena, *C jourdani*, within this grouping although no live specimens have been observed and it is categorised as DD. With three CR, 11 EN and 11 VU species, representing 25.5% of the Eastern Atlantic species, and a further 17 species NT (Table 1), 42.9% of Eastern Atlantic

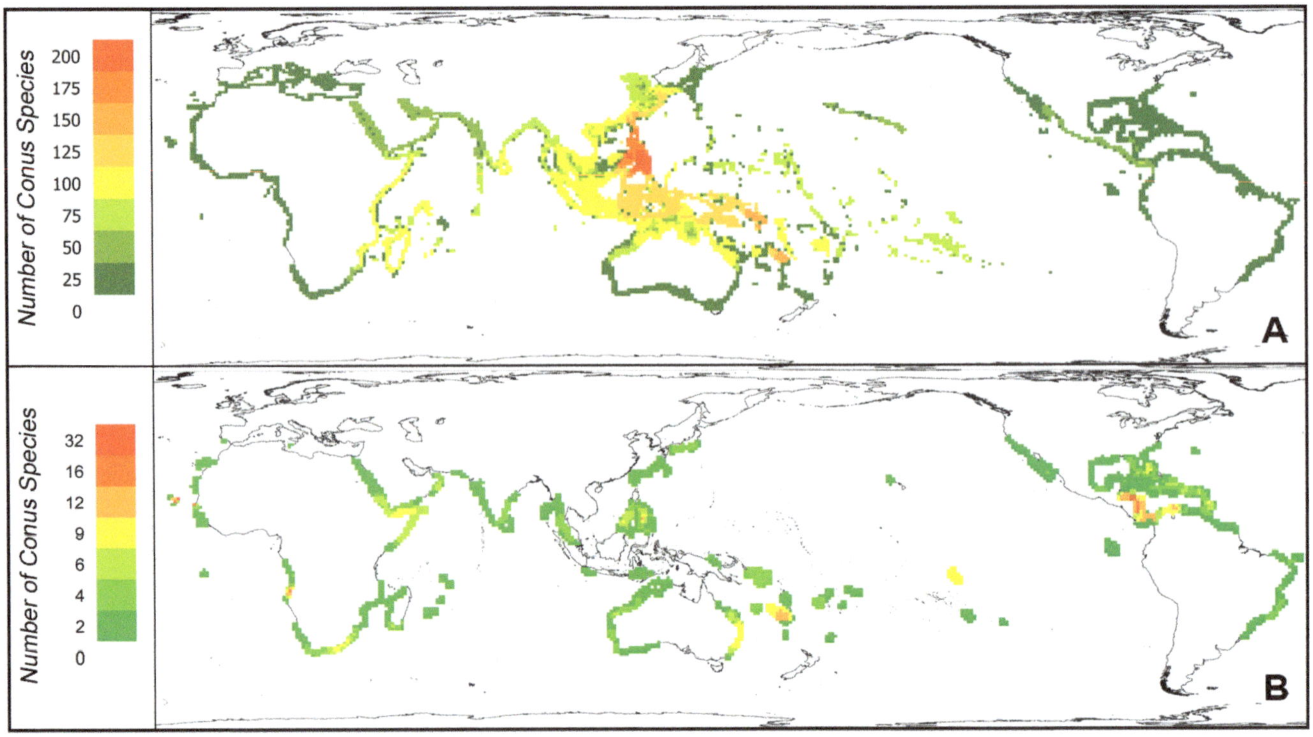

Figure 2. *Conus* species distribution. Species richness from a composite of individual species maps (Fig. 2A), and only species with mapped area less than the median indicating regions of potential endemism (Fig. 2B).

Conus are considered at risk of extinction or liable to become so. This exceptional concentration of threatened species is found nowhere else across the genus' wide distribution and the disproportionate contribution of species from Cape Verde and Senegal demands further explanation.

Cape Verde is home to 8.9% of all *Conus* species. With 53 species endemic from a total of 56 present in the archipelago, endemism is exceptionally high at 94.6%. Forty-three species are each restricted to a single island. All three CR species are found in Cape Verde, *C. lugubris*, *C. mordeirae*, and *C. salreiensis*, together with four EN and five VU (Table 1). There are also 12 NT species. With 24 species in either a threatened or near-threatened category, Cape Verde has 45.3% of its *Conus* diversity at risk compared to 7.4% for the remainder of the world. Angola and Senegal contribute the next largest numbers of endemic *Conus* species with 22 and 13 respectively, which together with 53 species endemic to Cape Verde account for 89.8% of all 98 species within the Eastern Atlantic. Senegal contributes seven EN and three VU species with Angola contributing three VU (Table 1).

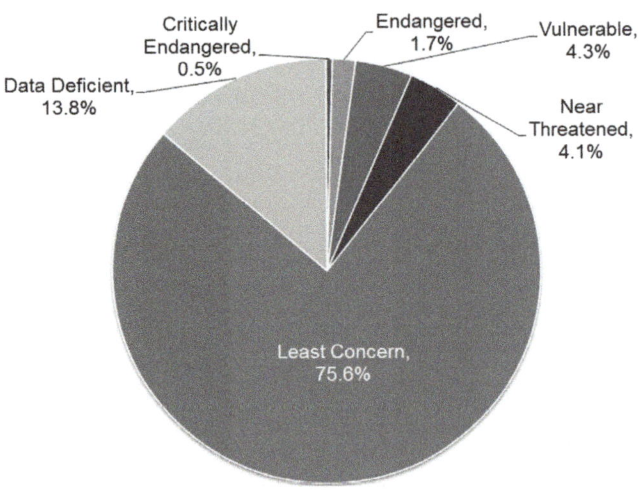

Figure 3. Global extinction risk to *Conus*. The percentage contribution for each assessed category to the global diversity of 632 *spp* of *Conus*. These are represented by 3 Critically Endangered species; 11 Endangered; 27 Vulnerable; 26 Near Threatened; 87 Data Deficient, and 478 of Least Concern.

Western Atlantic. We assessed 113 species of *Conus* from the Western Atlantic where they occur from the Carolinas and Bermuda south to Brazil and throughout the Gulf of Mexico and Caribbean. Species are widely variable in their distribution across the region. There are six threatened species, all categorised as VU (see Table 2), representing 5.3% of the total and a further four NT, together resulting in 8.8% of *Conus* species within this region considered at immediate or potential risk.

Indo-Pacific. We assessed 390 species of *Conus* from the Indo-Pacific where they occur across the tropics and subtropics, from East Africa south to South Africa and north to the Red Sea and the Persian Gulf and across the whole of the Indian Ocean and the Western and Central Pacific, south to Australia and New Zealand, north to Japan, east to French Polynesia and Easter Island and northeast to Hawaii.

Only nine species were found to be VU. All occur within the Indian Ocean with six species from the western flank: two from Oman, one from the southern Red Sea, two from the Mascarenes and one from South Africa. From the eastern flank there is one species from Thailand and two from Western Australia. There are also four NT species including one from Oman with the other

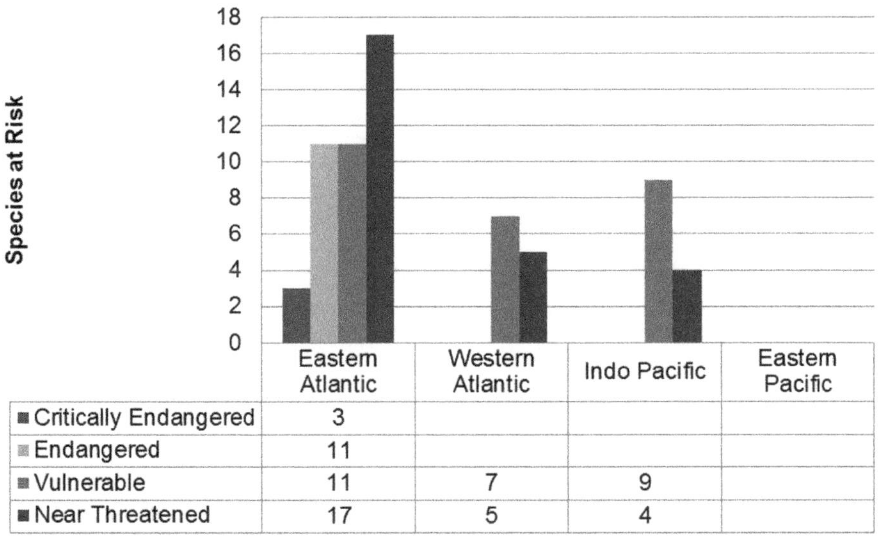

	Eastern Atlantic	Western Atlantic	Indo Pacific	Eastern Pacific
Critically Endangered	3			
Endangered	11			
Vulnerable	11	7	9	
Near Threatened	17	5	4	

Figure 4. Number of *Conus* species at risk by ocean basin. The number of species at risk by ocean basin for each threatened category. There are no species at risk in the Eastern Pacific.

three being the only *Conus* species potentially at risk in the Pacific – one each from Queensland, the Philippines and the Marquesas.

Eastern Pacific. We assessed 31 species of *Conus* from the Eastern Pacific where they occur from Southern California south along the Pacific coast of Meso-America to Southern Ecuador including the Galapagos and other island groups of the region.

No species were assessed as threatened or near threatened in the Eastern Pacific.

Threats

The nature of threats to those species of *Conus* at risk of extinction are varied and depend primarily, but not exclusively, on the proximity and nature of human habitation and development adjacent to coastlines where the molluscs occur. This alone, however, will not normally create a scenario for species extinction. Wide-ranging species are capable of maintaining their viability through resilience from multiple sub-populations. Although most threatened *Conus* species are range-restricted, this is not always the case: two species from the USA, *C. anabathrum* and *C. stearnsii* occur along the west coast of Florida where their ranges are substantially fragmented by shoreline development. However, restricted range, coupled with shallow water habitat, magnifies the impact of stressors such as coastal development or pollution. Of the 41 *Conus* species globally assessed as threatened with extinction, 32 (78.0%) occur within an AOO of 250 km^2 and a minimum depth of 5 m or less. In the Eastern Atlantic, of the 25 threatened species, this rises to 100%.

Threats to those *Conus* species assessed within one of the three threatened categories can be classified into four causal groups: 1. pollution, either from proximity to actual or potential petrochemical spills, or urban and industrial effluent; 2. disturbance to habitat from coastal development either resulting from human population increases, e.g. sea defences, residential and commercial structures, including aquaculture facilities, and port construction, or tourism infrastructure. Also included in this group is damage to habitat caused by damaging and extensive demersal fishing; 3. shell gathering, and 4. environmental change e.g. elevated sea-surface temperatures (Fig. 6). There will frequently be a

Table 1. Threatened *Conus* of the Eastern Atlantic (EA).

Critically Endangered (CR)		**Endangered (EN)**		**Vulnerable (VU)**	
Cape Verde	*C. lugubris*	Cape Verde	*C. ateralbus*	Angola	*C. allaryi*
Cape Verde	*C. mordeirae*	Cape Verde	*C. crotchii*	Angola	*C. cepasi*
Cape Verde	*C. salreiensis*	Cape Verde	*C. cuneolus*	Angola	*C. xicoi*
		Cape Verde	*C. fernandesi*	Cape Verde	*C. decoratus*
		Senegal	*C. belairensis*	Cape Verde	*C. felitae*
		Senegal	*C. bruguieresi*	Cape Verde	*C. fontonae*
		Senegal	*C. cloveri*	Cape Verde	*C. regonae*
		Senegal	*C. echinophilus*	Cape Verde	*C. teodorae*
		Senegal	*C. hybridus*	Senegal	*C. cacao*
		Senegal	*C. mercator*	Senegal	*C. guinaicus*
		Senegal	*C. unifasciatus*	Senegal	*C. tacomae*

Table 2. Threatened *Conus* of the Western Atlantic (WA).

Vulnerable (VU)	
Aruba	*C. hieroglyphus*
Florida	*C. anabathrum*
Florida	*C. stearnsii*
Bahamas	*C. richardbinghami*
Brazil	*C. henckesi*
Martinique	*C. hennequini*
Venezuela	*C. duffyi*

Table 3. Threatened *Conus* of the Indo-Pacific (IP).

Vulnerable (VU)	
Oman	*C. ardisiaceus*
Oman	*C. melvilli*
S Red Sea	*C. cuvieri*
Mascarenes	*C. julii*
Réunion	*C. jeanmartini*
SE South Africa	*C. immelmani*
W Thailand	*C. rawaiensis*
Australia	*C. compressus*
Australia	*C. thevenardensis*

combination of causes, for example tourism infrastructure may also increase shell gathering. Similarly, the proximity of shanty towns devoid of planning regulations poses an elevated risk of effluent discharge into the marine environment. Finally habitat destruction from sand removal, beach nourishment works and recreational use of the sea may all result in disturbance to local mollusc populations.

Cape Verde is experiencing a major structural change from a largely services and fisheries based economy supported by development aid and remittances from its diaspora to one of beach tourism [37]. This is accompanied by a myriad of threats from road and resort construction, unlawful removal of beach sand for cement [38] and casual shell gathering by tourists. All three CR species occur in Cape Verde where their populations are already reduced. *C. lugubris* and *C. mordeirae* live in areas where habitat has already been lost to development and *C. salreiensis* which is restricted to a single bay has had observable declines in population since a harbour was constructed. Each is found in an area along a shallow coastal strip of less than 11 km in length. Harbour expansion and the accidental discharge of engine fuel increase the pressures on small, range-restricted *Conus* populations such as *C. fernandesi, C. fontonae* and *C. regonae*. With so many *Conus* species occupying highly restricted ranges within the archipelago, modest threats such as these could have a profound impact.

Around the Dakar peninsula, Senegal, it has been observed that species restricted to its highly polluted coastal waters are showing a marked decline in abundance coupled with an overall diminution of shell size including *C. echinophilus, C. hybridus, C. mercator* and *C. unifasciatus*. In common with many maritime cities in developing countries, Dakar suffers from a burgeoning population with largely inadequate waste-processing infrastructure. South in Angola, *Conus* species categorised as at risk face similar threats to those in Senegal.

In the Western Atlantic some disturbance to *Conus* can be traced to human migration to the Florida coast. Tourism and retirement have driven large-scale construction projects for condominiums and other coastal infrastructure leading to significant loss of habitat for *C. anabathrum* and *C. stearnsii*. Tourism also represents the underlying risk to the vulnerable species *C. hennequini* in Martinique and *C. hieroglyphus* in Aruba. Shell collecting in the Bahamas threaten *C. richardbinghami*. General coastal development in Bahia, Brazil threatens *C. henckesi* where it occurs only off two small islands. The Venezuelan government has voiced plans for substantial development on the islands of Los Roques which will place the shallow water species, *C. duffyi* at risk.

The *Conus* species of the Indo-Pacific are at less risk. In the north-western Indian Ocean, the Persian Gulf, the southern Red Sea including the Gulf of Aden and the Horn of Africa, civil wars, poverty, piracy and the security situation offer some degree of protection from coastal development. However, there are still concerns from oil spillage in the region and two scarce species from Oman, *C. ardisiaceus* and *C. melvilli* together with *C. cuvieri* from Djibouti are categorised as VU. In Southern Natal and the Mascarene islands of Mauritius and Réunion respectively, *C. immelmani* and *C. julii*, have both declined in numbers almost certainly from over-collecting, with *C. jeanmartini* also from Réunion being subject to intensive trawling in its deep-water habitat.

In the Eastern Indian Ocean, *C. rawaiensis* occurs only in an area estimated at less than 35 km^2 in a single location off the western shores of Thailand in a region zoned for tourism. In Western Australia, an extreme localized warm-water event in 2011 from La Niña, in the region around Geraldton to Shark Bay including the Abrolhos Islands, resulted in a catastrophic decline of marine molluscs including *Conidae*. *C. compressus*, a restricted range species, possibly suffered a 50% decline in abundance. Also in Western Australia, *C. thevenardensis*, already rare, is subject to a range of threats including a large oil installation, tourism and dredging.

Other Red List categories

The results for the three threatened categories paint an incomplete picture. There are also 87 species assessed to be Data Deficient and 26 as Near Threatened, together representing 17.9% of the global diversity. Many of the Data Deficient species are considered to be scarce in the wild even though the causes and extent of the threats they face cannot yet be determined with sufficient accuracy. Twice as many of the 87 DD species, (39.1%) occur only at depths of 50 m or more, i.e. below gleaning, scuba, and (many) artisanal fishing gears, compared to 19.4% for the remaining 545 species globally. Specimens may be brought to the surface from these depths as by-catch from fisheries. However, demersal gear such as dredges may also contribute substantially to the endangerment of the species recovered through destruction of their habitat, especially for those that are also of restricted range. Occurrence in deep water does not automatically result in a DD categorisation. Despite paucity of data, taxa with a known distribution greater than 2,000 km^2 but with no known threat would normally be assessed as Least Concern. At the other extreme, there are also a number of DD species where there is an almost total absence of recent sightings but extinction cannot be proven, i.e. there is reasonable doubt that the last individual has died [39]. This is exemplified by species such as *C. jourdani* from St Helena which is only known from 'dead' shells washed onto the beach in one small bay; *C. bellulus* and *C. luteus* which have not

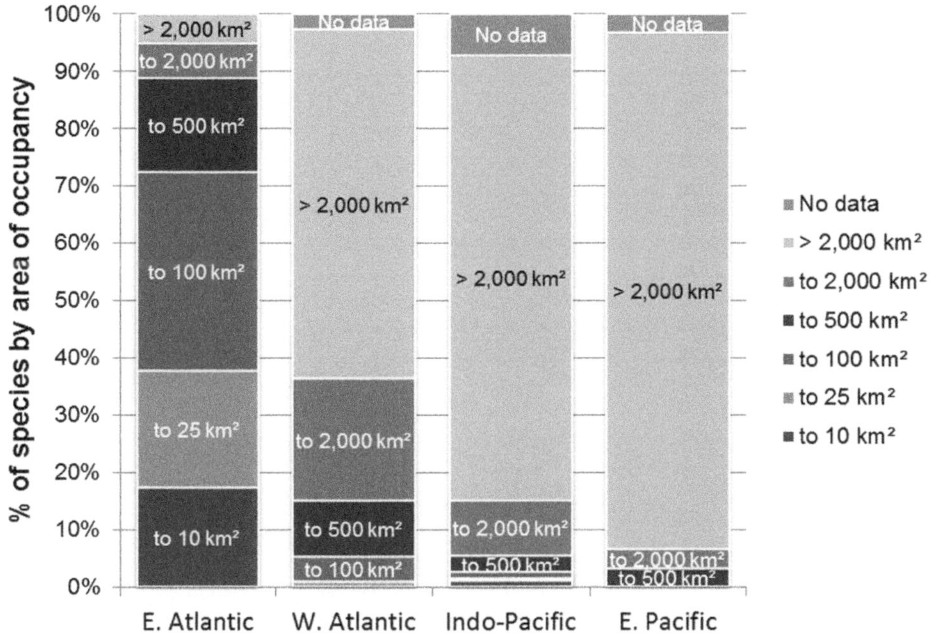

Figure 5. Contribution of range-restricted species to *Conus* biodiversity within each ocean basin. This illustrates by region the percentage of total species by area of occupancy, with wide-ranging species, i.e. >2,000 km^2 being minimal in the Eastern Atlantic but the major contributor to the Indo-Pacific and Eastern Pacific *Conus*. The abbreviated key describes the band sizes, e.g. to 10 km^2 = 0–10 km^2, to 25 km^2 = 11–25 km^2, to 100 km^2 = 26–100 km^2, etc.

been reported since the 1970s; *C. splendidulus* which has not been seen in 20 years and *C. sauros* which is possibly extinct.

Threatened species frequently have small ranges. Of 478 *Conus* species categorised as LC, 103 have an AOO of 2,000 km^2 of less (21.5%). However, of the 67 species with threatened and near-threatened categories, all except four occur within an AOO of 2,000 km^2 or less (94.0%), with 40 (59.7%) restricted to a range of 100 km^2 or less (Fig. 7). By comparison, 27 of the 59 DD species (45.8%) where it had been possible to approximate their AOO, were recorded at 2,000 km^2 or less (Fig. 7). Furthermore, many of

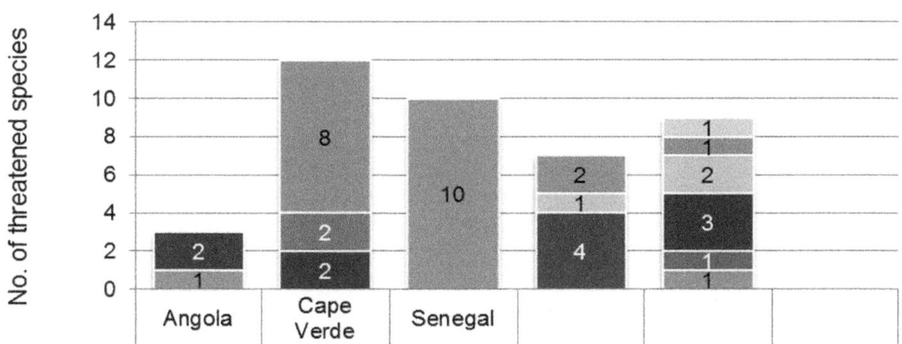

	Angola	Cape Verde	Senegal	Western Atlantic	Indo-Pacific	Eastern Pacific
	Eastern Atlantic					
Dem Fishing					1	
Tourism		8		2	1	
Shell Coll				1	2	
Ports etc		2				
Petro-chem	2	2			3	
Elev SST					1	
Effluent	1		10		1	
Coast Dev				4		

Figure 6. Main threats to *Conus* by ocean basin. The number of *Conus* species at risk (consolidation of CR, EN and VU) indicating primary causes of endangerment, being demersal fishing, tourism, shell collecting, ports and harbours, petro-chemical spills, elevated sea-surface temperatures, effluent discharge and runoff, and coastal development.

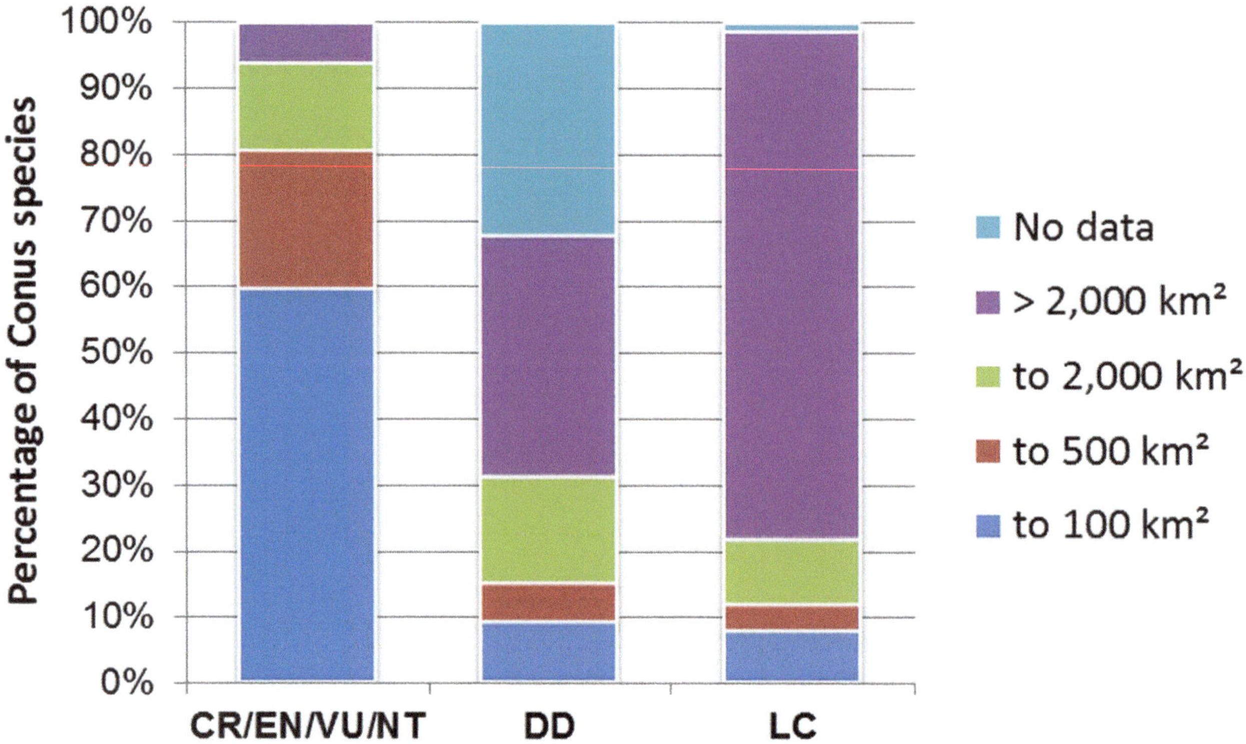

Figure 7. Contribution of range-restricted *Conus* species within assessment categories. This shows the percentage of species by assessment category, with DD species having a higher proportion of taxa with small ranges compared to LC species and also with a large number of species with no distribution data. All threatened and near-threatened categories have been grouped. Key: CR Critically Endangered, EN Endangered, VU Vulnerable, NT Near Threatened, DD Data Deficient and LC Least Concern

the 36.8% of DD species that are wider-ranging (AOO >2,000 km^2) are based on infrequent sightings. It would therefore be reasonable to suppose that a substantial proportion of the DD species are potential candidates for listing as threatened.

DD species are overwhelmingly found in the Indo-Pacific. There are three in the Eastern Atlantic, eight in the Western Atlantic, 75 in the Indo-Pacific and one in the Eastern Pacific. Outside the Indo-Pacific there is just a single species with a known AOO of more than 2,000 km^2.

Discussion

It is widely believed that extinction risk in the sea is less likely than in the terrestrial environment and that this is supported by the fossil record [2,3]. This view is based largely on perceived high fecundity, greater dispersal ability and geographic range size [40]. With 6.5% of *Conus* species at risk globally this would appear to follow this perception, however, in regions offering reduced dispersal opportunity, such as the whole of the Eastern Atlantic, 25.5% of species are threatened. Cone snails here have a similar level of extinction risk to species in well-assessed terrestrial taxa, such as freshwater invertebrates (34% of 7,784 species assessed at risk), lepidoptera (from 8.5% of butterflies in Europe to 17% in the U.S. at risk), European terrestrial molluscs (20% at risk) [41,42] and bryophyte flora from the Canaries (21% at risk) [43]. Contributing to the pattern seen, many cone snails have limited dispersal ability, small geographic ranges and/or are rare. The level of extinction risk is similar in other well assessed marine taxa, including corals (27% of species at risk) [7,44] and scombrid and billfish (11% of 61 species at risk) [45]. Given the rapid escalation of threats to the marine environment [46], if the pattern seen in these groups is typical of marine species generally, then there is a high risk that extinctions will soon become common in the sea, just as they now are on land.

Our global assessment of the conservation status of all 632 cone snails shows that three-quarters (75.6%) of species are classified as Least Concern under IUCN Red List standards. However, beneath this relatively optimistic result lies a picture of substantial regional variations with indicators signalling wider concerns. In the Eastern Atlantic along the shores of Senegal, Cape Verde and Angola, species restricted in their range and subject to the effects of industrialisation and urbanisation face an elevated risk of extinction. Endemism for marine species occurs most commonly in isolated island groups where the original dispersal was assisted by a pelagic larval stage or by transport on rafting matter [47]. Endemics may also be found where there may be non-reversing currents transporting water away from the tropics towards higher latitudes [40]. All Cape Verde endemic *Conus* have a non-planktonic larval stage having lost the ability during speciation to feed during larval dispersal [48]. This conforms to the hypothesis that non-planktonic, i.e. lecithotrophic, species of *Conus* commonly originate from planktotrophic species [49]. All three species assessed as Critically Endangered occur in the waters off Cape Verde where they are exposed to habitat degraded through coastal development primarily driven by tourism. Similarly, of the 11 Endangered species, four are found in Cape Verde with the remaining seven occurring off the coast of Senegal, in particular the Dakar peninsula, where high levels of pollution from industrial and residential effluent is thought to be the driver of declining abundance and observable reductions in body size. A further 11 species (40.7%) of the 27 assessed as Vulnerable occur in Cape Verde, Senegal and Angola, making West Africa home to 61% of the 41 *Conus* species threatened with extinction. Of the remaining 16 species categorised as Vulnerable, seven are found in the Western Atlantic where they are primarily exposed to coastal development, tourism and shell collecting. The remaining nine occur in the Indian Ocean where petrochemicals, shell-collecting and elevated sea-surface temperatures represent the principal causes of decline.

Threats

Overfishing. The effect of overfishing on the abundance of fish stocks has been extensively reported in both the scientific and general press over many years [50,51]. However, threats to invertebrates from fishing are seldom equated with extinction, especially marine molluscs. Although extremely unusual, near-extinctions in this group have occurred in the recent past; for example in the white abalone *Haliotus sorenseni* from southern California and Baja California by the mid-1990s had been fished to the edge of extinction [52]. Once counted in the millions there are now probably less than 1,600 individuals remaining.

Amongst marine molluscs, most species are sought by shell collectors [12]. Although this does not threaten the survival of the vast majority of molluscs, shell collecting has undoubtedly caused the decline and endangerment of some species, particularly 'trophy' shells. Throughout the Indo-Pacific, the spectacular giant triton (*Charonia tritonis*), has been extensively fished and in many areas has been extirpated [53]. Similarly, although primarily removed for its adductor muscles, the giant clam (*Tridacna gigas*) the largest of all bivalve molluscs, has met the same fate [54]. In Zanzibar, East Africa, the cowries *Cypraea tigris*, *C. histrio* and *C. lynx* were found to be up to 18 times less abundant in exploited tourist areas [8]. We identified three rare species of *Conus* threatened by shell collecting: *C. richardbinghami* from the Bahamas and *C. immelmani* and *C. julii* from the Mascarenes. Taxa already facing pressures from factors such as pollution may be pushed further towards extinction by gathering for shells, yet warning indicators such as sudden price inflation on the shell market may not alone warrant inclusion to a threatened category.

Bioprospecting. *Conus* is exceptionally important to biomedical science, although there is dispute about the number of animals taken for their bioactive compounds. To protect their intellectual property, pharmaceutical companies are silent on the issue, but researchers are adamant that volumes are negligible. In their dialogue in *Science* Chivian et al. (2003; 2004) raised important concerns about the quantity of cone snails taken from the wild, indicating that thousands were then collected to satisfy research demands [55,56]. This was forcefully rebutted by Duda et al. (2004) who reviewed recent conotoxin research from which they determined that a maximum of 20 research groups were working on *Conus* toxins at that time, and that any single characterisation required fewer than 21 animals to be sacrificed [57]. Regardless of where the true determinant lies, balancing the legitimate needs of medical research without further compromising natural resources is essential. Fortunately, alternative, more sustainable options are now available including milking venom without killing the animal [58], polymerase chain reaction (PCR) sequencing of DNA fragments that requires just one specimen [59] and more recently digital marine bioprospecting using massive parallel deep sequencing of transcriptomes that requires only minute samples of bioactive material [60].

Habitat loss. It has been shown that habit loss leads to declines in species richness, reduced biomass and loss of complexity [61,62], often accompanied by colonisation by species that inhibit recovery [63]. Virtually all of the world's 'trawlable' area of continental shelf has already been altered, and about half the area of all the continental shelves is hit by trawls every year [64], changing the structure and function of habitats, destroying assemblages and resulting in homogenisation of the seabed [65]. Of 133 marine species that have been recorded as having gone extinct either regionally or globally, 37% were attributed entirely or in part to habitat loss [66]. Extinctions of marine gastropod molluscs from loss of habitat are set to continue and include the horn snail *Cerithidea fuscata* from southern California last seen in 1935, the eelgrass limpet *Lottia alveus alveus* from the northwest USA last collected in 1929, and from the 19th century the rocky shore limpet '*Colisella*' *edmitchelli* also from southern California and the periwinkle *Littoraria flam*mea from China; all driven to extinction through loss of habitat from anthropogenic causes, with the possible exception of the eelgrass limpet that lost its habitat from a slime mould that may have been introduced from ships' ballast [2,34].

Our assessment found that with the exception of three species made vulnerable by shell collecting (see above), all 38 other *Conus* species threatened with extinction are impacted to some degree by habitat loss, either directly from coastal and port development or indirectly from pollution or from human exacerbated natural occurrences such as El Niño/La Niña–Southern Oscillation (ENSO) warm-water events (Fig. 6).

Red List comparatives

Our *Conus* assessment is the first global study for the IUCN Red List for any marine gastropod mollusc genus and one of the few for marine invertebrates. Other marine invertebrates that have been the subject of a global assessment include 845 reef-building corals, 247 lobsters and 195 cuttlefishes [44]. Data Deficiency is a common thread throughout each of these studies with 17%, 35% and 76% of species for each respective grouping [44] compared to 14% for *Conus*. Preliminary results available for oceanic squid show that 57% of this group are of Least Concern with the remaining 43% Data Deficient. As with the data deficient cone snails, many of these cephalopods are deep-water species that have only been captured on a few occasions [44].

Of the 845 corals that have been globally assessed 27.3% fall into a threatened category with a further 20.8% near threatened, although prior to the massive bleaching event of 1998 it has been estimated that 95.3% of non-DD species would have been categorised as Least Concern [7]. The exceptional ENSO event which resulted in this bleaching largely devalues any post-event comparison, although it has been shown that La Niña can impact some mollusc assemblages through stress, changes in productivity and availability of dietary preferences [67]. In Australia, the La Niña event of 2010-11 gave the highest monthly Southern Oscillation Index values on record accompanied by elevated sea-surface temperatures in Western Australia [68]. In the region around Geraldton to Shark Bay including the Abrolhos Islands, this coincided with an estimated 50% mortality in molluscs that included *Conus compressus* (H. Morrison pers. comm. 2011). Scleractinian corals and molluscs, including *Conus*, also share the threat of ocean acidification with the prospect of arrested development in their aragonite-forming structures [69,70]. Of all the threats faced by these fauna this is the most intractable and one that could even determine their continued existence.

For freshwater molluscs, Red List assessments have been completed for 1,500 of the 5,000 described species [6,42]. Results show that out of 7,784 freshwater invertebrates assessed to date, gastropods are the most threatened group with a threat range of 33% (if no DD species are threatened) to 68% (if all DD species are threatened) [42]. In common with *Conus*, the threatened species include range-restricted habitat specialists that are particularly at risk from loss of habitat and pollution.

Further Research and Conservation Priorities

As a global assessment for conservation has not been undertaken on any other marine gastropod it is not possible to explore relationships between different gastropod genera to identify commonality of risks. Further research is urgently needed to address this issue.

Data Deficient *Conus* species can normally be characterised by minimal sightings and a lack of data on distribution. This may result from their bathymetric profile (39% of DD species occur only at 50 m or more) and/or genuine rarity. However, DD species share a common trait with those categorised as threatened in having a higher percentage with restricted range than those from the general population, with the implication that there could be a significant proportion of DD species at risk. Bearing the title 'Data Deficient' or 'Near Threatened' places these 113 species jointly away from the spotlight afforded the 41 species in the three threatened categories. However, with the potential to double the number of species at risk it is essential that the taxa that make up these categories should not be ignored but instead benefit from further research into their true status.

One of the primary sources of information on species distribution, habitats, populations and threats for our Red List assessment has been the specimen shell trade. In many parts of the developing world, trade in shells provides valuable additional revenue to some of the poorest families living along tropical coastlines. Research is needed to assess the threat from rare shell collecting towards mollusc population decline to determine what measures should be taken to enable this activity to continue sustainably while at the same time allowing for protection of vulnerable species.

The need to identify conservation strategies for all species at risk is compelling, although for developing nations, snail conservation is unlikely to become a driver for environmental improvements. In the absence of in situ conservation measures, captive breeding programmes may ultimately be necessary, such as those undertaken for tree snails of the genus *Partula* from the Pacific Islands [71]. At present, except possibly for species such a *C. pennaceus* and *C. textile* that emerge as mature veligers, this is not a viable option, as the complexity of plankton essential for developing larvae cannot be easily replicated [28].

Over half (53.6%) of all *Conus* species occur at depths of 5 m or less where they are susceptible to gleaning, and nearly three-quarters (74.5%) occur at or above recreational SCUBA diving depths of 30 m or less. Marine Protected Areas (MPAs) offer one of the few sanctuaries, but regional authorities need encouragement to strengthen enforcement and to erect prominent signage against shell gathering within MPAs.

Cape Verde presents a special case in *Conus* conservation. With 45.3% of its species at risk there is a strong argument for legislating against export of both animals and shells, and with manageable borders the country is ideally suited to export controls. This small archipelago also signals a warning to other nations developing their coastal infrastructure: new roads bring visitors to areas previously protected by their isolation, and illegal sand removal for construction from beaches and shallow water of the littoral zone [72] pose a constant threat to habitat. Regional authorities should be required to undertake environmental impact assessments that take account of these issues when planning new developments.

The toxins that make *Conus* so successful are generally unique to each species [59] and any extinction in the genus could in turn deprive science of a potential pharmacological resource. The extraordinary number of species and the global distribution of these tropical snails make them an important contributor to marine biodiversity, and with the appeal of their shells they help support some of the world's poorest people.

Finally, there exists a well-defined community of cone snail aficionados who together are highly influential in the trade in cone shells. This includes leading academics as well as collectors and dealers. A positive first step from our Red Listing is that following a preliminary presentation of our findings at their international convention, a core of members has been motivated to explore a voluntary embargo in trade of critically endangered species and to consider this also for other *Conus* species at risk [73].

Acknowledgments

We would like to thank: Mark Westneat, Audrey Aronowsky, Sarah Kim and Beth Sanzenbacher at the Biodiversity Synthesis Center, Chicago for their organisation of the *Conus* synthesis workshop at the Field Museum in Chicago; Philippe Bouchet, José Coltro, Tom Duda, Alan Kohn, Eric Monnier, Hugh Morrison, Ed Petuch, Guido Poppe, Gabriella Raybaudi-Massilia, Sheila Tagaro, Manuel Jiménez Tenorio, Stephan Veldsman and Fred Wells for volunteering their time and expertise during the assessment and at the synthesis workshop; Monika Böhm, Heather Harwell, Andrew Hines, Suzanne Livingstone, Jonnell Sanciangco and Mary Seddon for facilitating at the synthesis workshop; Mike Filmer for helping to resolve the many taxonomic issues; Hannah Cubaynes, Zarozinia Sheriff and all the interns who assisted with species research; Klaus & Christina Groh of ConchBooks for use of images and maps for the Red List; Mia Theresa Comeros and Angela Goodpaster for their work on the Red List maps; and Caroline Pollock and Janet Scott of the IUCN Red List Unit for bringing the assessment to publication.

Author Contributions

Conceived and designed the experiments: HP BCO JPH CMR. Analyzed the data: HP BCO CMR. Contributed reagents/materials/analysis tools: KEC. Wrote the paper: HP JPH CMR. Revised the manuscript: JPH KEC CMR. Supported the synthesis workshop: HP KEC CMR.

References

1. Harnik PG, Simpson C, Payne JL (2012) Long-term differences in extinction risk among the seven forms of rarity. Proc Biol Sci 279: 4969–4976. doi:10.1098/rspb.2012.1902.
2. Roberts CM, Hawkins JP (1999) Extinction risk in the sea. Trends Ecol Evol 14: 241–246.
3. McKinney ML (1998) Is marine biodiversity at less risk? Evidence and implications. Divers Distrib 4: 3–8.
4. McManus JW (1997) Tropical marine fisheries and the future of coral reefs: a brief review with emphasis on Southeast Asia. Coral Reefs 16: S121–S127. doi:10.1007/s003380050248.
5. Jamieson GS (1993) Marine Invertebrate Conservation: Evaluation of Fisheries. Am Zool 33: 551–567.
6. IUCN (2013) IUCN Red List of Threatened Species. IUCN Red List Threat Species Version 20122. Available: http://www.iucnredlist.org. Accessed 8 February 2013.
7. Carpenter KE, Abrar M, Aeby G, Aronson RB, Banks S, et al. (2008) One-third of reef-building corals face elevated extinction risk from climate change and local impacts. Science 321: 560–563. doi:10.1126/science.1159196.
8. Newton LC, Parkes EVH, Thompson RC (1993) The Effects of Shell Collecting on the Abundance of Gastropods on Tanzanian Shores. Biol Conserv 63: 241–245.
9. Queensland Government Dept of Primary Industries and Fisheries (2007) Annual status report 2007 Queensland Marine Specimen Shell Collection Fishery. Available: http://www.daff.qld.gov.au/documents/Fisheries_SustainableFishing/AnnualStatusReport-QLDMarineSpecimen-ShellCollectionFishery-2007.pdf.
10. Bouchet P (1990) Turrid genera and mode of development: the use and abuse of protoconch morphology. Malacologia: 69–77.
11. Röckel D, Korn W, Kohn AJ (1995) Manual of the Living Conidae, Vol 1. Verlag Christa Hemmen.

12. Rice T (2007) A Catalog of Dealers' Prices for Shells: Marine, Land and Freshwater, 23rd edition. Of Sea and Shore Publications.
13. Chadwick A, Olivera BM (2009) Cone Shells and Human Culture. Cone Collect 12: 16–21. Available: http://www.theconecollector.com/.
14. Garber K (2005) Peptide leads new class of chronic pain drugs. Nat Biotechnol 23: 399. doi:10.1038/nbt0405-399.
15. Taylor JD, Kantor YI, Sysoev AV (1993) Foregut anatomy, feeding mechanisms, relationships and classification of the Conoidea (= Toxoglossa) (Gastropoda). Bull Nat Hist Museum London (Zoology) 59: 125–170.
16. Kohn AJ, Perron FE (1994) Life History and Biogeography – Patterns in Conus. Oxford Science Publications.
17. Monteiro A, Tenorio MJ, Poppe GT (2004) A Conchological Iconography. The Family Conidae. The West African and Mediterranean Species of Conus. Germany: ConchBooks, Hackenheim.
18. Tenorio MJ, Tucker JK, Chaney HW (2012) The Families Conilithidae and Conidae – The Cones of the Eastern Pacific. In: Poppe GT, Groh K, editors. A Conchological Iconography. ConchBooks, Hackenheim. p. 200.
19. Franklin JB, Fernando SA, Chalke BA, Krishnan KS (2007) Radular morphology of Conus (Gastropoda: Caenogastropoda: Conidae) from India. Molluscan Res 27: 111–122.
20. Zou S, Li Q, Kong L, Yu H, Zheng X (2011) Comparing the Usefulness of Distance, Monophyly and Character-Based DNA Barcoding Methods in Species Identification: A Case Study of Neogastropoda. PLoS One 6: e26619. doi:10.1371/journal.pone.0026619.
21. Kohn AJ (1990) Tempo and Mode of Evolution in Conidae. Malacologia 32: 55–67.
22. Duda TF, Kohn AJ (2005) Species-level phylogeography and evolutionary history of the hyperdiverse marine gastropod genus Conus. Mol Phylogenet Evol 34: 257–272. doi:10.1016/j.ympev.2004.09.012.
23. Kohn AJ (1983) Marine Biogeography and Evolution in the Tropical Pacific: Zoological Perspectives. Bull Mar Sci 33: 528–535.
24. Kohn AJ (1968) Microhabitats, Abundance and Food of Conus on Atoll Reefs in the Maldive and Chagos Islands. Ecology 49: 1046–1062.
25. Kohn AJ (1959) The Ecology of Conus in Hawaii. Ecol Monogr 29: 47–90.
26. Duda TF, Kohn AJ, Palumbi SR (2001) Origins of diverse feeding ecologies within Conus, a genus of venomous marine gastropods. Biol J Linn Soc 73: 391–409. doi:10.1006/bijl.2001.0544.
27. Olivera BM, Walker C, Cartier GE, Hooper D, Santos AD, et al. (1999) Speciation of cone snails and interspecific hyperdivergence of their venom peptides. Potential evolutionary significance of introns. Ann N Y Acad Sci 870: 223–237.
28. Livett BG, Gayler KR, Khalil Z (2004) Drugs from the sea: conopeptides as potential therapeutics. Curr Med Chem 11: 1715–1723.
29. Remigio EA, Duda TF (2008) Evolution of ecological specialization and venom of a predatory marine gastropod. Mol Ecol 17: 1156–1162. doi:10.1111/j.1365-294X.2007.03627.x.
30. Terlau H, Olivera BM (2004) Conus venoms: a rich source of novel ion channel-targeted peptides. Physiol Rev 84: 41–68. doi:10.1152/physrev.00020.2003.
31. Olivera BM (1997) E.E. Just Lecture, 1996. Conus venom peptides, receptor and ion channel targets, and drug design: 50 million years of neuropharmacology. Mol Biol Cell 8: 2101–2109.
32. Craig AG, Bandyopadhyay P, Olivera BM (1999) Post-translationally modified neuropeptides from Conus venoms. Eur J Biochem 264: 271–275.
33. Skilleter GA, Warren S (2000) Effects of habitat modification in mangroves on the structure of mollusc and crab assemblages. J Exp Mar Bio Ecol 244: 107–129. doi:10.1016/S0022-0981(99)00133-1.
34. Carlton JT (1993) Neoextinctions of Marine Invertebrates. Am Zool 33: 499–509.
35. Zuschin M, Hohenegger J, Steininger FF (2001) Molluscan assemblages on coral reefs and associated hard substrata in the northern Red Sea. Coral Reefs 20: 107–116. doi:10.1007/s003380100140.
36. IUCN Standards and Petitions Subcommittee (2010) Guidelines for Using the IUCN Red List Categories and Criteria. Version 8.0.: 85. Available: http://www.iucnssg.org/tl_files/Assets/pdf/RL Docs/RedListGuidelines.pdf.
37. AfDB OECD, UNDP UNECA (2012) African Economic Outlook 2012, Western African Countries. African Development Bank. Dakar, Senegal. Available: http://www.africaneconomicoutlook.org/en/countries/west-africa/cape-verde/.
38. Irwin A, Wilson C (2011) Cape Verde. Bradt Travel Guides Ltd.
39. IUCN (2012) IUCN Red List Categories and Criteria: Version 3.1. Second edition. Gland, Switzerland and Cambridge, UK: IUCN. iv.
40. Roberts CM, McClean CJ, Veron JEN, Hawkins JP, Allen GR, et al. (2002) Marine biodiversity hotspots and conservation priorities for tropical reefs. Science 295: 1280–1284. doi:10.1126/science.1067728.
41. Gerlach J, Hoffman Black S, Hochkirch A, Jepsen S, Seddon M, et al. (2012) Terrestrial invertebrate life. In: Collen B, Böhm M, Kemp R, Baillie JEM, editors. Spineless: Status and trends of the world's invertebrates. Zoological Society of London, United Kingdom. pp. 46–57.
42. Darwall W, Seddon M, Clausnitzer V, Cumberlidge N (2012) Freshwater invertebrate life. In: Collen B, Böhm M, Kemp R, Baillie JEM, editors. Spineless: Status and trends of the world's invertebrates. Zoological Society of London, United Kingdom. pp. 26–33.
43. González-Mancebo JM, Dirkse GM, Patiño J, Romaguera F, Werner O, et al. (2012) Applying the IUCN Red List criteria to small-sized plants on oceanic islands: conservation implications for threatened bryophytes in the Canary Islands. Biodivers Conserv 21: 3613–3636. doi:10.1007/s10531-012-0385-0.
44. Kemp R, Peters H, Allcock L, Carpenter KE, Obura D, et al. (2012) Marine invertebrate life. In: Collen B, Böhm M, Kemp R, Baillie JEM, editors. Spineless: Status and trends of the world's invertebrates. Zoological Society of London, United Kingdom. pp. 34–45.
45. Collette BB, Carpenter KE, Polidoro BA, Juan-Jorda MJ, Boustany A, et al. (2011) High Value and Long Life — Double Jeopardy for Tunas and Billfishes. Science (80-) 333: 291–292. doi:10.1126/science.1208730.
46. Roberts CM (2012) Ocean of Life. Allen Lane/Penguin Books.
47. Devantier LM (1992) Rafting of tropical marine organisms on buoyant coralla. Mar Ecol Prog Ser 86: 301–302.
48. Cunha RL, Castilho R, Rüber L, Zardoya R (2005) Patterns of cladogenesis in the venomous marine gastropod genus Conus from the Cape Verde islands. Syst Biol 54: 634–650. doi:10.1080/10635159100747l.
49. Duda TF, Palumbi SR (1999) Developmental shifts and species selection in gastropods. Proc Natl Acad Sci U S A 96: 10272–10277.
50. Myers RA, Hutchings JA, Barrowman NJ (1997) Why do fish stocks collapse? The example of cod in the Atlantic Canada. Ecol Appl 7: 91–106.
51. Thurstan RH, Brockington S, Roberts CM (2010) The effects of 118 years of industrial fishing on UK bottom trawl fisheries. Nat Commun 1: 15. doi:10.1038/ncomms1013.
52. National Marine Fisheries Service (2008) White Abalone Recovery Plan (Haliotis sorenseni). Available: http://www.nmfs.noaa.gov/pr/pdfs/recovery/whiteabalone.pdf.
53. Moore A, Ndobe S (2008) Reefs at risk in Central Sulawesi, Indonesia – status and Outlook. Proceedings of the 11th International Coral Reef Symposium, Ft. Lauderdale, Florida, 7-11 July 2008 Session number 18. pp. 840–844. Available: http://nova.edu/ncri/11icrs/proceedings/files/m18-35.pdf.
54. Wells S (1997) Giant clams: Status, trade and mariculture, and the role of CITES in management. IUCN—the World Conservation Union (Gland, Switzerland).
55. Chivian E, Roberts CM, Bernstein AS (2003) The threat to cone snails. Science 302: 391. doi:10.1126/science.302.5644.391b.
56. Chivian E, Roberts CM, Bernstein AS (2004) Response to: How much at risk are cone snails? Science 303: 955–957.
57. Duda TF, Bingham J-P, Livett BG, Kohn AJ, Raybaudi Massilia G, et al. (2004) How much at risk are cone snails? Science 303: 955–957.
58. Hopkins C, Grilley M, Miller C, Shon KJ, Cruz LJ, et al. (1995) A new family of Conus peptides targeted to the nicotinic acetylcholine receptor. J Biol Chem 270: 22361–22367.
59. Livett BG, Sandall DW, Keays D, Down J, Gayler KR, et al. (2006) Therapeutic applications of conotoxins that target the neuronal nicotinic acetylcholine receptor. Toxicon 48: 810–829. doi:10.1016/j.toxicon.2006.07.023.
60. Urbarova I, Karlsen BO, Okkenhaug S, Seternes OM, Johansen SD, et al. (2012) Digital marine bioprospecting: mining new neurotoxin drug candidates from the transcriptomes of cold-water sea anemones. Mar Drugs 10: 2265–2279. doi:10.3390/md10102265.
61. Airoldi L, Balata D, Beck MW (2008) The Gray Zone: Relationships between habitat loss and marine diversity and their applications in conservation. J Exp Mar Bio Ecol 366: 8–15. doi:10.1016/j.jembe.2008.07.034.
62. Munday PL (2004) Habitat loss, resource specialization, and extinction on coral reefs. Glob Chang Biol 10: 1642–1647. doi:10.1111/j.1365-2486.2004.00839.x.
63. Thrush SF, Dayton PK (2002) Disturbance to Marine Habitats by Trawling and Dredging: Implications for Marine Biodiversity. Annu Rev Ecol Syst 33: 449–473. doi:10.1146/annurev.ecolsys.33.010802.150515.
64. Watling L, Norse EA (1998) Effects of Mobile Fishing Gear on Marine Benthos. Conserv Biol 12: 1178–1179. doi:10.1046/j.1523-1739.1998.0120061178.x.
65. Gray JS, Dayton P, Thrush S, Kaiser MJ (2006) On effects of trawling, benthos and sampling design. Mar Pollut Bull 52: 840–843. doi:10.1016/j.marpolbul.2006.07.003.
66. Dulvy NK, Sadovy Y, Reynolds JD (2003) Extinction vulnerability in marine populations. Fish Fish 4: 25–64. doi:10.1046/j.1467-2979.2003.00105.x.
67. Riascos JM, Heilmayer O, Laudien J (2007) Population dynamics of the tropical bivalve Cardita affinis from Málaga Bay, Colombian Pacific related to La Niña 1999–2000. Helgol Mar Res 62: 63–71. doi:10.1007/s10152-007-0083-6.
68. Australian Government (2012) Record breaking La Niña events; An analysis of the La Niña life cycle and the impacts and significance of the 2010–11 and 2011–12 La Niña events in Australia. Melbourne, Australia: Bureau of Meteorology.
69. Doney SC, Fabry VJ, Feely RA, Kleypas JA (2009) Ocean Acidification: The Other CO2 Problem. Ann Rev Mar Sci 1: 169–192. doi:10.1146/annurev.-marine.010908.163834.
70. Rodolfo-Metalpa R, Houlbrèque F, Tambutté É, Boisson F, Baggini C, et al. (2011) Coral and mollusc resistance to ocean acidification adversely affected by warming. Nat Clim Chang 1: 308–312. doi:10.1038/nclimate1200.
71. Tonge S, Bloxam Q (1991) A review of the captive-breeding programme for Polynesian tree snails. Int Zoo Yearb 30: 51–59.
72. Lopes EP (2010) Recent data on marine bivalves (Mollusca, Bivalvia) of the Cape Verde Islands, with records of six species new to the archipelago. Zool Caboverdiana 1: 59–70.
73. Monteiro A, Malcolm G, Herndl G (2012) Endangered cone species. Cone Collect (Ed A Monteiro): 36. Available: http://www.theconecollector.com/. Accessed 1 January 2013.

11

Quantitative and Functional Characterization of the Hyper-Conserved Protein of *Prochlorococcus* and Marine *Synechococcus*

Caroline E. Whidden[1]*, Katrina G. DeZeeuw[1]*, Jackie K. Zorz[1], Andrew P. Joy[2], David A. Barnett[2], Milo S. Johnson[3], Olga Zhaxybayeva[3,4]*, Amanda M. Cockshutt[1]*

1 Department of Chemistry & Biochemistry, Mount Allison University, Sackville, NB, Canada, **2** Atlantic Cancer Research Institute, Moncton, NB, Canada, **3** Department of Biological Sciences, Dartmouth College, Hanover, New Hampshire, United States of America, **4** Department of Computer Science, Dartmouth College, Hanover, New Hampshire, United States of America

Abstract

A large fraction of any bacterial genome consists of hypothetical protein-coding open reading frames (ORFs). While most of these ORFs are present only in one or a few sequenced genomes, a few are conserved, often across large phylogenetic distances. Such conservation provides clues to likely uncharacterized cellular functions that need to be elucidated. Marine cyanobacteria from the *Prochlorococcus*/marine *Synechococcus* clade are dominant bacteria in oceanic waters and are significant contributors to global primary production. A Hyper Conserved Protein (PSHCP) of unknown function is 100% conserved at the amino acid level in genomes of *Prochlorococcus*/marine *Synechococcus*, but lacks homologs outside of this clade. In this study we investigated *Prochlorococcus marinus* strains MED4 and MIT 9313 and *Synechococcus* sp. strain WH 8102 for the transcription of the PSHCP gene using RT-Q-PCR, for the presence of the protein product through quantitative immunoblotting, and for the protein's binding partners in a pull down assay. Significant transcription of the gene was detected in all strains. The PSHCP protein content varied between 8±1 fmol and 26±9 fmol per ug total protein, depending on the strain. The 50 S ribosomal protein L2, the Photosystem I protein PsaD and the Ycf48-like protein were found associated with the PSHCP protein in all strains and not appreciably or at all in control experiments. We hypothesize that PSHCP is a protein associated with the ribosome, and is possibly involved in photosystem assembly.

Editor: Christos A. Ouzounis, Hellas, Greece

Funding: This work was supported by the Marjorie Young Bell Faculty Fellow Award, the Mount Allison President's Research and Creative Activity Fund (www.mta.ca) and the Natural Science and Engineering Research Council (NSERC) (http://www.nserc-crsng.gc.ca) Discovery Team grant to AMC. JKZ and CEW received Undergraduate Summer Research Awards from NSERC (http://www.nserc-crsng.gc.ca) and Mount Allison University (www.mta.ca). OZ was supported in part by a research stipend from Mount Allison University and by start-up funds from Dartmouth College. The study was in part supported by Dartmouth College Dean of Faculty start-up funds and Walter and Constance Burke Research Initiation Award to OZ. The funders had no role in the study design, data collection and analysis, decision to publish, or preparation of the manuscript.

Competing Interests: Amanda M. Cockshutt has read the journal's policy and has the following conflict: Amanda M. Cockshutt is a part owner of Environmental Proteomics NB Inc. who developed with Agrisera AB the antibodies used in this work. The authors confirm that co-author Amanda M. Cockshutt is a PLOS ONE Editorial Board Member. This does not alter the authors' adherence to PLOS ONE Editorial policies and criteria.

* Email: olga.zhaxybayeva@dartmouth.edu (OZ); acockshutt@mta.ca (AMC)

These authors contributed equally to this work.

Introduction

Cyanobacteria are ubiquitous oxygenic phototrophs on Earth, colonizing all environments reached by solar light and largely responsible for the fifty percent of global net primary production that occurs in the world's oceans [1,2]. Numerically dominating the oceanic waters are marine free-living unicellular picocyanobacteria from *Prochlorococcus* and Cluster 5.1 *Synechococcus* genera [3]. *Prochlorococcus* is the smallest known phototroph that inhabits subtropical, oligotrophic open ocean environments, where it thrives under extremely limiting concentrations of essential nutrients and contributes significantly to chlorophyll (Chl) biomass and primary production [4]. Marine *Synechococcus* spp. are better adapted to coastal water environments, although they do occur in the open ocean and their geographic distribution overlaps with that of *Prochlorococcus* [5–7]. On ribosomal RNA phylogeny *Prochlorococcus* and marine *Synechococcus* form a monophyletic clade [8]. The adaptations of *Prochlorococcus* and marine *Synechococcus* to their complementary ecological niches have been the subject of many investigations (e.g., [6,7,9–12]). There is emerging evidence that picocyanobacteria specifically may become even more implicated in the global carbon cycle with the escalating effects of climate change on the world's oceans. Ongoing climate warming has been correlated with a decline of total phytoplankton biomass [13,14], and a shift towards a major dominance of picoplankton in phytoplankton communities is anticipated [15].

Among cyanobacteria *Prochlorococcus* and marine *Synechococcus* have the smallest genomes [4]. Minimization of a genome by selective reduction and deletion of non-essential genes is hypothesized to be an evolutionary strategy to decrease the amount of energy and nutrients needed to maintain cellular processes in a resource-limited environment [16]. Hence, reduced genome size may explain the ecological success of picocyanobacteria in oligotrophic areas of the ocean [1]. Comparative analyses of genomes of *Prochlorococcus* and marine *Synechococcus* reveal a substantial gene flow among *Prochlorococcus* and marine *Synechococcus* spp. [17], as well as the presence of strain-specific genomic islands [4]. Despite high similarity among their 16S rRNA gene sequences, the rest of their genomes are surprisingly divergent and exhibit variable GC content [18]. This divergence is likely due to *Prochlorococcus*' high mutation rates in combination with the lack of several DNA repair enzymes in the genome, as a direct result of the aggressive genome reduction [10]. A notable exception from the observed sequence divergence patterns is a protein-coding gene of unknown function, termed PSHCP for *Prochlorococcus/Synechococcus* Hyper Conserved Protein [18]. The PSHCP ORF encodes a polypeptide of 62 amino acid residues that exhibits 100% amino acid sequence conservation and is universally present in this clade of marine picocyanobacteria.

Such remarkable amino acid sequence conservation and consistent inclusion within the clade suggests that the PSHCP gene product performs an essential role in a picocyanobacterial cell. The DNA sequence of the ORF has accumulated numerous synonymous substitutions and its nucleotide composition mimics GC content of the host genome, indicating that the protein has been under purifying selection since the last common ancestor of the *Prochlorococcus* and marine *Synechococcus*, and has not been recently transferred among the members of the clade [18]. The gene synteny in the immediate genomic region surrounding the *PSHCP* gene is conserved across all examined picocyanobacteria: the gene is flanked on its 5′ end by *rpl19*, encoding the large ribosomal subunit protein 19 and the tryptophanyl-tRNA, and on its 3′ end by the aspartyl-tRNA and *gltX*, encoding a glutamyl tRNA synthetase (Fig. 2 in [18]). Since in prokaryotes the genes of functionally interacting proteins tend to be present in operons [19], the functional assignments of neighboring genes suggest that PSHCP could have a role in protein translation. Predicted features of the protein provide additional clues to its function. PSHCP is highly basic with a predicted isoelectric point (pI) of 11.3 and a high proportion (16%) of positively charged amino acids, suggestive of PSHCP's role in RNA/DNA-interactions [18].

At the time of initial discovery, PSHCP lacked homologs in any other cyanobacteria or other organisms for which sequence data was available in GenBank [18]. Presence of the PSHCP gene is now reported in four freshwater *Synechococcus* spp. strains isolated from Laurential Great Lakes and in a photosynthetic endosymbiont of the amoeba *Paulinella chromatophora* [20]. Similar to their marine counterparts, freshwater *Synechococcus* spp. dominate the euphotic zone of oligotrophic lakes, both in numbers and in overall productivity. In the freshwater *Synechococcus* spp. strains PSHCP retained its 100% amino acid sequence identity to the marine homologs and is flanked by the same genes in the genome [20]. Constitutive transcription of *PSHCP* in the freshwater strains was observed under various irradiances and under limiting conditions of nitrogen and phosphorus, and PCR experiments suggested that *rpl19* and *PSHCP* are co-transcribed [20]. In the highly reduced genome of the endosymbiont (chromatophore) of *Paulinella chromatophora*, the *PSHCP* protein is 89% identical to its picocyanobacterial homologs and is also flanked by *rpl19* and *gltX* genes [20]. Although it is no longer 100% conserved, it is remarkable, however, that the PSHCP gene is retained in the chromatophore genome that underwent a three fold size reduction from the genome of its ancestral free-living cyanobacterium, again pointing to the potentially essential role of PSHCP. The closest relative of chromatophore is a freshwater cyanobacterium from the *Cyanobium* clade, a sister clade to the clade of *Prochlorococcus* and marine *Synechococcus*, suggesting that PSHCP has originated in the common ancestor of marine and freshwater picocyanobacteria. In this study, we used a combination of bioinformatic and laboratory approaches to evaluate expression of the PSHCP gene, to quantify the PSHCP protein abundance, and to suggest possible PSHCP functions.

Materials & Methods

Promoter Prediction

Homologs of the PSHCP gene were retrieved from GenBank using BLAST [21] and *Synechococcus* sp. WH 8102 as a query (since PSHCP amino acid sequences are identical in all picocyanobacteria, the choice of strain will not affect the results). To predict potential promoters, WebLOGOs [22] were constructed for the DNA sequences of the region upstream of the PSHCP gene and compared with the consensus sequence for the experimentally determined −10 promoter element from *Prochlorococcus* sp. MED4 [23].

Culturing

Cultures of *Synechococcus* sp. WH 8102 and *Prochlorococcus* spp. MED4 and MIT 9313 obtained from the Provasoli-Guillard National Center for Marine Algae and Microbiota were grown at 22°C in a 1:4 dilution in PCR S11 and Pro99 medium, respectively [24]. MED4 was exposed to 40 μmol photons $m^{-2}s^{-1}$ of light, and WH 8102 and MIT 9313 were exposed to 20 μmol photons $m^{-2}s^{-1}$. Cultures were harvested at a quantum yield of photosynthesis between 0.4 and 0.6 determined by the ratio of the variable fluorescence (F_V) to the maximum fluorescence (F_M) using a Xenon-PAM fluorometer (Walz), as described previously [25].

RT-PCR Data

Cell pellets were obtained by centrifugation and RNA was extracted using the Trizol Kit (Life Technologies) as per the manufacturer's instructions. RNA extracts were treated with DNase I (Life Technologies) and quantified using the Quant-it RNA Broad Range Assay Kit (BR Kit, Life Technologies). Total RNA (60 ng) was reverse transcribed with random hexamer primers using the iScript cDNA Synthesis Kit (Bio-Rad), with reverse transcriptase omitted from the –RT negative control samples. Primers were designed to amplify specific regions surrounding the *pshcp* gene (Table S1 and Figure 1A). Calibrated standards were produced by amplifying the entire region of interest for all species (869 bp for WH 8102, 835 bp for MIT 9313, and 817 bp for MED4) using genomic DNA as a template. The PCR products were purified using a MinElute PCR Purification Kit (Qiagen) and quantified using the Quant-it DNA BR Kit (Life Technologies). Triplicate samples were prepared for RT-PCR using iQ SYBR Green Supermix (Bio-Rad) with a final cDNA content of 1.5 ng and 40 nM of forward and reverse primers. Samples were amplified using the following program in an iCycler thermal cycler (Bio-Rad): 95°C for 2 min; 40 cycles of 95°C for 30 s, 54°C for 30 s, 72°C for 30 s; 5 min at 72°C.

Cloning of *pshcp* gene

Cell pellets were obtained from WH 8102 cell culture by centrifugation, and genomic DNA was extracted using the Ultraclean Microbial DNA Isolation Kit (MoBio). Primers were designed to specifically amplify the *pshcp* gene with incorporated restriction enzyme cut sites to facilitate downstream subcloning. The 199 bp *pshcp* insert was amplified using PCR, purified using a

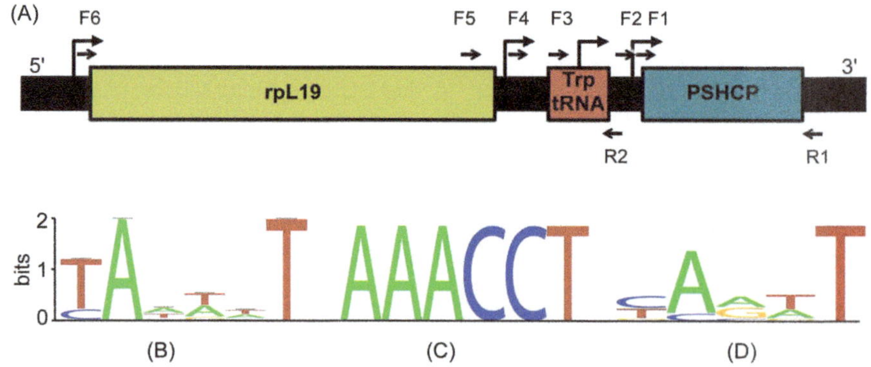

Figure 1. Schematic presentation of the genomic region around the PSHCP gene. Panel A. Mapping of promoters and primers onto the region. In all examined picocyanobacterial genomes the upstream region consists of the ribosomal protein L19 (rpL19) and tryptophanyl-tRNA genes. The locations of predicted promoters are shown by bent arrows, while the locations of the reverse (R1) and forward (F1 to F6) primers used for RT-Q-PCR analysis are indicated by half arrows (see Table S1 for further details). Panels B–D. Conservation of predicted and experimentally determined promoters across homologous regions from 18 *Prochlorococcus marinus* and *Synechococcus* spp. genomes. The regions were visualized using the WebLOGO web site (http://weblogo.berkeley.edu/). Panel B. Conservation of experimentally determined promoters from *Prochlorococcus marinus* MED4 [23]. Panel C. Conservation of predicted promoter located 37 bases into the *trp-tRNA* gene. Panel D. Conservation of the predicted promoter in the 5′ region 12–14 bases upstream of the *PSHCP* start codon. This promoter is present in all analyzed genomes except *Synechococcus* sp. RCC307. For promoters predicted upstream of the *trp-tRNA* and *rpl19* genes (panel A) WebLOGOs are not shown.

HiYield PCR DNA Extraction Kit (Real Genomics) and quantified as described above. The insert was cloned into the pCR2.1 vector and transformed into One Shot competent *E. coli* cells using the TOPO Cloning Kit (Life Technologies). Plasmid DNA was isolated from selected *E. coli* colonies using the Wizard Plus SV Miniprep DNA Purification System (Promega) and quantified as described above. The presence and integrity of the insert were confirmed by DNA sequencing by McGill University and Genome Quebec Innovation Centre.

The *pshcp* gene was subcloned into the pIVEX 2.4d vector following sequential restriction enzyme digestion by NcoI and XmaI (BioLabs), cleaning of the fragment (MinElute Gel Extraction Kit, Qiagen) and ligation with T4 DNA ligase reaction (Life Technologies) and transformation into competent DH5α *E. coli* cells.

Overexpression, purification, and preparation of PSHCP for antibody production

The Rapid Translation System (RTS) 500 *E. coli* HY Kit (Roche) was used for large scale overexpression of the PSHCP protein for antibody production. Purified overexpressed PSHCP was separated by gel electrophoresis on 2-well 1.0 mm Novex 4–12% Bis-Tris NuPAGE gels (Life Technologies) and stained with SimplyBlue SafeStain (Life Technologies). Bands corresponding to PSHCP were excised and used as immunogen for antibody elicitation in rabbits (Agrisera AB).

Antibody Validation

The specificity of the antibody was determined by comparing the immune serum to the pre-immune and by performing a competition assay, pre-incubating the immune serum with the immunogen protein to block reactivity (data not shown).

Whole Cell Protein Extraction and Quantitative Immunoblotting

Cell pellets were harvested from 50 mL of each culture and resuspended in 200 µL of protein extraction buffer (1X AEBSF, Roche and 1X PSB, Agrisera). Samples were then homogenized using the CY:24×2 rotor of the FastPrep-24 Instrument (MPBIO) for two 1 min periods with 1 min of cooling on ice after each treatment. The protein extracts were then centrifuged at 10,000×g for 3 min to remove insoluble material and unbroken cells. Protein concentrations of extracts were quantified using the microplate DC protein assay (Bio-Rad) with bovine gamma globulin (BGG) as a comparative protein standard.

The sample preparation, SDS-PAGE, immunoblotting and quantitation protocols were followed according to those described previously [26]. Immunoblotting of PSHCP required a shorter transfer of 45 min. onto nitrocellulose membranes. Primary antibodies (Agrisera) were diluted in 2% ECL advance blocking agent (GE Healthcare) in TBS-T, RbcL 1:10,000 and PSHCP 1:2,000. HRP-conjugated goat anti-rabbit (Agrisera) was employed as the secondary antibody at a dilution of 1:40,000. Immunoblots were developed with ECL Advance (GE Healthcare) and images were captured on a VersaDoc Imager (BioRad) and analyzed using Quantity One and Image Lab 3.0 software (BioRad).

Pull Down Assay: Bait Preparation

Recombinant His_6-PSHCP was overexpressed and purified with the NiNTA Purification System (Life Technologies) under native conditions, and employed as bait protein in the pull down assay. Genomic *pshcp* was PCR amplified using specific primers designed to facilitate its insertion into a pET303/C-terminal His_6-tag vector, and the presence and proper sequence of the *pshcp* insert was subsequently confirmed as described above. The ligated pET303 vector was transformed into BL21 *E. coli* overexpression cells (Life Technologies). The overexpression of *pshcp* was induced with IPTG for 3 hr according to the expression guidelines, and cells were harvested by centrifugation.

The lysate was prepared under native conditions, according to the manufacturer's protocol for bacterial cell lysates with some alterations. Buffer concentrations of imidazole were increased in the lysate preparation (25 mM) and in the subsequent purification of His_6-PSHCP in order to generate a more pure product. Lysis was achieved using a Branson Digital Sonifier at 70% amplification with 3×10 s, 3×35 s bursts on ice. The lysate was centrifuged at 13,000×g, for 15 min, at 4°C to remove cellular debris. The manufacturer's instructions for purification under native conditions were followed. The column was washed 2X with 35 mM imidazole native binding buffer (NBB), 2X with 50 mM, and eluted with 500 mM. 100 µg bait protein was brought to a final volume of 1–1.5 mL in 25 mM imidazole NBB using an Amicon 3K Ultra-15 Centrifugal Filter Unit.

Pull Down Assay: Prey Preparation

A whole cell native extraction of *Prochlorococcus* and marine *Synechococcus* was performed on cell pellets harvested from 200 mL of culture of significant biomass (day 7 of growth) by centrifugation. The pellet was resuspended in a solution of 1X EBA (110 mM KOAc, 0.5% Triton X-100, 100 mM NaCl, 1 mM AEBSF, pH 7.4) with 0.05% Nonidet P-40 at a maximum total volume of 300 µL. The cells were subject to two flash-freeze-thaw (FT) cycles in liquid nitrogen to allow for partial breakage, and four subsequent rounds of sonication/FT, each round consisting of 4×10 s bursts on ice at 70% amplification followed by FT. The final solution was centrifuged at 10,000×g, for 5 min, at room temperature. The remaining pellet was re-treated with the same procedure in order to extract all possible protein. The amount of protein in the combined extracts was determined using the microplate DC protein assay with BGG used as a protein standard. A portion of the extract was diluted with 1X EBA to obtain 500 µg of protein in 500 µL (1 mg/mL).

Pull Down Assay

NiNTA resin was resuspended and 50 µL each was placed into sterile 1.5 mL microcentrifuge tubes. Resins were equilibrated according to manufacturer's native purification protocol, using 500 µL sterile ddiH_2O and 25 mM imidazole NBB. To experimental columns, prepared bait protein was added (100 µg in 1–1.5 mL). To control columns, an equal volume of 25 mM imidazole NBB was added. Tubes were rotated end-over-end for 1 hr at room temperature. The non-binding bait solutions were aspirated and 500 µg prepared prey sample was added to all tubes (1:5 bait to prey mass ratio). Tubes were rotated end-over-end for 2 hr at room temperature to allow proteins to bind. The non-binding prey solutions were aspirated and resins were washed 3–4X with 25 mM imidazole NBB. Proteins were eluted in 5×100 µL fractions of 500 mM imidazole NBB.

In-solution Tryptic Digest

The pull down products, experimental and control, were subject to an in-solution tryptic digest in preparation for mass spectrometry analysis. To each sample, TRIS (1 M), pH 8.3 was added to a final concentration of 50 mM. DTT was added to a final concentration of 10 mM and the samples were incubated at 50°C for 30 min. Iodoacetamide (1 M) was added to a final concentration of 40 mM. The samples were vortexed briefly and incubated in the dark at room temperature for 30 min. A 1 µg/µL trypsin solution was added at a trypsin: protein ratio of approximately 1:40. The samples were incubated overnight (16–18 h) at 37°C. The reaction was stopped by adding 1/9th volume of 1% TFA solution. Samples were frozen at −20°C until mass spectrometry analysis.

Mass Spectrometry Analysis

Protein tryptic digests were analyzed by gradient nanoLC-MS/MS using a Quadrupole Orbitrap (Q-Exactive, Thermo-Fisher Scientific) mass spectrometer interfaced to a Proxeon Easy Nano-LC. Samples were adjusted to 1% aqueous acetic acid and injected (10 µL) onto a narrow bore (20 µm long ×100 µm inner diameter) C18 pre-column packed with 5 µm ReproSil-Pur resin (Thermo-Fisher Scientific). High resolution chromatographic separation was then achieved on a Thermo-Scientific Easy C18 analytical column with dimensions of 100 mm by 75 µm i.d. using 3 µm diameter ReproSil-Pur particles. Peptide elution was achieved using an acetonitrile/water gradient system, with LC-MS grade water and acetonitrile (VWR). Solvent A consisted of 0.1% formic acid (Sigma-Aldrich) in water and solvent B was made up of 90/9.9/0.1 acetonitrile/water/formic acid. A linear acetonitrile gradient was applied to the C18 column from 5–30% solvent B in 120 min followed by 100% B for 10 min at a flow rate of 300 nL/min.

The outlet diameter of the nano-flow emitter on the Q-Exactive (15 µm) was biased to +1.9 kV and positioned approximately 2 mm from the heated (250°C) transfer capillary. The S-lens of the mass spectrometer was maintained at 100 V. The Q-Exactive mass spectrometer was calibrated in positive ion mode with mass standards (caffeine, MRFA peptide and Ultramark) every three days as recommended by the instrument manufacturer. Mass spectrometric data was acquired in data dependent acquisition (DDA) mode, whereby a full mass scan from 350–1200 Th was followed by the acquisition of fragmentation spectra for the five most abundant precursor ions with intensities above a threshold of 20,000. Precursor ion spectra were collected at a resolution setting of 35,000 and an AGC (automatic gain control) value of 1×10^6. Peptide fragmentation was performed using high energy collision induced dissociation in the HCD cell and MS/MS spectra were collected in the Orbitrap at a resolution of 17,500 and an AGC setting of 1×10^5. Peptide precursors were selected using a repeat count of 2 and a dynamic exclusion period of 20 s.

Mass spectrometric protein identification data was analyzed using Proteome Discoverer version 1.3 (Thermo-Fisher Scientific) employing the Sequest scoring algorhithm. FASTA databases were obtained from Uniprot [27] for *Escherichia coli* (sp. BL21, 2041 kb), *Prochlorococcus marinus* (strains MED4, 774 kb and MIT 9313, 1028 kb) and *Synechococcus* sp. (strain WH 8102, 1010 kb). Searches were performed with the following settings: (a) enzyme specificity of trypsin with 2 allowed missed cleavages, (b) precursor and fragment tolerances were 10 ppm and 0.8 Da, respectively, (c) a variable modification of methionine oxidation (+15.99 Da), and (d) a fixed modification of cysteine carbamidomethylation (+57.02 Da). Proteome Discoverer 1.3 calculated a strict false discovery rate (FDR) of 0.1% based on the results of a decoy (reverse) database search. Peptides were scored with Sequest using minimal Xcorr values of 1.62 for +2 ions and 1.79 for +3 ions. Proteins were assigned a positive identification if at least one peptide was identified with high confidence or at least two unique peptides were identified with medium confidence.

Survey of PSHCP presence in GenBank and draft cyanobacterial genomes

Using the PSHCP protein from *Prochlorococcus marinus* str. CCMP1986 as a query in BLASTP searches, homologs of the PSHCP were retrieved from GenBank and the IMG portal [28]. The homologs were examined for ORF length and for the presence of amino acid substitutions. Amino acid composition of the PSHCP and pI were predicted using the EMBOSS package [29]. Conserved domains were predicted using CD search [30].

Information Content of Ribosomal RNA Genes in PSHCP-containing Cyanobacteria

110 completed genomes of free-living cyanobacteria available in GenBank were screened for the presence of PSHCP using BLAST searches and the PSHCP protein in *Prochlorococcus marinus* AS9601 as a query [21]. 5 S, 16 S and 23 S ribosomal RNA genes

from these cyanobacterial isolates, as well as from *Escherichia coli* K-12 substr. MG1655 (used as a reference), were downloaded from the SILVA database [31] and aligned using the SINA program [32]. Alignments were edited manually for several misaligned regions and the alignment positions with>50% gaps were eliminated. The alignments were screened for "signature" positions, defined as a nucleotide in an alignment site present in ≥90% of PSHCP-containing cyanobacteria and in ≤10% of the remaining cyanobacteria (analogously to [33]). The variation in the alignment sites in cyanobacteria that have or do not have PSHCP was also summarized using Shannon entropy [34]. More specifically, frequencies f of each nucleotide N (A, C, U and G) at each site s of 23 S rRNA alignment were calculated for the two groups of cyanobacteria and uncertainty H at site s (in bits) was calculated using Shannon's formula [34]:

$$H(s) = -\sum_{N=A,U,G,C} f(N,s)\log_2 f(N,s).$$

Total information content of the alignment site $T(s)$ was calculated on a 0–1 scale, and then a gap penalty was applied:

$$T(s) = \frac{H_{\max} - H(s)}{H_{\max}}(1 - f(gap)),$$

where $H_{max} = 2$ bits. Average information across all sites T_{avg} was calculated for each group, to compensate for the fact that two cyanobacterial groups have different level of overall sequence divergence. The relative information for each site $R(s)$ was defined as

$$R(s) = \frac{T(s)}{T_{avg}}.$$

The $R(s)$ values were mapped onto a secondary structure model of *Escherichia coli* described by Petrov *et al.* [35]. Distances of *E.coli*'s rRNA residues to the nearest amino acid residue of the ribosomal protein L2 were determined using the 3D structure information from the 3R8T record of the Protein Data Bank [36].

Results

PSHCP is universally present in one cyanobacterial clade

A survey of newly available cyanobacterial genomes reveals that PSHCP continues to be present in all *Prochlorococcus marinus* and marine *Synechococcus* genomes, as well as in freshwater strains of *Synechococcus* [20], in *Cyanobium* spp. and in the chromatophore genome of *Paulinella chromatophora* (Table S2). On an rRNA phylogenetic tree of cyanobacteria, these lineages form a monophyletic clade [3]. As noted in [18], the observed length variants of the *pshcp* genes are due to a misannotation of the start codon and a two amino acid variation at the C-terminus. Additionally, amino acid substitutions were observed in the PSHCP homologs from the genomes of *Synechococcus* sp. strains RCC307 and WH7803, and of the *Paulinella*'s chromatophore. The latter genome has undergone significant size reduction and exhibit an increased mutation rate [37], therefore it is not surprising that its PSHCP homolog does not show the extreme amino acid sequence conservation seen in the free-living cyanobacteria. Overall, the survey of the PSHCP homologs preserves our earlier observation that this protein is under strong purifying selection [18].

Amino Acid Composition and Predicted Biochemical Properties of PSHCP

PSHCP has a predicted isoelectric point of 11.3 and contains 10 (16%) positively charged amino acids (8 arginines and 2 lysines). Moreover, PSHCP contains seven glycines. The predicted secondary structure consists of an alpha helix and four small beta strands (Fig. 3 in [18]). These properties are consistent with those of many ribosomal proteins, which have stretches of sequences rich in arginine and lysine, as well as high glycine content in the "extensions" [38]. A few ribosomal proteins contain SH3-like domains, and PSHCP has a weak similarity to an SH3-like domain, as detected in a Conserved Domain Database [39] search (data not shown). Tertiary structures of very small proteins (including those of several ribosomal proteins) are often stabilized by disulfide bonds or binding to zinc [40]; however, PSHCP's amino acid sequence lacks the indicative cysteines or histidine residues that would be required for Zn-binding fold to function properly.

Transcriptional Analyses Identify Likely Transcription Start Sites for PSHCP Gene

To confirm PSHCP expression and to determine if PSHCP gene is expressed polycistronically with neighboring genes, we performed RT-Q-PCR experiments on *Prochlorococcus marinus* strains MED4 and MIT9313, and on marine *Synechococcus* sp. strain WH 8102. Utilizing the experimentally determined promoter sequence in *Prochlorococcus marinus* strain MED4 [23] (Figure 1B), we predicted promoters in the region upstream of the PSHCP genes (Figure 1C and D). To determine which, if any, of the predicted promoters are used to transcribe the *pshcp* gene, we performed RT-Q-PCR with a series of forward primers ranging from the beginning of the *pshcp* coding region through to the beginning of the *rpl19* gene. The inclusion of a standard curve allowed for absolute quantitation and ultimately a comparison between the strains. Table 1 presents the number of copies of each transcript per ng of RNA template. *Prochlorococcus marinus* strain MED4 has the highest number of copies of *pshcp* transcripts per ng of total RNA. In this strain, 42.8% of the transcripts appear to arise from the promoter predicted between the *tryptophanyl-tRNA* gene and the *pshcp* gene, whereas there is little to no apparent transcription from this promoter in the other strains. The bulk of the *pshcp* transcripts in MIT 9313 (74.9%) and WH 8102 (81.1%) strains arise from the strong predicted promoter within the *tryptophanyl-tRNA* gene. Larger transcripts are detected in all three strains, but to a greater degree in MED4, indicating some co-transcription with the *tryptophanyl-tRNA* and *rpl19*. It is possible that some of the shortest *pshcp* transcripts detected in MED4 arise from the polycistronic processing of mRNA rather than from transcription from the predicted promoter between *pshcp* and the *trp-tRNA*. Our results are largely in agreement with the earlier report [20] of co-transcription of *PSHCP* with *rpl19*, the *trp-tRNA* and the downstream *asp-tRNA*, but not with the downstream *gltX* gene in both the freshwater *Synechococcus* sp. strain ARC-11 and the marine *Synechococcus* strain WH 8102. We, however, measure the pool of long transcripts containing both *rpl19* and *pshcp* to represent less than 2% of the total transcript pool. Our transcriptional observations for the region near *pshcp* gene are in general consistent with findings of complex transcriptional patterns of bacterial genomes (e.g., [41,42]).

Table 1. Quantitative Transcript Analysis.

RT-Q-PCR Product	Copies per ng of RNA template (% of the total *pshcp* transcripts)		
	MED4	MIT 9313	WH 8102
hcp only	2826 (42.8)	96 (5.4)	0 (0)
hcp +5′ region	629 (9.5)	1329 (74.9)	679 (81.1)
hcp + *Trp-tRNA*	1566 (23.7)	201 (11.3)	13 (1.6)
hcp + *Trp-tRNA* +5′ region	900 (13.6)	0 (0)	43 (5.1)
hcp + *Trp-tRNA* +3′ end *rpl19*	684 (10.4)	121 (6.8)	93 (11.1)
hcp + *Trp-tRNA* + *rpl19*	0 (0)	29 (1.6)	8 (1.0)
Total	6606	1776	837

Examination of the global gene expression patterns in *Prochlorococcus* MED4 and MIT 9313 show down-regulation of *pshcp* in both strains under nitrogen deprivation conditions [43] and down-regulation of *pshcp* in MIT 9313 (but not MED4) in response to phosphorus starvation [44]. Under both nitrogen and phosphorus limitation the transcription of *pshcp* changes in parallel with the expression of genes involved in protein translation [43,44]. Constitutive expression of *pshcp* reported in [20] did not vary with light level or nutrient limitation. Since the latter study used a single end-point RT-PCR method to measure the level of transcription as opposed to RT-Q-PCR or microarray, subtle differences in expression such as those reported by Tolonen et al. [43] and Martiny et al. [44] may have gone undetected.

PSHCP is a relatively abundant protein

The level of PSHCP protein in whole cell extracts of *Prochlorococcus* and marine *Synechococcus* cultures was on the order of that of the RuBisCO large subunit protein (RbcL), indicating that PSHCP is a relatively abundant protein (Table 2). No correlation was observed between the number of *pshcp* transcripts per ng total RNA and the fmoles of PSHCP protein per ug total cell protein in the cell extracts.

Potential Binding Partners of PSHCP

A large number of proteins were identified by the pull down assay (Table 3 and Table S3). However, three proteins (the 50S ribosomal protein Rpl2, the Photosystem I protein PsaD and the Ycf48-like protein) were pulled down from extracts of *all* strains in the presence of PSHCP, but few or none in its absence. Given that PSHCP is so conserved in all *Prochlorococcus* and marine *Synechococcus* sequenced to date, thus its essential function and binding partners should also be conserved across strains, we have focused our analysis on proteins that were pulled down only in the presence of PSHCP in all three strains examined. Both Rpl2 and PsaD have basic isoelectric points (11.1–11.6 and 7.0–8.7, respectively), and therefore their interactions with PSHCP unlikely to be artifactual consequences of non-specific electrostatic interactions between PSHCP and a prey protein. Since Rpl2 and PsaD would both be highly positively charged at physiological pH, they would also not be expected to interact non-specifically *in vivo*.

Substitution patterns in the ribosomal RNA genes of PSHCP-containing cyanobacteria reveal non-compensatory substitutions

If PSHCP is interacting with Rpl2 protein (and hence likely be part of the ribosome), it may be recruited to stabilize ribosomal RNA structure. In the latter case, we would expect to observe non-compensatory substitutions in rRNA of PSHCP-containing cyanobacteria. Examination of 5 S, 16 S and 23 S rRNA nucleotide substitution patterns in 110 cyanobacterial isolates indeed revealed a few faster evolving sites in the 23 S rRNA of the PSHCP-containing cyanobacteria (Figure S1). Notable non-compensatory substitutions are located in the helical stems 18, 64, 75, 84 and 94 (Figure S1). Helical stems 75 and 76 are in proximity to the ribosomal protein L2 (Figure S2), with which PSHCP protein interacts (Table 3), suggesting that these helical stems could correspond to the regions of 23 S rRNA stabilized by the interactions of PSHCP with the RNA.

Discussion & Conclusions

The modern ribosome is an RNA/protein complex that catalyzes protein synthesis in all living organisms. RNA molecules of the ribosome are catalytically active [45], while the role of the ribosomal proteins is to stabilize the structure of RNAs, often by "filling in the cracks between RNA helices" [38]. In light of these observations, it has been suggested that the primordial ribosome likely consisted of RNA alone and ribosomal proteins were recruited later. A few ribosomal proteins are universally present in the three domains of life and therefore were likely present already at the time of the Last Universal Common Ancestor (LUCA).

Table 2. Quantitative Protein Analysis.

Protein Detected (fmoles/ug total protein)	Complex	*Prochlorococcus* MED4	*Prochlorococcus* MIT 9313	*Synechococcus* WH 8102
RbcL	RuBisCO	59±15	20±8	71±8
PSHCP	unknown	19±8	26±9	8±1

Table 3. Proteins of Unique Peptides Identified in Experimental Pull Down.

Protein Name	Accession Number	Strain	# of Unique Peptides	Molecular Weight (kDa)	Isoelectric point (pI)
50 S Ribosomal Protein L2	Q7U4J7	WH 8102	3	31.6–31.7	11.1–11.6
50 S Ribosomal Protein L2	Q7V539	MIT 9313	2	31.6–31.7	11.1–11.6
50 S Ribosomal Protein L2	Q7UZV0	MED4	8	31.6–31.7	11.1–11.6
Photosystem I reaction center subunit PsaD	Q7U4M3	WH 8102	1	15.7	7.0–8.7
Photosystem I reaction center subunit PsaD	Q7V564	MIT 9313	3	15.7	7.0–8.7
Photosystem I reaction center subunit PsaD	Q7UZS7	MED4	4	15.7	7.0–8.7
Ycf48-like protein	Q7U9P8	WH 8102	1	36.4–37.2	4.7–5.3
Ycf48-like protein	Q7V4Q3	MIT 9313	1	36.4–37.2	4.7–5.3
Ycf48-like protein	Q7V301	MED4	1	36.4–37.2	4.7–5.3

Genes for these universal ribosomal proteins are usually clustered in the genomes into the universal ribosomal operons. More than half of the ribosomal proteins, however, are either domain-specific or patchily distributed across various taxonomic groups. Non-universal ribosomal proteins are likely newer additions to the ribosome [45], and genes for these ribosomal proteins are usually scattered across the genome.

Predicted features of the amino acid sequence of PSHCP and its observed interaction with the ribosomal protein Rpl2 favour a hypothesis that PSHCP is a group-specific ribosomal protein. It is not located within one of the universal r-protein operons, but it is encoded near Rpl19, a Bacteria-specific ribosomal protein, although it does not appear to be significantly co-transcribed with Rpl19 (see Table 1). Rpl2 is a ribosomal protein that plays a central role in the ribosome and is located near the peptidyl transferase center [45]. According to SH3-hunter [46], Rpl2 may interact with SH3-containing proteins. It is possible that PSHCP interacts with Rpl2 via its predicted SH3-like domain. Additionally, PSHCP may stabilize rRNA structure at positions where non-compensatory substitutions are observed.

Our pull-down assays also reveal an association of PSHCP with two proteins in the photosynthetic complex, which suggests a role linking synthesis of the photosystem subunits and their assembly into native complexes. The Ycf48-like protein has been shown to play a role in Photosystem II assembly in the freshwater cyanobacterium *Synechocystis* sp. strain PCC 6803 [47]. More recently Komenda *et al.* [48] demonstrated that the Psb27 assembly factor, which we found associated with PSHCP in the examined *Prochlorococcus* strains (Table S3), can interact with Photosystem I as well as Photosystem II under certain conditions. This is further suggestive of a role for PSHCP linking the ribosome to photosystem assembly.

Further structural and biochemical studies of PSHCP are needed to firmly establish the functional role of this protein in the picocyanobacterial cell. To further explore the association of PSHCP with the ribosome and/or the photosystems, picocyanobacterial ribosomes and photosystems should be isolated and probed with the PSHCP specific antibody reported here. Examination of the suggested protein-protein interactions in a bacterial two-hybrid system would determine whether PSHCP and these identified binding partners are indeed interacting in the cell. The evolutionary studies of the ribosome suggest that ribosomes have over time recruited ribosomal proteins [45]. If PSHCP's role as a ribosomal protein is confirmed, its further study may provide additional clues into the evolution of the ribosome.

Supporting Information

Figure S1 Nucleotide conservation of 23 S rRNA in PSHCP-containing cyanobacteria in comparison to other cyanobacteria. The conservation is measured as a relative information content (see Materials & Methods for details). The difference in the relative information content between PSHCP-containing cyanobacteria and other cyanobacteria is color coded (see color bar) and mapped onto a predicted structure of 23 S rRNA of *Escherichia coli* [35]. Nucleotides on the red side of the spectrum correspond to the residues evolving faster in the PSHCP-containing cyanobacteria. Nucleotides shown in black correspond to sites that are absent in> 50% of cyanobacteria. Different domains of 23 S rRNA are designated by 0 and roman numerals I through VI. Helices are designated by numbers 1–101 in blue. Residue numbers (in brown) correspond to the nucleotides in *Escherichia coli* sequence.

Figure S2 Visualization of possible interactions of the ribosomal protein L2 with 23 S rRNA. Distances (in angstroms) of each nucleotide (P atom in the phosphate group) of the 23 S rRNA to the nearest residue (α-C atom of the amino acid) of the ribosomal protein L2 were calculated using crystal structure of the bacterial ribosome (PDB ID 3R8T). The distances were color coded (see scale bar on the figure) and mapped onto a predicted structure of 23 S rRNA of *Escherichia coli* [35]. Residues in black correspond to the nucleotides that were not resolved in the crystal structure. For notations of the 23 S rRNA features see legend to the Figure S1.

Table S1 Nucleotide sequences of primers used in the RT-Q-PCR experiments.

Table S2 Presence of PSHCP homologs in genomes.

Table S3 Results of proteomics analysis from PSHCP pull down assay. The spreadsheet contains the complete results of tandem mass spectrometric analysis of the pull down experiment with PSCHP. Analysis of each of the three strains is presented in a separate sheet (labeled with the strain name). Proteins discussed in the text are highlighted in green.

Column notations are as follows:

- *Column A:* UNIPROT accession number for the protein.
- *Column B:* Protein Score. The Sequest algorithm calculates a cross-correlation (Xcorr) value for every identified peptide. The protein score is the sum of Xcorr values for all peptides linked to a particular protein.
- *Column C:* Score for the control pull down (i.e. minus PCHCP). For definition of the score see description of column B.
- *Column D:* UNIPROT annotation of the protein.
- *Column E:* Coverage of the protein by the observed peptides.
- *Column F:* Number of unique peptides per protein.
- *Column G:* Molecular weight of the protein in kDa.
- *Column H:* Predicted isoelectric point of the protein.

Acknowledgments

The authors thank Natalie Donaher and Zakir Hossain for technical assistance, Laurel McIntyre, Miranda Corkum and Maria Thistle for help with cyanobacterial culturing, and Kelly Milton for pilot experiments.

Author Contributions

Conceived and designed the experiments: AMC OZ. Performed the experiments: CEW KGD JKZ APJ DAB MSJ. Analyzed the data: CEW KGD JKZ APJ DAB MSJ OZ AMC. Contributed reagents/materials/ analysis tools: AMC. Wrote the paper: CEW OZ AMC. Conceived the study, interpreted results and wrote the manuscript: AMC OZ. Performed transcriptional analyses, initial bioinformatics and cloning of PSHCP for antibody generation: KGD. Performed pull down assays and composed much of the manuscript: CEW. Performed quantitative immunoblotting experiments: JKZ. Performed the mass spectrometry experiments: AJ DB. Performed computational analyses of the evolution of rRNA genes in cyanobacteria: MSJ.

References

1. Partensky F, Garczarek L (2010) *Prochlorococcus*: Advantages and limits of minimalism. Annu Rev Mar Sci 2: 305–331. doi: 10.1146/annurev-marine-120308-081034
2. Ting CS, Rocap G, King J, Chisholm SW (2002) Cyanobacterial photosynthesis in the oceans: The origins and significance of divergent light-harvesting strategies. Trends Microbiol 10: 134–142.
3. Garrity G, editor-in-chief (2001). Bergey's Manual of Systematic Bacteriology. Springer-Verlag, 2nd edition, Vol. 1.
4. Scanlan DJ, Ostrowski M, Mazard S, Dufresne A, Garczarek L, et al. (2009) Ecological genomics of marine picocyanobacteria. Microbiol Mol Biol Rev MMBR 73: 249–299. doi: 10.1128/MMBR.00035-08
5. Zwirglmaier K, Jardillier L, Ostrowski M, Mazard S, Garczarek L, et al. (2008) Global phylogeography of marine *Synechococcus* and *Prochlorococcus* reveals a distinct partitioning of lineages among oceanic biomes. Environ Microbiol 10: 147–161. doi: 10.1111/j.1462-2920.2007.01440.x
6. Waterbury JB, Watson SW, Guillard RRL, Brand LE (1979) Widespread occurrence of a unicellular, marine, planktonic, cyanobacterium. Nature 277: 293–294. doi: 10.1038/277293a0
7. Chisholm SW, Olson RJ, Zettler ER, Goericke R, Waterbury JB, et al. (1988) A novel free-living prochlorophyte abundant in the oceanic euphotic zone. Nature 334: 340–343. doi: 10.1038/334340a0
8. Wilmotte A, Herdman M. (2001) Phylogenetic Relationships Among the Cyanobacteria Based on 16S rRNA Sequences. In Bergey's Manual of Systematic Bacteriology. Garrity G, editor-in-chief. Springer-Verlag, 2nd ed. Vol. 1. pp. 487–493.
9. Johnson ZI, Zinser ER, Coe A, McNulty NP, Woodward EMS, et al. (2006) Niche partitioning among *Prochlorococcus* ecotypes along ocean-scale environmental gradients. Science 311: 1737–1740. doi: 10.1126/science.1118052
10. Rocap G, Larimer FW, Lamerdin J, Malfatti S, Chain P, et al. (2003) Genome divergence in two *Prochlorococcus* ecotypes reflects oceanic niche differentiation. Nature 424: 1042–1047. doi: 10.1038/nature01947
11. Gierga G, Voss B, Hess WR (2012) Non-coding RNAs in marine *Synechococcus* and their regulation under environmentally relevant stress conditions. ISME J 6: 1544–1557. doi: 10.1038/ismej.2011.215
12. Pittera J, Humily F, Thorel M, Grulois D, Garczarek L, et al. (2014) Connecting thermal physiology and latitudinal niche partitioning in marine *Synechococcus*. ISME J 8: 1221–1236. doi: 10.1038/ismej.2013.228
13. Behrenfeld MJ, O'Malley RT, Siegel DA, McClain CR, Sarmiento JL, et al. (2006) Climate-driven trends in contemporary ocean productivity. Nature 444: 752–755. doi: 10.1038/nature05317
14. Boyce DG, Lewis MR, Worm B (2010) Global phytoplankton decline over the past century. Nature 466: 591–596. doi: 10.1038/nature09268
15. Finkel ZV, Beardall J, Flynn KJ, Quigg A, Rees TAV, et al. (2010) Phytoplankton in a changing world: Cell size and elemental stoichiometry. J Plankton Res 32: 119–137. doi: 10.1093/plankt/fbp098
16. Luo H, Friedman R, Tang J, Hughes AL (2011) Genome reduction by deletion of paralogs in the marine cyanobacterium *Prochlorococcus*. Mol Biol Evol.28: 2751–60. doi: 10.1093/molbev/msr081
17. Zhaxybayeva O, Doolittle WF, Papke RT, Gogarten JP (2009) Intertwined evolutionary histories of marine *Synechococcus* and *Prochlorococcus marinus*. Genome Biol Evol 1: 325–339. doi: 10.1093/gbe/evp032
18. Zhaxybayeva O, Gogarten JP, Doolittle WF (2007) A hyperconserved protein in *Prochlorococcus* and marine *Synechococcus*. Fems Microbiol Lett 274: 30–34. doi: 10.1111/j.1574-6968.2007.00806.x
19. Huynen MA, Snel B, von Mering C, Bork P (2003) Function prediction and protein networks. Curr Opin Cell Biol 15: 191–198.
20. Kutovaya OA, McKay RML, Bullerjahn GS (2010) Expression of hcp in freshwater *Synechococcus* spp., a gene encoding a hyperconserved protein in picocyanobacteria. J Basic Microbiol 50: 227–231. doi: 10.1002/jobm.200900337
21. Altschul SF, Madden TL, Schäffer AA, Zhang J, Zhang Z, et al. (1997) Gapped BLAST and PSI-BLAST: a new generation of protein database search programs. Nucleic Acids Res 25: 3389–3402. doi: 10.1093/nar/25.17.3389
22. Crooks GE, Hon G, Chandonia J-M, Brenner SE (2004) WebLogo: A sequence logo generator. Genome Res 14: 1188–1190. doi: 10.1101/gr.849004
23. Vogel J, Axmann IM, Herzel H, Hess WR (2003) Experimental and computational analysis of transcriptional start sites in the cyanobacterium *Prochlorococcus* MED4. Nucleic Acids Res 31: 2890–2899.
24. Moore (1995) Comparative physiology of *Synechococcus* and *Prochlorococcus*: Influence of light and temperature on growth, pigments, fluorescence and absorptive properties. Mar Ecol-Prog Ser 116: 259–275. doi: 10.3354/meps116259
25. Six (2004) Photophysiology of the marine cyanobacterium *Synechococcus* sp. WH8102, a new model organism. Aquat Microb Ecol 35: 17–29. doi: 10.3354/ame035017
26. Brown CM, MacKinnon JD, Cockshutt AM, Villareal TA, Campbell DA (2008) Flux capacities and acclimation costs in *Trichodesmium* from the Gulf of Mexico. Mar Biol 154: 413–422. doi: 10.1007/s00227-008-0933-z
27. The UniProt Consortium (2012) Update on activities at the Universal Protein Resource (UniProt) in 2013. Nucleic Acids Res 41: D43–D47. doi: 10.1093/nar/gks1068
28. Markowitz VM, Chen I-MA, Palaniappan K, Chu K, Szeto E, et al. (2011) IMG: the integrated microbial genomes database and comparative analysis system. Nucleic Acids Res 40: D115–D122. doi: 10.1093/nar/gkr1044
29. Rice P, Longden I, Bleasby A (2000) EMBOSS: the European Molecular Biology Open Software Suite. Trends Genet 16: 276–277.
30. Marchler-Bauer A, Bryant SH (2004) CD-Search: protein domain annotations on the fly. Nucleic Acids Res 32: W327–331. doi: 10.1093/nar/gkh454
31. Quast C, Pruesse E, Yilmaz P, Gerken J, Schweer T, et al. (2012) The SILVA ribosomal RNA gene database project: improved data processing and web-based tools. Nucleic Acids Res 41: D590–D596. doi: 10.1093/nar/gks1219
32. Pruesse E, Peplies J, Glöckner FO (2012) SINA: Accurate high-throughput multiple sequence alignment of ribosomal RNA genes. Bioinformatics 28: 1823–1829. doi: 10.1093/bioinformatics/bts252
33. Roberts E, Sethi A, Montoya J, Woese CR, Luthey-Schulten Z (2008) Molecular signatures of ribosomal evolution. Proc Natl Acad Sci U S A 105: 13953–13958. doi: 10.1073/pnas.0804861105
34. Shannon CE (1948) A mathematical theory of communication. ATT Tech J 27: 623–656.
35. Petrov AS, Bernier CR, Hershkovits E, Xue Y, Waterbury CC, et al. (2013) Secondary structure and domain architecture of the 23 S and 5 S rRNAs. Nucleic Acids Res 41: 7522–35. doi: 10.1093/nar/gkt513
36. Dunkle JA, Wang L, Feldman MB, Pulk A, Chen VB, et al. (2011) Structures of the bacterial ribosome in classical and hybrid states of tRNA binding. Science 332: 981–984. doi: 10.1126/science.1202692
37. Nowack ECM, Melkonian M, Gloeckner G (2008) Chromatophore genome sequence of *Paulinella* sheds light on acquisition of photosynthesis by eukaryotes. Curr Biol 18: 410–418. doi: 10.1016/j.cub.2008.02.051

38. Klein DJ, Moore PB, Steitz TA (2004) The Roles of Ribosomal Proteins in the Structure Assembly, and Evolution of the Large Ribosomal Subunit. J Mol Biol 340: 141–177. doi: 10.1016/j.jmb.2004.03.076
39. Marchler-Bauer A, Zheng C, Chitsaz F, Derbyshire MK, Geer LY, et al. (2013) CDD: conserved domains and protein three-dimensional structure. Nucleic Acids Res 41: D348–D352. doi: 10.1093/nar/gks1243
40. Krishna SS, Majumdar I, Grishin NV (2003) Structural classification of zinc fingers: survey and summary. Nucleic Acids Res 31: 532–550.
41. Mitschke J, Georg J, Scholz I, Sharma CM, Dienst D, et al. (2011) An experimentally anchored map of transcriptional start sites in the model cyanobacterium *Synechocystis* sp. PCC6803. Proc Natl Acad Sci U S A 108: 2124–2129. doi: 10.1073/pnas.1015154108
42. Sharma CM, Hoffmann S, Darfeuille F, Reignier J, Findeiß S, et al. (2010) The primary transcriptome of the major human pathogen *Helicobacter pylori*. Nature 464: 250–255. doi: 10.1038/nature08756
43. Tolonen AC, Aach J, Lindell D, Johnson ZI, Rector T, et al. (2006) Global gene expression of *Prochlorococcus* ecotypes in response to changes in nitrogen availability. Mol Syst Biol 2: 53. doi: 10.1038/msb4100087
44. Martiny AC, Coleman ML, Chisholm SW (2006) Phosphate acquisition genes in *Prochlorococcus* ecotypes: Evidence for genome-wide adaptation. Proc Natl Acad Sci U S A 103: 12552–12557. doi: 10.1073/pnas.0601301103
45. Fox GE (2010) Origin and Evolution of the Ribosome. Cold Spring Harb Perspect Biol 2: a003483. doi: 10.1101/cshperspect.a003483
46. Ferraro E, Peluso D, Via A, Ausiello G, Helmer-Citterich M (2007) SH3-Hunter: discovery of SH3 domain interaction sites in proteins. Nucleic Acids Res 35: W451–454. doi: 10.1093/nar/gkm296
47. Komenda J, Nickelsen J, Tichy M, Prasil O, Eichacker LA, et al. (2008) The Cyanobacterial Homologue of HCF136/YCF48 Is a Component of an Early Photosystem II Assembly Complex and Is Important for Both the Efficient Assembly and Repair of Photosystem II in *Synechocystis* sp. PCC 6803. J Biol Chem 283: 22390–22399. doi: 10.1074/jbc.M801917200
48. Komenda J, Knoppova J, Kopecna J, Sobotka R, Halada P, et al. (2011) The Psb27 Assembly Factor Binds to the CP43 Complex of Photosystem II in the Cyanobacterium *Synechocystis* sp. PCC 6803. Plant Physiol 158: 476–486. doi: 10.1104/pp.111.184184

A Hierarchical Classification of Benthic Biodiversity and Assessment of Protected Areas in the Southern Ocean

Lucinda L. Douglass[1,2]*, Joel Turner[2], Hedley S. Grantham[1,3], Stefanie Kaiser[4,5], Andrew Constable[6,7], Rob Nicoll[8], Ben Raymond[6,7], Alexandra Post[9], Angelika Brandt[4], Daniel Beaver[1,2]

1 Centre for Biodiversity and Conservation Science, School of Biological Sciences, The University of Queensland, Brisbane, Queensland, Australia, **2** Centre for Conservation Geography, Sydney, New South Wales, Australia, **3** Betty and Gordon Moore Centre for Science and Oceans, Conservation International, Arlington, Virginia, United States of America, **4** Biocentre Grindel and Zoological Museum, University of Hamburg, Hamburg, Germany, **5** German Centre for Marine Biodiversity Research, Wilhelmshaven, Germany, **6** Australian Antarctic Division, Department of the Environment, Australian Government, Kingston, Tasmania, Australia, **7** Antarctic Climate and Ecosystems Cooperative Research Centre, University of Tasmania, Hobart, Tasmania, Australia, **8** WWF Australia, Ultimo, New South Wales, Australia, **9** Marine and Coastal Environment Group, Geoscience Australia, Canberra, Australian Capital Territory, Australia

Abstract

An international effort is underway to establish a representative system of marine protected areas (MPAs) in the Southern Ocean to help provide for the long-term conservation of marine biodiversity in the region. Important to this undertaking is knowledge of the distribution of benthic assemblages. Here, our aim is to identify the areas where benthic marine assemblages are likely to differ from each other in the Southern Ocean including near-shore Antarctica. We achieve this by using a hierarchical spatial classification of ecoregions, bathomes and environmental types. Ecoregions are defined according to available data on biogeographic patterns and environmental drivers on dispersal. Bathomes are identified according to depth strata defined by species distributions. Environmental types are uniquely classified according to the geomorphic features found within the bathomes in each ecoregion. We identified 23 ecoregions and nine bathomes. From a set of 28 types of geomorphic features of the seabed, 562 unique environmental types were classified for the Southern Ocean. We applied the environmental types as surrogates of different assemblages of biodiversity to assess the representativeness of existing MPAs. We found that 12 ecoregions are not represented in MPAs and that no ecoregion has their full range of environmental types represented in MPAs. Current MPA planning processes, if implemented, will substantially increase the representation of environmental types particularly within 8 ecoregions. To meet internationally agreed conservation goals, additional MPAs will be needed. To assist with this process, we identified 107 spatially restricted environmental types, which should be considered for inclusion in future MPAs. Detailed supplementary data including a spatial dataset are provided.

Editor: Roberto Pronzato, University of Genova, Italy

Funding: Research funding was provided by WWF (wwf.panda.org). The funders had no role in data collection and analysis or decision to publish the manuscript.

Competing Interests: Rob Nicoll is an employee of the non-governmental advocacy organisation, WWF-Australia. This employment does not alter the authors' adherence to all the PLoS ONE policies on sharing data and materials.

* Email: douglass.lucinda@gmail.com

Introduction

The high-latitude Southern Ocean, south of the Polar Front, is globally significant for conservation due to its unique species and environments. The presence of high levels of endemism shows it to be a distinct 'realm' within the Earth's oceans [1–3]. The region is remote and has been altered less by human pressures than most other parts of the world [4]. However, exploitation has occurred across all its major marine ecosystems [5]. This has prompted the establishment of international agreements, in particular, the Convention on the Conservation of Antarctic Marine Living Resources (CCAMLR) to conserve the biodiversity while, among other roles, managing exploitation of marine living resources in the region [5–7].

An international effort is underway to establish a representative system of marine protected areas (MPAs) for the Southern Ocean and CCAMLR has the primary responsibility for developing this system [8,9]. CCAMLR aims to complete this task to meet marine protection objectives set by The United Nations World Summit on Sustainable Development [8]. CCAMLR has identified the development of a broad-scale classification of the Southern Ocean and a finer-scale, classification of its biodiversity as necessary to achieve this aim [10]. Our aim is to complement this process by identifying the areas where benthic marine assemblages are likely to differ in near-shore Antarctica and the Southern Ocean. We achieve this by using a hierarchical spatial classification of the benthic environment and assessing the extent to which the existing system of MPAs encompasses the benthic diversity.

Developing a hierarchical classification of the benthic environment in the Southern Ocean

Regional classifications are important to MPA planning and aim to identify biogeographic patterns by spatially subdividing an area, such that assemblage compositions within each subdivision are expected to be homogeneous relative to adjacent regions [2,11]. Classifications aim to separate different assemblages and, in particular, identify areas within which species that are endemic or have restricted ranges will reside. Importantly, this process also aims to identify ecological separation (poor connectivity) for species common across assemblages. For example, some widespread species show genetic differentiation suggesting that gene flow is restricted between isolated sub-populations [12–15]. Therefore, at a finer-scale, higher heterogeneity is likely to be present than is often considered in the historical biogeography in the region e.g. [16]. The focus of the classification developed here is to identify possible general patterns of assemblages and of more restricted benthic species and populations at a circumpolar scale.

Physical environmental data can be used as surrogates for biodiversity where biological data are scarce [17,18]. Ideally, the process of subdivision would be done largely, if not entirely, on the basis of biological data. However, biological data paucity is prevalent across marine ecosystems globally leading to a necessity to use surrogates [19,20]. On a circum-Antarctic scale, biological data are limited, particularly for the Amundsen Sea, eastern Antarctica, under ice shelves and below 1500 m [21–23]. While this paucity of biological data introduces uncertainty into regionalisation or MPA planning processes, environmental data can be used to assess the similarity or differences between regions for the purposes of these planning processes [17,18]. In fact, additional biological data does not always greatly increase the efficiency of protected area planning [24].

When utilizing physical environmental data in this way, a classification method should be chosen that enhances the biological relevance of the results, for example by incorporating known relationships between the physical environment and biodiversity. Recent research within the Southern Ocean has primarily focussed on using statistical clustering methods to identify areas with relatively dissimilar attributes, examples include: [11,25,26]. This approach has particular utility when synoptic data are available on abiotic factors that are known to be drivers or correlates of biological patterns but few data are available on actual biological patterns [27]. Where the relationships between the distribution of benthos and abiotic factors can be reasonably inferred, they can be manually incorporated to delineate regions rather than relying on an unsupervised classification e.g. clustering, [11]. For instance, the continental shelf usually supports distinct species assemblages when compared to the deeper ocean [22,28,29]; this has also been observed on seamounts [30]. Furthermore, the utility of biogeographical classifications to MPA planning is improved by a hierarchical (nested) classification, which shows the subdivision at increasingly finer scales [2,20]. This approach was used by Spalding et al. [2] and Last et al. [31] to identify distinct regions based on taxonomic composition and the degree to which separation might occur as a result of patterns of dispersal, life histories of taxa, and isolation.

While a circumpolar classification of the pelagic environment has been completed and adopted by CCAMLR [25], no equivalent, accepted benthic bioregionalisation currently exists [26], although there are sufficient data to develop one. Databases of molluscs, bryozoans and zooplankton are being compiled and analysed, albeit they are limited in their geographic extent [1,32–35]. The SCAR-MarBIN database is improving data availability at a circumpolar scale [36]. The relationships between depth, geomorphology and species distributions have been investigated [28,37–39]. Furthermore, the connectivity between populations has been explored using genetics [14,40,41]. The Biogeographic Atlas of the Southern Ocean, when published, will contribute to bringing together much of this research (see www.atlas.biodiversity.aq). However at the time of writing, no synthesis of this recent research exists for the Southern Ocean.

Here we present a circumpolar hierarchical classification of the benthic environment based on Last et al., [31] and intended for general use in spatial planning and management. The classification includes the identification of ecoregions, bathomes and environmental types, the latter of which are based on geomorphic features. Similar to the Delegations of Australia and France, [17], ecoregions were delineated by accounting for recent biogeographic research, patterns of endemism and the influence of environmental drivers as potential barriers to dispersal (Fig. 1). Environmental types represent the lowest order in the hierarchy and we used these to assess the representativeness of currently designated MPAs. We also identified environmental types which have restricted distributions and are probable locations for the focus of future MPA design.

Environmental drivers

Environmental drivers are the physico-chemical processes and other factors that set the habitat conditions and influence the distribution and abundance of taxa, including their connectivity between similar habitats. Reduced connectivity can give rise to divergent evolution of similar taxa. Two major environmental drivers are depth and geomorphology [28,29,42–44]. Other important drivers include seabed temperature, icebergs and sea ice coverage, sea-surface productivity and ocean currents [45–48].

Depth. The Southern Ocean contains a smaller proportion of depth-restricted species than other oceans, probably due to the presence of a deeper shelf, deep-water formation (Antarctic Bottom Water), lack of thermocline and the intermittent glaciations of the shelf, creating less difference between environments on the shelf and the deep sea [49–51]. Wide depth ranges in Antarctic species have been typically associated with survival in deep-sea refuges e.g. [49]. That is, sub-populations of eurybathic shelf species may have endured past glaciations on the slope or deeper, which could have then served as a potential source for shelf (re-) colonization [28,49,52]. Thatje et al. [50] suggested, that life on the Antarctic continental slope during the last glacial maximum may have been extremely difficult due to turbidity currents cascading down the slope. Sedimentation processes would not have occurred everywhere though, and were probably more pronounced along canyons compared to adjacent flanks and plateaus leaving some scope for slope fauna to survive [28]. Limited depth distributions of shelf species, on the other hand, could serve as evidence for survival through glaciations in ice-free refuges on the continental shelf [53].

Many benthic species do have restricted depth ranges and depth-related factors can be strong barriers to dispersal of benthic species [37,38,54,55,56 and see 'Barriers to dispersal' section]. For instance, the octopus genus *Paraledone,* the echinoid genus *Sterechinus*, the ostracod genus *Macroscapha* and the amphipod genus *Eurythenes* show niche separation by depth [40,57–59]. This depth partitioning gives rise to different assemblage structures in different depths [60], the ranges of which can be characterized as bathomes [31]. Furthermore, molecular studies show widely distributed species often represent species complexes, with each individual species having a much more restricted range size. For example, isopods: [56,61,62]; crinoids: [13]; and ophiuroids: [55].

Geomorphic features. Geomorphic features are a classification of the seabed based on its surface morphology, and have been shown to be significant for delineating benthic habitats [63]. The features used here are meso-scale habitat features related to the sedimentary habitat along with the near-bottom conditions of the ocean environment [31]. Major habitat characteristics include the availability of hard-rock surfaces and the erosion or deposition of sediment along with their physical attributes [42,64]. For

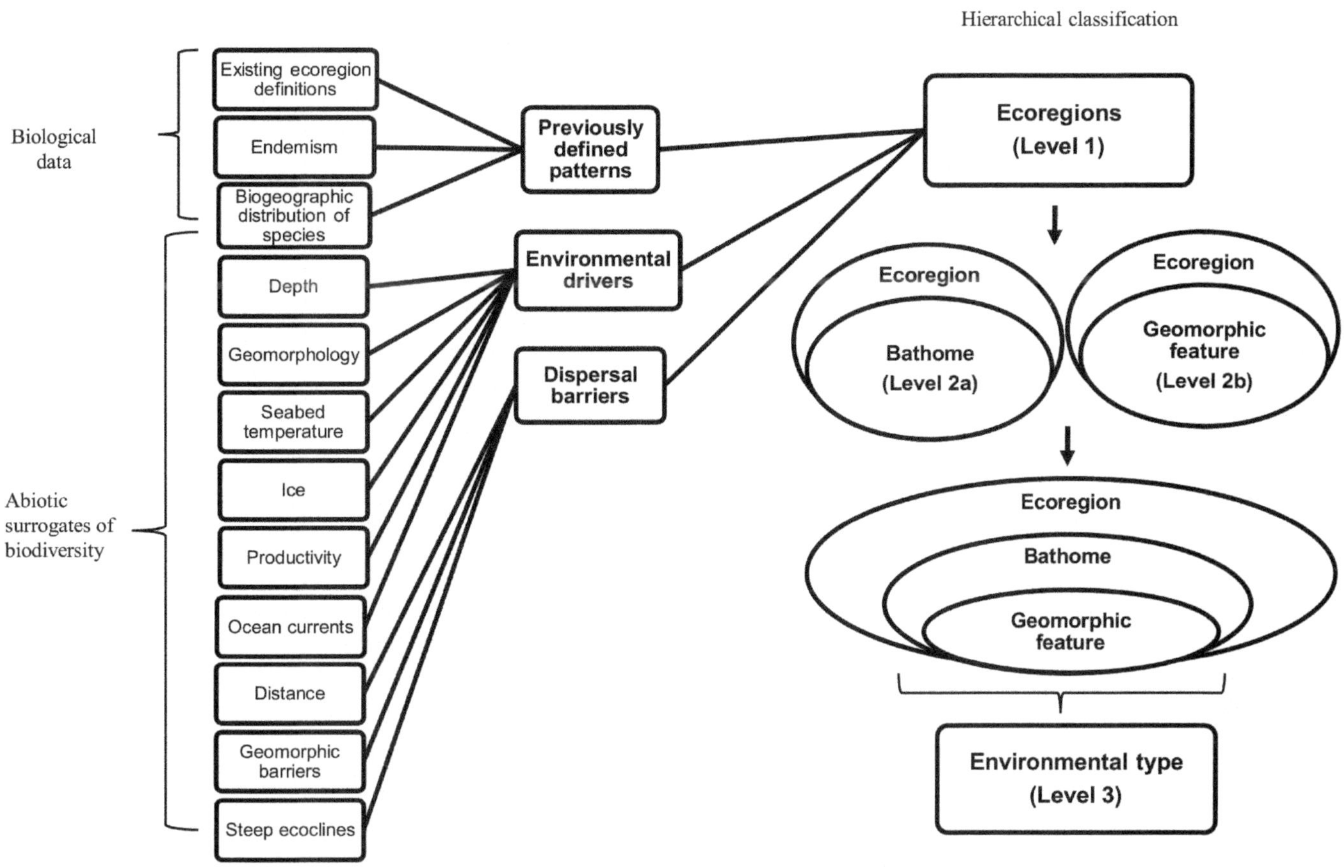

Figure 1. The framework used to classify ecosystems within the Southern Ocean. Ecoregions were defined on the basis of important environmental drivers and their potential to prevent dispersal. Biogeographic patterns identified in the literature were also incorporated. A hierarchical approach was then applied within each ecoregion, nesting the two main habitat types (geomorphic features within bathomes) to identify environmental types.

example, Antarctic shelf depressions eroded during glacial maxima now have low currents and fine sediments, providing appropriate habitats for mobile deposit feeder and infaunal communities [46,64,65]. The species rich coral-sponge communities found within some shelf-cutting canyons are postulated to be due to an abundance of food contained in channelled water sourced from productive shelf regions [43,66]. Elsewhere, canyons are shown to alter current flow causing upwelling and mixing which are attributed to enhanced local biological systems [67]. Seamounts have also been shown to be important for biodiversity and can have high endemism levels [30,68–70]. Hydrothermal vents are shown to contain distinct species and unique community structures [71]. Research in the Weddell Sea shows a clear distinction in faunas of the shelf, slope and abyss [28,53]. However, this may not be the case in all regions and for all species. For example, such patterns have not been found to exist within the Scotia Sea [28].

Seabed temperature. Seabed temperature has a major influence on the ecology of Antarctic benthic fauna; its spatial and temporal variability contributes to determining the biogeography of those fauna; and the composition of assemblages [27,45]. Seabed temperature is suspected to constrain the migration of benthic fauna and may lead to genetic variation and eventual speciation [27]. This hypothesis is supported by recent research on the Antarctic sea spider, Nymphon australe and the octopus genus, Pareledone [12,72]. Temperature can be biologically meaningful even where the spatial and temporal variation is narrow [73].

Ice. Ice regimes are a key structuring element in the ecology of the Antarctic benthos [48,74,75]. Icebergs scour the shelf seabed, impacting the distribution of benthic habitats and the diversity of assemblages [39,42,76]. Areas under glacier tongues and fast ice are deprived of inorganic sedimentation leading to dominance of sessile suspension feeders [76]. Benthic habitat conditions can also be affected by the influence of ice on light availability, water currents, temperature, biogeochemistry, algal growth and productivity [42,48,77–80]. These conditions may be modified depending on the concentration and duration of the sea ice [78,81].

Productivity. Sea surface productivity provides a vital food source to the benthos [82] and the inter-annual reliability and magnitude of productivity blooms are likely drivers of total productivity in the benthic system [75]. Phytoplankton blooms are highest (i) where frontal activity has created an upwelling of nutrient-rich water, (ii) down-stream from iron-rich landforms and (iii) within the ice-melt zones and polynyas [83–87]. Polynyas are regions of reduced ice concentration surrounded by heavy pack ice. Feeding especially within the near shore benthos can be seasonal [88–90]. Many benthic ecological processes though, appear not to be coupled with the summer sea surface productivity bloom due to the presence of persistent sediment food banks and the influence of bottom circulation on sediment distribution [46,75]. However, interaction between the sea surface and benthos can occur through sedimentation of organic matter from the surface [75], and through trophic transfer from herbivores at the surface to delivery to the benthos [91]. For example, in the latter case, salps feed on phytoplankton then vertically migrate to the demersal zone where they form the near-exclusive prey species of benthic octocorallian polyps [47]. In the former case, macrobenthic biomass in muddy sediments of the deep Antarctic shelf is correlated with regional primary production and sea-ice duration [75]. Similarly, organic matter is transferred within sinking water generated from rejected brine during ice formation and is suspected to act as an important source of food for benthic species creating locations of high species richness in deep-sea areas traversed by this organically laden bottom water [22,51].

Barriers to dispersal

The geographic distribution of a species and genetic connectivity between subpopulations is linked to the ability of individuals to disperse away from their parents. This ability will be constrained by dispersal barriers including the distance between suitable habitats, geomorphic barriers and steep environmental clines. Long-distance dispersal in benthic organisms is confined to passive oceanic transport of adults or via turbidites (i.e. rafting), and transport during pelagic larval stages [92].

Habitats may support different species assemblages where the distance between suitable habitats is greater than the distance over which a species can disperse. While the dispersal distances for most species in the Southern Ocean are unknown [93], it is likely that the dispersal range for many brooding species will be restricted; brooding species protect and feed their offspring without a pelagic larval stage [50]. Within the Southern Ocean, benthic communities have more species with either lecithotrophic (non-feeding) larval development or brooding than species with planktotrophic (feeding) larvae [50]. While the strong, circumpolar currents of the Southern Ocean may assist the dispersal of some species with pelagic larval stages [94,95], predation, parasitism and starvation will constrain the probability of moving great distances, especially for lecithotrophic larvae which have a limited food source [94,96]. Furthermore, strong genetic variation has been shown in some widely dispersive species suggesting higher heterogeneity at a finer-scale [12–14].

Geomorphic barriers, steep ecoclines and ocean currents can also cause ecological and genetic differentiation. For example, differences in seabed temperature across relatively short distances can cause genetic differentiation within a species [12]. Similarly, genetic variation can occur through the inhibition of adult migration by geomorphic features [97]. Ocean current and frontal systems, depth-related factors as well as habitat fragmentation can also be strong physical barriers to gene flow [14,81,94,95,98], indicating that the ecological dynamics of populations can be spatially restricted. Examples include studies comparing shelf dwelling benthic diversity between the Antarctic Peninsula and Scotia Arc [14,55]. Changes in oceanographic conditions are also important barriers to dispersal in the open ocean [14,29,94].

MPA Assessment

A gap analysis can identify where additional MPAs are required to ensure that all ecosystems and biodiversity are represented within a system of MPAs [99]. There are very few MPAs within the Southern Ocean. The largest is south of the South Orkney Islands [100,101]. Other MPAs include those within the Australian, French, Norwegian and South African territories and some small areas protected under the Environmental Protocol to the Antarctic Treaty (Madrid Protocol).

We use the fine-scale (<100's of km's) environmental types as surrogates for different assemblages of biodiversity to assess the representativeness of existing MPAs. A system of MPAs that captures each environmental type is assumed to also capture the range of biodiversity these habitats represent. We also identify those environmental types that have distributions restricted by geographic area and location. The environmental types within these restricted locations are not necessarily more important than others ecologically. However, they tend to be the focus of MPA selection since there are limited spatial options for protecting the biodiversity for which they are a surrogate.

Materials and Methods

The region that is managed by CCAMLR defined the study area. The northern boundary of this region is a line approximating the location of the Polar Front [25]. The southern boundary was defined as the northern edge of the permanent ice shelf of the Antarctic continent.

Data

Table 1 provides an overview of the circumpolar datasets used in this study, their spatial and temporal resolutions along with their data sources. Additional regional information on seabed currents was also geo-referenced including: [102–105]. The spatial resolutions of the environmental data sets varied but were generally around 10 km with the exception of the seafloor temperature. However, only the depth and geomorphic data were directly utilized to define the environmental types, with the remaining data used to elucidate general environmental patterns. Thus, the effects of these scale mismatches were minimal.

The datasets were not modified except for sea-surface chlorophyll *a*. For this dataset we used a time series to identify areas of consistently high and low productivity. For each available season t, the mean summer concentration in each grid cell $i(x_{it})$ was transformed to an anomaly a_{it} by subtracting the seasonal mean $\bar{x}_t$ for the entire study area (using log-transformed values):

$$a_{it} = \log_{10}(x_{it}) - \log_{10}(\bar{x}_t)$$

To distinguish areas of intermittent high productivity from those of consistent but moderate to high productivity, these anomalies were further transformed giving an index p of the consistency of productivity from season to season [17]:

$$p_i = u_i|u_i|\sigma_i$$

Where u_i and σ_i are the mean and standard deviation of $i(x_{it})$ calculated over all available seasons t.

We also developed two datasets of proposed and existing MPAs. The MPAs proposed from MPA planning analyses current at September 2013 were; 1) The seven MPAs proposed from a study of Eastern Antarctica south of 60°S and between 30°E and 150°E [17] and; 2) The MPAs proposed from a study of the Ross Sea region south of 60°S and between 150°E and 150°W [106]. Existing MPAs that have been designated or acknowledged by CCAMLR were included. These MPAs included; 1) The South Orkney Islands Southern Shelf MPA within the high seas [100]; 2) The Kerguelen and Crozet Islands within the French EEZ [107]; 3) The Prince Edward and Marion Islands within the South African EEZ [108]; 4) The Heard and MacDonald Islands within the Australian EEZ; 5) A twelve nautical mile buffer around Bouvet Island for the Bouvetøya Nature Reserve [109]; 6) One nautical mile exclusion zones around Vulnerable Marine Ecosystems notified in accordance with CCAMLR conservation measures that regulate bottom fishing (specifically measures 22-06 and

Table 1. Circumpolar datasets used within the classification.

Data	Spatial resolution	Temporal resolution	Source
Depth	1minute	Not applicable	Smith and Sandwell, [118]
Geomorphology	1–12 km	Not applicable	O'Brien *et al.* [64]
Seafloor temperature	1 degree	Annual mean over all years available in the World Ocean Atlas 2005 [127]	Clarke et al. [27]
Sea surface chlorophyll-a	9 km	Mean values for each austral summer season (20th Dec to 20th March) for years 1998–2010	Feldman and McClain, [128]
Sea ice concentration	6.25 km	The proportion of the year where sea ice concentration was at least 85% derived from daily estimates during the 1st January 2003 to 31st December 2009	Spreen et al. [129]
Frontal systems	20 km	Annual mean calculated across 1992–2007	Sokolov and Rintoul, [130]

22-07) [110–112] and; 7) Marine areas listed as designated under the Environmental Protocol to the Antarctic Treaty and which have management plans (available at www.ats.aq) approved by CCAMLR in accordance with Antarctic Treaty Consultative Meeting decision 9 in 2005 [113,114].

Analysis

Broad spatial boundaries, termed benthic ecoregions, were identified according to the primary environmental drivers and dispersal barriers that drive the distribution of Southern Ocean benthic biodiversity. Within these ecoregions, biodiversity patterns are further driven by finer-scale spatial processes. Therefore, we applied a hierarchical framework based on Last et al. [31] to identify each environmental type. Bathomes (broad-scale depth classes) were nested within ecoregions, and geomorphic features were nested within bathomes. Each environmental type is therefore a unique combination of geomorphic feature, bathome and ecoregion. The nesting of geomorphic features within bathomes avoids the potential for false within-class homogeneity that can arise if such features are not modified by depth [115]. The analysis was performed using both manual and automated processing within a geographic information system and is further described in the following six sections. Further detail of the analysis is provided as supplementary information including ecoregion boundary descriptions (Text S1), the spatial coverage of each level of the classification (Table S1) and a description of the methods used to generate geomorphic features (Text S2).

Benthic ecoregions. The primary environmental drivers used to identify benthic ecoregions were depth, geomorphology, seabed temperature, sea ice concentration and chlorophyll-a as a proxy for productivity. The three major types of barriers to dispersal we used were; 1) the distance between potential habitat types (i.e. bathomes and geomorphic features); 2) large geomorphic barriers and; 3) presence of steep ecoclines in temperature, ice or productivity. The influence of both frontal and seabed currents were considered as either dispersal barriers or assisting connectivity between habitats (as further described in Text S1).

We initially delineated the distinct environments of the continental shelf and slope from the deeper ocean and then applied bathomes based on depth-species relationship studies to further account for the depth structuring of benthic assemblages. The 4000 m bathymetric contour was the most appropriate to generally separate the geomorphic features associated with the shelf and slope of the Antarctic continent from the geomorphic features associated with the abyssal ocean. Exceptions to this contour included parts of the East Indian Abyssal, Pacific-Antarctic Ridge, Kerguelen and Pacific Basin ecoregions which were adjusted to better capture geomorphology. Where the structural slope of the Oates ecoregion extends past the 4000 m contour in the Indian Ocean sector, the boundary of the structural slope was used in preference to the 4000 m contour. The boundary between the Kerguelen ecoregion and the Antarctic continent was defined by the southern limit of the contourite drift (i.e. a geomorphic feature), the slope of the broader Kerguelen Plateau (i.e. including Banzare and Elan Banks) and the canyon at 77.5°E, 61.5°S. Similar to Spalding et al [2] we included the seamounts and the seamount chain near to 120°W within the Amundsen ecoregion.

Where possible, previously defined ecoregions were reviewed and incorporated especially: [1,2,17,25,32,33]. These studies have focussed primarily on the Antarctic continental shelf and slope where more data are available. The previously defined boundaries were upheld if the ecoregion contained levels of endemism greater than 10% as reported by Linse et al. [32] or Griffiths et al. [1]. In each case, these boundaries were adjusted to better reflect the distribution of bathomes and geomorphic features. For instance, geomorphic features can act as barriers to dispersal for some benthic species leading to genetic distinctions [97]. Therefore, if deemed more appropriate, features such as canyons were used to replace the previously defined boundaries. Also, where a previously defined boundary divided a feature such as a canyon, it was redrawn to capture the entire feature. Within the deeper ocean, ecoregion boundaries were defined by considering the effect of structuring agents on patterns of biodiversity.

For areas without previously defined ecoregions, we initially identified where areas of similar habitat types (predominantly bathomes and geomorphic features) were separated by more than 200 km. This distance is based on the general global maximum of 200 km for dispersal distance across multiple marine species even where uncertainly is accounted for [116,117]. Spatial separation is assumed to increase the likelihood of areas of the same habitat type supporting genetically dissimilar species due to isolation and therefore contributes to the separation of ecoregions. All data layers were then used to corroborate or refute the validity of these regions defined by the separation of similar types of areas by the minimum distance. For instance, where large geomorphic barriers, including ridges, could contribute to separating areas of deeper habitats this would be seen as an ecoregion boundary between these two areas of habitat. Similarly, rapid changes in seabed temperature, sea ice, persistent productivity and ocean currents were also considered as signifying potential barriers to dispersal or where an ecoregion boundary could be most meaningful ecologically.

Bathomes. Bathomes are broad-scale depth classes. They were derived by dividing the region into different depth divisions based on bathymetry data which are frequently updated from satellite altimetry and ship depth soundings [17,118,119]. The boundaries were based on the depths at which there are expected to be rapid transitions in assemblage composition. We identified those depths based on available depth-species relationship studies within the Southern Ocean. Depth-species relationship studies are summarised in Table 2.

Geomorphic features. We identified geomorphic features based on O'Brien et al. [64] (see Table 1 and Table 3). This dataset, however, did not cover all areas within the study region due to the absence of data in areas of persistent ice cover. All areas not covered were located on the Antarctic continental shelf and we classified them as 'Shelf'. We also modified some features. 'Fracture zone cliffs' and 'Canyon axes' had been mapped as lines by O'Brien et al [64]. We converted 'Fracture zone cliffs' to areas through interpretation of the bathymetry contours with contours located close together considered part of the cliff. 'Canyon axes' were converted to areas by applying a 5 km buffer around the axes to capture both the canyon and its surrounding zone of influence. We deemed a 5 km buffer to be the most appropriate distance to ensure connectivity within dendritic canyon systems and discontinuity between single channelled canyon systems. The geomorphic features classified by O'Brien et al. [64] as 'Wave affected banks' are identical to their 'Shelf banks', but at depths of less than 200 m. This depth modification of geomorphology is captured by our nested hierarchical analysis and therefore these features were combined and renamed 'Banks'. The 'Coastal terrane' of South Georgia Island was reclassed as 'Island coastal terrane' to be consistent with other island areas. The Antarctic ice shelf including areas under floating ice tongues termed 'ice shelf cavities' were excluded.

Nesting data layers. Bathomes were nested within ecoregions, and geomorphic features were nested within bathomes to identify environmental types except for the following seven geomorphic features. Four geomorphic features 'Shelf', 'Cliff', 'Coastal terrane' and 'Island coastal terrane' were not further modified by depth. Seamounts and Seamount ridges were classed by only their shallowest bathome. Canyons were classed by whether they commenced on the shelf or slope. The environmental types were reviewed to minimise the potential for false heterogeneity. Environmental types with a very small area or restricted distribution were checked and dissolved into the surrounding environmental type if deemed invalid. For example, small patches of shallow rugose ocean floor within the South Atlantic ecoregion initially appeared as a restricted or rare environmental type. Upon review, this environmental type was found to be a processing artefact caused by the presence of adjacent seamounts. It was therefore eliminated by reclassing it as the most appropriate surrounding environmental type.

Representation of MPAs. We condensed the geomorphic features into four broad classes (Table 3). The distributions of existing MPAs and environmental types were overlaid. For each geomorphic class within each ecoregion that contains an existing MPA, we calculated; 1) the proportion of area within an MPA and, 2) the proportion of the number of environmental types with at least part of their area within an MPA. The total number of environmental types and seafloor area of each geomorphic class was also calculated.

Restricted environmental types. We specifically identified those environmental types which had a restricted spatial distribution. We assumed environmental types to have restricted distribution if they were either small in number (3 or less) or are predominantly limited to a particular discrete location without replication within an ecoregion. Since our focus was identifying areas where additional MPAs are required to fill large gaps in representation, restricted environments were not identified within regions where proposed MPAs have been identified through fine-scale MPA planning processes. Therefore, the region east of 30°E to 150°W and south of 60°S was excluded from the restricted environments analysis. Instead, the proposed MPAs within these regions were compiled and incorporated [17,106].

Results and Discussion

Classification

We identified 23 benthic ecoregions and 9 bathomes (Fig. 2, Table 2). A general description of the ecoregions is provided (Table 4). Existing geomorphology mapping has identified 28 geomorphic feature types (Table 3, [64]). When these three datasets were combined within the hierarchical framework, 562 environmental types were identified (Fig. 3 and Table S1). The spatial dataset is available online (see http://dx.doi.org/10.4225/15/53A3760D4AFAA).

The ecoregion boundaries reflect many of the broad patterns of the underlying data used for the analysis (Fig. 3). Generally, ecoregion boundaries most closely resemble depth, geomorphology and seabed temperature with productivity and ice cover tending to contain stronger within ecoregion variation (e.g. around polynyas or persistent productivity hotspots, Fig. 3). The Southern Ocean is dominated by deeper environments with over 85% of the seabed deeper than 2000 m (Fig. 3). This is also reflected in the geomorphology where over 75% of the Southern Ocean is made up of three geomorphic types, Rugose Ocean Floor, Abyssal Plain and Lower Slope (Fig. 3). The Atlantic basin, the largest ecoregion, is dominated by depths greater than 4500 m.

Within the deeper environments of the Southern Ocean, the primary drivers of the classification were the distribution of rare and mostly isolated shallow habitats often associated with seamounts, seamount ridges, plateaus and islands. Also important was the presence of mid-ocean ridges and large plateaus that separate deeper habitats. On the shelf and slope, where survey effort is higher, biological data on species distributions (in the form of previous biogeographical research) were the key drivers. Therefore, our results broadly reflect that of Spalding et al. [2], the Delegations of Australia and France, [17] and Clarke et al. [33] with some boundaries adjusted to more closely reflect the distribution of bathomes and geomorphic features.

In contrast to most other recent benthic classifications in the region examples include: [2,17,32], the results of this study encompass the deeper environments of the Southern Ocean as well as the continental shelf and islands. While the Global Oceans and Deep Seabed (GOODS) biogeographic classification provides a global scale classification, the only previous benthic classification to include these deeper environments at a circum-Antarctic scale was exploratory and incomplete [26]. This exploratory classification identified 20 regions using the statistical relationships between three physical datasets of depth, slope and seabed temperature. The classification presented here incorporates known or inferred relationships between the benthic biology and the physical environment and their influence on barriers to dispersal. The seamount classification provided by Clark et al. [120] provides a biologically meaningful classification of seamounts and we also

Table 2. Benthic bathomes and the ecological and biological events used to define them.

Bathome (m) Depth Range	Ecological and Biological Events. Unless otherwise stated, the Bivalvia, Gastropoda, Isopoda and Polychaeta taxa discussed only include the species studied in Brandt et al. [37].
0–100	Seaweed availability limits depth of herbivores [131].
	Polychaete species richness is highest.
	12 paramunnid species and 8 genera of Isopod family Paramunnidae are restricted to the top 100 m (S. Kaiser, unpublished data).
	4 pycnogonid species have been recorded exclusively from the top 100 m [132]
100–200	200 m is likely to be the maximum extent of the influence of wave action and sunlight penetration [64].
	Polychaete species richness begins to rapidly decrease.
	Chlorophyll concentration is generally negligible below 200 m [83].
200–500	High gastropod species richness at approximately 200 m–300 m.
	The end of the depth range of *Channichthys rhinoceratus* which is located in depths less than 200 m [38].
	200–500 m is the depth range of *Champsocephalus gunnari* [38,133].
	Zanclorhynchus spinifer depth range begins [38].
	500 m is the approximate maximum depth of scouring by contemporary ice bergs [42,43,134].
	Lepidonotothen squamifrons exhibits some areas of high density close to the 500 m isobath [38].
	Species of the isopod family, Santiidae are only found in the top 500 m.
	Most hydroid species only occur above 500 m [135]
500–1000	Upper slope mollusc assemblage is present between 400 and 800 m [136].
	Many echinoid species can only be found in the top 1000 m see suppl. Material of [28] and [137].
	Many bryozoan species restricted to the top 1000 m [53].
	Seven fold less gastropod species than at 200–300 m.
	The bivalve families Arcidae and Vesicomyidae can now be found. Condylocardiidae, Nuculidae, Hiatellidae and Erycinidae are no longer found.
	The Polychaeta families; Dorvilleidae, Chaetopteridae, Lacydoniidae, Pectinariidae and Spintheridae are no longer found.
	Much lower densities of fish than depths <500 m [133]. Four fish-depth ranges begin (*Bathyraja eatonii, Bathyraja irrasa, Alepocephalus* cf. *Antipodianus* and *Etmopterus* cf. *granulosus*) [38]. *Z. spinifer* depth range ends [38]. *Sterechinus neuymeyeri* depth range is between 0–810 m [40].
	Most pycnogonid species (63.8%) occur above 1000 m [132].
1000–1500	Bivalves, gastropods and polychaete diversity decreases from the shelf to the slope then stabilises at low numbers. There is no observed replacement of diminished shelf polychaete species from the slope and rise community (A. Brandt, unpublished data). The bivalve famillies Arcidae (*Bathyarca sinuata*) and Vesicomyidae (*Vesicomya sirenkoi*) can now be found regularly down to the abyss.
	The number of isopod species increases. Isopod families, Macrostylidae, Ischnomesidae and Haploniscidae are mostly found below 1000 m depth (with the exception of a few species found at shallower depths).
	Chaetognaths become much less abundant [138].
1500–2000	The gastropod families; Acetonidae, Cancellariidae and Cerithiidae are no longer found.
	The bivalve families; Pectinidae, Lyonsiidae and Astartidae are mostly only found down to 2000 m.
	Sequenziidae (gastropod) and Cyamiidae (bivalve) are located in depths shallower than 1500 m and deeper than 2000 m.
	ANDEEP samples showed isopod species typical of the shelf to penetrate to a depth of 1500–2000 m [139].
	Presence of a lower slope mollusc assemblage between 800 to 2000 m [136].
2000–3000	Depth band in which the gastropod family, Marginellidae is located (also located at depths <800 m). Also, the family Propilidiidae can now be found.
	The shelf inhabiting bivalve family Philobryidae is generally found in shallow water (<2000 m). However, the genus *Adarcnarca* can be found deeper than 2000 m.
	Isopod species richness continues to increase. The Isopod family, Austrarcturellidae is no longer found and Gnathiidae is found above and below this range.
	The main depth range of the polychaete family Pectinariidae (beside <500 m). Also, the Sabellariidae begin to be located (beside <200 m) and Eunicidae are no longer found.
3000–4500	Isopod species richness is highest within this depth range.
	Number of gastropod and polychaete species per depth begins to decrease.
	Echinoid species *Echinosigra amphora* and *Pourtalesia debilis* only occur between 3000 m and 4000 m [137].
	Gastropod species richness becomes extremely low and the gastropod families, Volutomitridae, Trichotropidae, Pleurobranchiidae, Fissurellidae are no longer found.
	The Isopoda families, Bopyridae and Stenetriidae are no longer found.
	The polychaete families, Pectinariidae are no longer found and Apistobranchidae end close to 3000 m.
	The bivalve families, Malletiidae and a species of Kelliidae are found deeper than around 3000 m.
	Only six bryozoan species occur below 3000 m [53].

Table 2. Cont.

Bathome (m) Depth Range	Ecological and Biological Events. Unless otherwise stated, the Bivalvia, Gastropoda, Isopoda and Polychaeta taxa discussed only include the species studied in Brandt et al. [37].
	Polychaetes are now a deep-sea assemblage composed of genera considered typical within the deep sea worldwide. This shift begins at approximately 2500–3000 m and extends onto the abyssal plain [22]
4500+	Isopod and polychaete species richness drops rapidly. However this could also be due to sampling bias, as there are very few samples deeper than 5000 m. Isopoda families: Cirolanidae, Dendrotionidae, Gnathiidae, Paramunnidae and *Xostylus* incertae sedis are no longer found within samples.

consider two of the four criteria used (i.e. summit depth and seamount proximity). However, the Clark et al. [120] classification incorporates seamounts i.e. [121] identified using a technique we found to overestimate the number of seamounts and therefore preferred the hand drawn seamounts identified by O'Brien et al, [64]. For the continental shelf and islands, this work provides a physical underpinning and whole of Southern Ocean context for the patterns in biodiversity found by Clarke et al. [33], the Delegations of Australia and France, [17] and Spalding et al. [2]. In addition to the identification of the regions themselves, we also offer a classification of the seabed habitats contained within each ecoregion.

Environmental types were mostly identified by nesting together ecoregions, bathomes and geomorphic features to avoid the potential for false within-class homogeneity [115]. However, there were some exceptions. To add utility to MPA planning, we reflect the variety of three geomorphic features by classifying seamounts and seamount ridges by only their shallowest bathome and canyons by whether they commence on the shelf or slope. When an MPA is placed to protect the benthos of integral physical features (e.g. those likely to contain vulnerable marine ecosystems) that have a high level of connectivity among the habitats they contain, such as seamounts, seamount ridges or canyons, the feature should be included as a whole and therefore the internal bathome changes become less relevant to spatial MPA planning [122]. The coastal terrane of some islands and the Antarctic continent represent areas adjacent to rugged coastlines with highly irregular depth structuring which could not be fully resolved with the depth data used and were therefore not further classed by bathomes [64]. Future MPA planning within regions where finer-scale bathymetry is available should further class these environmental types by the bathomes they contain and could be guided by the framework we provide. A level 2 nesting (see Fig. 1) provides a broader spatial management unit. For instance, a level 2b nesting of geomorphic features within ecoregions identifies 210 types. However, caution should be used when not further nesting geomorphic features within bathomes (i.e. level 3 of Fig. 1) since this can result in false within-class homogeneity [115]. Some environmental types are described within the 'restricted environmental types' section below.

According to the advice of the Science Committee to CCAMLR, we treat the benthic and pelagic environments separately [26,123]. The assumption is that these environments are decoupled due to the extreme depths and complex topography and oceanography of the Southern Ocean [75]. However, there is increasing evidence of interaction, and coupling may be favoured by intense seasonality and quick and efficient cycling through the pelagic food web [47,75,81,91,124]. We partially account for this interaction by considering sea surface productivity, the duration of ice cover and frontal systems within the analysis.

The classification we provide aims to capture the larger scale patterns by incorporating a combination of physical and biological data and also taking into consideration probable barriers to dispersal. As new data become available they can be used to refine the classification. Increasing knowledge, especially of cryptic species (i.e. species that are morphologically similar but differ genetically) seems most likely to lead to adjustment of the boundaries and the identification of further within-region heterogeneity, and not to a fundamentally different view of the distribution of habitats. Biogeographic classifications aim to identify general patterns within complex and diverse systems and therefore false heterogeneity for some species is inevitable. However, for the purposes of MPA planning, incorporating the general geographic and depth distribution patterns will increase the chance of representing the range of species habitat requirements within the final MPA system. The ecoregions and the environmental types can be used to consider the representativeness of biodiversity in a system of MPAs across the whole of the Southern Ocean.

Assessment of Representativeness

We found the level of existing representation for each environmental type across geomorphic classes is low (figure 4). None of the 23 ecoregions contains a system of MPAs that is representative of the benthic biodiversity they contain, with 12 ecoregions having no MPAs. Of the remaining 11 ecoregions, none includes a system of MPAs that sample each of the environmental types present in the ecoregion. The three ecoregions which perform the best are 1) Del Cano where South Africa declared the Prince Edward Islands MPA in 2012, 2) the South Orkney Islands, where CCAMLR established its first entirely high seas MPA in 2009 and 3) the Kerguelen Plateau where Australia and France have established MPAs. For some geomorphic classes, the number of environmental types in MPAs especially within the Ross Sea, Oates, Deep Kerguelen and Antarctic Peninsula ecoregions, is greater than 20% (figure 4). However, the area of seabed in MPAs for these geomorphic classes is less than 1% suggesting that only a small portion of each environmental type is represented in MPAs.

We have provided an assessment of progress toward a representative system of MPAs. The primary aim of the analysis described here is to identify important gaps in the protected area system, particularly related to ecoregions and environmental types. We have not identified how the biodiversity within a region needs to be adequately included in MPAs in order to achieve viability of the protected biodiversity; requirements will vary between different ecoregions and environmental types.

There are a number of existing conservation planning processes directed towards resolving some of these gaps in the existing system of MPAs (Fig. 2), including Eastern Antarctica [17] and the Ross Sea for example: [106,125]. These proposals will help fill the gaps for achieving a Southern Ocean system of MPAs. Analyses such as the one we present here could be used to help assess progress towards a representative system of MPAs covering each

Table 3. The geomorphic features of the Southern Ocean classified according to the attributes of the seabed surface substratum.

Geomorphic class	Name (map code)	Definition adapted from O'Brien et al. [64]
Continental Shelf and related features	Bank (2)	Broad shallow regions typically at depths of 100–200 m. The boundary between the shelf bank and shelf depressions is set at a depth of around 500. Banks within both the South Georgia and Kerguelen Plateau ecoregions were classed as Oceanic Shallow features.
	Coastal (Rugged) Terrane (5)	Region of varying seafloor type and depth ranges along rugged coastlines
	Cross Shelf Valley (7)	Shelf depressions that are connected to the shelf edge via valleys.
	Shelf (22)	Unclassified regions within the continental shelf region.
	Shelf Deeps-Depressions (23)	Shelf region generally deeper than 550 m with closed contours.
	Volcano (28)	Distinguished from other islands and seamounts, where volcanic processes directly impact the marine environment. Mapped volcanoes within the Southern Ocean all occur on the shelf.
Oceanic Shallow features	Island Arc (9)	Islands formed from bow-shaped volcanic ridges adjacent to subduction zones.
	Island Coastal Terrane (10)	Similar to coastal (rugged) terrane representing a zone of high variability around islands.
	Margin Ridge (13)	Ridges formed from igneous or basement intrusions along the continental margin and protruding hundreds of meters above the (abyssal) sediment plain.
	Marginal Plateau (14)	Areas of relatively level sea floor at mid depth extending from continental margins and separated from the shelf by a saddle.
	Plateau (16)	Relatively flat regions elevated above the surrounding sea floor by more than a few hundred meters.
	Ridge (18)	Elongate ridges that may extend from a plateau or other feature.
	Seamount Ridges (20)	Elongate ridges that protrude hundreds to thousands of meters above the surrounding sea floor. Their shape has the potential to influence deep current activity.
	Seamounts (21)	Roughly circular areas which rise above the surrounding sea floor by at least 1000 m.
Slope and related features	Canyon (3)	A relatively narrow, deep gully with steep sides. Axes were traced along landward contour inflection points, particularly in the shelf edge region.
	Lower Slope (11)	Region on the continental slope of reduced gradient with a lower limit where slope canyons are no longer obvious (around 2500–3500 m below sea level).
	Plateau Slope (17)	Broad regions sloping from the margins of large plateaus to the surrounding deep ocean floor.
	Structural Slope Region (24)	Low relief topographic features formed from underlying structures, such as basement protrusions, that extend beyond the lower slope.
	Trough Mouth Fans (26)	Broad aprons of smooth to slightly gullied sediment on the Upper Slope extending from the shelf break to 2500–3000 m water depth.
	Upper Slope (27)	Seaward dipping slope extending from the continental shelf break which is defined as the position at which the rate of change in slope gradient is at a maximum.
Abyss and related features	Abyssal Plain (1)	Extensive, flat, gently sloping or nearly level region of sediment covered seafloor at abyssal depths.
	Cliff (4)	Very steep or near vertical features normally occurring at major crustal fractures or on the sides of glacial valleys on the shelf and are likely to expose hard substrates.
	Contourite Drift (6)	Sediment mounds constructed by strong bottom currents that rise gently above the surrounding sea floor.
	Fracture Zone (8)	Major oceanic crustal fracture zones.
	Mid-Ocean Ridge Rift Valley (12)	Elongate troughs created by seafloor spreading, extending several hundred meters below the rift shoulders and containing hydrothermal vents.
	Ocean Trough (15)	Closed elongate depressions (in the ocean floor) more than 4500 m deep and hundreds of kilometres long, generally associated with fracture zones.
	Rugose Ocean Floor (19)	Relatively young oceanic crust with rugged features protruding through the sediment.
	Trench (25)	Arcuate depressions, typically at depths of more than 5000 m and reaching 6000 m in places, formed by subduction of oceanic crust.

of the ecoregions and which areas may be most suitable for filling gaps. An accounting procedure to help assess the adequacy of MPAs would improve future assessments.

Restricted environmental types. In this study we contribute to Southern Ocean MPA planning processes by identifying environmental types with a restricted spatial distribution and within ecoregions largely lacking any existing or proposed representation in MPAs. We identify 107 restricted environmental types (Fig. 2). The distribution of restricted environments is driven by rare environmental units that arise when a component of the hierarchical classification is scarce. The island arc, a geomorphic feature within the Southern Sandwich Islands ecoregion, is the only habitat of its kind in the Southern Ocean. Therefore the ecosystems associated with the particular environments created where a ridge of volcanic islands is formed adjacent to a subduction zone can only be protected in this location. Other restricted environments arise from an uncommon combination of geomorphic features and bathomes. There are many seamounts

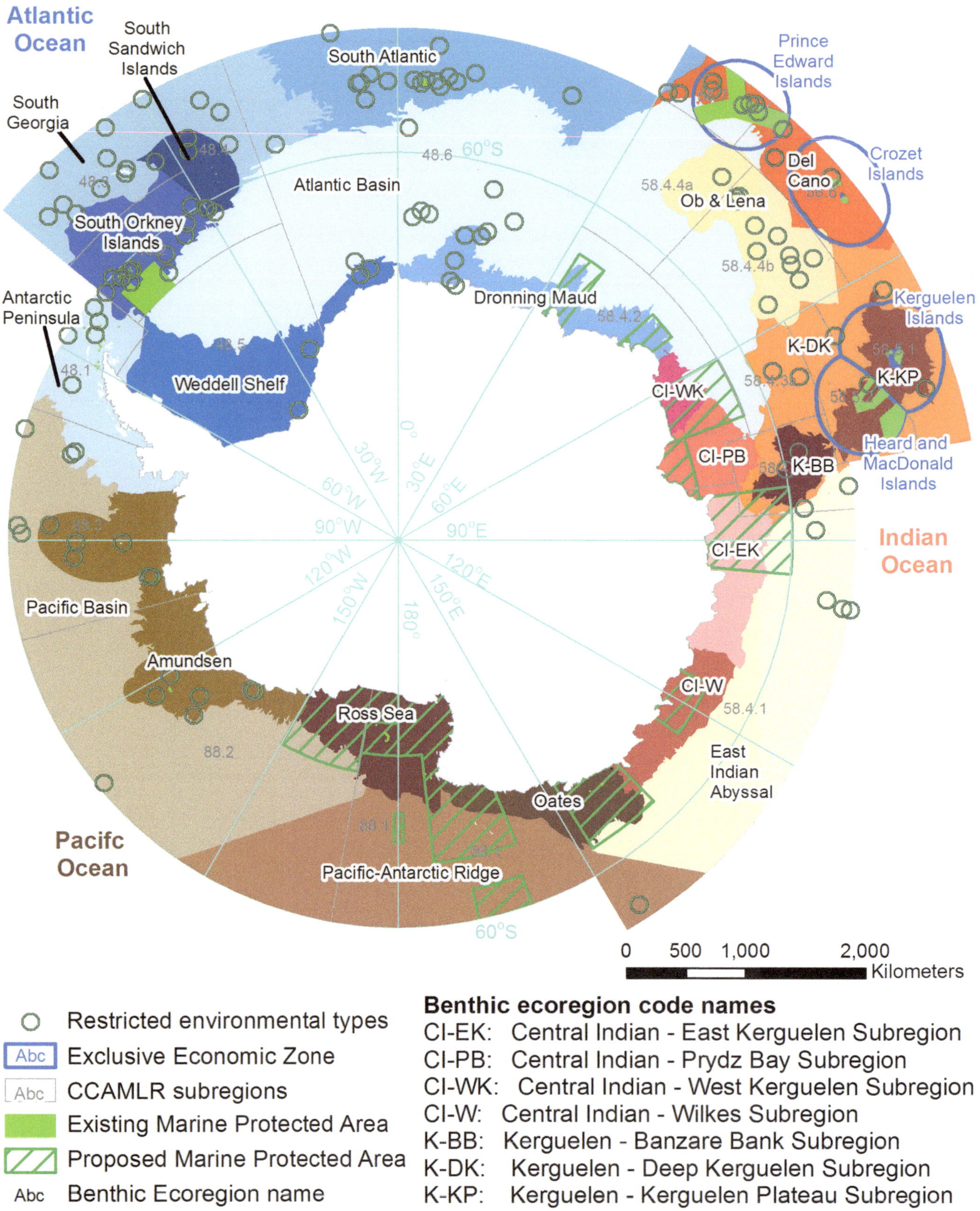

Figure 2. The benthic ecoregions, restricted environments and marine protected areas identified within the Southern Ocean. An environmental type is a unique combination of an ecoregion, bathome and geomorphic feature. Existing marine protected areas and regions where planning processes are underway to propose future representation, were identified. Where large gaps in existing and proposed representation were found, the locations of geographically restricted environmental types were identified. These restricted environments indicate areas of potential future marine protected area selection since there are limited spatial options for protecting the biodiversity for which these environments are a surrogate.

Table 4. Benthic ecoregions and their features (see Fig.2 for the location of each ecoregion).

Benthic ecoregion	General description
Amundsen	The productive shelf and polynyas of the Amundsen and Bellingshausen seas. The oceanic shallow environments of Peter I Island, De Gerlache Seamounts and the Marie Byrd Seamount group.
Antarctic Peninsula	The shallow, productive shelf of the west Antarctic Peninsula with a low duration of sea ice cover and warm seabeds relative to other Antarctic shelf areas. The island ecosystems of the South Shetland Islands. 13 endemic molluscs [11]. Greater than 10% of gastropods endemic [12].
Atlantic Basin	The very deep and very cold rugose ocean floor and abyssal plain of the South Atlantic Ocean Basin and Weddell Sea.
Central Indian-East Kerguelen Subregion	Central Indian region of East Antarctica that is influenced by the Kerguelen Plateau including downstream productivity from frontal activity across the Kerguelen Plateau [17].
Central Indian-Prydz Bay Subregion	Central Indian region of East Antarctica that contains the cold, productive waters of Prydz Bay and the Prydz Gyre which oceanographically separates the east Kerguelen and west Kerguelen Central Indian subregions [17].
Central Indian-West Kerguelen Subregion	Central Indian region of East Antarctica that is not influenced by the Kerguelen Plateau nor the Weddell Gyre [17].
Central Indian -Wilkes Subregion	Central Indian region of East Antarctica that is oceanographically separated from the Central Indian-East Kerguelen subregion [17,140].
Del Cano	The shallow, warm seabeds in the Subantarctic Frontal Zone including the South West Indian Ridge seamounts, the Del Cano Rise and the Crozet and Prince Edward Islands.
Dronning Maud	Maud Rise and associated open ocean polynya, Astrid Ridge, Gunnerus Ridge and the canyons offshore Dronning Maud Land. Easternmost extent of the Weddell Gyre. 20 endemic molluscs (19% of documented species) [11]. 21% of documented gastropods are endemic [12].
East Indian Abyssal	The very deep and cold seabeds of the rugose ocean floor and abyssal plains of the South Indian Ocean Basin.
Kerguelen-Banzare Bank Subregion	Shallower (mostly depths between 1000 to 3000 m), warmer seabeds of the Banzare Bank, south of the frontal activity of the Fawn Trough.
Kerguelen-Deep Kerguelen Subregion	Deep (mostly depths greater than 3000 m) ocean surrounding the Kerguelen Plateau and Banzare Bank.
Kerguelen-Kerguelen Plateau Subregion	Shallower (mostly depths between 200 m to 3000 m), warmer seabeds of the Kerguelen Plateau, north of the frontal activity of the Fawn Trough.
Oates	Oceanographically separated from the Central Indian-Wilkes subregion with wind and sea ice vectors diverging at its western border [17]. The eastern border is adjacent to the Ross Sea region [2,33,125].
Ob & Lena	Shallow, warm seabeds in the Polar Frontal Zone, including the Ob and Lena banks and the seamounts to their east.
Pacific Basin	The very deep rugose ocean floor and abyssal plains of the South Pacific Ocean Basin which is warmer than other deep ocean basin regions of the Southern Ocean.
Pacific-Antarctic Ridge	The Pacific-Antarctic Ridge region with large extents of shallower environments of depths less than 2000 m.
Ross Sea	Very cold seabed and high sea ice duration of the productive Ross Sea. 22 endemic molluscs (11.5% of documented species) [11]. 16% of documented gastropods endemic [12].
South Atlantic	Shallower environments of the Mid Atlantic Ridge and associated seamounts.
South Georgia	Productive, shallow environments in the Polar Frontal Zone including the island ecosystems of South Georgia Island and the seamounts of the North Scotia Ridge. 65 endemic molluscs (32.7% of documented species) [11]. 15% of documented Cheilostomata endemic [12]. 13% of documented bivalves endemic [12]. 36% of documented gastropods endemic [12].
South Orkney Islands	The island ecosystems of the South Orkney Islands and the seamounts and plateaus of the South Scotia Arc, many of which underlie the Southern Antarctic Circumpolar Current Frontal Zone. 22 endemic molluscs (19.6% of documented species) [11]. 25% of documented gastropods endemic [12]
South Sandwich Islands	Highly productive island ecosystems of the South Sandwich Islands and the deeper waters of the South Sandwich Trench.
Weddell Shelf	Very cold seabed and high sea ice duration of the productive Weddell shelf, usually rather deep, ~500 m, at places even down to 1000 m. 55 endemic molluscs (19.7% of documented species) [11]. 26% of documented gastropods endemic [12]

within the Pacific basin ecoregion however there are only two, the Belgica Guyot and the Lecointe Guyot, that have very shallow mounts of less than 200 m depth. Furthermore, there are only five discrete locations within the CCAMLR region where these shallow seamount environmental types occur.

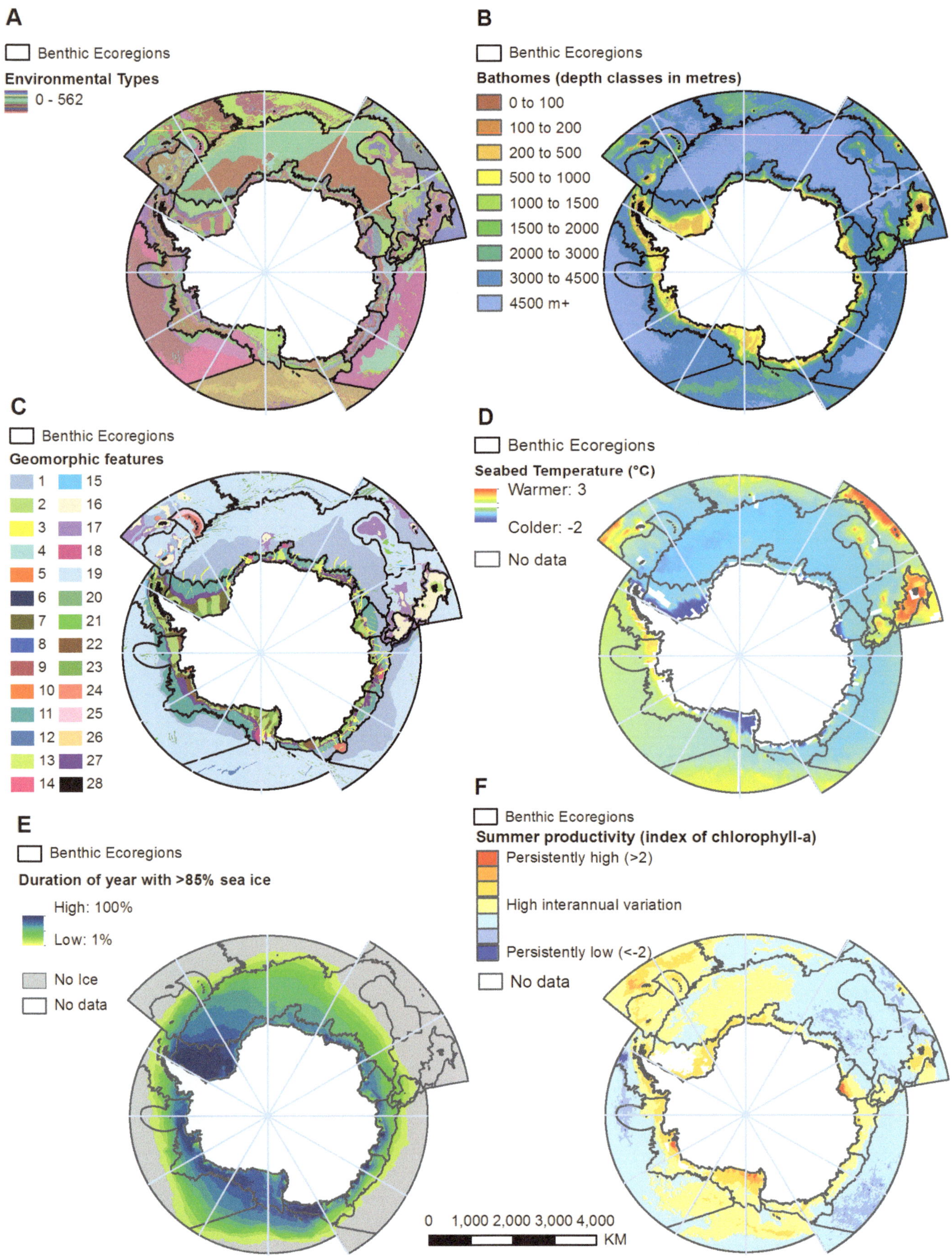
A
Benthic Ecoregions
Environmental Types
0 - 562
B
Benthic Ecoregions
Bathomes (depth classes in metres)
0 to 100
100 to 200
200 to 500
500 to 1000
1000 to 1500
1500 to 2000
2000 to 3000
3000 to 4500
4500 m+
C
Benthic Ecoregions
Geomorphic features
1
2
3
4
5
6
7
8
9
10
11
12
13
14
15
16
17
18
19
20
21
22
23
24
25
26
27
28
D
Benthic Ecoregions
Seabed Temperature (°C)
Warmer: 3
Colder: -2
No data
E
Benthic Ecoregions
Duration of year with >85% sea ice
High: 100%
Low: 1%
No Ice
No data
F
Benthic Ecoregions
Summer productivity (index of chlorophyll-a)
Persistently high (>2)
High interannual variation
Persistently low (<-2)
No data
0 1,000 2,000 3,000 4,000
KM

Figure 3. Environmental types and the bio-physical data used to drive the classification. A) The 562 environmental types (in colour) and ecoregion outlines (refer to figure 2 for names) broadly reflect the underlying data used within the classification **B)** Bathomes derived from bathymetry and species-depth relationships **C)** Geomorphic features; Abyssal Plain (1), Bank (2), Canyon (3), Cliff (4), Coastal (rugged) Terrane (5), Contourite Drift (6), Cross Shelf Valley (7), Fracture Zone (8), Island Arc (9), Island Coastal Terrane (10), Lower Slope (11), Marginal Ridge (12), Marginal Plateau (13), Mid-Ocean Ridge Rift Valley (14), Ocean Trough (15), Plateau (16), Plateau Slope (17), Ridge (18), Rugose Ocean Floor (19), Seamount Ridges (20), Seamount (21), Shelf (22), Shelf Deep (Depressions) (23), Structural Slope Region (24), Trench (25), Trough Mouth Fans (26), Upper (Continental) Slope (27), Volcano (28) **D)** Seabed temperature **E)** Duration of the year where more than 85% of the region is covered by sea ice **F)** High positive and negative values indicate areas of consistently high and low summer productivity respectively. Values approaching zero indicate areas that vary greatly between years.

Maud Rise is another example of a biologically important region containing restricted environmental types. At Maud Rise, oceanography, sediment characteristics and sea ice processes have been linked to high biodiversity throughout all trophic levels from pelagic predators to benthic species [30]. Maud Rise is an underwater plateau and slope region with two associated seamounts within the Atlantic Basin ecoregion. Four of the six environmental types of Maud Rise are restricted. These include: 1) the only seamount in the ecoregion with a mount in the −1500 to −2000 m bathome; 2) the only seamount ridge within the ecoregion with a mount in the −1000 to −1500 m bathome and; 3) the only plateau in the ecoregion with two environmental types resulting from plateau areas within the −2000 to −3000 m bathome and −3000 to −4500 m bathome. The Maud Rise seamount is suspected to contribute to the most recurring open-ocean polynya in the Southern Ocean which forms over Maud Rise [30,126]. These polynyas drive convection which influences interaction between surface productivity, the benthos and the upwelling of nutrients all of which may contribute to the rich and prospering food web in the region [30].

Conclusion

The international effort overseen by CCAMLR to establish a representative network of MPAs within the Southern Ocean has already made a significant step in declaring the world's first MPA within the high seas located outside national jurisdiction. The

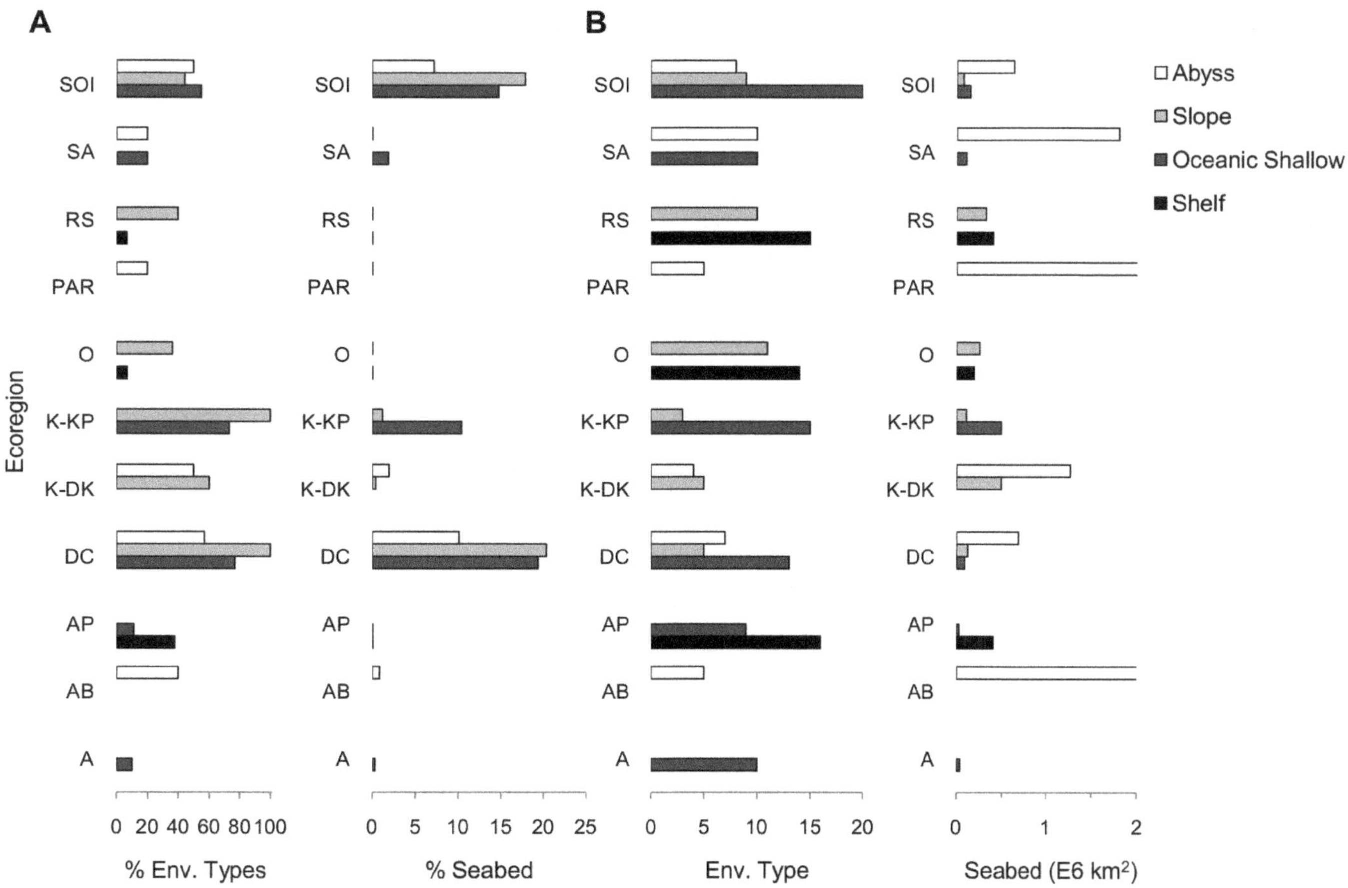

Figure 4. Marine protected areas and their spatial coverage of ecoregions and representation of environmental types. A) The proportion of the seafloor and the proportion of environmental types included either partly or wholly within existing MPAs are displayed within four broad geomorphic classes. These classes are further described in Table 3 and are represented within the plots from top to bottom as Abyss (white), Slope (light grey), Oceanic Shallow (dark grey) and Continental Shelf (black). **B)** The total number of environmental types and seafloor area for each geomorphic class are also shown. The incomplete bars within PAR and AB have values of 2.95 E6 km^2 and 6.59 E6 km^2 respectively. The ecoregion code names are: SOI (South Orkneys Islands), SA (South Atlantic), RS (Ross Sea), PAR (Pacific-Antarctic Ridge), O (Oates), K-KP (Kerguelen-Kerguelen Plateau Subregion), K-DK (Kerguelen-Deep Kerguelen Subregion), DC (Del Cano), AP (Antarctic Peninsula), AB (Atlantic Basin), A (Amundsen).

efforts to achieve the conservation commitments made by CCAMLR and parties to the Antarctic Treaty sets a precedent for conservation within other high seas regions of the global ocean [101]. However the current levels of representation clearly contain significant gaps that future conservation planning efforts need to address. The environmental types presented here are an improvement on previous statistical classifications because, rather than opting for a method that identifies statistically dissimilar areas within the data, we have identified environmental types using known and probable relationships between biogeographic patterns, environmental drivers, and knowledge of dispersal of benthic taxa. The classification of environmental types allows the gaps in the MPA network to be quantified. It also provides a new circum-Antarctic map of environmental types that can be used to support spatial management aimed at conserving benthic biodiversity across the entire Southern Ocean.

Acknowledgments

The authors thank Alan Williams, Ben Sharp, Brendan Brooke, Falk Huettmann, Grant Humphries, Hugh Possingham, Katrin Linse, Louise Emmerson, Matthew Pinkerton, Mark Spalding, Nic Bax, Peter Last, Philippe Koubbi, Piers Dunstan, Rudy Kloser, Steven Rintoul, Susie Grant, Vincent Lyne, Zhi Huang and two anonymous reviewers for their advice and comment and Bruno Danis, Philip E O'Brien, Rowan Romeyn, Steven Rintoul and Serguei Sokolov for their data provision.

Author Contributions

Conceived and designed the experiments: DB LLD. Performed the experiments: DB LLD JT AC. Analyzed the data: DB LLD JT AC BR. Contributed reagents/materials/analysis tools: AP SK BR AB. Wrote the paper: LLD HSG DB SK AC BR RN.

References

1. Griffiths H, Barnes D, Linse K (2009) Towards a generalized biogeography of the Southern Ocean benthos. Journal of Biogeography 36: 162–177.
2. Spalding M, Fox H, Allen G, Davidson N, Ferdana Z, et al. (2007) Marine ecoregions of the world: a bioregionalization of coastal and shelf areas. Bioscience 57: 573–583.
3. Kaiser S, Brandão SN, Brix S, Barnes DK, Bowden DA, et al. (2013) Patterns, processes and vulnerability of Southern Ocean benthos: a decadal leap in knowledge and understanding. Marine Biology 160: 2295–2317.
4. Halpern BS, Walbridge S, Selkoe KA, Kappel CV, Micheli F, et al. (2008) A global map of human impact on marine ecosystems. Science 319: 948.
5. Kock K (2007) Antarctic Marine Living Resources–exploitation and its management in the Southern Ocean. Antarctic Science 19: 231–238.
6. Agnew DJ (1997) Review–The CCAMLR Ecosystem Monitoring Programme. Antarctic Science 9: 235–242.
7. Constable A, de la Mare W, Agnew D, Everson I, Miller D (2000) Managing fisheries to conserve the Antarctic marine ecosystem: practical implementation of the Convention on the Conservation of Antarctic Marine Living Resources (CCAMLR). ICES Journal of Marine Science 57: 778.
8. SC-CAMLR (2009) Scientific Committee for the Conservation of Antarctic Marine Living Resources: Report of the twenty-eighth meeting 2009 26–30 October; North Hobart, Tasmania Australia.
9. SC-CAMLR (2009) Report of the Joint SC-CAMLR–CEP Workshop (Baltimore, USA, 3 and 4 April 2009) Science Committee for the Conservation of Antarctic Marine Living Resources (SC-CAMLR-XXVIII/6).
10. SC-CAMLR (2005) Report of the Scientific Committee workshop on Marine Protected Areas held at Silver Spring, MD, USA, 29 August to 1 September 2005. Annex 7 to the Convention for the Conservation of Antarctic Living Marine Resources.
11. Harris P, Whiteway T (2009) High seas marine protected areas: Benthic environmental conservation priorities from a GIS analysis of global ocean biophysical data. Ocean & Coastal Management 52: 22–38.
12. Arango C, Soler Membrives A, Miller K (2011) Genetic differentiation in the Circum-Antarctic sea spider *Nymphon australe* (Pycnogonida; Nymphonidae). Deep Sea Research Part II: Topical Studies in Oceanography 58: 212–219.
13. Wilson N, Hunter R, Lockhart S, Halanych K (2007) Multiple lineages and absence of panmixia in the "circumpolar" crinoid *Promachocrinus kerguelensis* from the Atlantic sector of Antarctica. Marine Biology 152: 895–904.
14. Hoffman J, Peck L, Linse K, Clarke A (2010) Strong Population Genetic Structure in a Broadcast-Spawning Antarctic Marine Invertebrate. Journal of Heredity 102: 55–66.
15. Hemery L, Eléaume M, Roussel V, Améziane N, Gallut C, et al. (2012) Comprehensive sampling reveals circumpolarity and sympatry in seven mitochondrial lineages of the Southern Ocean crinoid species *Promachocrinus kerguelensis* (Echinodermata). Molecular Ecology 21: 2502–2518.
16. Longhurst A (1998) Ecological geography of the sea. San Diego: Academic Press.
17. Delegations of Australia and France (2011) Proposal for a representative system of Marine Protected Areas (RSMPA) in the East Antarctica planning domain. Submitted to the scientific committee to the Commission for the Conservation of Antarctic Marine Living Resources by the Delegations of Australia and France, SC-CCAMLR-XXX/11.
18. Pressey RL, Bottrill MC (2009) Approaches to landscape-and seascape-scale conservation planning: convergence, contrasts and challenges. Oryx 43: 464–475.
19. McArthur M, Brooke B, Przeslawski R, Ryan D, Lucieer V, et al. (2010) On the use of abiotic surrogates to describe marine benthic biodiversity. Estuarine, Coastal and Shelf Science 88: 21–32.
20. Huang Z, Brooke BP, Harris PT (2010) A new approach to mapping marine benthic habitats using physical environmental data. Continental Shelf Research 31: S4–S16.
21. Griffiths H (2010) Antarctic Marine Biodiversity-What Do We Know About the Distribution of Life in the Southern Ocean? PLoS ONE 5: e11683.
22. Brandt A, Gooday A, Brandão S, Brix S, Brökeland W, et al. (2007) First insights into the biodiversity and biogeography of the Southern Ocean deep sea. Nature 447: 307–311.
23. Gutt J, Barnes D, Lockhart SJ, van de Putte A (2013) Antarctic macrobenthic communities: A compilation of circumpolar information. Nature Conservation 4: 1–13.
24. Grantham HS, Moilanen A, Wilson KA, Pressey RL, Rebelo TG, et al. (2008) Diminishing return on investment for biodiversity data in conservation planning. Conservation Letters 1: 190–198.
25. Grant S, Constable A, Raymond B, Doust S (2006) Bioregionalisation of the Southern Ocean: report of experts workshop, Hobart, September 2006: WWF Australia.
26. Penhale P, Grant S (2007) Commission of the Conservation of Antarctic Marine Living Resources (CCAMLR) Workshop on Bioregionalisation of the Southern Ocean (Brussels, Belgium, 13 to 17 August 2007) Annex 9: Report of the Workshop Executive Summary.
27. Clarke A, Griffiths H, Barnes D, Meredith M, Grant S (2009) Spatial variation in seabed temperatures in the Southern Ocean: Implications for benthic ecology and biogeography. Journal of Geophysical Research 114: G03003.
28. Kaiser S, Griffiths H, Barnes D, Brandão S, Brandt A, et al. (2011) Is there a distinct continental slope fauna in the Antarctic? Deep Sea Research Part II: Topical Studies in Oceanography 58: 91–104.
29. Koubbi P, Moteki M, Duhamel G, Goarant A, Hulley PA, et al. (2011) Ecoregionalisation of myctophid fish in the Indian sector of the Southern Ocean: Results from generalized dissimilarity models. Deep Sea Research Part II: Topical Studies in Oceanography 58: 170–180.
30. Brandt A, Bathmann U, Brix S, Cisewski B, Flores H, et al. (2011) Maud Rise – a snapshot through the water column. Deep Sea Research Part II: Topical Studies in Oceanography 58: 1962–1982.
31. Last P, Lyne V, Williams A, Davies C, Butler A, et al. (2010) A hierarchical framework for classifying seabed biodiversity with application to planning and managing Australia's marine biological resources. Biological Conservation 143: 1675–1686.
32. Linse K, Griffiths H, Barnes D, Clarke A (2006) Biodiversity and biogeography of Antarctic and sub-Antarctic mollusca. Deep Sea Research Part II: Topical Studies in Oceanography 53: 985–1008.
33. Clarke A, Griffiths H, Linse K, Barnes D, Crame J (2007) How well do we know the Antarctic marine fauna? A preliminary study of macroecological and biogeographical patterns in Southern Ocean gastropod and bivalve molluscs. Diversity and Distributions 13: 620–632.

34. McLeod DJ, Hosie GW, Kitchener JA, Takahashi KT, Hunt BPV (2010) Zooplankton Atlas of the Southern Ocean: The SCAR SO-CPR Survey (1991–2008). Polar Science 4: 353–385.
35. Barnes D, Griffiths H (2008) Biodiversity and biogeography of southern temperate and polar bryozoans. Global Ecology and Biogeography 17: 84–99.
36. Danis B, Griffiths H (2009) Polar science: bid for freely accessible biodiversity archive. Nature 458: 830–830.
37. Brandt A, Linse K, Schüller M (2009) Bathymetric distribution patterns of Southern Ocean macrofaunal taxa: Bivalvia, Gastropoda, Isopoda and Polychaeta. Deep Sea Research Part I: Oceanographic Research Papers 56: 2013–2025.
38. Duhamel G, Hautecoeur M (2009) Biomass, abundance and distribution of fish in the Kerguelen Islands EEZ (CCAMLR statistical division 58.5.1). CCAMLR Science 16: 1–32.
39. Koubbi P, Ozouf-Costaz C, Goarant A, Moteki M, Hulley P-A, et al. (2010) Estimating the biodiversity of the East Antarctic shelf and oceanic zone for ecoregionalisation: Example of the ichthyofauna of the CEAMARC (Collaborative East Antarctic Marine Census) CAML surveys. Polar Science 4: 115–133.
40. Díaz A, Féral JP, David B, Saucède T, Poulin E (2011) Evolutionary pathways among shallow and deep-sea echinoids of the genus *Sterechinus* in the Southern Ocean. Deep Sea Research Part II: Topical Studies in Oceanography 58: 205–211.
41. González-Wevar CA, David B, Poulin E (2011) Phylogeography and demographic inference in *Nacella (Patinigera) concinna* (Strebel, 1908) in the western Antarctic Peninsula. Deep Sea Research Part II: Topical Studies in Oceanography 58: 220–229.
42. Beaman RJ, Harris PT (2005) Bioregionalization of the George V Shelf, East Antarctica. Continental Shelf Research 25: 1657–1691.
43. Post A, O'Brien P, Beaman R, Riddle M, De Santis L (2010) Physical controls on deep water coral communities on the George V Land slope, East Antarctica. Antarctic Science 22: 371–378.
44. Barry JP, Grebmeier JM, Smith J, Dunbar RB (2003) Oceanographic versus seafloor-habitat control of benthic megafaunal communities in the SW Ross Sea, Antarctica. Antarctic Research Series 78: 327–354.
45. Hall S, Thatje S (2011) Temperature-driven biogeography of the deep-sea family Lithodidae (Crustacea: Decapoda: Anomura) in the Southern Ocean. Polar Biology 34: 363–370.
46. Gutt J (2007) Antarctic macro-zoobenthic communities: a review and an ecological classification. Antarctic Science 19: 165–182.
47. Gili J, Rossi S, Pagès F, Orejas C, Teixidó N, et al. (2006) A new trophic link between the pelagic and benthic systems on the Antarctic shelf. Marine Ecology Progress Series 322: 43–49.
48. Gutt J (2001) On the direct impact of ice on marine benthic communities, a review. Polar Biology 24: 553–564.
49. Brey T, Dahm C, Gorny M, Klages M, Stiller M, et al. (1996) Do Antarctic benthic invertebrates show an extended level of eurybathy? Antarctic Science 8: 3–6.
50. Thatje S, Hillenbrand C, Larter R (2005) On the origin of Antarctic marine benthic community structure. Trends in Ecology & Evolution 20: 534–540.
51. Clarke A (2003) The polar deep seas. In: Tyler PA, editor. Ecosystems of the Deep Oceans. Amsterdam: Elsevier. 239–260.
52. Brandt A (1992) Origin of Antarctic Isopoda (Crustacea, Malacostraca). Marine Biology 113: 415–423.
53. Barnes DKA, Kuklinski P (2010) Bryozoans of the Weddell Sea continental shelf, slope and abyss: did marine life colonize the Antarctic shelf from deep water, outlying islands or in situ refugia following glaciations? Journal of Biogeography 37: 1648–1656.
54. France SC, Kocher T (1996) Geographic and bathymetric patterns of mitochondrial 16S rRNA sequence divergence among deep-sea amphipods, *Eurythenes gryllus*. Marine Biology 126: 633–643.
55. Hunter RL, Halanych KM (2008) Evaluating connectivity in the brooding brittle star *Astrotoma agassizii* across the Drake Passage in the Southern Ocean. Journal of Heredity 99: 137–148.
56. Held C, Wägele JW (2005) Cryptic speciation in the giant Antarctic isopod *Glyptonotus antarcticus* (Isopoda, Valvifera, Chaetiliidae). Scientia marina 69: 175–181.
57. Allcock A (2005) On the confusion surrounding *Pareledone charcoti* (Joubin, 1905) (Cephalopoda: Octopodidae): endemic radiation in the Southern Ocean. Zoological Journal of the Linnean Society 143: 75–108.
58. Brandão SN, Sauer J, Schön I (2010) Circumantarctic distribution in Southern Ocean benthos? A genetic test using the genus *Macroscapha* (Crustacea, Ostracoda) as a model. Molecular Phylogenetics and Evolution 55: 1055–1069.
59. Havermans C, Sonet G, d'Acoz CdU, Nagy ZT, Martin P, et al. (2013) Genetic and Morphological Divergences in the Cosmopolitan Deep-Sea Amphipod *Eurythenes gryllus* Reveal a Diverse Abyss and a Bipolar Species. PLoS ONE 8: e74218.
60. Barry JP, Dayton PK (1991) Physical Heterogeneity and the Organization of Marine Communities. In: Kolasa J, Pickett STA, editors. Ecological Heterogeneity. New York: Springer-Verlag. 270–320.
61. Held C (2003) Molecular evidence for cryptic speciation within the widespread Antarctic crustacean Ceratoserolis trilobitoides (Crustacea, Isopoda). Antarctic biology in a global context: 135–139.
62. Leese F, Kop A, Wägele J-W, Held C (2008) Cryptic speciation in a benthic isopod from Patagonian and Falkland Island waters and the impact of glaciations on its population structure. Frontiers in zoology 5: 19.
63. Harris PT, Baker EK (2012) Seafloor Geomorphology as Benthic Habitat: GeoHab Atlas of seafloor geomorphic features and benthic habitats. Amsterdam: Elsevier. 936 p.
64. O'Brien PE, Post AL, Romeyn R (2009) Antarctic-wide Geomorphology as an aid to habitat mapping and locating Vulnerable Marine Ecosystems. Science Committee to the Commission of Antartctic Marine Living Resources (SC-CAMLR-XXVIII/10) Workshop on Vulnerable Marine Ecosystems. La Jolla, CA, USA 3–7th August 2009: GeoScience Australia. Conference paper: WS-VME-09/10. Available: data.aad.gov.au with search ID 'ant_seafloor_geomorph.' Accessed: 2013 Nov.
65. Post AL, Beaman RJ, O'Brien PE, Eléaume M, Riddle MJ (2011) Community structure and benthic habitats across the George V Shelf, East Antarctica: Trends through space and time. Deep Sea Research Part II: Topical Studies in Oceanography 58: 105–118.
66. Schlacher TA, Schlacher-Hoenlinger MA, Williams A, Althaus F, Hooper JNA, et al. (2007) Richness and distribution of sponge megabenthos in continental margin canyons off southeastern Australia. Marine Ecology Progress Series 340: 73–88.
67. Harris PT, Whiteway T (2011) Global distribution of large submarine canyons: Geomorphic differences between active and passive continental margins. Marine Geology 285: 69–86.
68. Richer de Forges BR, Koslow JA, Poore G (2000) Diversity and endemism of the benthic seamount fauna in the southwest Pacific. Nature 405: 944–947.
69. Bowden DA, Schiaparelli S, Clark MR, Rickard GJ (2011) A lost world? Archaic crinoid-dominated assemblages on an Antarctic seamount. Deep Sea Research Part II: Topical Studies in Oceanography 58: 119–127.
70. Gjerde KM (2006) Ecosystems and Biodiversity in Deep Waters and High Seas. UNEO Regional Seas Report and Studies 178: 1–60.
71. Rogers AD, Tyler PA, Connelly DP, Copley JT, James R, et al. (2012) The discovery of new deep-sea hydrothermal vent communities in the Southern Ocean and implications for biogeography. Plos Biology 10: e1001234.
72. Allcock AL, Barratt I, Eléaume M, Linse K, Norman MD, et al. (2011) Cryptic speciation and the circumpolarity debate: A case study on endemic Southern Ocean octopuses using the COI barcode of life. Deep Sea Research Part II: Topical Studies in Oceanography 58: 242–249.
73. Morley S, Martin S, Bates A, Clark M, Ericson J, et al. (2012) Spatial and temporal variation in the heat tolerance limits of two abundant Southern Ocean invertebrates. Marine Ecology-Progress Series 450: 81–92.
74. Clarke A, Aronson R, Crame J, Gili J, Blake D (2004) Evolution and diversity of the benthic fauna of the Southern Ocean continental shelf. Antarctic Science 16: 559–568.
75. Smith CR, Mincks S, DeMaster DJ (2006) A synthesis of bentho-pelagic coupling on the Antarctic shelf: food banks, ecosystem inertia and global climate change. Deep Sea Research Part II: Topical Studies in Oceanography 53: 875–894.
76. Clarke A, Arntz WE (2006) An introduction to EASIZ (Ecology of the Antarctic Sea Ice Zone): An integrated programme of water column, benthos and bentho-pelagic coupling in the coastal environment of Antarctica. Deep Sea Research Part II: Topical Studies in Oceanography 53: 803–814.
77. Barry J, Dayton P (1988) Current patterns in McMurdo Sound, Antarctica and their relationship to local biotic communities. Polar Biology 8: 367–376.
78. Massom RA, Stammerjohn SE (2010) Antarctic sea ice change and variability-Physical and ecological implications. Polar Science 4: 149–186.
79. Wilson P, Ainley D, Nur N, Jacobs S, Barton K, et al. (2001) Adélie penguin population change in the pacific sector of Antarctica: relation to sea-ice extent and the Antarctic Circumpolar Current. Marine Ecology Progress Series 213: 301–309.
80. Baines PG (2007) Coastal and Regional Currents of Antarctica. In: Riffenburgh B, editor. Encyclopedia of the Antarctic. New York: Routledge. 1146.
81. Grebmeier JM, Barry JP (1991) The influence of oceanographic processes on pelagic-benthic coupling in polar regions: A benthic perspective. Journal of Marine Systems 2: 495–518.
82. Falkowski P, Barber R, Smetacek V (1998) Biogeochemical controls and feedbacks on ocean primary production. Science 281: 200.
83. Knox GA (2007) Biology of the Southern Ocean. Boca Raton, United States: CRC Press.
84. Arrigo K, van Dijken G, Bushinsky S (2008) Primary production in the Southern Ocean, 1997–2006. Journal of Geophysical Research 113: C08004.
85. Sokolov S, Rintoul S (2007) On the relationship between fronts of the Antarctic Circumpolar Current and surface chlorophyll concentrations in the Southern Ocean. Journal of Geophysical Research 112: C07030.
86. Arrigo KR, van Dijken GL (2003) Phytoplankton dynamics within 37 Antarctic coastal polynya systems. Journal of Geophysical Research 108: 3271.
87. Fichefet T, Goosse H (1999) A numerical investigation of the spring Ross Sea polynya. Geophysical Research Letters 26: 1015–1018.
88. Clarke A, Meredith MP, Wallace MI, Brandon MA, Thomas DN (2008) Seasonal and interannual variability in temperature, chlorophyll and macronutrients in northern Marguerite Bay, Antarctica. Deep Sea Research Part II: Topical Studies in Oceanography 55: 1988–2006.
89. Barnes DK, Clarke A (1995) Seasonality of feeding activity in Antarctic suspension feeders. Polar Biology 15: 335–340.

90. Brockington S, Clarke A, Chapman A (2001) Seasonality of feeding and nutritional status during the austral winter in the Antarctic sea urchin *Sterechinus neumayeri*. Marine Biology 139: 127–138.
91. Suhr SB, Pond D (2006) Antarctic benthic foraminifera facilitate rapid cycling of phytoplankton-derived organic carbon. Deep Sea Research Part II: Topical Studies in Oceanography 53: 895–902.
92. Bradbury IR, Snelgrove PVR (2001) Contrasting larval transport in demersal fish and benthic invertebrates: the roles of behaviour and advective processes in determining spatial pattern. Canadian Journal of Fisheries and Aquatic Sciences 58: 811–823.
93. Potthoff M, Johst K, Gutt J (2006) How to survive as a pioneer species in the Antarctic benthos: minimum dispersal distance as a function of lifetime and disturbance. Polar Biology 29: 543–551.
94. Thornhill DJ, Mahon AR, Norenburg JL, Halanych KM (2008) Open ocean barriers to dispersal: a test case with the Antarctic Polar Front and the ribbon worm *Parborlasia corrugatus* (Nemertea: Lineidae). Molecular Ecology 17: 5104–5117.
95. Matschiner M, Hanel R, Salzburger W (2009) Gene flow by larval dispersal in the Antarctic notothenioid fish *Gobionotothen gibberifrons*. Molecular Ecology 18: 2574–2587.
96. Cowen RK, Sponaugle S (2009) Larval dispersal and marine population connectivity. Annual Review of Marine Science 1: 443–466.
97. Shaw P, Arkhipkin A, Al-Khairulla H (2004) Genetic structuring of Patagonian toothfish populations in the Southwest Atlantic Ocean: the effect of the Antarctic Polar Front and deep water troughs as barriers to genetic exchange. Molecular Ecology 13: 3293–3303.
98. White C, Selkoe KA, Watson J, Siegel DA, Zacherl DC, et al. (2010) Ocean currents help explain population genetic structure. Proceedings of the Royal Society B: Biological Sciences 277: 1685.
99. Scott JM, Davis F, Csuti B, Noss R, Butterfield B, et al. (1993) Gap analysis: A geographic approach to protection of biological diversity. Wildlife monographs 123: 3–41.
100. CCAMLR (2009) Conservation Measure 91-03 (2009): Protection of the South Orkney Islands southern shelf: Commission to the Conservation of Antarctic Marine Living Resources. Available: http://www.ccamlr.org/en/measure-91-03-2009. Accessed: 2013 May.
101. Christiansen S (2010) Background document for the high seas MPAs regional approaches and experiences: Side event at the 12th UNEP global meeting of the regional seas conventions and action plans, 20th of September 2010. WWF-International.
102. McCartney M, Donohue K (2007) A deep cyclonic gyre in the Australian-Antarctic Basin. Progress in Oceanography 75: 675–750.
103. Robertson R, Visbeck M, Gordon A, Fahrbach E (2002) Long-term temperature trends in the deep waters of the Weddell Sea. Deep Sea Research Part II: Topical Studies in Oceanography 49: 4791–4806.
104. Fukamachi Y, Rintoul SR, Church A, Aoki S, Sokolov S, et al. (2010) Strong export of Antarctic BottomWater east of the Kerguelen plateau. DOI:10.1038/NGEO842. Nature Geoscience 3: 327–331.
105. Meijers AJS, Klocker A, Bindoff NL, Williams GD, Marsland SJ (2010) The circulation and water masses of the Antarctic shelf and continental slope between 30 and 80°E. Deep Sea Research II 57: 723–737.
106. N.Z and U.S (2012) Rev 1 Conservation Measure 91-XX (2012) Ross Sea Region Marine Protected Area. Submitted to the scientific committee to the Commission for the Conservation of Antarctic Marine Living Resources by the delegations of New Zealand and the United States. Available: www.mfat.govt.nz/ross-sea-mpa/docs/ Accessed: 2013 May 5.
107. MEDD (2006) Décret no 2006-1211 du 3 octobre 2006: Portant création de la réserve naturelle des Terres australes françaises: Ministère De L'écologie Et Du Développement Durable (MEDD).
108. Lombard A, Reyers B, Schonegevel L, Cooper J, Smith-Adao L, et al. (2007) Conserving pattern and process in the Southern Ocean: designing a marine protected area for the Prince Edward Islands. Antarctic Science 19: 39–54.
109. Njaastad B (2011) Personal communication 28th March. Norway: Head of environmental management section, Norwegian Polar Institute.
110. CCAMLR (2010) Conservation Measure 22-07. Interim measure for bottom fishing activities subject to Conservation Measure 22-06 encountering potential vulnerable marine ecosystems in the Convention Area. Commission for the Conservaton of Antarctic Marine Living Resources. Available: http://www.ccamlr.org/sites/drupal.ccamlr.org/files//22-07.pdf. Accessed: 2013 Sep.
111. CCAMLR (2012) Conservation Measure 22-06. Bottom fishing in the Convention Area.: Commission for the Conservaton of Antarctic Marine Living Resources. Available: http://www.ccamlr.org/sites/drupal.ccamlr.org/files//22-06_3.pdf. Accessed: 2013 Sep.
112. CCAMLR Secretariat (2013) Vulnerable Marine Ecosystem Registry. Accessed: 2013 Jul 26.
113. CCAMLR (2012) Conservation Measure 91-02: Protection of the values of Antarctic Specially Managed and Protected Areas. Commission for the Conservation of Antarctic Marine Living Resources. Available: www.ccamlr.org/sites/drupal.ccamlr.org/files//91-02.pdf. Accessed: 2013 May 6.
114. ERA (2011) Antarctic Protected Areas. Data source: Environmental Research & Assessment. Available: www.ats.aq/devPH/apa/ep_protected.aspx. Accessed: 2013 Apr 10.
115. Williams A, Bax N, Kloser R, Althaus F, Barker B, et al. (2009) Australia's deep-water reserve network: implications of false homogeneity for classifying abiotic surrogates of biodiversity. ICES Journal of Marine Science 66: 214.
116. Palumbi S (2004) Marine reserves and ocean neighborhoods: the spatial scale of marine populations and their management. Annual Review of Environment and Resources 29: 31.
117. Halpern B, Regan H, Possingham H, McCarthy M (2006) Accounting for uncertainty in marine reserve design. Ecology Letters 9: 2–11.
118. Smith W, Sandwell D (2010) Measured and Estimated Seafloor Topography Version 13.1. World Data Center-A for Marine Geology and Geophysics. Available: http://www.ngdc.noaa.gov/mgg/fliers/97mgg03.html. Accessed: 2010 Nov.
119. Smith W, Sandwell D (1997) Global sea floor topography from satellite altimetry and ship depth soundings. Science 277: 1956.
120. Clark MR, Watling L, Rowden AA, Guinotte JM, Smith CR (2011) A global seamount classification to aid the scientific design of marine protected area networks. Ocean & Coastal Management 54: 19–36.
121. Kitchingman A, Lai S, Morato T, Pauly D (2007) How many seamounts are there and where are they located. Seamounts: Ecology, Fisheries and Conservation: 26–40.
122. UQEC (2009) Scientific Principles for Design of Marine Protected Areas in Australia: A Guidance Statement: The University of Queensland Ecology Centre. Available: http://ecology.uq.edu.au/mpa-principals-guidance-statement. Accessed: 2014 Jun. 29 p.
123. SC-CAMLR. (2010) Pre-release version: Report of the Twenty-ninth Meeting of the Scientific Committee Hobart, Australia 25–29 October 2010. Science Committee for the Conservation of Antarctic Marine Living Resources.
124. Schnack-Schiel SB, Isla E (2005) The role of zooplankton in the pelagic-benthic coupling of the Southern Ocean. Scientia marina 69.
125. Ainley DG, Ballard G, Weller J (2010) Ross Sea Bioregionalisation Part I: Validation of the 2007 CCAMLR Bioregionalization Workshop Results Towards Including the Ross Sea in a Representative Network of Marine Protected Areas in the Southern Ocean. CAMLR reference number: CCAMLR WG-EMM-10/11.
126. Martin S (2001) Polynyas. In: John HS, editor. Encyclopedia of Ocean Sciences. Oxford: Academic Press. 2241–2247.
127. Locarnini RA, Mishonov A, Antonov J, Boyer T, Garcia H, editors (2006) World Ocean Atlas 2005. Vol. 1, Temperature. Washington, D.C.: National Oceanic and Atmospheric Administration Atlas NESDIS 61, United States Government. Available: www.nodc.noaa.gov/OC5/indprod.html. Accessed:2011 Sep. 182 p.
128. Feldman GC, McClain CR, editors (2010) SeaWiFS Reprocessing: NASA Goddard Space Flight Center. Available: http://oceancolor.gsfc.nasa.gov/.
129. Spreen G, Kaleschke L, Heygster G (2008) Sea ice remote sensing using AMSR-E 89 GHz channels. J Geophys Res 113: 3481–3484.
130. Sokolov S, Rintoul SR (2009) The circumpolar structure and distribution of the Antarctic Circumpolar Current fronts. Part 1: Mean circumpolar paths. Journal of Geophysical Research-Oceans 114: 1–19.
131. Wiencke C, Roleda MY, Gruber A, Clayton MN, Bischof K (2006) Susceptibility of zoospores to UV radiation determines upper depth distribution limit of Arctic kelps: evidence through field experiments. Journal of Ecology 94: 455–463.
132. Munilla León T (2001) Synopsis of the pycnogonids from Antarctic and Subantarctic waters. Polar Biology 24: 941–945.
133. Williams R, De La Mare W (1995) Fish distribution and biomass in the Heard Island zone (Division 58.5. 2). CCAMLR Science 2: 1–20.
134. O'Brien PE, Leitchenkov G (1997) Deglaciation of Prydz Bay, East Antarctica based on echo sounder and topographic features. In: Barker PF, Cooper AK, editors. Geology and seismic stratigraphy of the Antarctic Margin, 2: American Geophysical Union, Antarctic Research Series. 109–126.
135. Pena Cantero A (2004) How rich is the deep-sea Antarctic benthic hydroid fauna? Polar Biology 27: 767–774.
136. Aldea C, Olabarria C, Troncoso JS (2008) Bathymetric zonation and diversity gradient of gastropods and bivalves in West Antarctica from the South Shetland Islands to the Bellingshausen Sea. Deep Sea Research Part I: Oceanographic Research Papers 55: 350–368.
137. David B, Choné T, Festeau A, Mooi R, De Ridder C (2003) Antarctic Echinoids. An Interactive Database on CD-ROM. Version 2.0. Dijon: Biogeosciences Publisher, University of Burgundy.
138. Kruse S, Bathmann U, Brey T (2009) Meso-and bathypelagic distribution and abundance of chaetognaths in the Atlantic sector of the Southern Ocean. Polar Biology 32: 1359–1376.
139. Brandt A, Brökeland W, Choudhury M, Brix S, Kaiser S, et al. (2007) Deep-sea isopod biodiversity, abundance and endemism in the Atlantic sector of the Southern Ocean-results from the ANDEEP I–III expeditions. Deep Sea Research II 54: 1760–1775.
140. Nicol S, Pauly T, Bindoff N, Wright S, Thiele D, et al. (2000) Ocean circulation off east Antarctica affects ecosystem structure and sea-ice extent. Nature 406: 504–507.

A Connectivity-Based Eco-Regionalization Method of the Mediterranean Sea

Léo Berline[1,3,4]*, Anna-Maria Rammou[2], Andrea Doglioli[2], Anne Molcard[1], Anne Petrenko[2]

1 Université du Sud Toulon-Var, Aix-Marseille Université, CNRS/INSU/IRD, Mediterranean Institute of Oceanography (MIO), La Garde, France, **2** Aix-Marseille Université, CNRS/INSU/IRD, Mediterranean Institute of Oceanography (MIO), Marseille, France, **3** CNRS, Laboratoire d'Océanographie de Villefranche, Villefranche-sur-Mer, France, **4** Université Pierre et Marie Curie, Paris 6, Laboratoire d'Océanographie de Villefranche, Villefranche-sur-Mer, France

Abstract

Ecoregionalization of the ocean is a necessary step for spatial management of marine resources. Previous ecoregionalization efforts were based either on the distribution of species or on the distribution of physical and biogeochemical properties. These approaches ignore the dispersal of species by oceanic circulation that can connect regions and isolates others. This dispersal effect can be quantified through connectivity that is the probability, or time of transport between distinct regions. Here a new regionalization method based on a connectivity approach is described and applied to the Mediterranean Sea. This method is based on an ensemble of Lagrangian particle numerical simulations using ocean model outputs at 1/12° resolution. The domain is divided into square subregions of 50 km size. Then particle trajectories are used to quantify the oceanographic distance between each subregions, here defined as the mean connection time. Finally the oceanographic distance matrix is used as a basis for a hierarchical clustering. 22 regions are retained and discussed together with a quantification of the stability of boundaries between regions. Identified regions are generally consistent with the general circulation with boundaries located along current jets or surrounding gyres patterns. Regions are discussed in the light of existing ecoregionalizations and available knowledge on plankton distributions. This objective method complements static regionalization approaches based on the environmental niche concept and can be applied to any oceanic region at any scale.

Editor: Brian R. MacKenzie, Technical University of Denmark, Denmark

Funding: The research leading to these results has received funding from the European Community's Seventh Framework Programme (FP7/2007–2013) under Grant Agreement No. 287844 for the project "Towards COast to COast NETworks of marine protected areas (from the shore to the high and deep sea), coupled with sea-based wind energy potential" (COCONET). The funders had no role in study design, data collection and analysis, decision to publish, or preparation of the manuscript.

Competing Interests: The authors have declared that no competing interests exist.

* Email: leo.berline@univ-amu.fr

Introduction

The ecoregionalization of the ocean is useful for scientific research, conservation and management of the marine environment and marine resources. For instance, ecoregionalization is needed to extrapolate punctual or transect data to broader areas and to target specific regions for interdisciplinary research (as in the Mediterranean Sea, [1]). Conservation and management goals range from selecting areas to protect [2] to defining fisheries zones or zones for monitoring and mitigating marine pollution.

To date, several approaches of ecoregionalization were used depending on the data at hand [3]. The taxonomic approach is based on species distributions and identifies areas of broadly similar assemblage of species [4–6]. The ecological approach is based on habitat characteristics; it separates areas of similar seasonal cycles of physical and biogeochemical variables [7–10]. This approach benefited from the nearly continuous coverage of satellite data. Lastly, the integrative approach is a combination of both taxonomic and ecological approaches that takes into account both the habitat and the species inhabiting it [11].

However, in the marine environment the species distribution not only results from selection by the local environment but also from dispersal of propagules and adults organisms (e.g. the metapopulation concept of Levins [12,13]). Therefore an ecoregionalization based on dispersal by ocean circulation is needed; recent studies start taking into account dispersal in defining management units [14]. However it was never achieved quantitatively at basin scale. Today this is possible, as widely available ocean circulation models provide 3 dimensional, time varying, realistic and consistent depictions of oceanic currents at basin scale. The goal of this paper is to present a regionalization method based on connectivity, assessed from ensemble Lagrangian simulations using ocean circulation model velocity outputs.

This method is applied to the Mediterranean basin, which is a target region for spatial planning owing to its high level of endemism and high biodiversity [15]. Surface circulation shows a complex pattern of larger and smaller gyres, driven by the entrance of Atlantic water at Gibraltar Strait [1], local meteorology and bathymetry. The oligotrophy increases toward the East, but productive spots also exist over shelves and deep mixing areas, thus creating a significant heterogeneity in ecosystem functioning and habitats.

Materials and Methods

The general outline of the method is as follow (Fig. 1): Lagrangian trajectories are computed from ocean circulation model velocity outputs for particles seeded over the whole model domain at three depths (0.5 m, 50 m and 100 m). The domain is divided into a regular grid (hereinafter connectivity grid) and the trajectories are used to derive the mean connection time between every pair of grid cell. In this way a mean connection time matrix is obtained and then transformed into an oceanographic distance matrix, used as input to a hierarchical clustering algorithm. Finally clustering produces a partition of the domain.

Daily outputs velocity fields for four years (2007–2010) were taken from the configuration PSY2V3 of the operational system MERCATOR OCEAN [16]. The PSY2V3 configuration covers the North Atlantic ocean and Mediterranean Sea, and is based on the NEMO-OPA primitive equations code [17] with assimilation of observed data (satellite and in situ). Here, only the domain subset covering the Mediterranean Sea was used. Daily surface forcing are provided by ECMWF [18]. The velocity components are distributed in an Arakawa C type grid [19]. The horizontal resolution is 1/12° (~8 km) and there are 50 fixed vertical levels with higher resolution at the surface. The vertical mixing is described by a TKE closure scheme [20] and the advection by a TVD 2nd order centered scheme [21].

The trajectories followed by numerical particles were calculated offline with the Lagrangian diagnostic tool ARIANE [22]. The trajectories only result from the horizontal advection at three depths (0.5 m hereinafter called surface, 50 m and 100 m) chosen to represent the transport in the epipelagic layer. No vertical velocity was considered to keep particles in the 0–100 m range. The one year integration time was chosen to allow particles to cover the whole basin and therefore quantify basin scale connectivity and to keep computation time reasonable. Particles were seeded every 10 km on a regular square grid covering the whole domain, totaling 25,646 initial positions for surface depth and 23,770 for depths 50 and 100 m because the domain is smaller. Particles were seeded every 3 days from the 1st to the 25th of every month, from January 1st 2007 to December 25st 2009 in order to fully sample the variability of the circulation. This represents a total of 8,309,304 particles for surface depth, respectively 7,701,480 for depths 50 and 100 m. The choice of 10 km and 3 days is a compromise between matching the horizontal resolution of the model, taking into account mesoscale processes and keeping an affordable computing time of resulting trajectories. We thus obtained three ensembles of trajectories, one per depth.

In order to quantify the connections over the model domain, the domain was divided into grid cells of 50 km×50 km on a regular square grid, the connectivity grid, with a total of 1095 cells covering only regions with depths greater than 100 m. The 50 km resolution is sufficient to keep a reasonably realistic coastline while being suitable with the seeding density chosen. Thus each connectivity grid cell contains 5*5 = 25 particles for each initial seeding date, except grid cells including land that contains less particles.

To quantify the connectivity between each grid cell, we used the Mean Connection Time, hereinafter MCT. Defining T(i,j) as the transit time from grid cell i to grid cell j, MCT(i,j) was computed as

$$MCT(i,j) = \frac{1}{M}\sum_{n=1}^{n=M} T_n(i,j)$$

M being the number of particle transitioning from i to j. Note that for each trajectory, all intermediate transitions were used to compute the MCT. The sensitivity of MCT to the number of particles was tested. The suite of MCT matrices converged when the number of particles was greater than 6,000,000, therefore we considered that 8,309,304 particles and respectively 7,701,480 particles for depths 50 and 100 m were sufficient to obtain a robust MCT matrix. Moreover, to keep MCT robust, it was computed only when M was greater or equal to 50. Four MCT matrices of size 1095×1095 were computed: one MCT matrix from each ensemble of trajectories (MCT_0, MCT_{50}, MCT_{100} for 0.5, 50 and 100 m depths trajectories respectively) and also one MCT matrix using the three ensembles together ($MCT_{3depths}$).

Not all grid cells of the domain were connected within one year, especially remote cells (e.g. Northern Aegean and Gibraltar Strait). Thus the resulting MCT matrices had gaps (from 37% to 56%). These gaps are a problem for the steps of computing the oceanographic distance and applying hierarchical clustering on it. Therefore a gap filling procedure was introduced as follows (see Appendix S1):

○ For each unconnected pair of grid cells i-j, we looked for grid cells k so that i-k and k-j pairs are connected. There must be at least 50 grid cells k as for M.

○ Then we computed MCT(i,j) for pair i-j as the sum of the MCT(i,k) and MCT(k,j), averaged on all existing cells k, and filled the MCT(i,j) value in the matrix.

After 3 iterations of this procedure, each MCT matrix was filled. The resulting MCT values ranged from 10 days to 3000 days. This gap filling procedure avoided the very long integration time (>8 years) needed if we were to fill the whole MCT matrices from original trajectories alone.

This led to four full MCT matrices, which are asymmetric since the time to go from i to j is not equal to the time to go from j to i. Then the oceanographic distance (OD) was defined after [23] as the minimum of the two MCT values associated to each pair of grid cells i and j (travel from i to j and return travel from j to i). We chose the minimum value as it corresponds to the fastest route of transport which is also the shortest in length.

$$OD(i,j) = \min(MCT(i,j), MCT(j,i))$$

This gave four symmetric matrices, (OD_0, OD_{50}, OD_{100}, $OD_{3depths}$) where all diagonal terms (autoconnection time) were set to zero.

Finally hierarchical clustering analysis was applied on each of the oceanographic distance matrix. This method has proved to be robust in the classification of atmospheric wind data (e.g. [24]) and hydrological data (e.g. [25]). Hierarchical clustering assigns grid cells to different clusters in a way that each grid cell belongs to only one cluster [26], and each cluster belongs to a larger cluster (Fig. 2). The grid cells are grouped according to their similarity, which here is the oceanographic distance. Thus there is no distance metric applied as in usual clustering exercises. During each sequence of the clustering algorithm, the distances between the new clusters formed and the other grid cells are computed. This step requires a linkage criterion to be defined. Here we used the flexible [27] and Ward linkages [28]. WPGMA linkage was also tested ([27]) but flexible and Ward best balanced the dendrogram. For a given cut-off level of the dendrogram, we obtained a partition of the grid cells in a certain number of clusters, which is, in the spatial domain, a regionalization. Each cluster corresponded to a region on the connectivity grid whose contours were identified. Finally for each cluster, the within-cluster MCT was computed and plotted as a function of the number of clusters from 2 to 31 (Fig. 3).

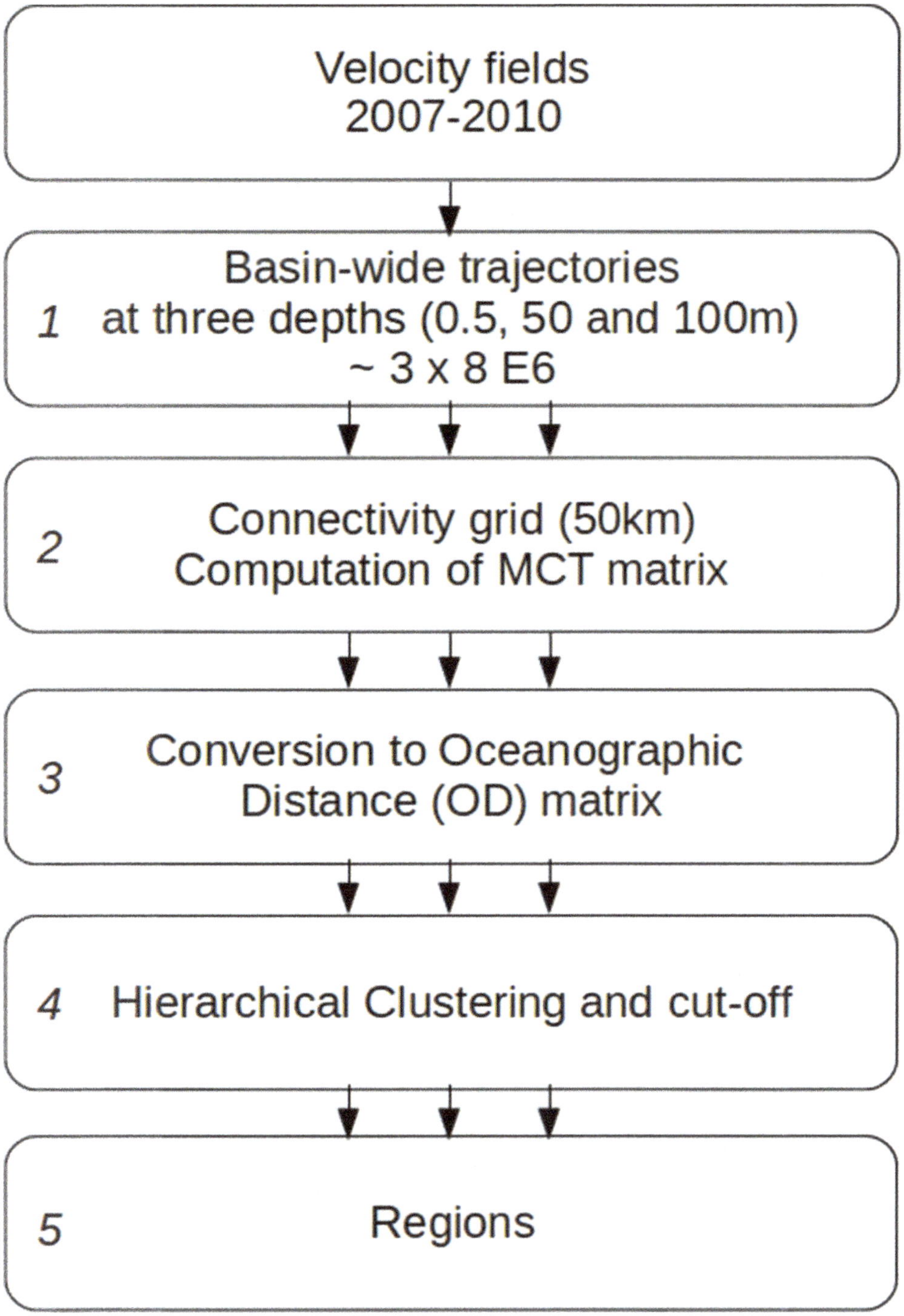

Figure 1. Schematic of the steps of the regionalization method. Note that steps 2 to 5 are repeated using trajectories at the 3 depths separately, shown with the three arrows, and then using them altogether.

Our "best estimate" regionalization was computed using flexible link and the matrix $OD_{3depths}$, built from the complete ensemble of trajectories (Fig. 4). We also computed one regionalization for each of the three depths and two linkages (6 cases). To assess the sensitivity of the regionalization results to the linkage and depth used, we computed the boundary stability, which is simply the local frequency of occurrence of a boundary in the spatial domain among the 6 cases, as defined in [29].

The choice of the optimal cut-off level and number of cluster is not straightforward here, because the distance matrix (OD) is not computed with a distance metric applied to a given dataset. Thus, usual criteria based on dataset variance within clusters cannot be used (e.g. [30]) because there is no dataset. Instead we took a simple approach comparing results from Ward and flexible linkage. For each partition into n clusters, we compute the proportion of cells classified in the same cluster with Ward and flexible (see Appendix S2). This proportion increases from 82%, to 88% from for $n=2$ to $n=6$ clusters, then drops to values $<70\%$ for $n>6$. Therefore we consider that the optimal cluster number is 6 as it gives more information while keeping consistent results

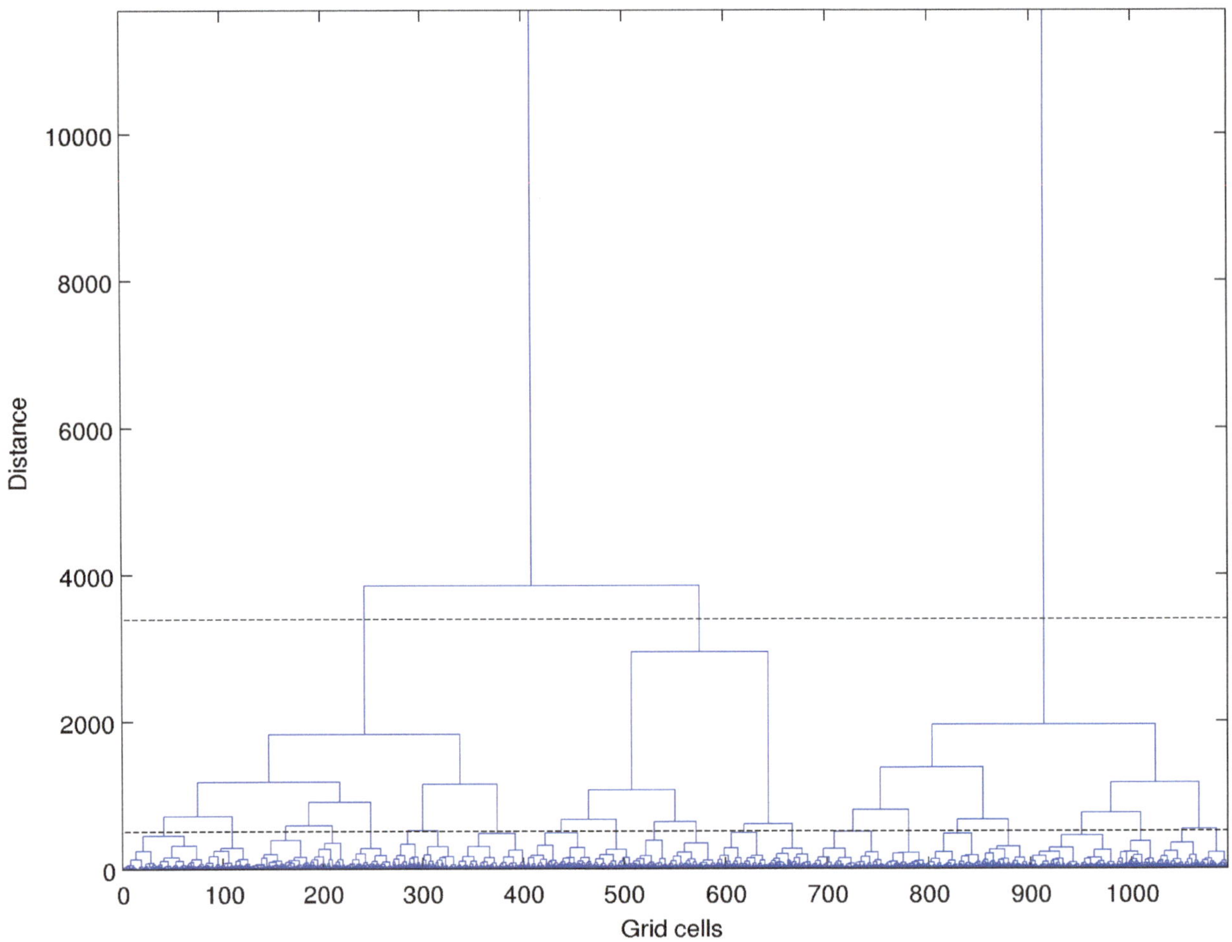

Figure 2. Cluster dendrogram of the oceanographic distance matrix $OD_{3depths}$ using the flexible linkage. Horizontal black lines show the cut-off values for 3 and 22 clusters.

among the two linkages. However, as no absolute criterion is available, we show the maximum number of clusters that we can interpret, which is 22 clusters. The clusters above 22 require detailed regional information to be interpreted, which is beyond the scope of this study.

Results

When the number of clusters increases, the within-cluster MCT diminishes, as well as the size of each region (Figs. 3 and 4). The average MCT ranges from 188 days for 2 clusters to ca. 90 days for 22 clusters.

On the basis of our interpretation of regions with respect to circulation, we retained 22 clusters (Fig. 4). The boundaries of each region were identified and colored according to the number of cluster obtained varying the cutoff distance from 10,000 (2 clusters) to 507 (22 clusters). The boundary #1 partly cuts the Sicily Strait (Fig. 4) and separates the Western and Eastern basins. The boundary #2 isolates Levantine basin from Ionian Sea and Adriatic Sea. The boundary #3 isolates the northern Ionian and Adriatic Sea from Southern Ionian. Then boundary #4 separates the Western basin into a western and an eastern part. The boundary #5 isolates the Levantine basin plus a part of AW current off Lybia from the Aegean Sea. The boundary stability map (Fig. 5) shows that some of the boundaries shown on Fig. 4 are stable (e.g. boundary #7, 11, 16) while others are variable in position or occurrence (e.g. boundary #4). Also, some boundaries (e.g. #2, 6, 8) have only a portion that is stable.

Then considering the 22 regions, the Western basin is separated into eight regions; regions A and B in the Northern part of the basin, G, F and E in the South and C, D that contains the Tyrrhenian sub-basin, region H at the center. In the Eastern basin, the Adriatic Sea is one region I. The Ionian Sea is separated into regions J, V, T at the center and K to the east, with U and S along the coasts of Libya and Tunisia. The Aegean Sea is divided into two regions, M in the East, L in the West. The Levantine basin has four regions: two coastal regions N and O, one southern region P and one center region Q. Considering only stable boundaries, the Western basin only has 5 regions. The Eastern basin has few continuous boundaries, only 4 regions are delimited (Adriatic, South of Sicily Strait and regions U and O).

Discussion

The boundary stability shows that the majority but not all boundaries are robust to changes in linkages and depths. Often, linkages or depth changes can produce minor shifts in boundary position, hence reducing boundary stability as defined here. When a boundary is not stable, it means that either the circulation is variable, either it is located in a region where the distance (OD) among grid cells is small thus the boundary position varies according to the overall content of each cluster. Thus the

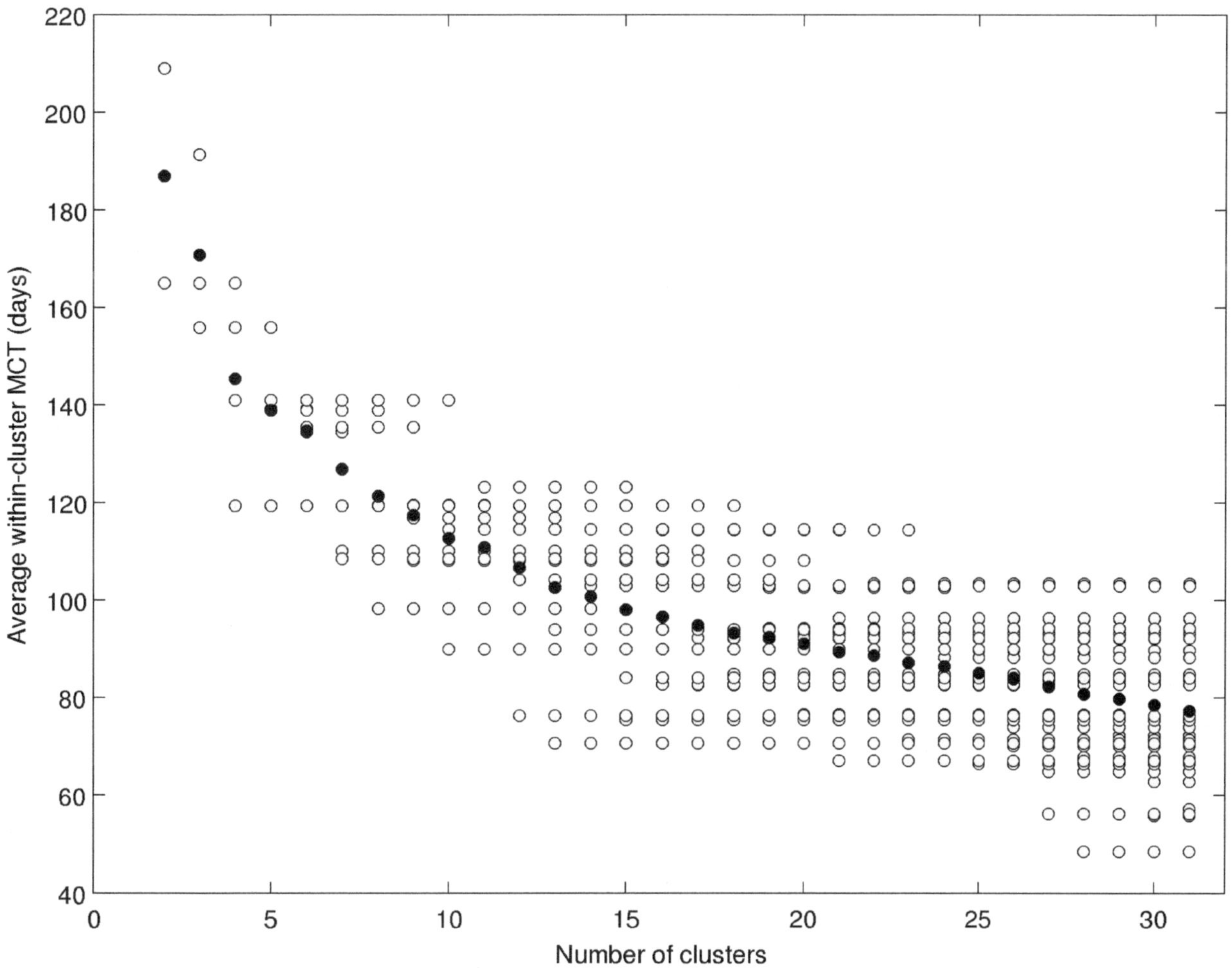

Figure 3. Within cluster mean connection time as a function of the cluster number for $MCT_{3depths}$. White dots are the mean for each cluster, black dots are the mean over all clusters.

boundary map must be analyzed jointly with the boundary stability to assess our regionalization.

1. Regions reveal circulation patterns

First the meaning of the regions obtained needs to be explained. One region contains grid cells that are connected at shorter time scale with each other than they are to the grid cells of the other regions. In the following, the relationship between the clusters boundaries, their stability and the circulation is examined in detail in comparison with the model average velocity fields (Fig. 4) and literature.

The hierarchy of cluster boundary is in good agreement with the surface general circulation scheme proposed by Millot et al. [31], their figure 2. Boundary #1 separates the Western and Eastern basins at the Sicily strait, boundary #2 isolates the Eastern Levantine, then boundary #3 the Adriatic Sea together with the northern part of the Ionian. Boundaries are often parallel to the mean velocity field. For instance boundary #16 is parallel to the Northern Current, boundary #11 parallel to the Asia Minor Current. Boundary can also separate two currents branches (the ATC along Tunisia and the AIS along Sicily, for part of boundaries #1 and 10, see [32]). This illustrates the barrier role of semi-permanent jets in the ocean. However, this is not always the case (e.g. boundary #1 at the Sicily Strait, boundary #18 at Oranto Strait). This can occur as the MCT matrix was computed from the time varying flow field, not from the mean field shown here and because each cluster is separated according to its overall distance with other clusters.

In the Western basin, boundary #16 is associated to the path of the Northern Current [31] and is the most stable. The boundary #6 from Spain to the Baleares follows approximately the Balearic front and is also rather stable. The Tyrrhenian Sea contains regions B, C, D with partly stable boundaries. Region C east of the Strait of Bonifacio contains the wind induced cold recirculation identified by [31], which is a potential dense water formation zone [33]. The Southern region G is restricted to the Alboran Sea.

In the Eastern basin, the Ionian Sea has two Southern regions U and S. Boundary #10 follows the Sicilian current of AW and region U contains the area of accumulation of eddies of the Ionian Sea [31]. The region V can correspond to the meandering stream identified by [34] or considered as interannual variability by [31]. The South-eastern Levantine has a region O with a stable boundary #7. Region O corresponds to the eddy accumulation zone $\sum$LE following [35]. The Asia Minor current along the Southern coasts of Turkey is captured in region N and has a stable boundary #11. Finally, the Aegean Sea is divided into an Eastern

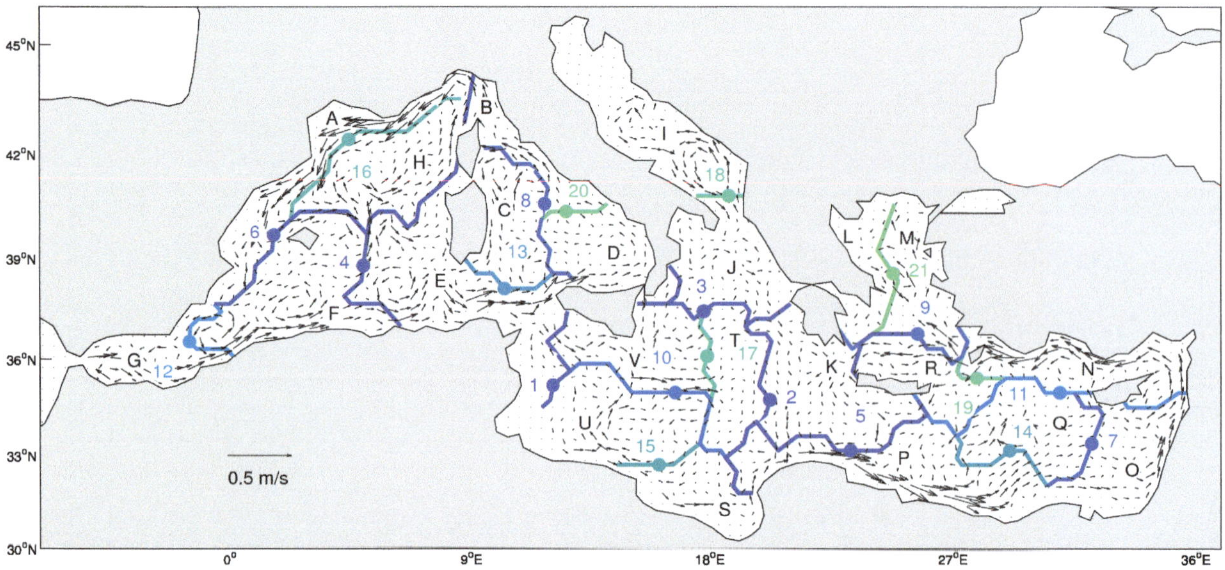

Figure 4. Map of the 21 clusters boundaries obtained from clustering of the oceanographic distance matrix $OD_{3depths}$ using the flexible link. Each boundary is colored and numbered according to the cut-off distance on the dendrogram (from blue – high distance- to green-low distance). Each region is identified by a letter from A to V. The velocity from the circulation model, averaged for the 4-year (2007–2010) and the 3 depths is overlaid as black vectors.

region M fed by AW and a North-Western region L fed by Black Sea outflow waters.

Some regions are virgin of any boundaries (Fig. 5), like the center of Gulf of Lion, the Alboran Sea, the Eastern Tyrrhenian Sea, the Northern Adriatic Sea, South of Greece, the South-East of the Levantine basin. This means that these regions are intraconnected at a time scale of less than ca 90 days (see Fig. 3).

Thus this regionalization reveals known circulation patterns and summarizes them in a way that complements the simple average velocity field analysis. It can be used to quantitatively compare the circulation patterns from contrasted periods or from different models.

2. Some boundaries coincide with major environmental boundaries and range limits of zooplankton assemblages

The identification of regions close to each other, not geographically but in terms of oceanographic connections, should help understanding the spatial distribution of properties that are passively transported by currents, such as conservative physical properties, or planktonic organisms living in the surface layer (epipelagic).

First, boundaries emerging from circulation alone often match major discontinuities in variables describing the environment. For instance a strong latitudinal salinity gradient exists near the Balearic Islands, close to our boundary #6. However, our

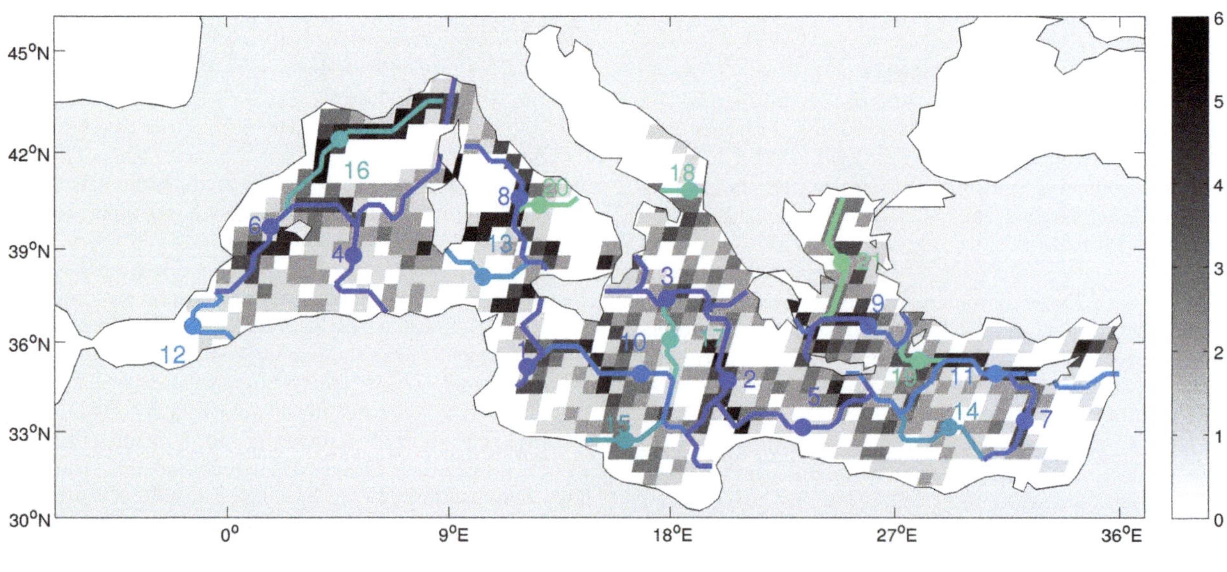

Figure 5. Map of the boundary stability (gray scale) derived from the 6 cases of clustering (3 depths×2 linkages). Boundary stability is defined as the number of occurrence of a boundary in each grid cell among the 6 cases. Boundaries are overlaid as in figure 4.

boundary #6 coincides with the Balearic Current but not to the Balearic salinity front, located more to the South [36]. Our boundary #16 coincides with a temperature and salinity front in the Ligurian Sea, and also in phytoplankton biomass (Fig. 1 in [10]). Off the Catalan coast, boundary #16 is consistent with the alongshore distribution of fish larvae [37], although located more offshore. Also, boundary #18 south of Adriatic Sea coincides with salinity fronts as seen in MEDATLAS [38]. This results from the dynamic links between density gradients and surface currents. The boundary #21 found in the Aegean Sea parallels the front in phytoplankton biomass [10]. At the Sicily strait, corresponding to our boundary #1, a boundary was also found by [9] (their figure 2) based on a clustering of sea surface temperature and ocean color data.

Within our regions, planktonic organisms are connected at shorter time scales than between regions. Thus hydrodynamical boundaries can become faunistic boundaries as suggested by Gaylord and Gaines [39] for larvae of benthic organisms. Given the spatial resolution, the MCT can correctly resolve connections of plankton organisms with a life cycle greater than 10 days, such as most zooplankton species [40]. Indeed, consistent with boundary #6 north of the Balearic Islands, a boundary exists between Atlantic zooplankton species to the South and Mediterranean species to the North [41,42]. Also, consistent with our boundaries #1 and #2, differences in zooplankton species composition between Eastern and Western basin were reported by several authors ([43] and references therein, [44]) although the spatial resolution of zooplankton data is generally not sufficient for accurately locating boundaries.

Ecoregions drawn qualitatively from expert knowledge of species assemblages ([45] their figure 2) also distinguish Atlantic-water regions including our region G, a Northern Current region including our region A, three Adriatic regions, one Aegean Sea region including our regions L and M, and two large zonal Eastern basin regions mostly consistent with boundaries #5 and #11.

However, for living organisms such as zooplankton, circulation alone is not sufficient to explain the distribution of a given species as it is adapted to its environment, in particular to a temperature range, e.g. [46]. Thus within our connected regions environmental conditions will restrict a species distribution to its specific preferendum, i.e. its ecological niche. Moreover, we deal with particles in the 0–100 m layer, which only properly represent epipelagic zooplankters dispersal.

3. How to use this regionalization?

To use this regionalization, the question of the number of clusters to retain will arise. With our approach, no existing criterion is available to define the optimal number. However the number of clusters can be chosen based on the time scale we are interested in, as regions isolated at a given time scale become connected at a larger time scale. Therefore the time scale of interest defines the appropriate cut-off distance and the resulting cluster number and sizes (Fig. 3). For instance, one can look for the scale of dispersal of planktonic larvae and hence consider the Pelagic Larval Duration (PLD) time scale. A PLD of 120 days (e.g. a crustacean as spiny lobster *Palunirus elephas* [47]) gives an adequate cluster number of ca 8. For a PLD of ca. 70 days (e.g. a labridae fish as *Lipophrys trigloides* [48] the adequate cluster number is ca 30. The lower bound time scale we can address with the present regionalization (~10 days) is set by the spatial resolution of our connectivity grid. Shorter time scales could be achieved with a finer connectivity grid.

Few existing studies can be compared to our regionalization because the approach is original. In the Mediterranean Sea, Andrello et al. [49] obtained clusters of coastal marine protected areas (MPA) based on their connectivity assessed by Lagrangian simulations. Although the velocity fields, Lagrangian simulations set up and clustering method are different, we can compare the overall grouping obtained (their figure 5-A). Considering only clusters containing several MPAs (8 clusters out of 38), their clusters are mostly contained within single regions and do not spread across several regions. Exceptions occur in the Northern Ligurian Sea and Ionian Sea with MPAs located very close or even onto our regions' boundaries. This probably results from the difference in the input velocity fields and subsequent connectivity quantification.

This new regionalization method quantifies the dispersal range of organisms, This dispersal dimension was shown to explain species distribution (e.g. [50]) and is thus critically needed [51]. This approach complements the usual regionalization methods rooted in the environmental niche concept (e.g. [9,10]). For instance, the Chl-a based regionalization from [10] reflects the regime of nutrients inputs and stratification, thus they are not directly linked to surface circulation patterns. Adding our connectivity-based regionalization helps understanding the types of environment that plankton is facing, through passive horizontal transport, vertical mixing and production processes. Practically, our OD matrices could be used as a constraint during the clustering of Chl-a, as for chronological clustering [52].

Also, our regions illustrate why plankton organisms may be encountered outside their optimum range (plankton expatriates, e.g. [53]) and where transport-driven fluctuations of plankton communities are expected. Indeed fluctuations of region boundaries may produce large biogeographic fluctuations noticeable at fixed points (e.g. [54,55]). Regions can also help tracking invasions of exotic organisms, for instance the so-called lessepsian species coming from the Suez Canal [56]. Apart from living organisms, our regions could be used to quantify areas of dispersion of pollutants coming from ships or land sources [57].

Finally this regionalization is useful as a framework to interpret the genetic differentiation of a given species sampled throughout the Mediterranean (e.g. [58]). Further, our approach could be used to define a priori units for grouping existing MPA or set up new MPA (e.g. [59] for the Gulf of Lions), as envisioned in the EU Integrated Project COCONET (www.coconet-fp7.eu).

4. Perspectives

The regionalization proposed here will eventually be compared to an ongoing biogeochemistry-based regionalization [60], and to zooplankton species distribution as available in database COPEPODS [61].

Concerning the methods, several points can be made. With a similar approach but shorter simulations, we can explore the seasonal variability of clusters boundaries that may be significant [62]. Here we used hierarchical clustering to extract clusters from the oceanographic distance, but clusters could also be computed with other methods such as graph theory that uses the asymmetry of the connectivity matrix (e.g. [49,63]). Finally, this method was applied to the Mediterranean Sea but it can be applied anywhere, at any spatial scales as long as accurate and long term model velocity outputs are available.

Acknowledgments

Mercator-ocean (www.mercator-ocean.fr) is thanked for providing velocity outputs for configuration PSY2V3. Fabien Lombard is thanked for useful feedback on an earlier version of the manuscript. The two reviewers are thanked for their constructive comments.

Author Contributions

Conceived and designed the experiments: LB AMR AD AM. Performed the experiments: AMR LB. Analyzed the data: LB AMR. Wrote the paper: LB AMR AD AM AP.

References

1. Durrieu de Madron X, Guieu C, Sempere R, Conan P, Cossa D, et al. (2011) Marine ecosystems' responses to climatic and anthropogenic forcings in the Mediterranean. Progress in Oceanography 91: 97–166.
2. Giakoumi S, Sini M, Gerovasileiou V, Mazor T, Beher J, et al. (2013) Ecoregion-Based Conservation Planning in the Mediterranean: Dealing with Large-Scale Heterogeneity. PLoS ONE 8(10): e76449. doi:10.1371/journal.pone.0076449.
3. UNESCO (2009) Global Open Oceans and Deep Seabed (GOODS) Biogeographic Classification. Paris. UNESCO-IOC. IOC Technical Series 84 92pp.
4. Forbes E (1856) Map of the distribution of marine life. in Johnston AK, editor. The Physical Atlas of Natural Phenomena. Edinburgh (Scotland): William Blackwood and Sons. pp. 99–102.
5. Ekman S (1953) Zoogeography of the Sea. London: Sidgwick and Jackson. 417 p.
6. Briggs JC (1995) Global Biogeography. Amsterdam, Elsevier. 472 p.
7. Longhurst A (1995) Seasonal cycles of pelagic production and consumption. Progress in Oceanography 36: 77–167.
8. Longhurst A (1998) Ecological geography of the sea Academic Press, London. 560 p.
9. Oliver MJ, Irwin AJ (2008) Objective global ocean biogeographic provinces. Geophysical Research Letters, 35, L15601, doi:101029/2008GL034238.
10. D'Ortenzio F, Ribera d'Alcalà M (2009) On the trophic regimes of the Mediterranean sea: a satellite analysis. Biogeosciences 6: 139–148.
11. Koubbi P, Moteki M, Duhamel G, Goarant A, Hulley PA, et al. (2011) Ecoregionalization of myctophid fish in the Indian sector of the Southern Ocean: Results from generalized dissimilarity models. Deep Sea Research Part II: Topical Studies in Oceanography 58: 170–180.
12. Levins R (1969) Some demographic and genetic consequences of environmental heterogeneity for biological control. Bulletin of the Entomological Society of America 15: 237–240.
13. Obura D (2012) The Diversity and Biogeography of Western Indian Ocean Reef-Building Corals. PLoS ONE doi:10.1371/journal.pone.0045013.
14. Casale P, Mariani P (2014) The first 'lost year' of Mediterranean sea turtles: dispersal patterns indicate subregional management units for conservation. Marine Ecology Progress Series 498: 263–274.
15. Coll M, Piroddi C, Steenbeek J, Kaschner K, Lasram F B, et al. (2010). The Biodiversity of the Mediterranean Sea: Estimates, Patterns, and Threats. PLoS ONE, 5(8).
16. Bahurel P (2006) MERCATOR OCEAN global to regional ocean monitoring and forecasting. In: Chassignet, E P and Verron, J editors, Ocean Weather Forecasting Springer, Netherlands, pp. 381–395.
17. Madec G (2008) "NEMO ocean engine". Note du Pole de modélisation, Institut Pierre-Simon Laplace (IPSL), France, No 27 ISSN No 1288–1619, 367 p.
18. Uppala S, Kallberg PW, Simmons AJ, Andrae U, Da Costa V, et al. (2005) The ERA-40 re-analysis. Q J R Meteorol Soc, 131: 2961–3012.
19. Mesinger F, Arakawa A, Kasahara A (1976) Numerical methods used in atmospheric models, vol 1 World Meteorological Organization, International Council of Scientific Unions 37 p.
20. Blanke B, Delecluse P (1993) Variability of the tropical Atlantic ocean simulated by a general circulation model with two different mixed layer physics. Journal of Physical Oceanography 23: 1363–1388.
21. Zalesak ST (1979) Fully multidimensional flux corrected transport algorithms for fluids. J Comput Phys 31: 335–362.
22. Blanke B, Raynaud S (1997) Kinematics of the Pacific Equatorial Undercurrent: An Eulerian and Lagrangian approach from GCM results. Journal of Physical Oceanography 27: 1038–1053.
23. Alberto F, Raimondi PT, Reed DC, Watson JR, Siegel DA, et al. (2011) Isolation by oceanographic distance explains genetic structure for Macrocystis pyrifera in the Santa Barbara Channel. Molecular Ecology, 20: 2543–2554.
24. Burlando M, Antonelli M, Ratto C (2008) Mesoscale wind climate analysis: identification of anemological regions and wind regimes. Int J Climatol 28: 629–641.
25. Hjelmervik KT, Hjelmervik K (2013) Estimating temperature and salinity profiles using empirical orthogonal functions and clustering on historical measurements. Ocean Dynamics 63: 809–821.
26. Legendre P, Rogers DJ (1972) Characters and clustering in taxonomy: a synthesis of two taximetric procedures. Taxon 567–606.
27. Lance GN, Williams WT (1967). A general theory of classificatory sorting strategies 1. Hierarchical systems. The computer journal, 9: 373–380.
28. Ward JH Jr (1963) Hierarchical Grouping to Optimize an Objective Function. Journal of the American Statistical Association 58: 236–244.
29. Oliver MJ, Glenn S, Kohut JT, Irwin AJ, Schofield OM, et al. (2004), Bioinformatic approaches for objective detection of water masses on continental shelves, J Geophys Res, 109, C07S04, doi:10.1029/2003JC002072.
30. Calinski RB, Harabasz J (1974) A dendrite method for cluster analysis. Commun. Stat. 3: 1–27.
31. Millot C, Taupier-Letage I (2005) Circulation in the Mediterranean sea. In: The Mediterranean Sea, Springer, pp. 29–66.
32. Poulain PM, Zambianchi E (2007) Surface circulation in the central Mediterranean Sea as deduced from Lagrangian drifters in the 1990s. Continental Shelf Research 27(7): 981–1001.
33. Fuda JL, Etiope G, Millot C, Favali P, Calcara M, et al. (2002) Warming, salting and origin of the Tyrrhenian Deep Water (2002) Geophys. Res. Letters 29, doi:10.1029/2001GL014072.
34. Malanotte-Rizzoli P, Artale V, Borzelli-Eusebi GL, Brenner S, Crise A, et al. (2014) G.: Physical forcing and physical/biochemical variability of the Mediterranean Sea: a review of unresolved issues and directions for future research, Ocean Sciences 10: 281–322, doi:10.5194/os-10-281-2014.
35. Hamad N, Millot C, Taupier-Letage I (2005) A new hypothesis about the surface circulation in the eastern basin of the Mediterranean sea. Progress in Oceanography 66: 287–298.
36. Mariani P, MacKenzie BR, Iudicone D, Bozec A (2010) Modelling retention and dispersion mechanisms of bluefin tuna eggs and larvae in the northwest Mediterranean Sea. Progress in Oceanography 86(1): 45–58.
37. Sabatés A, Olivar MP, Salat J, Palomera I, Alemany F (2007) Physical and biological processes controlling the distribution of fish larvae in the NW Mediterranean. Progress in Oceanography, 74(2): 355–376.
38. Rixen M, Beckers JM, Brankart JM, Brasseur P (2001) A numerically efficient data analysis method with error map generation. Ocean Modelling 2: 45–60.
39. Gaylord B, Gaines SD (2000) Temperature or transport ? range limits in marine species mediated solely by flow. The American Naturalist 155: 769–789.
40. Carlotti F, Poggiale JC (2010) Towards methodological approaches to implement the zooplankton component in "end to end" food-web models. Progress in Oceanography, 84: 20–38.
41. Fernandez de Puelles ML, Pinot J-M, Valencia J (2003) Seasonal and interannual variability of zooplankton community in waters off Mallorca island (Balearic Sea, Western Mediterranean): 1994–1999. Oceanologica Acta 26: 673–686.
42. Fernandez de Puelles ML, Molinero JC (2007) North Atlantic climate control on plankton variability in the Balearic Sea, western Mediterranean. Geophysical Research Letters 34, L04608 doi:101029/2006GL028354.
43. Siokou-Frangou I, Christaki U, Mazzocchi MG, Montresor M, Ribera d'Alcalá M, et al. (2010) Plankton in the open Mediterranean Sea: a review. Biogeosciences 7: 1543–1586, doi:105194/bg-7-1543-2010.
44. Nowaczyk A, Carlotti F, Thibault-Botha D, Pagano M (2011) Distribution of epipelagic metazooplankton across the Mediterranean Sea during the summer BOUM cruise. Biogeosciences, 8: 2159–2177, doi:105194/bg-8-2159-2011.
45. Bianchi C, Morri C (2000) Marine biodiversity of the Mediterranean Sea: situation, problems and prospects for future research. Marine Pollution Bulletin 40: 367–376.
46. Beaugrand G (2005) Monitoring pelagic ecosystems using plankton indicators. ICES Journal of Marine Science 62: 333–338.
47. Queiroga H, Blanton J (2004) Interactions between behavior and physical forcing in the control of horizontal transport of decapods crustacean larvae. Advances in Marine Biology 47: 198–214.
48. McPherson E, Raventos N, (2006) Relationship between pelagic larval duration and geographic distribution of Mediterranean littoral fishes. Marine Ecology Progress Series 327: 257–265.
49. Andrello M, Mouillot D, Beuvier J, Albouy C, Thuiller W, et al. (2013) Low connectivity between mediterranean marine protected areas: a biophysical modeling approach for the dusky grouper *Epinephelus marginatus*. Plos One 8(7): e68564 doi:101371/journalpone0068564.
50. Wernberg T, Thomsen MS, Connell SD, Russell BD, Waters JM, et al. (2013) The Footprint of Continental-Scale Ocean Currents on the Biogeography of Seaweeds. PLoS ONE doi:10.1371/journal.pone.0080168.
51. Guisan W, Thuiller W (2005) Predicting species distribution: offering more than simple habitat models. Ecology Letters 8: 993–1009.
52. Legendre P, Dallot S, Legendre L (1985) Succession of species within a community: chronological clustering, with applications to marine and freshwater zooplankton. American Naturalist 125: 257–288.
53. Olli K, Wassmann P, Reigstad M, Ratkova TN, Arashkevich E, et al. (2007) The fate of production in the central Arctic Ocean – top-down regulation by zooplankton expatriates. Progress in Oceanography 72: 84–113.
54. Chiba S, Di Lorenzo E, Davis A, Keister JE, Taguchi B, et al. (2013) Large-scale climate control of zooplankton transport and biogeography in the Kuroshio Oyashio Extension region. Geophysical Research Letters 40: 5182–5187.

55. Berline L, Zakardjian B, Molcard A, Ourmières Y, Guihou K (2013) Modelling transport and stranding of jellyfish Pelagia noctiluca in the Mediterranean Marine Pollution Bulletin, 70: 90–99.
56. Jribi I, Bradai MN (2012) First record of the Lessepsian migrant species *Lagocephalus sceleratus* (Gmelin,1789) (Actinopterygii: Tetraodontidae) in the Central Mediterranean. BioInvasions Records 1: 49–52 doi:10.3391/bir.2012.1.1.11.
57. Olita A, Cucco A, Simeone S, Ribotti A, Fazioli L, et al. (2012) Oil spill hazard and risk assessment for the shorelines of a Mediterranean coastal archipelago. Ocean & Coastal Management 57: 44–52.
58. Serra IA, Innocenti AM, Di Maida G, Calvo S, Migliaccio M, et al. (2010) Genetic structure in the Mediterranean seagrass *Posidonia oceanica*: disentangling past vicariance events from contemporary patterns of gene flow. Molecular Ecology 19: 557–68.
59. Guizien K, Belharet M, Marsaleix P, Guarini JM (2012) Using larval dispersal simulations for marine protected area design: Application to the Gulf of Lions (northwest Mediterranean). Limnology and Oceanography, 57: 1099–1112 doi:10.4319/lo.2012.57.4.1099.
60. Reygondeau G, Irisson J-O, Guieu C, Gasparini S, Ayata S-D, et al. (2013) Toward a dynamic biogeochemical division of the Mediterranean Sea in a context of global climate change. EGU General Assembly 2013, Vienna, Austria, 07–12 April 2013. Available at: http://meetingorganizer.copernicus.org/EGU2013/EGU2013-10011.pdf.
61. O'Brien TD, Wiebe PH, Hay S (2010) ICES Zooplankton Status Report 2008/2009 ICES Cooperative Research Report No 307 152.
62. Pizzigali C, Rupolo V, Lombardi E, Blanke B (2007) Seasonal probability dispersion maps in the Mediterranean Sea obtained from the Mediterranean Forecasting System Eulerian velocity fields. Journal of Geophysical Research 112, C05012.
63. Treml EA, Halpin PN, Urban DL Pratson LF (2008) Modeling population connectivity by ocean currents, a graph-theoretic approach for marine conservation. Landscape Ecology 23: 19–36.

14

Identification and Expression Profiling of MicroRNAs in the Brain, Liver and Gonads of Marine Medaka (*Oryzias melastigma*) and in Response to Hypoxia

Karen Lau[1,2]*, Keng Po Lai[1,2], Jessie Yun Juan Bao[3], Na Zhang[3], Anna Tse[1,2], Amy Tong[3¤], Jing Woei Li[4], Si Lok[3¤], Richard Yuen Chong Kong[2], Wing Yee Lui[1], Alice Wong[1], Rudolf Shiu Sun Wu[1,2]*

1 School of Biological Sciences, Kadoorie Biological Sciences Building, The University of Hong Kong, Hong Kong SAR, China, **2** The State Key Laboratory in Marine Pollution, Hong Kong, Hong Kong SAR, China, **3** Genome Research Centre, The Hong Kong Jockey Club Building for Interdisciplinary Research, The University of Hong Kong, Pokfulam, Hong Kong SAR, China, **4** School of Life Sciences, Hong Kong Bioinformatics Centre, The Chinese University of Hong Kong, Hong Kong SAR, China, **5** Department of Biology and Chemistry, City University of Hong Kong, Kowloon, Hong Kong SAR, China

Abstract

The marine medaka (*Oryzias melastigma*) has been increasingly used as a fish model for detecting environmental stresses and chemical contaminants in the marine environment. Recent mammalian studies have shown that environmental stresses can alter the expression profiles of microRNAs (miRNAs), leading to transgenerational effects. Here, we use high-throughput Illumina RNA sequencing (RNA-Seq) for miRNA transcriptome analysis of brain, liver, and gonads from sexually mature male and female marine medaka. A total of 128,883,806 filtered sequence reads were generated from six small RNA libraries, identifying a total of 2,125,663 non-redundant sequences. These sequences were aligned and annotated to known animal miRNAs (miRBase) using the BLAST method. A total of 223 distinct miRNA types were identified, with the greatest number expressed in brain tissue. Our data suggested that 55 miRNA types from 34 families are common to all tested tissues, while some of the miRNAs are tissue-enriched or sex-enriched. Quantitative real-time PCR analysis further demonstrated that let-7a, miR-122, and miR-9-3p were downregulated in hypoxic female medaka, while miR-2184 was specifically upregulated in the testis of hypoxic male fish. This is the first study to identify miRNAs in *O. melastigma* using small RNA deep sequencing technology. Because miRNA expression is highly conserved between marine medaka and other vertebrates, marine medaka may serve as a good model for studies on the functional roles of miRNAs in hypoxia stress response and signaling in marine fish.

Editor: Christophe Antoniewski, CNRS UMR7622 & University Paris 6 Pierre-et-Marie-Curie, France

Funding: These authors have no support or funding to report.

Competing Interests: The authors have declared that no competing interests exist.

* Email: rudolfwu@hku.hk (RSSW); karenytlau@hotmail.com (KL)

¤ Current address: The Chinese University of Hong Kong, Li Ka Shing Institute of Health Sciences, Prince of Wales Hospital, Shatin, New Territories, Hong Kong SAR, China

Introduction

MicroRNAs (miRNAs) are small non-coding RNA molecules (~22 nucleotides in length) encoded in the genome. They have emerged as one of the most important and abundant group of gene expression regulator in a wide range of species spanning from mammals to plants and viruses [1]. MiRNAs are first transcribed into primary miRNAs (pri-miRNAs) and then processed in the nucleus into precursor miRNAs (pre-miRNAs) with hairpin stem-loop structures. In animals, pre-miRNAs are then exported to the cytoplasm, where they are processed into mature miRNAs [2]. Mature miRNAs preferentially target 3′-untranslated regions (3′-UTRs) of mRNAs with which they share partial complementary sequence, leading to post-transcriptional gene silencing through translational repression. They can also lead to degradation of the target mRNA when complete sequence complementarity exists. Each miRNA may have multiple gene targets, and each gene target may also be regulated by more than one miRNA [3,4].

Because of the important biological functions of miRNA, considerable efforts have been made to establish miRNA databases such as miRBase (http://www.mirbase.org), which detail all published miRNA sequences, annotations, and predicted gene targets [5]. The number of miRBase entries has increased exponentially since it was established in [6]. The latest version of miRBase comprises 30,424 distinct mature miRNA products from 206 species, including over 2,578 mature human miRNAs. Considerably fewer mature miRNAs have been reported in fish: 255 have been reported from zebrafish (*Danio rerio*), 108 from pufferfish (*Fugu rubripes*) and 146 from the Japanese freshwater medaka (*Oryzias latipes*) [5,7–8]. Many of these fish miRNAs are conserved and have homologs in the human and mouse genomes [9–10].

Mammalian studies have shown that prenatal exposure to certain environmental stresses (e.g., hypoxia) and chemical contaminants (e.g., endocrine disrupting chemicals) can lead to alterations in phenotypes (e.g., delayed physical and sexual development, infertility, increased mortality and behavioral changes) in successive generations, despite the offspring have never been directly exposed to the stress and contaminants before [11,12]. Such transgenerational effects may not be caused by alterations in DNA sequence but by changes in the epigenome. For example, changes may occur in DNA methylation, covalent modification of histones, or the activation or silencing of genes by miRNAs [13]. MiRNAs have been shown to modulate cellular differentiation and development, and various hypoxia-responsive miRNAs (HRMs) have been reported in human breast and colon cancers [14–16]. Although the functional roles of several HRMs have been extensively studied in mammals [17,18], the biological roles of miRNAs in fish remain poorly understood. Small freshwater fish models (e.g. zebrafish and freshwater medaka) have been used extensively to identify and study biological responses to stresses in the freshwater environment [19,20]. The marine medaka (which is very closely related to *O. latipes*) has been developed as a small fish model for similar studies in the marine environment [21]. However, the miRNA profile of this species is unknown.

New technologies such as high throughput RNA sequencing (RNA-Seq) have rapidly increased the rate of miRNA discovery. The high sensitivity of RNA-Seq allows for both the detection of species-specific, conserved miRNAs and the detection of weakly expressed miRNAs. MiRNA discovery has often relied on the mapping of small RNA sequence reads to a reference genome. Because a reference genome is currently not available for *O. melastigma*, predicting miRNAs in this species is difficult. However, with the development of alignment-type methods such as BLAST [22] for identifying homologs of known miRNAs, it has become possible to predict candidate miRNAs in the marine medaka. This is the first report on the identification and expression profiles of miRNAs in male and female *O. melastigma,* and also the differential responses of selected miRNAs to hypoxic stress. The results presented here will be invaluable for future transgenerational studies and risk assessment using this model marine fish species.

Methods

Medaka maintenance and RNA Isolation

All animal research procedures were approved by the Committee on the Use of Live Animals in Teaching and Research (CULATR, #2714-12) at the University of Hong Kong. The stock of marine medaka used in our experiment was obtained from Interocean Industries (Taiwan) and has been reared in our laboratory for over 10 generations. Marine medaka were maintained under optimal growth and breeding conditions as described in Kong *et al*, (5.8 mg O_2 L^{-1}, 28±2°C, pH 7.2 in a 14-h light: 10-h dark cycle) [21]. The fish were fed with hormone-free Aquatox Feed (Zeigler Bros. Inc. Gardners, PA) and live artemia. For hypoxia study, fifty juvenile male and fifty female marine medaka were reared under normoxia (5.8+0.2 mg O_2/L) or hypoxia (1.5+0.2 mg O_2/L) for three months. At 120 days post fertilization, the fish were anesthetized on an ice bath. Then, brain, liver, ovary and testis tissue were dissected from randomly selected male ($n = 10$ from normoxia and $n = 10$ from hypoxia) and female ($n = 10$ from normoxia and $n = 10$ from hypoxia) fish. Total RNA from pooled tissue samples was extracted using the mirVana miRNA isolation kit (Applied Biosystems) and treated with DNase (Ambion) to remove contaminating genomic DNA. RNA quality was assessed using the Agilent 2100 Bioanalyzer system and samples with a RNA Integrity Number (RIN) greater than 9 were used for small RNA library construction.

Illumina small RNA library construction and sequencing

Six small RNA (cDNA) libraries were constructed (brain, liver and ovaries from female; brain, liver and testis from male), each prepared from 1 µg of total RNA. Small RNAs were gel purified to enrich for molecules less than 30 nucleotide in length, and Illumina adaptors (Integrated DNA technology) were ligated to their 5′ and 3′ ends to enable the construction of strand-specific cDNA libraries [23]. The ligation products were used as a template for cDNA synthesis using SuperScript II Reverse Transcriptase (Invitrogen), and an oligonucleotide primer with sequence complementary to the 3′ adapter was used to initiate cDNA synthesis (5′-CTCGGCATTCCTGCTGAACCGCTC-3′). cDNAs were amplified in 12 cycles with AccuPrime *Pfx* DNA Polymerase (Life Technologies) using PCR primer A (5′ - AATGATACGGCGACCACCGAGATCTACACTCTTTCCC-TACACGACGCTCTTCCGATCT 3′) and PCR primer B (5′-CAAGCAGAAGACGGCATACGAGATCGGTCTCGGCAT-TCCTGCTGAACCGCTCTTCCGATCT 3′), and products were purified on Novex 8% TBE polyacrylamide gels (Invitrogen). Amplified DNA of 140–190 nt was extracted from the gel and purified using the QiaQuick Gel Extraction Kit (Qiagen). Sequencing was performed on the Illumina Genome Analyzer GAIIx. Sequences were extracted from image files using the Illumina pipeline, set at default parameters to yield sequence reads of 58 base lengths. Low-quality sequences, homopolymers and adaptor sequences were removed. The filtered reads were processed into different read lengths for further analysis.

Illumina sequencing data analysis

Because the libraries were constructed strand-specifically, only reads in the sense orientation were considered in our analysis. The filtered reads were binned into non-redundant sets of unique small RNAs to tabulate the expression level profile for each library. The relative abundance of each unique miRNA was given as the number of tabulated reads per million filtered reads (RPM). Non-redundant miRNAs of between 20 and 23 bp having RPM≥6 were annotated using BLAST (Basic Local Alignment Search Tool) against miRBase version 17 [5], a database of known mature miRNA and mature star sequences. For unique species, only the highest-scoring alignments were considered for further analysis. Candidates were required to have at least 15 nucleotide matches to reference miRNAs with no more than 3 mismatches and to have an E-value lower than 0.06. In cases where a single reference miRNA from the database might match more than one sequence species from our *O. melastigma* small RNA library, the most abundant miRNA having the highest homology to the reference miRNA sequence was selected as the canonical candidate miRNA [24].

qPCR analysis and identification of hypoxia-responsive miRNAs

TaqMan MicroRNA Assays (primer and probe sets PN: 4427975 and 4440886, Applied Biosystems) were used to detect and quantify five representative miRNA genes (let-7a, miR-9-3p, miR-27a, miR-122 and miR-2184) from the RNA tissue samples collected from male ($n = 10$ from normoxia and $n = 10$ from hypoxia) and female ($n = 10$ from normoxia and $n = 10$ from hypoxia) fish, according to the manufacturer's protocol. U6 small nuclear RNA and 5S ribosomal RNA (5S rRNA) were used as an endogenous control for the analysis. All target sequences are shown in Table S1. cDNA was synthesized from total RNA using gene-specific primers. Reverse transcriptase reactions contained 10 ng of total RNA, 50 nM stem-loop RT primer, 1X RT buffer, 0.25 mM dNTPs, 3.3 U/μl MultiScribe reverse transcriptase, and 0.25 U/μl RNase inhibitor (TaqMan MicroRNA Reverse Transcription Kit, Applied Biosystems). The RT reactions were incubated in a C1000 Thermal Cycler (Bio-Rad) at 16°C for 30 min, then at 42°C for 30 min, at 85°C for 5 min and then held at 4°C. qRT-PCR was performed using the StepOnePlus Real-Time PCR system (Applied Biosystems). The 20 μl PCR included 1.3 μl RT product, 10 μl TaqMan Universal PCR Master Mix II (2X) without UNG (Applied Biosystems), 1 μl TaqMan Small RNA Assay, and 7.7 μl nuclease-free water. Reactions were incubated in a 96-well optical plate at 95°C for 10 min, followed by 40 cycles of 95°C for 15 sec and 60°C for 1 min. Reactions were run in triplicate and included a no-template control for each gene. Expression levels were determined based on the threshold cycle values (C_t) of each target miRNA normalized to the 5S rRNA and U6 snRNA. Statistical significance was assessed using Wilcoxon–Mann–Whitney test.

Availability of sequence data

The sequence data from this study have been submitted to the NCBI SRA (http://www.ncbi.nlm.nih.gov/sra) under the accession number SRP041922.

Results

Construction of small RNA libraries and Illumina sequencing

To identify miRNA from marine medaka (*O. melastigma*), six small RNA libraries were generated and sequenced using the Illumina GAIIx. A schematic description of the workflow for *O. melastigma* miRNA discovery is shown in Figure 1. A total of 128,883,806 high quality filtered sequence reads were generated, and 2,125,663 non-redundant RNAs were identified. Between 18 and 25 million usable pre-processed reads were obtained for each library. The number of unique species following removal of redundant sequences is shown in Table 1.

The length distributions of sequenced small RNAs in the six libraries were analyzed (Figure S1). In brain and liver, small RNAs of 21–23 nt in length were the most abundant species. However, in male and female gonads, the most abundant species are those between 26 and 28 nt. This distribution might reflect the presence of piRNAs in germ cells. It is consistent with the distribution that Houwing *et al* reported for zebrafish, in which ovary- and testis-specific piRNAs were detected [25]. The identification of piRNAs in ovaries and testis in the medaka will be assessed in future studies.

A stringent screening process for the most promising miRNA candidates was implemented, using sequence length (20–23 bp),

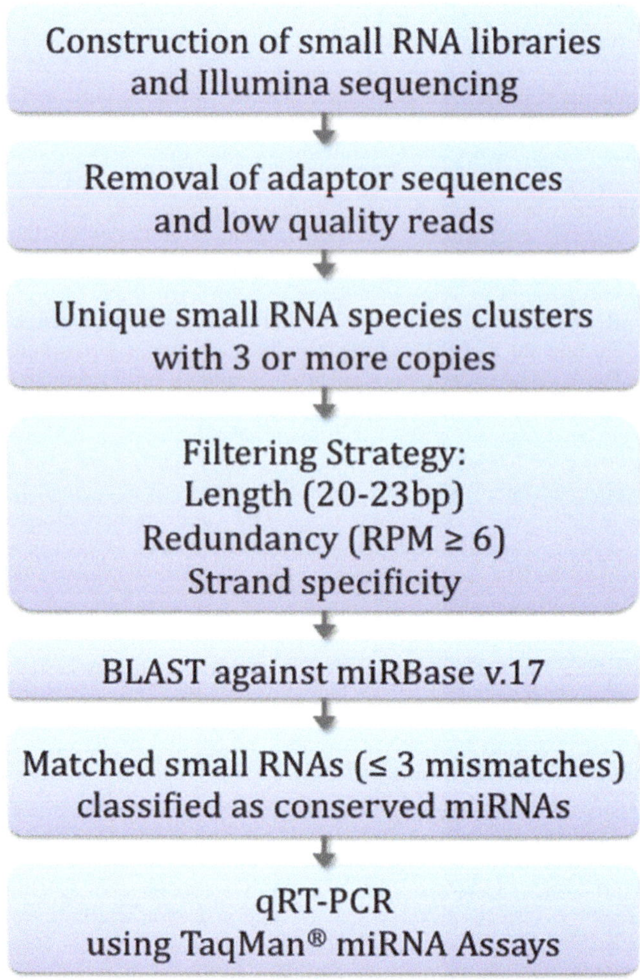

Figure 1. Schematic diagram of the workflow for *O. melastigma* miRNA discovery. Brain, liver and gonadal (ovary and testis) tissues of male and female marine medaka were submitted to small RNA libraries preparation and were sequenced using Illumina GAIIX. After the removal of low quality reads and filtering, high quality sequencing reads were blasted against miRBase v.17 to identify the conserved miRNAs of marine medaka.

Table 1. Statistics of pre-processed sequencing reads in six small RNA libraries of *O. melastigma*.

Small RNA Library	Number of Usable Reads	Total Number of Unique miRNA (All sizes)	Unique miRNA (20–23 bp)
Female brain	22,651,811	82,780	22,419
Female liver	18,898,991	54,580	8,763
Female ovary	22,890,085	745,265	55,506
Male brain	20,247,293	65,818	15,968
Male liver	18,660,890	68,024	10,463
Male testis	25,534,736	1,109,196	111,644

expression of RPM≥6, and strand specificity as screening criteria. A total of 3,750 miRNAs were found to satisfy the criteria for potential miRNA (Table 2). The highest complexity was found in brain tissue, with 1,150 and 1,127 screened miRNAs detected in female and male brains, respectively. In contrast, liver, ovary and testis samples each yielded between 300 and 400 miRNAs.

Identification of miRNAs in marine medaka

To identify miRNAs in marine medaka tissues, all screened miRNAs were mapped against the current release of miRBase at the time of analysis [5]. Approximately 20–30% of the screened unique sequences from each library could be annotated as miRNA candidates based on homology to miRBase entries. Some canonical sequences from our libraries differed by only one or two bases from miRBase entries, indicating they are the likely marine medaka orthologs of those miRNAs. The annotated miRNA sequences for the six individual libraries are shown in Tables S2–S7. We also observed potential miRNA paralogs in our libraries: sequence variants related to canonical miRNA sequences. At this time, we cannot rule out the possibility that these are sequencing or amplification errors and not true paralogs encoded by related miRNA genes or arising by differential processing. However, the great sequencing depth of our study ensures a high level of accuracy. To simplify our dataset, potential paralogs were assigned the same miRNA number pending independent analysis. Using BLAST searches to reveal matches to known sequences, a total of 223 known miRNA types from 109 miRNA families were identified (Table S8), 183 of which were known miRNAs and 40 of which were known miRNAs* (all previously named star forms have been re-assigned the −5p/−3p nomenclature).

Tissue- and sex-enriched miRNA expression in marine medaka

Tissue-enriched miRNAs in brain, ovary, testis and liver was elucidated. The highest numbers of genes were expressed in brain tissue (Table 3). In females, 198, 91, and 100 annotated miRNAs were identified in brain, liver and ovary, respectively. In males, 195, 103, and 83 annotated miRNAs were identified in brain, liver and testis, respectively. Comparison of expression profiles showed that 47.5% of identified miRNAs were tissue-enriched under the criteria used in this study: 39.9% of were brain-enriched, while only 3.6% were liver-enriched and 4% were gonad-enriched (Figure 2). The distribution of miRNAs across all three tissues in female and male medaka is shown in Figure 3; 69 miRNAs are found in female tissues and 60 miRNAs are found in male tissues, with only 55 miRNAs being common to all tissues in both males and females (24.7% of total miRNAs). A list of all commonly conserved miRNAs is shown in Table S9.

Further analysis showed that a small number of miRNA candidates are sex-enriched (Figure 4, Table 3). In the brain tissue, 89 miRNAs were identified, 9 of which (miR-20b, miR-83, miR-124-5p, miR-133, miR-184, miR-203a, miR-205a, miR-725, miR-962-3p) are only found in female brain and 3 (let-7a-2-3p, miR-17-3p, miR-124-3p) are enriched in male brain (Figure 4a). Of the 8 liver-enriched miRNAs, only 1 (miR-4448) is enriched in female liver and 2 (miR-749, miR-199b-3p) are only expressed in male liver (Figure 4b). We also observed 9 gonadal enriched miRNAs; 2 are only found in the female gonad (miR-27a, miR-1692) and 3 are enriched in the male gonad (miR-2184, miR-2895, and miR-4682) (Figure 4c).

Highly expressed miRNAs in marine medaka tissues

Among the 223 orthologous miRNA types, miR-9a-3p was the most highly expressed in brain tissue, with RPM values of 23,080 and 26,939 in female and male, respectively (Tables S2–S3). Other highly expressed miRNAs in brain included let-7a, miR-29b and miR-153, each having RPM values greater than 6,000. In liver tissues, the most highly expressed miRNAs were miR-122 with RPM>1000 and miR-194 with RPM>6000 (Tables S4–S5). In gonads, the most highly expressed miRNA was let-7a, with RPM values of 3,563 in ovary and 2,351 in testis (Tables S6–S7). However, the expression levels of let-7a in gonads were fairly low compared to those in brain tissue: 13,722 RPM in females and 10,299 RPM in males. Other orthologous miRNAs that were abundantly expressed in ovary were miR-21, miR-26a, miR-26b, miR-216b, miR-202-5p and miR-143, and in testis miR-21, miR-29b, miR-216b, miR-202-5p, miR-2476, and miR-143.

Interspecies comparison of miRNA expression in marine medaka tissues

The 223 identified miRNA types from 109 miRNA families were found to be homologous to miRNAs identified in 21 different species. Brain miRNAs showed a higher level of species conservation (21 species) than those identified in liver (15 species) and gonads (14 species). For a detailed species conservation analysis, *O. melastigma* miRNAs were compared with those identified in different model organisms including human (*Homo sapiens*), mouse (*Mus musculus*), nematode (*Caenorhabditis elegans*), fruit fly (*Drosophila melanogaster*) and four fish species: zebrafish (*Danio rerio*), tiger blowfish (*Fugu rubripes*), green-spotted pufferfish (*Tetraodon nigroviridis*) and Japanese freshwater medaka (*O. latipes*). BLAST searches revealed that the majority of identified *O. melastigma* miRNAs aligned well with human and zebrafish miRNA sequences (Table S10). A total of 144 of the identified *O. melastigma* miRNAs have previously been identified in zebrafish, and 92 have been identified in freshwater medaka.

Table 2. Annotation of canonical sequences based on known reference (miRBase).

Small RNA Library	Number of screened miRNA	Number of annotated canonical sequences	Number of annotated miRNA types
Female brain	1,150	243	198
Female liver	372	107	91
Female ovary	380	113	100
Male brain	1,127	241	195
Male liver	400	126	103
Male testis	321	89	83

The screened miRNAs satisfy four criteria: (1) sequencing length between 20 and 23 bp; (2) redundancy ≥6 RPM; and (3) strand specificity.

Table 3. Tissue- and sex-enriched miRNA expression in *O. melastigma*.

Tissue Type	Identified *O.melastigma* miRNA
Brain	let-7, let-7a-1-3p, let-7i-3p, miR-99-3p, miR-18a, miR-106, miR-10d, miR-124, miR-124a, miR-125b-1-3p, miR-129-1-3p, miR-129-3p, miR-129-5p, miR-130b, miR-132, miR-132-5p, miR-212, miR-135, miR-135b, miR-137, miR-137-5p, miR-137b, miR-138, miR-138-1-3p, miR-138-2-3p, miR-138b, miR-142-5p, miR-150, miR-153-5p, miR-153b, miR-22-5p, miR-22b, miR-23b-3p, miR-24-1-5p, miR-92a, miR-27b, miR-27b-5p, miR-365, miR-34a, miR-7-1-3p, miR-9a, miR-9a-3p, miR-9-3p, miR-181a-3p, miR-182, miR-183, miR-187, miR-190a-3p, miR-200b, miR-203b-5p, miR-216, miR-218, miR-218b, miR-219, miR-219-2-3p, miR-221-5p, miR-222a-5p, miR-301a, miR-301c, miR-375, miR-454, miR-455, miR-456, miR-458, miR-460b-3p, miR-489, miR-723, miR-724, miR-727, miR-727-5p, miR-728, miR-734, miR-737, miR-338, miR-2187, miR-2188-3p, miR-2188
Brain (Female only)	miR-20b, miR-124-5p, miR-133, miR-83, miR-184, miR-203a, miR-205a, miR-725, miR-962-3p
Brain (Male only)	let-7a-2-3p, miR-17-3p, miR-124-3p
Liver	miR-122, miR-122-3p, miR-148, miR-192, miR-199
Liver (Female only)	miR-4448
Liver (Male only)	miR-199b-3p, miR-749
Gonads	miR-196, miR-196a, miR-202, miR-202-5p
Ovary only	miR-27a, miR-1692
Testis only	miR-2895, miR-4682, miR-2184

Only 3 *O. melastigma* miRNAs (Let-7, miR-1 and miR-124) were highly conserved across all model animals, 52 were conserved in all of the vertebrate models, and 31 were found to be teleost-specific (Table 4).

qRT-PCR and stress response of marine medaka miRNAs

To validate miRNAs discovered during high-throughput sequencing, five representative miRNA candidates were selected for qRT-PCR analysis using RNA from the brain, liver and gonads of male and female *O. melastigma*. TaqMan MicroRNA Assays specific for mature miRNAs were performed with stem-loop RT primers. MiRNAs that were highly expressed in both sexes or that were tissue-enriched were chosen: let-7a (non-tissue enriched), miR-9-3p (brain-enriched), miR-122 (liver-enriched), miR-27a (ovary-enriched), and miR-2184 (testis-enriched). Consistent with our RNA sequencing data, qRT-PCR analysis (Figure S2) showed that let-7a was amplified in all of the male and female tissues examined, with the highest expression level observed in brain. Likewise, miR-9-3p was expressed only in the brain, while miR-122 was expressed only in the liver. As shown in our RNA-seq results, miR-2184 was most highly expressed in testis but was also found at extremely low levels in all other tissues. However, testicular expression of miR-2184 was significantly greater than that of all other tissues ($p<0.05$ for all). The highest level of miR-27a expression was observed in fish ovary. However, in contrast to the sequencing data, RT-PCR analysis indicated that miR-27a expression is not ovary-enriched. This discrepancy could be due to the presence of iso-miRNAs (multiple variants of miRNA genes) miR-27a, b, b-5p, c and e, which qRT-PCR may not be able to distinguish. Interestingly, RNA sequencing identified the presence of miR-27b and b-5p in fish brain only, while miR-27c expression was detected in all tissues and miR-27e in brain and gonads.

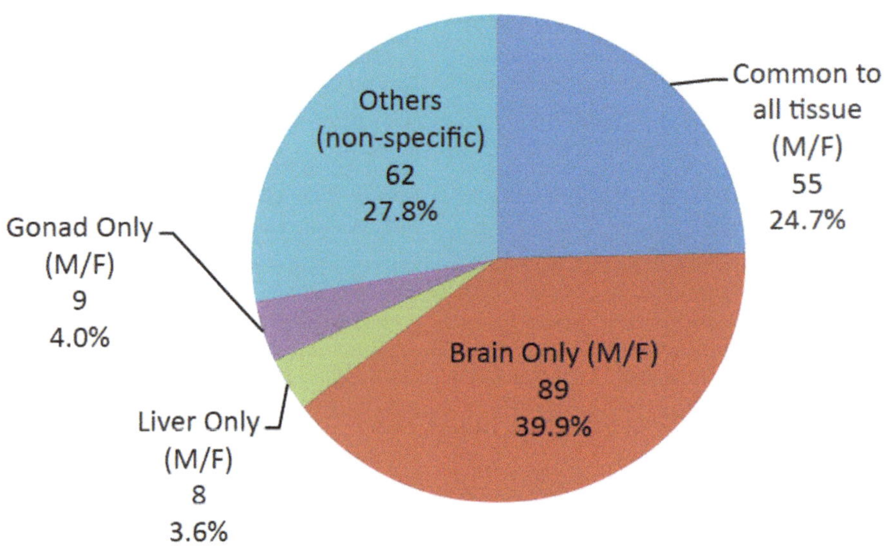

Figure 2. Pie chart of tissue-enriched miRNAs and conserved miRNAs common to all tissues in marine medaka. Diagram demonstrated brain-enriched, liver-enriched and gonad-enriched miRNAs of marine medaka. 'Others' are representative of miRNAs that are not tissue-enriched and not common to all male and female tissues.

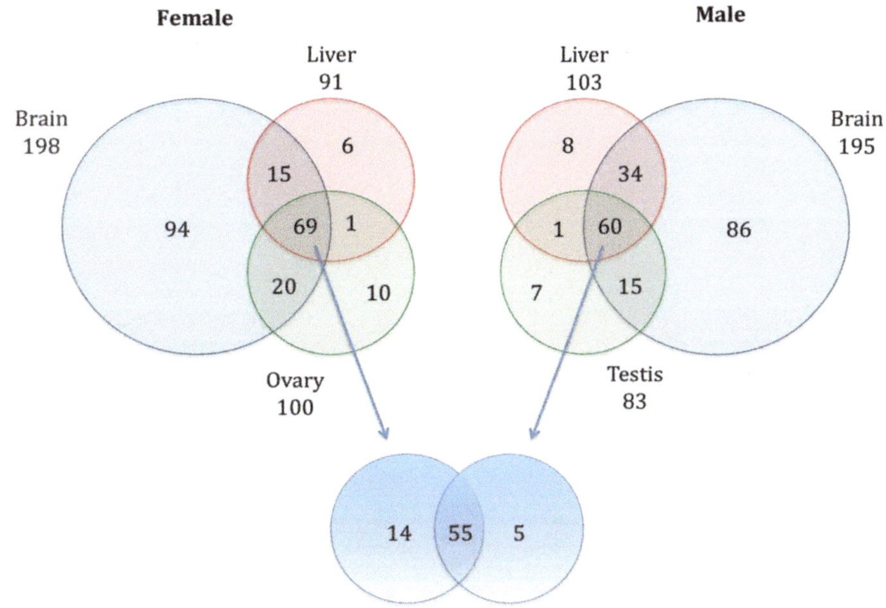

Figure 3. Venn diagrams of distribution of conserved miRNAs across tissues in female and male marine medaka. Diagram showed the distribution of conserved miRNAs in brain (grey), liver (red) and gonads (green) of marine medaka. Bottom diagram (blue) demonstrated the overlapped miRNAs between female and male medaka.

Globally, hypoxia is an important environmental stressor affecting every area in the marine environment [26]. Experiments were therefore further carried out to test the response of five selected miRNA (let-7a, miR-9-3p, miR-27a, miR-122 and miR-2184) to hypoxic stress. qRT-PCR analysis was performed on the brain, liver and gonadal tissues of male and female medaka fish that have been exposed to normoxia or hypoxia for 3 months. Our results demonstrated a significant suppression of let-7a in hypoxic liver and brain tissues of female marine medaka as compared to normoxic tissues, and a similar reduction of let-7a was also found in male liver (Figure 5A–C). However, there is no significant change of let-7a in gonadal tissues and male brain of marine medaka (data not shown). Moreover, a profound inhibition of miR-122 was observed in hypoxic female and male liver (Figure 5D–E). A significant reduction of mi-9-3p was also demonstrated in hypoxic female brain, but not in male brain tissue (Figure 5F and data not shown). In gonadal tissues, a 2-fold induction of miR-2184 was found in testis in response to hypoxia (Figure 5G). However, there is no significant change of miR-27a in ovary under hypoxic exposure (data not shown). Taken together, our data indicated that hypoxia would deregulate the expression profile of tissue-enriched miRNAs in marine medaka, but the biological implications of these changes have yet to be elucidated.

Discussion

The complete genome sequence of the *O. melastigma* has yet to be reported; thus, the established methods for novel miRNA discovery that involve mapping to the genome are not yet possible. The primary aim of the present study is to identify the known miRNAs in the organs along the brain-pituitary-gonad axis and the center of detoxification, liver in marine medaka. All of these organs are important for ecotoxicological studies. BLAST-based method was utilized to identify 223 distinct miRNA candidates in the brain, ovary, testis and liver of marine medaka based on homology to reported orthologous miRNAs. Using high-throughput sequencing, we identified unique miRNAs with expression levels ranging from 6 to nearly 27,000 RPM. Thus the methodology was highly sensitive and had sufficient dynamic range to detect miRNAs with both high and low expression levels. Our qRT-PCR analysis further confirmed that the identified *O. melastigma* miRNAs (let-7a, miR-9-3p, miR-122, miR-27a and miR-2184) are authentic miRNAs, which are differentially expressed by specific tissue and sex. The miRNAs were also highly tissue specific, with the majority brain-enriched. Many identified *O. melastigma* miRNAs in the present study are found to be highly conserved between species and taxa, and orthologous to miRNAs from as many as 21 other species, ranging from nematodes to humans. The majority of these miRNAs aligned well with miRNAs from humans and zebrafish and was also conserved in most vertebrates. Our analysis showed that 92% of *O. melastigma* miRNAs identified in the present study have orthologs previously found in zebrafish. Thus, species in the same taxonomic group appear to express similar miRNAs. The specific functions and gene targets of these teleost miRNAs are not well understood. Our qRT-PCR analysis further identified the hypoxia-responsive miRNAs: let-7a & miR-122 (female liver); let-7a & miR-9-3p (female brain) and miR2184 (testis) in hypoxia-exposed marine medaka.

In mammals, miRNAs are of great importance in the regulation of many fundamental biological processes such as cell proliferation, differentiation, apoptosis, signal transduction and organ development [27–29]. Several known classes of miRNAs have been predicted to regulate specific biological pathways. For example, the first known miRNAs, lin-4 and let-7, were found to play a major role in developmental timing [30,31]. Highly expressed miRNAs are likely to have essential and broad regulatory functions. For example, mature let-7 regulates cell proliferation and differentiation and is also highly conserved across animal species [32]. Let-7 is one of the most highly expressed across all tissues in our data. Thirteen members of the let-7 family were identified in *O. melastigma*, 8 of which were common to the

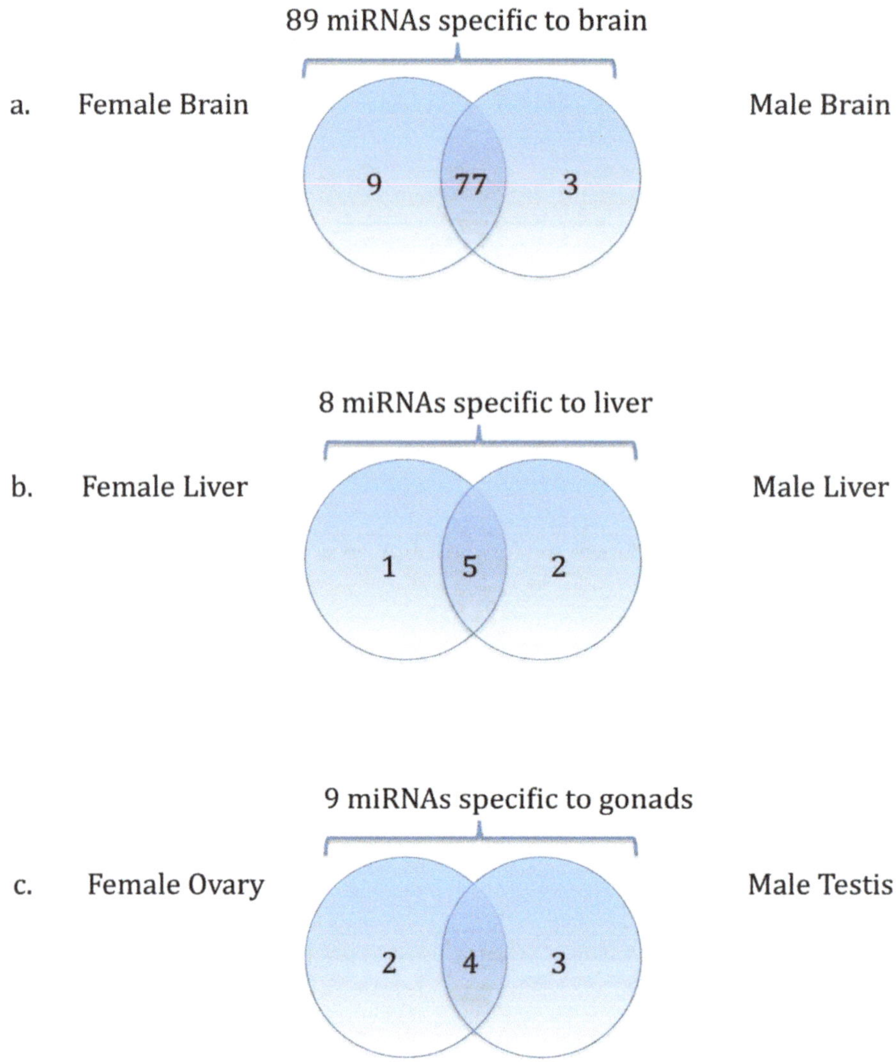

Figure 4. Venn diagrams of sex-dependent distribution of the tissue-enriched miRNAs. Diagram showed the overlapping of tissue-enriched miRNAs in a) brain, b) liver, and c) gonads of female and male medaka.

male and female tissue investigated (let-7a, let-7b, let-7d, let-7e, let-7g, let-7h, let-7i and let-7j). All let-7 members share a similar seed region which regulates the interaction between miRNA and its target genes, so it is generally believed that let-7 family imposes similar biological functions among different species. There is ample reports that demonstrated the importance of let-7 in different biological functions of brain, liver and gonads [33–35]. Sabrina *et al* reported that the introduction of let-7 caused the activation of Toll-like receptor 7, resulting in neurodegeneration in mouse's brain [33]. Also, let-7 was found to control glucose homeostasis and insulin sensitivity in liver [34]. Furthermore, let-7 regulated ageing of testis stem cell niche in drosophila [35].

Other highly expressed miRNAs identified in *O. melastigma* include miR-9, miR-21, miR-29, miR-122, miR-124, miR-143 and miR-202-5p. The orthologous miRNAs identified in the present study have diverse annotated regulatory functions. The brain-enriched miR-9 and miR-124 are both known to have a crucial role in neurogenesis and neuronal development, particularly in the regulation of neural differentiation, proliferation and cell migration [36–38]. The expression of miR-122 was found to be liver-enriched. This is consistent with the results of previous studies and is consistent with its known important role in cholesterol and fatty acid metabolism [39–41]. Recently, let-7, miR-122 and miR-143 have been shown to have tumor suppressive activity, and de-regulated let-7 expression has been associated with cancer [14,16,41]. Some miRNAs have been implicated in the regulation of apoptosis including miR-21 and miR-29, which we found to be expressed in all medaka tissues. MiR-21 is thought to be anti-apoptotic and oncogenic, while miR-29 is thought to be pro-apoptotic, suppressing tumor growth in mice [42,43]. We also observed gonad-enriched expression of miR-202-5p in medaka. This miRNA has been linked to the regulation of gonadal sex-differentiation and together with miR-21 appears to be estrogen-responsive [44,45]. Thus, there is potential for highly expressed medaka miRNAs to respond to endocrine disruptors. However, the effects of such alterations are completely unknown.

Our data also demonstrated the suppression of let-7a in liver and brain tissues under hypoxic condition. This response to oxygen depletion is concordant to the previous finding that hypoxia caused reduction of let-7a in nasopharyngeal carcinoma [46]. Moreover, let-7 transgenic mice study indicated an

Table 4. Fish-enriched miRNAs identified in *O. melastigma*.

O. melastigma miRNA	Sequence	Species Conservation
let-7h	UGAGGUAGUAAGUUGUGUUGUU	dre, fru, tni
let-7j	UGAGGUAGUUGUUUGUACAGUU	dre, fru, tni
miR-10c	UACCCUGUAGAUCCGGAUUUGU	dre, fru, tni
miR-10d	UACCCUGUAGAACCGAAUGUGU	dre, fru, tni, ola
miR-16b	UAGCAGCACGUAAAUAUUGGAG	dre
miR-457a	AAGCAGCACAUCAUUACUGGUA	dre
miR-19d	UGUGCAAACCCAUGCAAAACUG	dre, ola
miR-22b	AAGCUGCCAGUUGAAGAGCUGU	dre, fru, tni
miR-27c	UUCACAGUGGUUAAGUUCUGC	dre, fru, tni, ola
miR-27e	UUCACAGUGGCUAAGUUCAGU	dre, fru, tni
miR-130c	CAGUGCAAUAUUAAAAGGGCAUU	dre, ola
miR-135c	UAUGGCUUUCUAUUCCUAUGUG	dre
miR-301c	CAGUGCAAUAGUAUUGUCAUA	dre
miR-456	CAGGCUGGUUAGAUGGUUGUCU	dre
miR-458	AUAGCUCUUUAAAUGGUACU	dre, fru, tni, ola
miR-460	CCUGCAUUGUACACACUGUGC	dre, fru, tni, ola
miR-462	UAACGGAACCCAUAAUGCAGCUG	dre, ola
miR-722	UUUUGCAGAAACGUUUCAGAUU	dre
miR-723	AGACAUCAGAAAAAUCUGUGCU	dre
miR-724	UUAAAGGGAAUUUGCGACUGUU	dre
miR-725	UUCAGUCAUUGUUUCUGGUCGU	dre
miR-727	UUGAGGCGAGUUGAAGACUUCA	dre
miR-728	AUACUAAGUAUACUACGUUUAC	dre
miR-730	UCCUCAUUGUGCAUGCUGUGUG	dre
miR-731	AAUGACACGUUUUCUCCCGGAUU	dre, ola
miR-734	UAAAUGCUGCAGAAUUGUGC	dre
miR-737	AAAUCAAAGCCUAAAGAAAAUA	dre
miR-1388	AUCUCAGGUUCGUCAGCCCAUG	dre, ola
miR-2184	AACAGUAAGAGUUUAUGUGCUG	dre
miR-2187	UUACAGGCUAUGCUAAUCUGU	dre
miR-2188	AAGGUCCAACCUCACAUGUCCU	dre

dre: *Danio rerio*; fru: *Fugu rubripes*; tni: *Tetraodon nigrovirdis*; ola: *Oryzias latipes*.

impairment of glucose tolerance through the deregulation of insulin receptor (*Insr*) and insulin receptor substrate 2 (*Irs2*) in liver [34], suggesting that the hypoxia-mediated let-7a suppression may result in the dysfunction of medaka liver under hypoxia. But, it has been reported that let-7 can be induced by HIF-1α in response to hypoxia in vascular endothelial cell (HUVECs) [18]. Also, we found the suppression of miR-9-3p in brain tissue of hypoxic fish, which is contradictory to the findings in pulmonary artery smooth muscle cells where hypoxia caused the upregulation of miR-9 [47]. The opposite response of miRNAs to hypoxia treatment suggested that the differential expression of miRNAs in response to stresses may vary among different models. Moreover, we demonstrated the reduction of miR-122 in hypoxic liver tissues. It is similar to the finding in hypoxia-exposed Atlantic salmon [48]. The testis-enriched miRNA, miR-2184 is found to be upregulated in response to hypoxia. So far, there is no report to demonstrate the biological roles of miR-2184. But H-cadherin (CDH13), one of the predicted gene targets of miR-2184 reported in TargetScan-Fish, has been shown to be downregulated in testicular germ cell tumors and associated with differentiation of seminoma [49]. Thus, it would be interesting to study the functional roles of miR-2184-CDH13 pathway in testis in response to hypoxia.

Our previous report has demonstrated that hypoxia is an endocrine disruptor [50]. And, recent studies in mammals have suggested that endocrine-disrupting chemicals can modify the epigenome of the germ line via miRNAs, producing phenotypes in subsequent generations [13,51]. The miRNA profiling and identification of hypoxia-responsive miRNAs of *O. melastigma* in this study serves as an important platform, not only for understanding fundamental biological processes but also provides new insight into the functional roles of miRNAs in responses to environmental stresses such as hypoxia. Conceivably, changes in miRNA profile and specific miRNAs may be employed as biomarkers for transgenerational effects of environmental stress such as hypoxia, which is likely to be an important emerging environmental problem.

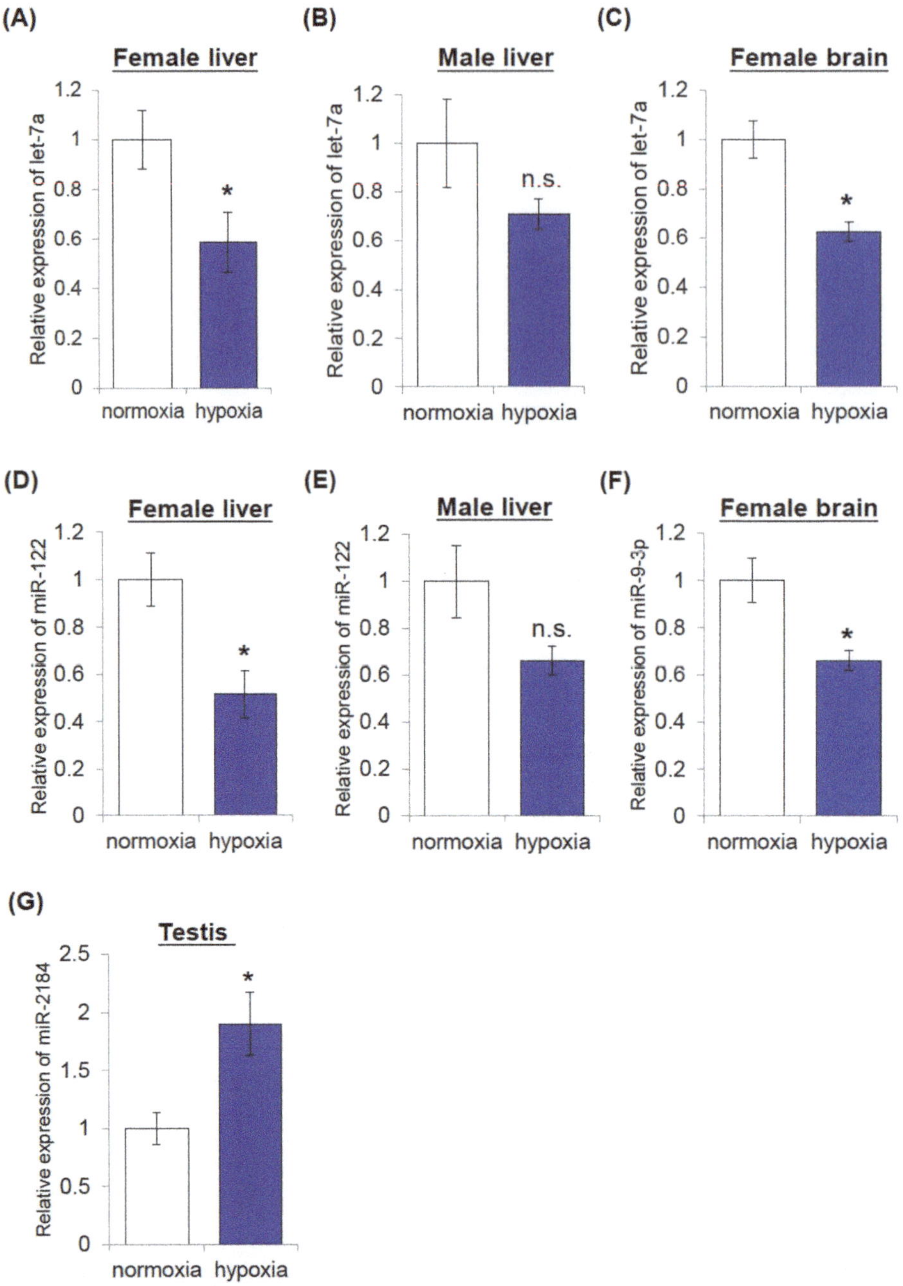

Figure 5. Identification of hypoxia-responsive miRNAs in marine medaka using qRT-PCR analysis. Female and male marine medaka were exposed to either hypoxia or normoxia for 3 months. The differentiation expressions of miRNAs in response to hypoxic stress were analyzed by quantitative real-time PCR. qRT-PCR analysis demonstrated the suppression of let-7a in female liver, male liver and female brain under hypoxia, respectively (A–C). Diagram showed the downregulation of miR-122 in female liver (D) and male liver (E). (F) Demonstrated the reduction of miR-9-3p in female brain. Diagram showed the upregulation of miR-2184 in testis of marine medaka (G). Data are presented as the means ± SEM (**p*<0.05; n.s.: not significant).

Supporting Information

Figure S1 Length (nt) distribution of small RNAs in male and female brain, liver and gonads of *O. melastigma*. Diagram showed distributions of small RNAs in female brain (upper left), male brain (upper right), female liver (middle left), male liver (middle right), female ovary (bottom left) and male testis (bottom right) of marine medaka.

Figure S2 The qRT-PCR analysis of five representative miRNA candidates using TaqMan MicroRNA Assays. (A) let-7a was amplified in all of the male and female tissues, (B) miR-9-3p was expressed in brain only, (C) miR-122 was expressed in liver only, (D) miR-2184 was most highly expressed in testis, and (E) miR-27a was most highly expressed in the ovary.

Table S1 Sequence information for all TaqMan MicroRNA Assays.

Table S2 Annotated miRNA sequences and expression levels in female brain.

Table S3 Annotated miRNA sequences and expression levels in male brain.

Table S4 Annotated miRNA sequences and expression levels in female liver.

Table S5 Annotated miRNA sequences and expression levels in male liver.

Table S6 Annotated miRNA sequences and expression levels in female ovary.

Table S7 Annotated miRNA sequences and expression levels in male testis.

Table S8 Total conserved miRNAs identified in *O. melastigma*.

Table S9 MiRNAs common to all *O. melastigma* tissues.

Table S10 Highly conserved miRNAs identified in *O. melastigma*.

Author Contributions

Conceived and designed the experiments: KL RSSW. Performed the experiments: KPL A. Tse KL. Analyzed the data: AW WYL JWL. Contributed reagents/materials/analysis tools: JYJB NZ SL A. Tong. Contributed to the writing of the manuscript: KL RSSW KPL RYCK.

References

1. Bartel DP (2004) MicroRNAs: genomics, biogenesis, mechanism, and function. Cell 116: 281–297.
2. Zeng Y (2006) Principles of micro-RNA production and maturation. Oncogene 25: 6156–6162.
3. Lim LP, Lau NC, Garrett-Engele P, Grimson A, Schelter JM, et al. (2005) Microarray analysis shows that some microRNAs downregulate large numbers of target mRNAs. Nature 433: 769–773.
4. Rajewsky N (2006) microRNA target predictions in animals. Nat Genet 38: S8–S13.
5. miRBase website (2013) http://www.mirbase.org. Accessed 3 Dec 2013.
6. Kozomara A, Griffiths-Jones S 11(2011) miRBase: integrating microRNA annotation and deep-sequencing data. Nucleic Acids Res 39: D152–D157.
7. Lewis BP, Shih IH, Jones-Rhoades MW, Bartel DP, Burge CB (2003) Prediction of mammalian microRNA targets. Cell 115: 787–798.
8. Lagos-Quintana M, Rauhut R, Meyer J, Borkhardt A, Tuschl T (2003) New microRNAs from mouse and human. RNA 9: 175–179.
9. Li SC, Chan WC, Ho MR, Tsai KW, Hu LY, et al. (2010) Discovery and characterization of medaka miRNA genes by next generation sequencing platform. BMC Genomics 11(Suppl 4): S8.
10. Chen PY, Manninga H, Slanchev K, Chien M, Russo JJ, et al. (2005) The developmental miRNA profiles of zebrafish as determined by small RNA cloning. Genes Dev 19: 1288–1293.
11. Anway MD, Cupp AS, Uzumcu M, Skinner MK (2005) Epigenetic transgenerational actions of endocrine disruptors and male fertility. Science 308: 1466–1469.
12. Dunaeva TY, Trofimova LK, Graf AV, Maslova MV, Maklakova AS, et al. (2008) Transgeneration effects of antenatal acute hypoxia during early organogenesis. Bull Exp Biol Med 146: 385–387.
13. Skinner MK, Manikkam M, Guerrero-Bosagna C (2011) Epigenetic transgenerational actions of endocrine disruptors. Reprod Toxicol 31: 337–343.
14. Esquela-Kerscher A, Slack FJ (2006) Oncomirs - microRNAs with a role in cancer. Nat Rev Cancer 6: 259–269.
15. Kulshreshtha R, Ferracin M, Wojcik SE, Garzon R, Alder H, et al. (2007) A microRNA signature of hypoxia. Mol Cell Biol 27: 1859–1867.
16. Boyerinas B, Park SM, Hau A, Murmann AE, Peter ME (2010) The role of let-7 in cell differentiation and cancer. Endocr Relat Cancer 17: F19–F36.
17. Huang X, Ding L, Bennewith KL, Tong RT, Welford SM, et al. (2009) Hypoxia-inducible mir-210 regulates normoxic gene expression involved in tumor initiation. Mol Cell 35: 856–867.
18. Chen Z, Lai TC, Jan YH, Lin FM, Wang WC, et al. (2013) Hypoxia-responsive miRNAs target argonaute 1 to promote angiogenesis. J Clin Invest 123: 1057–1067.
19. Volz DC, Bencic DC, Hinton DE, Law JM, Kullman SW (2005) 2,3,7,8-Tetrachlorodibenzo-p-dioxin (TCDD) induces organ- specific differential gene expression in male Japanese medaka (*Oryzias latipes*). Toxicol Sci 85: 572–584.
20. Kissling GE, Bernheim NJ, Hawkins WE, Wolfe MJ, Jokinen MP, et al. (2006) The utility of the guppy (*Poecilia reticulata*) and medaka (*Oryzias latipes*) in evaluation of chemicals for carcinogenicity. Toxicol Sci 92: 143–156.
21. Kong RY, Giesy JP, Wu RS, Chen EX, Chiang MW, et al. (2008) Development of a marine fish model for studying *in vivo* molecular responses in ecotoxicology. Aquat Toxicol 86: 131–141.
22. Altschul SF, Gish W, Miller W, Myers EW, Lipman DJ (1990) Basic local alignment search tool. J Mol Biol 215: 403–410.
23. Hafner M, Landgraf P, Ludwig J, Rice A, Ojo T, et al. (2008) Identification of microRNAs and other small regulatory RNAs using cDNA library sequencing. Methods 44: 3–12.
24. Bar M, Wyman SK, Fritz BR, Qi J, Garg KS, et al. (2008) MicroRNA discovery and profiling in human embryonic stem cells by deep sequencing of small RNA Libraries. Stem Cells 26: 2496–2505.
25. Houwing S, Kamminga LM, Berezikov E, Cronembold D, Girard A, et al. (2007) A role for Piwi and piRNAs in germ cell maintenance and transposon silencing in zebrafish. Cell 129: 69–82.
26. Wu RS (2002) Hypoxia: from molecular responses to ecosystem responses. Mar Pollut Bull 45: 35–45.
27. Karp X, Ambros V (2005) Developmental biology. Encountering microRNAs in cell fate signaling. Science 310: 1288–1289.
28. Miska EA (2005) How microRNAs control cell division, differentiation and death. Curr Opin Genet Dev 15: 563–568.
29. Kloosterman WP, Plasterk RH (2006) The diverse functions of microRNAs in animal development and disease. Dev Cell 11: 441–450.
30. Reinhart BJ, Slack FJ, Basson M, Pasquinelli AE, Bettinger JC, et al. (2000) The 21-nucleotide let-7 RNA regulates developmental timing in *Caenorhabditis elegans*. Nature 403: 901–906.
31. Lee RC, Feinbaum RL, Ambros V (1993) The *C. elegans* heterochronic gene lin-4 encodes small RNAs with antisense complementarity to lin-14. Cell 75: 843–854.
32. Roush S, Slack FJ (2008) The let-7 family of microRNAs. Trends Cell Biol 18: 505–516.
33. Lehmann SM, Krüger C, Park B, Derkow K, Rosenberger K, et al. (2012) An unconventional role for miRNA: let-7 activates Toll-like receptor 7 and causes neurodegeneration. Nat Neurosci 15: 827–835.
34. Frost RJ, Olson EN (2011) Control of glucose homeostasis and insulin sensitivity by the Let-7 family of microRNAs. Proc Natl Acad Sci U S A 108: 21075–21080.
35. Toledano H, D'Alterio C, Czech B, Levine E, Jones DL (2012) The let-7-Imp axis regulates ageing of the Drosophila testis stem-cell niche. Nature 485: 605–610.
36. Delaloy C, Liu L, Lee JA, Su H, Shen F, et al. (2010) MicroRNA-9 coordinates proliferation and migration of human embryonic stem cell–derived neural progenitors. Cell Stem Cell 6: 323–335.
37. Gao FB (2010) Context-dependent functions of specific microRNAs in neuronal development. Neural Dev 5: 25.
38. Kawahara H, Imai T, Okano H (2012) MicroRNAs in neural stem cells and neurogenesis. Front Neurosci 6: 30.
39. Wienholds E, Kloosterman WP, Miska E, Alvarez-Saavedra E, Berezikov E, et al. (2005) MicroRNA expression in zebrafish embryonic development. Science 309: 310–311.
40. Esau C, Davis S, Murray SF, Yu XX, Pandey SK, et al. (2006) miR-122 regulation of lipid metabolism revealed by *in vivo* antisense targeting. Cell Metab 3: 87–98.
41. Jopling C (2012) Liver-specific microRNA-122: Biogenesis and function. RNA Biol 9: 137–142.
42. Chan JA, Krichevsky AM, Kosik KS (2005) MicroRNA-21 is an anti-apoptotic factor in human glioblastoma cells. Cancer Res 65: 6029–6033.
43. Xiong Y, Fang JH, Yun JP, Yang J, Zhang Y, et al. (2010) Effects of microRNA-29 on apoptosis, tumorigenicity, and prognosis of hepatocellular carcinoma. Hepatology 51: 836–845.
44. Wickramasinghe NS, Manavalan TT, Dougherty SM, Riggs KA, Li Y, et al. (2009) Estradiol downregulates miR-21 expression and increases miR-21 target gene expression in MCF-7 breast cancer cells. Nucleic Acids Res 37: 2584–2595.
45. Bannister SC, Smith CA, Roeszler KN, Doran TJ, Sinclair AH, et al. (2011) Manipulation of estrogen synthesis alters MIR202* expression in embryonic chicken gonads. Biol Reprod 85: 22–30.

46. Hua Z, Lv Q, Ye W, Wong CK, Cai G, et al. (2006) MiRNA-directed regulation of VEGF and other angiogenic factors under hypoxia. PLoS One 1: e116.
47. Shan F, Li J, Huang QY (2014) HIF-1 Alpha-Induced Up-Regulation of miR-9 Contributes to Phenotypic Modulation in Pulmonary Artery Smooth Muscle Cells During Hypoxia. J Cell Physiol 229(10): 1511–1520.
48. Kure EH, Sæbø M, Stangeland AM, Hamfjord J, Hytterød S, et al. (2013) Molecular responses to toxicological stressors: profiling microRNAs in wild Atlantic salmon (Salmo salar) exposed to acidic aluminum-rich water. Aquat Toxicol 138–139: 98–104.
49. Lind GE, Skotheim RI, Fraga MF, Abeler VM, Esteller M, et al. (2006) Novel epigenetically deregulated genes in testicular cancer include homeobox genes and SCGB3A1 (HIN-1). J Pathol 210: 441–449.
50. Wu RS, Zhou BS, Randall DJ, Woo NY, Lam PK (2003) Aquatic hypoxia is an endocrine disruptor and impairs fish reproduction. Environ Sci Technol 37: 1137–1141.
51. Brevik A, Lindeman B, Brunborg G, Duale N (2012) Paternal benzo[a]pyrene exposure modulates microRNA expression patterns in the developing mouse embryo. Int J Cell Biol 2012: 407431.

How Can We Identify and Communicate the Ecological Value of Deep-Sea Ecosystem Services?

Niels Jobstvogt[1,2]*, Michael Townsend[3], Ursula Witte[1], Nick Hanley[4]

1 Oceanlab, University of Aberdeen, Aberdeen, United Kingdom, **2** Aberdeen Centre for Environmental Sustainability (ACES), University of Aberdeen, Aberdeen, United Kingdom, **3** National Institute of Water and Atmospheric Research (NIWA), Hamilton, New Zealand, **4** Department of Geography and Sustainable Development, University of St Andrews, St Andrews, United Kingdom

Abstract

Submarine canyons are considered biodiversity hotspots which have been identified for their important roles in connecting the deep sea with shallower waters. To date, a huge gap exists between the high importance that scientists associate with deep-sea ecosystem services and the communication of this knowledge to decision makers and to the wider public, who remain largely ignorant of the importance of these services. The connectivity and complexity of marine ecosystems makes knowledge transfer very challenging, and new communication tools are necessary to increase understanding of ecological values beyond the science community. We show how the Ecosystem Principles Approach, a method that explains the importance of ocean processes via easily understandable ecological principles, might overcome this challenge for deep-sea ecosystem services. Scientists were asked to help develop a list of clear and concise ecosystem principles for the functioning of submarine canyons through a Delphi process to facilitate future transfers of ecological knowledge. These ecosystem principles describe ecosystem processes, link such processes to ecosystem services, and provide spatial and temporal information on the connectivity between deep and shallow waters. They also elucidate unique characteristics of submarine canyons. Our Ecosystem Principles Approach was successful in integrating ecological information into the ecosystem services assessment process. It therefore has a high potential to be the next step towards a wider implementation of ecological values in marine planning. We believe that successful communication of ecological knowledge is the key to a wider public support for ocean conservation, and that this endeavour has to be driven by scientists in their own interest as major deep-sea stakeholders.

Editor: James P. Meador, Northwest Fisheries Science Center, NOAA Fisheries, United States of America

Funding: This research project was funded by MASTS (Marine Alliance for Science and Technology for Scotland; URL: www.masts.ac.uk). MASTS is funded by the Scottish Funding Council (grant reference HR09011) and contributing institutions. Townsend's involvement was funded by NIWA (National Institute of Water and AtmosphericResearch; URL: www.niwa.co.nz) under the Coasts and OceansResearch Programme 3 (2013/14 SCI). The funders had no role in study design, data collection and analysis, decision to publish, or preparation of the manuscript.

Competing Interests: The authors have declared that no competing interests exist.

* Email: niels.jobstvogt@abdn.ac.uk

Introduction

The concept of ecosystem services (ES) has inspired a movement away from conservation for the sake of nature's inherent value to one that explicitly identifies, links and communicates the benefits of conservation to human wellbeing [1–3]. The endeavour of describing, quantifying and valuing the economic benefits that nature provides to society through ES has been identified as a powerful tool to make ecosystems count in cost-benefit analysis for environmental decision making [1,4]. Throughout this paper, however, the term 'value' is used in a broader sense, as a holistic concept which can include social, ecological and economic values. This broadening of the concept of value is needed because for the remotest places on earth like the deep sea, it is particularly challenging to make direct links between changes in system functioning and effects on the delivery of final ES (and thus on human well-being) [5,6].

The deep sea accounts for nearly 91% of the world's oceans with depths ranging from 200 m to almost 11,000 m. Despite its remoteness and size, its ecosystems are far from being unaffected by anthropogenic impacts such as fishing, climate change, and pollution [7–10]. To date many knowledge gaps remain around the functioning of deep-sea ecosystems. This is partially explained through the high costs, difficulties, and risks that are associated with deep-sea research. This lack of ecological knowledge means that we also know very little about the social and economic value of protecting the deep sea. By identifying and quantifying the ES benefits provided by the deep sea it is likely that appreciation for these benefits will change. This should lead to a larger emphasis on mitigating anthropogenic impacts in the oceans.

The major challenges of accounting for deep-sea ES stem from most people's lack of awareness about the deep-sea environment, and from the prevalence of intermediate services relative to easier-to-appreciate final services. Intermediate services in this paper refer to the indirect services that the ecosystem provides, such as habitat provision and nutrient cycling (the Millennium Ecosystem Assessment [1] refers to this category as supporting services). Intermediate services are the functional basis of the final services supplied by the system (Figure 1). The final services are considered as the ecosystem's contribution to human well-being [11] and include the ES categories of provisioning (e.g. commercial fish species), regulating (e.g. waste absorption and detoxification) and

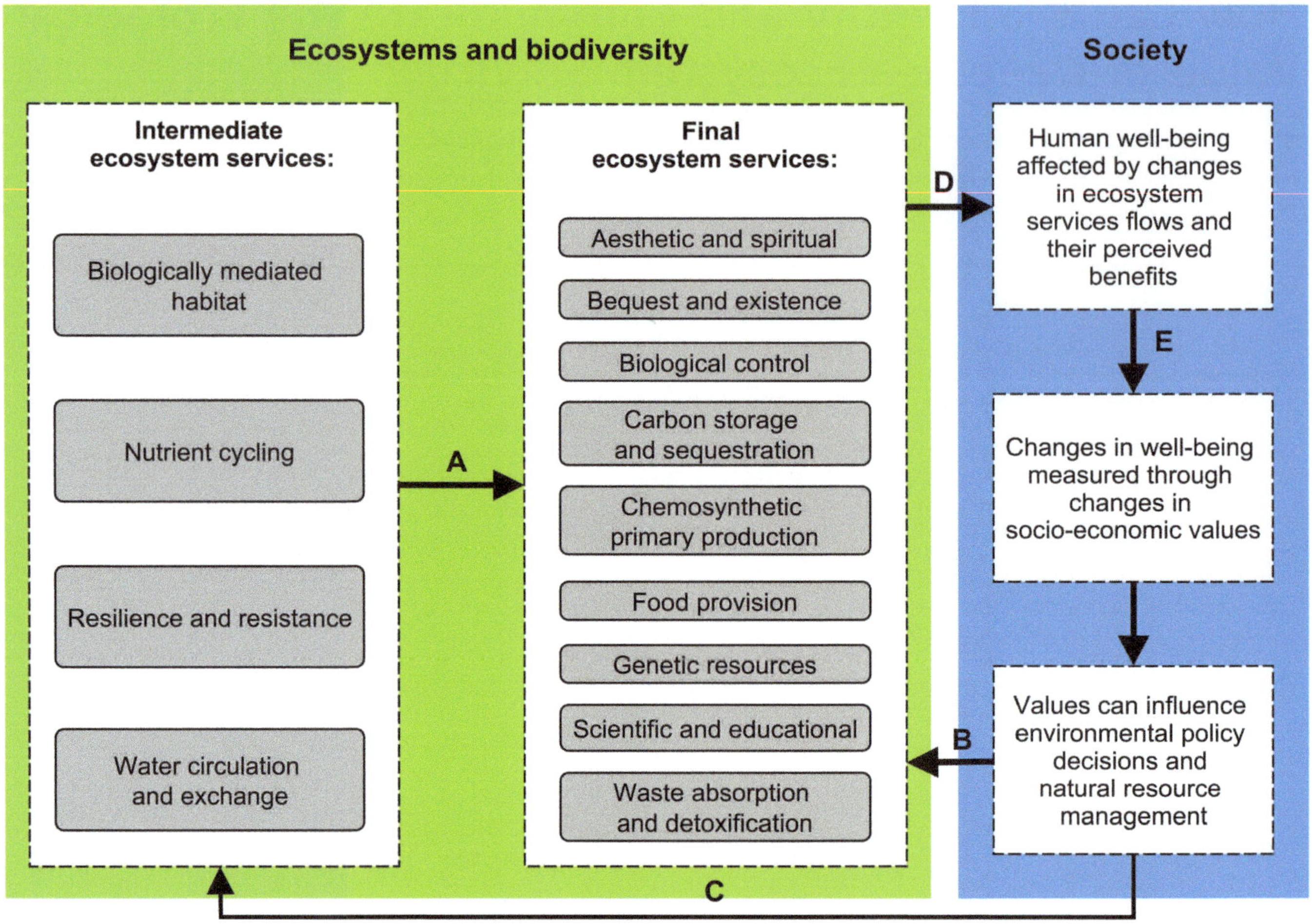

Figure 1. The ecosystem services framework for the example of deep-sea ecosystem services (ES). Environmental policies can either influence the management of final ES directly (arrow B) or indirectly via the intermediate ES (arrow C). The latter requires a sufficient understanding of the dependencies between intermediate and final ES (arrow A). Our understanding for the benefits provided by deep-sea ES (arrow D) and the values associated with them (arrow E) is currently very limited. The framework was simplified from [53] and adapted to the ES used for this expert consultation.

cultural (e.g. aesthetic values) [1]. There is a need to improve the integration of intermediate services and the processes that sustain them into the way in which we assess ES [12], since this ecological understanding is essential for demonstrating how human well-being ultimately depends on ecological processes and biodiversity [13,14].

In this paper we analyse the links between ES and the underlying processes that sustain them. This analysis was undertaken with the help of an expert panel of deep-sea scientists, focussing on submarine canyons. Within the deep sea, submarine canyons are an important ecosystem, which host unique reservoirs of biodiversity [15–18]. Along with much of the deep sea, they remain mostly underexplored and unprotected [9,19,20]. We selected this deep-sea ecosystem to test the Ecosystem Principles Approach (EPA) [21,22]. The EPA has recently been developed as a way of incorporating and translating ecological knowledge into easily understandable 'units' of information ('ecosystem principles') suitable for a wide range of audiences and thus for use in an ecosystem management context. The focus is on known and broadly-accepted information, with scientists from a wide field of expertise condensing this knowledge into principles that explain the linkages between ES, and their dependencies on underlying processes. The ecosystem principles also provide marine managers with qualitative information on temporal, spatial, and causal dependencies of ES flows. In the New Zealand case study by Townsend and colleagues [22], the EPA highlighted the importance of accounting for intermediate services in marine management, which were often provided by different geographical areas relative to the location at which final services were taken into account in the ES assessment. Economic theory suggests that intermediate ES can and should be valued through the final services that they support and the resultant direct benefits to people [40]. Indeed, ES are often perceived as a purely economic concept [23], but they also have social and ecological values, which when integrated with economic values produce a more holistic ecosystem assessment that can better inform natural resource management decisions [24]. This dominance of economic approaches and monetary valuation of ES stems from the often-felt pressure among the nature conservation sector to "speak the same language" as business and policy sectors in order to make conservation count [25–29].

In contrast, the EPA's advantage lies where economic valuation reaches its limits, by offering a more holistic picture which might help non-experts to better understand the high ecological value that scientists associate with the deep sea. Links between ecosystem processes and ES are less well known for the deep sea. For this reason it may be more helpful to highlight the links between deep-sea ES and processes themselves, rather than presenting decision makers with a set of economic values that are likely to underestimate the ecosystems' holistic value due to the omission of important ecological aspects of the deep sea [23]. In this paper we test the applicability of the EPA to little known and remote deep-sea ecosystems, such as submarine canyons, and demonstrate how the approach can provide decision makers with an accessible knowledge base for conservation decisions, despite some deficiencies in scientific data and associated uncertainty. We further discuss the approach's utility for expert consultation and cross-disciplinary knowledge transfer.

Methodology

2.1 Case study area: The Nazaré Canyon

The Nazaré Canyon on the Portuguese continental margin (Figure 2), also described as "Europe's Grand Canyon" [16], was chosen as the case study area of this paper to test the applicability of the EPA for deep-sea ecosystems. The Nazaré Canyon has attracted scientific interest due to its habitat heterogeneity and is considered to be a biodiversity hotspot [16]. Like other submarine canyon ecosystems it plays an important role in transportation processes at the continental margins [30,31]. The canyon is shallowest at 1 km off the Portuguese coast (50 m depth) and with a total length of 210 km it extends into the Iberian abyssal plain where it reaches depths of over 4,900 m [32].

2.2.1 Sampling and survey structure. Twenty-three researchers with knowledge of the Nazaré Canyon, covering a broad range of disciplines such as ecology, biology, microbiology, biogeochemistry, geography, geomorphology, geology, sedimentology and oceanography, were invited to participate in a HERMIONE (Hotspot Ecosystem Research and Man's Impact On European Seas project; URL: www.eu-hermione.net; last access July 2013) workshop in September 2012 and in an online pre-workshop survey (Figure 3). These opportunities were used to gather ideas and feedback for the two main surveys that followed. We used an email-based Delphi process to gather structured information by consensus from the invited expert panel in two consecutive rounds of surveying (post-workshop survey I and II; Figure 3). The Delphi process was originally developed as an interactive forecasting technique, where an expert panel goes through iterative survey rounds. The group results of each round are fed back to participants who are able to adapt their responses in the next survey round. The main idea of the Delphi process is to lead the group towards a consensus through the indirect exchange of information via a process coordinator. This process allowed us to subsequently include experts' requests for changes and additional information into the post-workshop surveys (Figure 3). The Delphi process makes it less likely that some researchers dominate the discussion and the outcomes, by maintaining anonymity throughout the communication process, thus avoiding the potential peer pressure of an expert workshop setting (further detail on the Delphi process in [33]). The survey questionnaires are available on request from the corresponding author.

2.2.2 Ethics statement. The nature of this research did not require ethical approval according to the University of Aberdeen Research Ethics Framework (*University of Aberdeen Research Ethical Review Checklist, Appendix A,* pp. 26–28; URL: www.abdn.ac.uk/documents/research-governance-framework-appendix4.pdf; last access March 2013). All study participants were recruited using an opt-in strategy and therefore consent was not explicitly recorded. Workshop and surveys did not include any sensitive personal questions. We asked participants to state their age, years of research experience, and field of studies; answering these questions was optional. Participants were provided with information on study objectives, sponsors, the participants' role, survey and workshop durations, potential benefits to the participants, summarised methodology, destination of gathered data and research results, the potential science impact and a contact address for further questions. Throughout the post-workshop surveys participants were identifiable via their email addresses. However, data on stated opinions and personal information was stored anonymously and kept confidential at all times. We chose email as the preferred communication method to facilitate the exchange of information during the Delphi phase of the expert consultation.

2.3 Submarine canyon ecosystem services

The experts helped to identify ES that were either perceived as less important or not relevant for the submarine canyon based on Table 1. Subsequently, the ES 'genetic resources', 'biological control', 'aesthetic and spiritual', 'scientific and educational', and 'chemosynthetic primary production' were excluded from Table 1 as less important relative to the other deep-sea ES. Accordingly eight ES (Table 1) were taken forward as a focus for the development of ecosystem principles.

2.4 Ecosystem Principles Approach

One of the main goals of the expert consultation was to develop a list of submarine canyon ecosystem principles, which could then be linked to ES. As an initial step, a review on the submarine canyon literature identified relatively well-explored ecosystem processes and relationships. The review findings were then discussed in the expert workshop and principles added or refined according to experts' suggestions. The following paragraph shows how the concept of ecosystem principles was described to canyon experts: "[An ecosystem principle] explicitly defines a key element

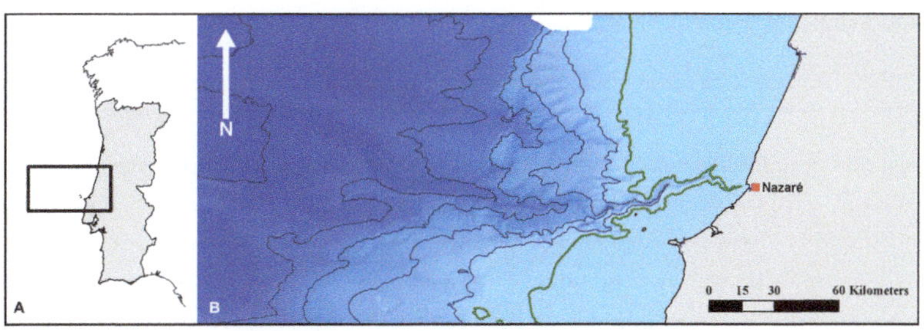

Figure 2. Nazaré Canyon. (**A**) Overview map of Portugal and the Nazaré Canyon area. (**B**) Nazaré Canyon bathymetry map with the Portuguese coastline to the east. Contour lines (blue) at 1000 m intervals; the 200 m depth contour, indicating the shelf edge, is marked in green. Data courtesy of Instituto Hidrografico, Lisbon and National Oceanography Centre, Southampton.

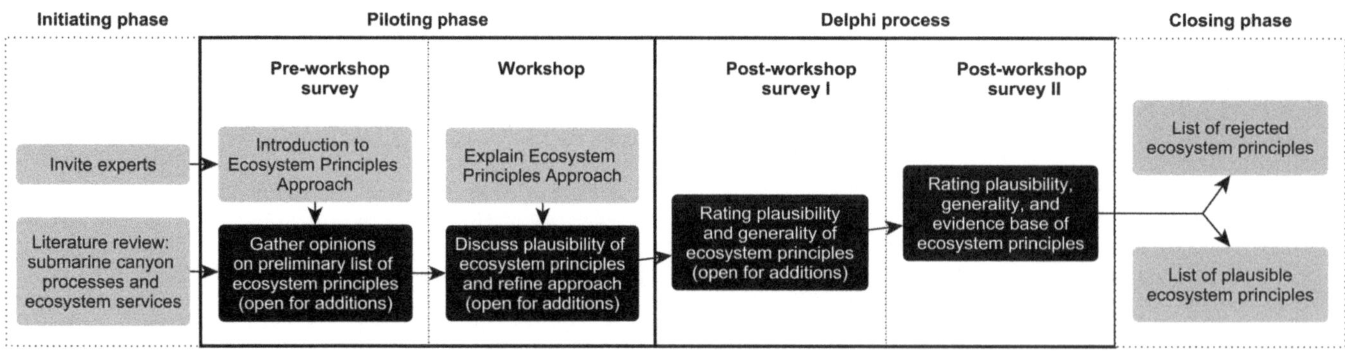

Figure 3. The survey phases of the submarine canyon expert consultation. Survey steps where experts were directly involved are highlighted as black boxes.

of how we expect the ecological system to operate" [22]. The workshop invitation summarised the EPA and included the methodology paper by Townsend and colleagues [22] as preparation for the workshop. During the workshop we provided further information in the form of a presentation that helped to increase participants' familiarity with the EPA. Principles were excluded from the initial list after each of the two consecutive post-workshop surveys (Delphi process; Figure 3) when fewer than 50% of the experts agreed with the plausibility of the principle. Experts were also able to propose new principles or suggest changes to the list of principles that was identified from the literature and subsequently refined throughout the piloting and Delphi phases (Figure 3).

We also asked the expert panel to categorise ecosystem principles according to their level of generality. The following categories were available: (i) general deep-sea principle, (ii) general canyon principle, (iii) shelf-incising canyon specific principles, to (iv) Nazaré Canyon-specific principle. The option with the highest frequency was then presented as the group vote in the subsequent survey.

Only in the second stage of the Delphi process were experts asked to distinguish their rating based on evidence on the one side and their expert view (as individuals) on the other side. The evidence base was rated on a five-point scale from 'very poor' to 'very good'. For the presentation of the group result, this was then divided into three categories of good, intermediate or poor evidence, according to the average group scores. These evidence scores had no influence on the decision to include or exclude any principle, but were introduced to separate personal opinion from the levels of evidence that existed in support of the principle.

During the workshop experts stressed the importance of the connectivity function of submarine canyons at the continental

Table 1. Submarine canyon ecosystem services.

Ecosystem services	Descriptions
Provisioning services:	
Carbon sequestration and storage	The uptake, storage, and burial of organic material within the canyon.
Food provision	The provision of marine organisms for human consumption.
Genetic resources and chemical compounds*	The use of canyon organisms in biotechnological, pharmaceutical, or industrial applications.
Regulating services:	
Biological control*	The control of diseases and invasive species.
Waste absorption and detoxification	The burial, decomposition and transformation of waste within the canyon ecosystem.
Cultural services:	
Aesthetic and spiritual*	The canyon ecosystem aesthetic and spiritual or inspirational source for religion, arts, movies, documentaries, books and folklore.
Bequest and existence	Safeguarding the canyon ecosystem for future generations and for the existence of marine species.
Scientific and educational*	The cognitive use of the canyon ecosystem for science and education.
Intermediate services:	
Biologically mediated habitat	Canyon habitats formed by marine organisms that provide nursery and refuge sites for other marine life.
Nutrient cycling	The storage and recycling of nutrients by canyon organisms.
Chemosynthetic primary production*	Primary productivity that is not dependent on energy from the sun.
Resilience and resistance	The amount of disturbance that the canyon ecosystem can cope with and its ability to regenerate after disturbance.
Water circulation and exchange	The currents, such as up-and down-welling, dense shelf water cascading, and mixing of water masses.

Services are grouped into four categories: provisioning, regulating, cultural and intermediate.
Listed items taken from [5,7,36,52] with alterations.
*Deep-sea ES that were not taken forward for the development of submarine canyon principles.

margin and we therefore chose 'water circulation and exchange' as an example to demonstrate how ecosystem principles link the supply of ES with their underlying processes. In the final results section, those ecosystem principles which in their description indicated a relationship between the ES 'water circulation and exchange' and other services were linked. The ecosystem principles relevant to this particular ES were assigned by the authors of this paper.

Results

3.1 Sample characteristics

The workshop was attended by 14 deep-sea researchers, 11 of whom had completed the pre-workshop pilot survey (Figure 3). All 14 workshop attendees were invited to participate in the two stages of the Delphi process (Figure 3), which 11 did for the first round and 10 for the second round. The average survey participant had 21 years of research experience and the survey covered academics from senior professors to PhD researchers, with male and female researchers equally represented.

3.2 Ecosystem principles

Over the course of the Delphi process, 21 ecosystem principles were identified from the literature and were then assessed and refined by the expert group (Tables 2 and 3). To highlight the nature of ecosystem principles, we present principle P1 as an example, which was rated to be plausible by all experts: 'canyons host a large number of different habitats and as a result increase species diversity at a regional scale' (for further principle descriptions we refer the reader to Tables 2 and 3). Four principles were discarded, whereas 17 principles were rated as plausible. Ten ecosystem principles fell into the category 'general submarine canyon principles', five into category 'general deep-sea principles', and two into 'shelf-incising canyon specific principles' (Tables 2 and 3). This indicated that the majority of ecosystem principles were at an appropriate level to describe processes and linkages between ES for submarine canyons in general and that they can be readily transferred to other canyons. The Delphi process had the expected effect of driving opinions closer towards consensus. Seven of the ten experts who participated in Delphi-rounds I and II (Figure 3) were closer to the group rating after the second Delphi-round.

Comparing the ratings of evidence and plausibility, we recognised that the existence of supporting evidence was not necessarily a requisite for an ecosystem principle to be plausible. Seven principles obtained intermediate evidence scores and for P16 and P17 evidence was rated as poor. However, this lack of evidence did not translate into a lack of plausibility. It was therefore an advantage to separate the two ratings from each other to distinguish between the experts' opinions and their evidence-based judgments. However, plausibility was clearly lower overall when evidence ratings were poor (Tables 2 and 3).

Developing principles to link food availability with biodiversity was challenging, and none were rated as plausible (P18 and P20). Experts had strong concerns of oversimplification when it came to the type, quality and amount of organic matter as a source of food and how changes of those parameters affected biodiversity. For biodiversity there was again a concern of over-simplification by omitting information on the spatial scale of biodiversity. In the same way, geographical scale mattered to experts, and lack of information on depth ranges and exact geographical position was criticised. The rating on generality provided a preliminary solution for implementing information on the geographical transferability of principles. Giving experts the chance to express uncertainty about the generality of principles as well as disentangling opinions about the rating of evidence allowed them to express their expectations for submarine canyons based on their research experience, and to transfer widely accepted knowledge from other ecosystems.

3.3 Linking principles and services

Many of the principles in Tables 2 and 3 have the capacity to provide information on *where* and *when* principles are likely to operate: for the principles included in Figure 4, particularly P8 and P7 reflect these spatio-temporal components. Other principles such as P3 and P12 explain *how* certain ES are provided and go into more detail on the processes involved. Principles like P17 and P16 that address effects of high biodiversity on ecosystem processes are capable of linking a broader set of ES such as 'carbon storage', 'food provision', 'bequest and existence' and 'waste absorption'. The ecosystem principles associated with 'biologically mediated habitat' were mainly thought to have an effect on biodiversity (e.g. P1 and P13) and to indirectly affect final ES such as 'food provision' and 'bequest and existence values' (P5).

For sustainable ecosystem management it can be equally important to understand the processes and principles that are involved in the provision of ES, as it is to understand the social and economic benefits of those services. We used 'water circulation and exchange' as an example to showcase how ecosystem principles explain links between ecological processes and ES (Figure 4; see also Table S1 for further details). 'Water circulation and exchange' has an important connectivity function in the submarine canyon (P11) and upwelling effects can lead to enhanced 'nutrient cycling' and as a result enhance productivity (P7 and P6; Figure 4). Further, 'nutrient cycling' might be important as an intermediate service for two different final ES, 'food provision' and 'bequest and existence values', because it can enhance fish abundance (P7). The 'bequest and existence' value can arise through the value that people tend to hold for iconic species (including fish), whereas 'food provision' relies on the abundance of commercially important fish as a consumptive resource. Trophic relationships, enhanced biomass, maintenance of deep-sea organisms (including non-iconic and non-commercial species) are important processes that sustain 'bequest and existence' as well as 'food provision' (P3, P4, P9; Figure 4) and should therefore be considered for management purposes. For 'carbon storage' and 'waste absorption' ecosystem management might be more concerned with other processes such as the transportation of organic and inorganic material, means of transportation, sedimentation rates, storage time, and burial processes that are important in parts of the submarine canyon (P3, P8, P12; Figure 4). How 'water circulation and exchange' is linked to 'resilience and resistance' as well as 'biologically mediated habitats' could not be resolved through the ecosystem principles developed in our workshop. This might be an indication that either too little evidence exists to support any ecosystem principles or that the links with processes that sustain these two ES are too complex to be described in the simplified form of ecosystem principles.

Discussion

The deep-sea case study for the Nazaré Canyon resulted in new insights on how to address the difficulties of assessing marine ES for ecosystem management purposes, especially when uncertainty is high due to lack of scientific data. In times where the demands on deep-sea resources are increasing, and scientific data on the potential impacts on marine biodiversity is scarce [5,34–36],

Table 2. Submarine canyon ecosystem principles with expert ratings on their plausibility and evidence base.

ID	Ecosystem principles	Plausibility	Evidence (mean score ± SE)
	General submarine canyon principles:		
P1	Canyons host a large number of different habitats and as a result increase species diversity at a regional scale.	100%	GOOD (3.8±0.2)
P2	The canyon topography tends to have a focusing or channelling effect for sediment and organic material.	100%	GOOD (3.8±0.2)
P3	The strength of large scale transportation events varies and occurrence ranges from a yearly to decadal pattern. They can be triggered by storms, high sediment load in the water column, cooling and increasing salinity of surface waters, or slope failures.	100%	GOOD (3.9±0.3)
P4	The transport of organic material from shallower waters to the deep seabed, which is mainly driven by large scale transportation events, is an important source of food for deep-sea organisms.	100%	MEDIUM (3.3±0.3)
P5	Canyons can serve as fish feeding ground, refuge and nursery area and therefore often show higher abundance of fish than their surroundings.	90%	MEDIUM (2.8±0.3)
P6	Canyons can enhance the mixing of water masses and as a result influence the exchange of nutrients, heat and salt between the shelf and the deep sea.	90%	MEDIUM (3.4±0.4)
P7	The canyon topography affects up- and down-welling of water masses at the continental margin. Upwelling events around the canyon head enhances productivity locally; as a result fish abundance can be higher.	90%	MEDIUM (3.3±0.3)
P8	By transporting large amounts of organic material from the shelf into deeper waters, canyons act as temporary stores of sediment and carbon. It can take decades or even centuries until the transported material reaches the abyssal plain, where it is then deposited on geological time scales.	80%	GOOD (3.8±0.2)
P9	Food quantity and quality tends to be higher within some canyon areas compared to the surrounding slope. This can enhance the biomass of the benthic and pelagic fauna.	80%	MEDIUM (3.3±0.4)
P10	Many species that are found in canyons are not found on the slope. They are therefore contributing to regional diversity.	80%	MEDIUM (3.1±0.4)

ID = principle identification number. The plausibility rating: ten experts participated in the full rating process (i.e. 100% = 10 experts). The evidence rating (1–5 from 'very good' to 'very poor'): poor (mean score <2.5), medium (2.5≤ mean score <3.5) and good (mean score ≥3.5); SE = standard error.

approaches such as the Ecosystem Principles Approach (EPA) are crucial to draw the link between the ecological and socio-economic dimensions of the ecosystem. Currently this linkage is poorly understood, contributing to the under-valuation of deep-sea ecosystems which is likely to undermine conservation efforts. We briefly outline the utility of expert consultation under these circumstances and reflect on the ability of the EPA to integrate more ecology into the assessment of ES and into the decision-making process. We first discuss how the EPA can help to communicate the overall importance of deep-sea ecosystems for the provision of ecosystem services. Further, we explain how the EPA can improve marine ecosystem-based management by promoting the inclusion of information on ecosystem processes into an ecosystem services assessment.

4.1 Communicating ecological values

The EPA has the ability to broaden access to ecological knowledge so that decision makers are not dependent on science advisors alone, but can take informed decisions on the basis of simplified ecological knowledge made available to them [37,38].

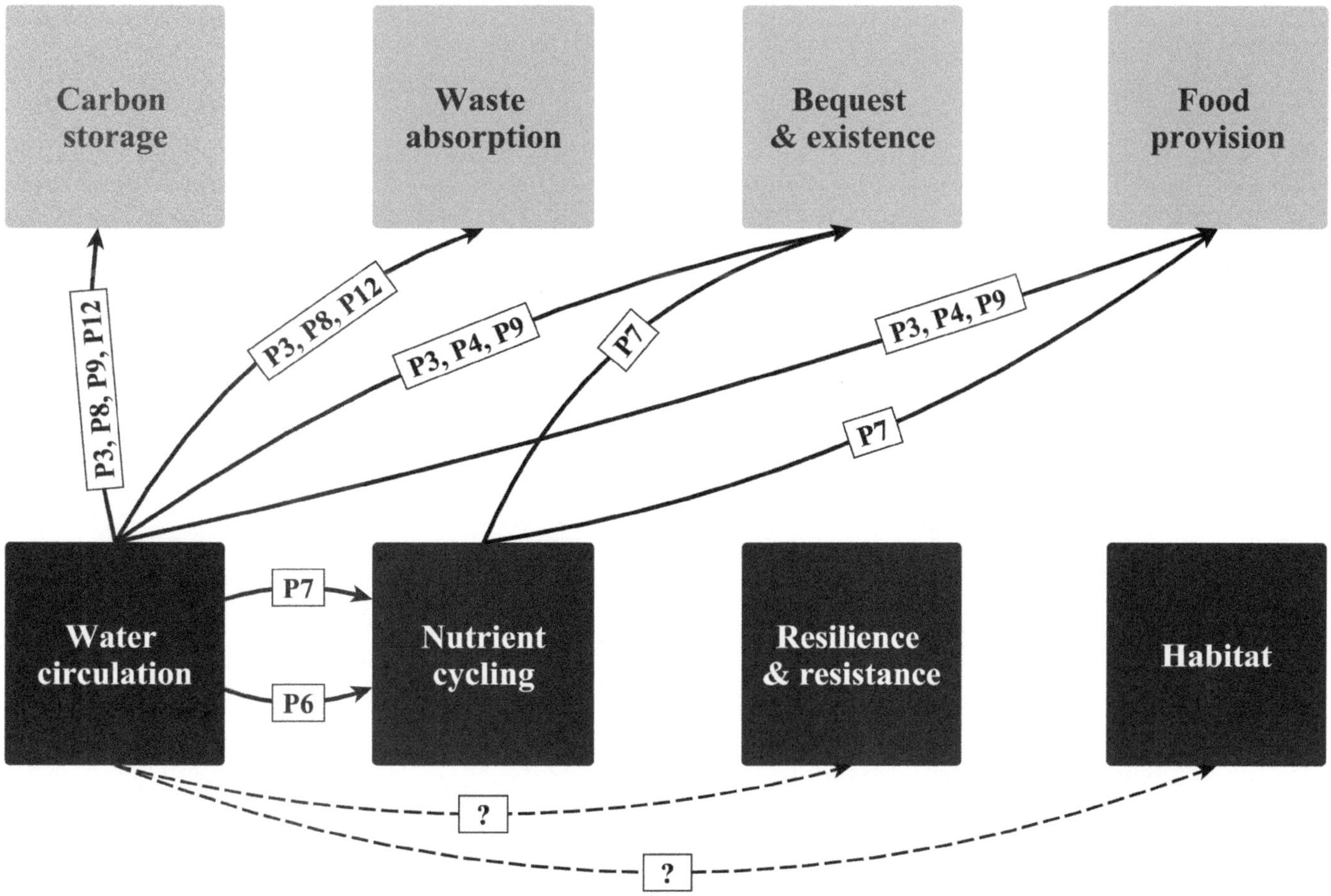

Figure 4. Links between 'water circulation' (black box) and other canyon ecosystem services explained through ecosystem principles. The intermediate services are in the lower half (dark grey and black boxes) and final services in the upper half (light grey boxes) of the diagram. Principles are indicated as arrows with their respective ID (cf. explanation in Table S1). Research gaps highlighted as question marks with dotted lines. Principles unrelated to 'water circulation' were omitted from this figure.

This broadening of access could also assist in increasing information about deep-sea ES amongst the general public. The process of developing ecosystem principles for the submarine canyon environment made clear that experts were able and willing to make predictions on connections between ES, generalisations on important canyon processes, and to link canyon characteristics to effects on ES. Also, while there remain research gaps and uncertainties, the list of ecosystem principles presented in Tables 2 and 3 provides the best available science knowledge to date, presented in easily understandable units of information. The list includes spatial and temporal information, as well as information on how the principles influence the supply of ES. We found that it was very important for the participating scientists to distinguish between their opinions (i.e. plausibility of the principles) and the (less subjective) rating of the existing scientific evidence base to inform principles. It was not imperative for principles to have a good evidence base, but rather to be generally accepted as plausible (cf. P5, P17 and P16). Herein lies a predictive strength of the EPA, in backing up the uncertainty associated with deep-sea science with a consensus-based approach, thereby decreasing uncertainty about ES linkages. The generality of principles was equally important, accounting for concerns that some ecosystem principles were valid on a larger scale than others i.e. 'general deep-sea principles' or 'shelf-incising canyon specific principles'. This additional type of information is crucial to highlight the ability of the EPA to transfer principles to other submarine canyons or even other deep-sea ecosystems. The majority of experts rated the principles as either very broadly applicable to the deep sea or to submarine canyons in general, irrespective of the type of canyon. The broad applicability of principles was thought to be an effect of reducing the complexity of ecological information.

We share the view of one workshop participant who stated that it will be difficult to determine when the list of ecosystem principles is complete. New evidence, the inclusion of researchers with different academic backgrounds, and assigning more time to the task might increase the number of principles on the list. Thus including a broad range of disciplines into the principle development process is crucial. Nonetheless, there exists an asymptotic relationship between effort expended and the number of principles, where spending more time on identifying and reviewing submarine canyon principles might increase the detail of principles, but not their utility for management decisions. The EPA's utility lies not in providing large amounts of detail, but in providing meaningful, concise information to better understand the overall functioning of the ecosystem in conjunction with the ES it provides. The EPA is based on what we know today and the ecosystem principles in this paper cover a broad range of topics, with further workshops or surveys being likely to provide diminishing returns of new principles to our established list.

There are many aspects of the deep sea that the science community remains uncertain about, but the EPA helped to focus and assimilate known information with the underlying ecosystem processes that are better understood and agreed on. However, the fact that some ecosystem principles were discarded (Table 3) indicated that there remain gaps in understanding on how canyon biodiversity is influenced by current regimes, and by different types and quality of organic material, and also the importance of recruitment processes between deeper and shallower waters. The high specificity and complexity of these and other processes might not allow us to develop these processes as ecosystem principles at the current time.

4.2 Utility of the Ecosystem Principles Approach

The deep sea is hard to sample and poorly understood, yet we were able to draw on experts' knowledge and condense what they know about submarine canyons, one of the deep-sea's biodiversity hotspots. We showed that the EPA, in combination with a Delphi process, can be a useful tool for working at the fringes of our current knowledge, using collective expert opinions to evaluate and arbitrate on the content of ecological understanding. Through this process we can also highlight where knowledge and hence research gaps lie, and where funding is needed. The EPA might be seen as a balancing act between the precautionary principle on the one side and economic reasoning on the other. The precautionary principle as framed in Rio in 1992 states that lack of scientific certainty shall not be used as an excuse to postpone actions that might prevent environmental degradation [39]. Lubchenco [37] lists guidance for decision-making under uncertainty as one of the roles that science should play in society. This might include reliance on more subjective approaches such as the EPA to support more holistic decision-making in marine resource management until we have a greater body of scientific evidence to prove or disprove what researchers have outlined as ecosystem principles.

The principles for 'water circulation and exchange' (Figure 4) demonstrate the EPA's ability to provide information on the ecological value of the ecosystem, and how these are linked to the kinds of final ES which economists are likely to value in monetary terms. The approach does not focus on final ES alone, but provides information at multiple levels without losing sight of the indirect impacts on ecosystems through intermediate services and underlying processes, and of the multiple connections between ecosystems. By linking processes with services through ecosystem principles, we draw the attention towards the network character of ecosystems. The highly interlinked nature of this network means that it is actually far from straightforward to categorise and separate services for ES valuation, especially given the multitude of connections within the marine environment. This is so irrespective of whether social, ecological, or economic definitions of 'value' are assessed. Presenting the information on final and intermediate services together with their underlying processes in a network style can better inform future management scenarios with more realistic ecological information than assessments that are limited to final ES alone.

Current ES valuation frameworks suggest that intermediate services should be valued only in terms of the final services that they support and not be included directly in a valuation of ES flows to avoid double-counting of their social or economic value [25,40–42]. The resulting requirement for effective management is that underlying processes and linkages are sufficiently well understood [23] and the ES they support are provided within the managed area [5]. However, the spatial and temporal distances between marine intermediate and final services can span millennia and act on a global scale, as is the case with the ocean nutrient cycle [35,43]. Marine ecosystems are highly connected systems with many processes being important for the provision of intermediate and final ES and crossing ecosystem boundaries [5]. Hence, if intermediate and final ES are spatially separated, the chances are that recommendations which focus only on final ES will be poor for marine resource management [22,44]. The field of ES valuation has its roots in terrestrial ecosystems where ecosystems with their services and processes are less open than in marine ecosystems [45,46]; valuation approaches might have to be adapted for the marine environment to account for its higher connectivity. Also, to capture the holistic value of intermediate services, we would have to successfully value all final ES (including cultural ES values and other non-marketed ES), which is still more of a research aspiration rather than a currently-achievable outcome. A failure to recognise the contribution of intermediate services for the ES valuation in ecosystems where they dominate, such as the deep sea, will likely lead to misguided policy decisions [47].

The EPA should be seen as an addition to baseline ecological research and economic ES valuation, rather than as a substitute for either. The EPA's advantage lies where monetary valuation reaches its limits, in highlighting links between ES and their underlying processes, and in linking intermediate services with final ES. While economic ES values can help set marine management priorities that are socially and economically desirable (Daily et al., 2009), the EPA focusses on the ecological 'value' of the ecosystem and can provide important information on how such management priorities can be achieved. Where economic values require empirical links to well-being and monetary quantification, ecological 'value' is more focussed on the importance of the ecosystem processes and characteristics that lead to such economically valued benefits being produced.

4.3 Future research opportunities and lessons learnt

The EPA might not only enhance the availability of ecological information and its uptake by decision makers, but can also improve how research results are shared across disciplines. Inter- as well as trans-disciplinary collaborations are complicated by the existence of language barriers. The use of different key terms or jargon restricts access to the pool of knowledge to only a small number of experts. The lack of frameworks that translate research findings into understandable and meaningful formats has been described as one of the major reasons why information might not reach beyond disciplinary boundaries [48,49]. Different methodologies, attitudes and perceptions between disciplines might further decrease the flow of scientific evidence [38,48]. Thus to allow economists, geologists, biologists, oceanographers and other disciplines involved in marine science to share information it would be beneficial to work on a global matrix of ecosystem principles similar to the ES valuation databases provided by the Marine Ecosystem Services Partnership (MESP; URL: www.marineecosystemservices.org; last access August 2013) which gathers studies on monetary ES values. In contrast to the MESP database, the EPA would be able to add to the evidence base not just on economic and social, but also on ecological values. Also, a more extensive dataset on ecosystem principles for marine ecosystems would increase the chances that more complex management scenarios could be developed, such as in Bayesian belief networks (BBNs), which depend heavily on the availability of baseline information on ecosystem processes, even though BBNs are able to deal with knowledge gaps when expert knowledge is available [50]. Other fields that are using approaches like habitat mapping as well as biological value mapping might benefit from

Table 3. Submarine canyon ecosystem principles continued from Table 2.

ID	Ecosystem principles	Plausibility	Evidence (mean score ± SE)
	Shelf-incising canyon specific principles:		
P11	Canyons function as major transport pathways between the shelf and the deep sea.	100%	GOOD (4.0±0.3)
P12	Sediment, organic material, and pollutants that are transported alongshore get trapped by the canyon and transported down the canyon slope.	90%	MEDIUM (3.4±0.3)
	General deep-sea principles:		
P13	Areas with reef forming or habitat creating organisms can support higher diversity than their surroundings. These habitats are most common on hard substrates, such as areas with steep slopes, rocks, boulders, vertical walls, or overhangs.	100%	GOOD (4.1±0.3)
P14	The biomass of invertebrates living in and on the seafloor can constitute an important food source for commercially important deep-sea species.	100%	GOOD (3.7±0.3)
P15	The organisms inhabiting soft substrates play a major role for the recycling of nutrients. The process is largely dominated by bacteria, but is to a smaller extent also attributed to the animals living in and on the sediment.	70%	GOOD (3.5±0.3)
P16	Higher biodiversity can support higher rates of ecosystem processes.	70%	POOR (2.4±0.5)
P17	Higher biodiversity increases the insurance value of an ecosystem by increasing the likelihood that the ecosystem is able to provide the same ecosystem functions after an ecosystem impact occurred.	60%	POOR (2.0±0.4)
	Rejected ecosystem principles*:		
P18	Diversity tends to be lower in areas with high food availability.	10%	Not assessed
P19	Space and resource occupancy by native species can decrease invasion risk.	30%	POOR (1.4±0.2)
P20	Where strong bottom currents are common, food availability and substrate characteristics become less important and current speed becomes the main driver for species abundance and diversity.	40%	Not assessed
P21	The disturbances caused by strong bottom currents keep species diversity and abundance at low levels.	40%	POOR (2.3±0.4)

ID = principle identification number. The plausibility rating: ten experts participated in the full rating process (i.e. 100% = 10 experts). The evidence rating (1–5 from 'very good' to 'very poor'): poor (mean score <2.5), medium (2.5≤mean score <3.5) and good (mean score ≥3.5); SE = standard error.
*Principles P18–P21 were rejected by the majority of experts, i.e. their plausibility was below 50%.

the EPA as well, given that spatial ecosystem principles were developed [51].

Using the EPA it should be possible to provide more precise temporal and spatial information on ecosystem principles, and to develop management strategies based on the list of principles and evidence on social and economic values and resulting management priorities. Showing the EPA's potential to improve people's understanding of ecosystem functioning was beyond the scope of this study, but future research involving the wider public and decision makers would be beneficial to test the effect of such simplified ecological knowledge on their decisions.

Three insights in particular emerge from our interdisciplinary workshop, which might improve future marine conservation initiatives and their acceptance:

(1) To further the field of marine ES valuation it would be beneficial to acknowledge that the traditional approaches to ES valuation, which have their roots in terrestrial research, might not be easily transferable to a highly linked marine

environment. Marine ecosystem boundaries are much less clearly defined than in terrestrial environments, and ES flows are less easily traceable. We might therefore need different approaches to valuing ES flows in the marine environment. The EPA is but one potential approach to improve integration of ecological values with social and economic values.

(2) The precautionary principle demands that we are cautious with our exploitation of the environment, but in the same time that management recommendations are made on a timely basis to the best of our knowledge, without postponing decisions for indefinite time until more certainty has been gained. The marine science community should more willingly embrace its important societal role in providing recommendations for nature conservation management with the support of social science approaches.

(3) We propose greater transparency in decisions on the conservation importance of marine areas. It should be possible to enhance understanding of the social, ecological as well as the economic values of certain areas, and to justify their protection, by providing easy understandable information on marine ES and how they relate to underlying ecosystem processes.

Acknowledgments

We are deeply thankful to the workshop and survey participants for their time and input, and to Dr Rene van der Wal, Dr Nicolas Krucien, and Prof Graham Pierce (University of Aberdeen) for constructive feedback on early drafts of this paper. We also like to thank Dr Veerle Huvenne (National Oceanography Centre) and the Instituto Hidrografico for the Nazaré Canyon bathymetry data, and the conference organizers for the workshop opportunity during the HERMIONE Annual Meeting 2012. We also thank two anonymous reviewers for their comments.

Author Contributions

Conceived and designed the experiments: NJ NH UW MT. Performed the experiments: NJ. Analyzed the data: NJ MT. Contributed reagents/ materials/analysis tools: MT. Wrote the paper: NJ MT NH UW.

References

1. MEA (2005) Millenium Ecosystem Assessment: ecosystems and human well-being synthesis report. Washington DC: Island Press, 141.
2. Salles J-M (2011) Valuing biodiversity and ecosystem services: Why put economic values on Nature? Comptes rendus biologies 334: 469–482.
3. Lele S, Springate-baginski O, Lakerveld R, Deb D, Dash P (n.d.) Ecosystem services: origins, contributions, pitfalls, and alternatives. Conservation & Society: 1–46.
4. Hanley N, Barbier E (2009) Pricing Nature: Cost-Benefit Analysis and Environmental Policy. Cheltenham: Edward Elgar, 353.
5. Armstrong CW, Foley NS, Tinch R, Van den Hove S (2012) Services from the deep: Steps towards valuation of deep sea goods and services. Ecosystem Services 2: 2–13.
6. Jobstvogt N, Hanley N, Hynes S, Kenter J, Witte U (2014) Twenty thousand sterling under the sea: Estimating the value of protecting deep-sea biodiversity. Ecological Economics 97: 10–19.
7. Van den Hove S, Moreau V (2007) Deep-sea biodiversity and ecosystems: A scoping report on their socio-economy, management and governance. Cambridge: UNEP-WCMC, 84.
8. Benn AR, Weaver PP, Billet DSM, Hove S Van Den, Murdock AP, et al. (2010) Human activities on the deep seafloor in the North East Atlantic: an assessment of spatial extent. PloS one 5(9): 1–15.
9. Ramirez-Llodra E, Tyler PA, Baker MC, Bergstad OA, Clark MR, et al. (2011) Man and the Last Great Wilderness: Human Impact on the Deep Sea. PLoS ONE 6(7): 1–25.
10. Jones DOB, Yool A, Wei C-L, Henson SA, Ruhl HA, et al. (2013) Global reductions in seafloor biomass in response to climate change. Global change biology: 1–12.
11. Haines-Young R, Potschin M (2013) Common International Classification of Ecosystem Services (CICES): Consultation on Version 4, August–December 2012. Nottingham, 34.
12. Turner RK, Paavola J, Cooper P, Farber S, Jessamy V, et al. (2003) Valuing nature: lessons learned and future research directions. Ecological Economics 46: 493–510.
13. Cardinale BJ, Duffy JE, Gonzalez A, Hooper DU, Perrings C, et al. (2012) Biodiversity loss and its impact on humanity. Nature 486: 59–67.
14. Mace GM, Norris K, Fitter AH (2012) Biodiversity and ecosystem services: a multilayered relationship. Trends in ecology & evolution 27: 19–26.
15. Stiles ML, Ylitalo-Ward H, Faure P, Hirshfield MF (2007) There's no place like home: Deep seafloor ecosystems of New England and the mid-Atlantic. Oceana, 40.
16. Tyler P, Amaro T, Arzola R, Cunha M, De Stigter H, et al. (2009) Europe's Grand Canyon: Nazaré Submarine Canyon. Oceanography 22: 46–57.
17. Danovaro R, Company JB, Corinaldesi C, D'Onghia G, Galil B, et al. (2010) Deep-sea biodiversity in the Mediterranean Sea: the known, the unknown, and the unknowable. PloS one 5(8): 1–25.
18. Auster PJ, Moore J, Heinonen KB, Watling L (2005) A habitat classification scheme for seamount landscapes: assessing the functional role of deep-water corals as fish habitat. In: Freiwald A, Roberts JM, editors. Cold-Water corals and ecosystems. Berlin Heidelberg: Springer, 761–769.
19. Glover AG, Smith CR (2003) The deep-sea floor ecosystem: current status and prospects of anthropogenic change by the year 2025. Environmental Conservation 30: 219–241.
20. Würtz M (2012) Mediterranean submarine canyons: Ecology and governance. Gland, Switzerland and Malaga, Spain: IUCN, 216.
21. Townsend M, Thrush S (2010) Ecosystem functioning, goods and services in the coastal environment. Prepared by the National Institute of Water and Atmospheric Research for Auckland Regional Council. Auckland Regional Council Technical Report 2010/033.
22. Townsend M, Thrush S, Carbines M (2011) Simplifying the complex: an "Ecosystem Principles Approach" to goods and services management in marine coastal ecosystems. Marine Ecology Progress Series 434: 291–301.
23. Cornell S (2010) Valuing ecosystem benefits in a dynamic world. Climate Research 45: 261–272.
24. De Groot RS, Wilson M a, Boumans RM. (2002) A typology for the classification, description and valuation of ecosystem functions, goods and services. Ecological Economics 41: 393–408.
25. Fisher B, Turner K, Zylstra M, Brouwer R, De Groot R, et al. (2008) Ecosystem services and economic theory: integration for policy-relevant research. Ecological Applications 18: 2050–2067.
26. Brink P, Mazza L, Badura T, Kettunen M, Withana S (2012) Nature and its role in the transition to a green economy. The Economics of Ecosystems & Biodiversity (TEEB), 72.
27. Kallis G, Gómez-Baggethun E, Zografos C (2013) To value or not to value? That is not the question. Ecological Economics 94: 97–105.
28. Balmford A, Fisher B, Green RE, Naidoo R, Strassburg B, et al. (2010) Bringing Ecosystem Services into the Real World: An Operational Framework for Assessing the Economic Consequences of Losing Wild Nature. Environmental and Resource Economics 48: 161–175.
29. Peterson MJ, Hall DM, Feldpausch-Parker AM, Peterson TR (2010) Obscuring ecosystem function with application of the ecosystem services concept. Conservation biology: the journal of the Society for Conservation Biology 24: 113–119.
30. De Leo FC, Smith CR, Rowden AA, Bowden DA, Clark MR (2010) Submarine canyons: hotspots of benthic biomass and productivity in the deep sea. Proceedings Biological sciences/The Royal Society 277: 2783–2792.
31. Harris PT, Whiteway T (2011) Global distribution of large submarine canyons: Geomorphic differences between active and passive continental margins. Marine Geology 285: 69–86.
32. Lastras G, Arzola RG, Masson DG, Wynn RB, Huvenne V a I, et al. (2009) Geomorphology and sedimentary features in the Central Portuguese submarine canyons, Western Iberian margin. Geomorphology 103: 310–329.
33. Linstone HA, Turoff M (2002) The Delphi Method: Techniques and Applications. New Jersey: New Jersey Institute of Technology, 616.
34. Mengerink KJ, Van Dover CL, Ardron J, Baker M, Escobar-Briones E, et al. (2014) A Call for Deep-Ocean Stewardship. Science 344: 696–698.
35. Thurber AR, Sweetman AK, Narayanaswamy BE, Jones DOB, Ingels J, et al. (2013) Ecosystem function and services provided by the deep sea. Biogeosciences Discussions 10: 18193–18240.
36. Armstrong CW, Foley N, Tinch R, Van den Hove S (2010) Ecosystem goods and services of the deep sea. Deliverable D6.2 HERMIONE Project, 68.

37. Lubchenco J (1998) Entering the Century of the Environment: A New Social Contract for Science. Science 279: 491–497.
38. Rudd MA, Lawton RN (2013) Scientists' prioritization of global coastal research questions. Marine Policy 39: 101–111.
39. Myers N (2002) The Precautionary Principle Puts Values First. Bulletin of Science, Technology & Society 22: 210–219.
40. UK NEA (2011) UK NEA: Synthesis of the key findings. UK National Ecosystem Assessment, Cambridge, UNEP-WCMC, 87.
41. Boyd J, Banzhaf S (2007) What are ecosystem services? The need for standardized environmental accounting units. Ecological Economics 63: 616–626.
42. Wallace KJ (2007) Classification of ecosystem services: Problems and solutions. Biological Conservation 139: 235–246.
43. Roberts CM, Andelman S, Branch G, Bustamante RH, Carlos Castilla J, et al. (2003) Ecological criteria for evaluating candidate sites for marine reserves. Ecological Applications 13: 199–214.
44. McLeod K, Leslie H (2009) Ecosystem-based management for the oceans. Washington, DC: Island Press, 368.
45. Hawkins SJ (2004) Scaling up: the role of species and habitat patches in functioning of coastal ecosystems. Aquatic Conservation: Marine and Freshwater Ecosystems 14: 217–219.
46. Raffaelli D (2006) Biodiversity and ecosystem functioning: issues of scale and trophic complexity. Marine Ecology Progress Series 311: 285–294.
47. Johnston RJ, Russell M (2011) An operational structure for clarity in ecosystem service values. Ecological Economics 70: 2243–2249.
48. Davies AL, Colombo S, Hanley N (2013) Improving the application of long-term ecology in conservation and land management. Journal of Applied Ecology 51: 63–70.
49. Peters DPC (2010) Accessible ecology: synthesis of the long, deep, and broad. Trends in Ecology & Evolution 25: 592–601.
50. Campbell D, Chilton S, Clark S, Fitzsimmons C, Gazzola P, et al. (2012) Bayesian Belief Networks as an Interdisciplinary Marine Governance and Policy Tool. Valuing Nature Network Report, 27.
51. Townsend M, Thrush SF, Lohrer AM, Hewitt JE, Lundquist CJ, et al. (2014) Overcoming the challenges of data scarcity in mapping marine ecosystem service potential. Ecosystem Services 8: 44–55.
52. Beaumont N, Townsend M, Mangi S, Austen MC (2006) Marine biodiversity: an economic valuation. Building the evidence base for the Marine Bill. Department for Environment Food and Rural Affairs, 64.
53. Mace GM, Bateman I, Albon S, Balmford A, Brown C, et al. (2011) UK NEA Technical Report: Conceptual Framework and Methodology. UK National Ecosystem Assessment: Technical Report. Cambridge: UK National Ecosystem Assessment, UNEP-WCMC, 11–26.

16

Projected Range Contractions of European Protected Oceanic Montane Plant Communities: Focus on Climate Change Impacts Is Essential for Their Future Conservation

Rory L. Hodd[1]*, David Bourke[2], Micheline Sheehy Skeffington[1]

1 Plant Ecology Research Unit (PERU), Botany and Plant Science, National University of Ireland Galway, Galway, Ireland, 2 Department of Botany, and Trinity Centre for Biodiversity Research, School of Natural Sciences, Trinity College, Dublin, Dublin, Ireland

Abstract

Global climate is rapidly changing and while many studies have investigated the potential impacts of this on the distribution of montane plant species and communities, few have focused on those with oceanic montane affinities. In Europe, highly sensitive bryophyte species reach their optimum occurrence, highest diversity and abundance in the north-west hyperoceanic regions, while a number of montane vascular plant species occur here at the edge of their range. This study evaluates the potential impact of climate change on the distribution of these species and assesses the implications for EU Habitats Directive-protected oceanic montane plant communities. We applied an ensemble of species distribution modelling techniques, using atlas data of 30 vascular plant and bryophyte species, to calculate range changes under projected future climate change. The future effectiveness of the protected area network to conserve these species was evaluated using gap analysis. We found that the majority of these montane species are projected to lose suitable climate space, primarily at lower altitudes, or that areas of suitable climate will principally shift northwards. In particular, rare oceanic montane bryophytes have poor dispersal capacity and are likely to be especially vulnerable to contractions in their current climate space. Significantly different projected range change responses were found between 1) oceanic montane bryophytes and vascular plants; 2) species belonging to different montane plant communities; 3) species categorised according to different biomes and eastern limit classifications. The inclusion of topographical variables in addition to climate, significantly improved the statistical and spatial performance of models. The current protected area network is projected to become less effective, especially for specialised arctic-montane species, posing a challenge to conserving oceanic montane plant communities. Conservation management plans need significantly greater focus on potential climate change impacts, including models with higher-resolution species distribution and environmental data, to aid these communities' long-term survival.

Editor: Han Y.H. Chen, Lakehead University, Canada

Funding: Research for this paper was funded by the Irish Research Council, the Irish National Parks and Wildlife Service and NUIG Thomas Crawford Hayes Fund. DB was funded by the Science, Technology, Research and Innovation for the Environment (STRIVE) Programme, financed by the Irish Government under the National Development Plan 2007-2013, administered on behalf of the Department of the Environment, Heritage and Local Government by the Irish Environmental Protection Agency, under grant 2007-633 CCRP-2.26 and a Royal Irish Academy Mobility Grant. The funders had no role in study design, data collection and analysis, decision to publish, or preparation of the manuscript.

Competing Interests: The authors have declared that no competing interests exist.

These authors contributed equally to this work.

* E-mail: rlhodd@gmail.com

Introduction

Mountain regions are thought to be especially susceptible to the effects of climate change [1]. Oceanic mountains, subject to both maritime and continental influences, present a number of difficulties for climatic change predictions and are likely to be significantly different from those of lowland regions, with altitudinal gradients and orographic effects adding to the local spatial and temporal variability [2], [3].

Climate change is projected to have far-reaching impacts on biodiversity and may lead to widespread changes in species distribution and community composition among a variety of taxa throughout the world [4], [5]. Among the many impacts of climate change are the observed and projected changes in species altitudinal [6], [7] and geographical range [8], [9]. Species vulnerable to higher temperatures either contract in range with increasing warmth, to occupy areas of higher altitudes, or move their range northwards, becoming extinct in more southerly regions [5]. Many factors influence a species' ability to alter its range in response to climate change, including dispersal ability and availability of suitable habitat. Species migration is also curtailed by natural barriers, habitat fragmentation and human impacts [10]. Therefore, it is likely that many species will be unable to disperse at rates necessary to keep up with rapidly changing climatic conditions [11]. On the other hand, there is a growing body of phylogenetic evidence to suggest that long-distance dispersal has occurred for a range or arctic-alpine plant species under past changing climates [12], [13], [14], indicating that dispersal may not be as limiting a factor to the future survival of some species as previously thought.

Montane plant species and communities are particularly vulnerable to climate change [15], [16] and have already sustained many alterations in distribution and species composition, including an increased abundance of generalist species, at the expense of specialised arctic-montane and arctic-alpine species [7], [17], [18], [19], [20]. Arctic-montane species belonging to oceanic mountain ranges are particularly adapted both physically and physiologically to grow in harsh, cold conditions [21]. Consequently they have poor competitive and dispersal ability and may encounter difficulties in adjusting rapidly and efficiently to changes affecting their environment [22] and, therefore, have an increased extinction risk [23]. This extinction risk is especially high in oceanic mountain ranges, as vulnerable species in these areas are likely to experience a physiologically-induced decrease in competitive ability in response to climate change [24]. Additionally, the oceanic mountain ranges of Europe tend to be lower than continental ranges, and therefore have no nival zone to accommodate upward migration of vulnerable species [25]. However, relatively few studies have been carried out to date on climate change effects on oceanic and coastal montane vegetation [26] and its responses may be different.

Bryophytes are ideal for use as predictors and indicators of past, present and future climate change [27], [28], as they often react rapidly to changes in climate [29], [30], [31]. Distributional changes have been detected in a number of British, Irish and European bryophytes, likely as a result of climate change [31], [32], including the northward and eastward shift of south-westerly distributed oceanic bryophytes and an increased vulnerability of northern, montane bryophyte species [32]. Oceanic bryophytes in Europe reach their optimum occurrence and greatest diversity in the hyperoceanic regions of western Scotland and western Ireland (Figure 1A); south-western Ireland has the highest diversity and abundance of Atlantic bryophytes [33]. These bryophyte assemblages, not found elsewhere in Europe, play an essential role locally in community structure and composition [34], [35]. In contrast to widespread lowland bryophytes [36], many oceanic species are particularly vulnerable to changes in climate, as they are not known to produce sporophytes in these areas, spreading only by vegetative means, resulting in slow and unpredictable establishment of new populations [37]. The distribution of these species is also limited by their highly specific microclimatic requirements, which are only met in areas where both climate and local topography are ideal [33], [35].

The montane oceanic species and vegetation of western Europe are of high conservation value at both a national and international scale [34]. They occur in a highly sensitive and vulnerable landscape in upland, temperate regions [38]. As these largely montane heath communities are of such restricted distribution, Ireland and the UK have special responsibility under the EU Habitats Directive to conserve them, particularly those supporting rare assemblages of oceanic bryophytes [34], [39]. These include montane moss heath, dominated by *Racomitrium lanuginosum* and a community of large leafy liverworts, known as 'mixed northern hepatic mat', which is restricted to shady north-facing mountain slopes in western Scotland and western Ireland [35]. The constituent species of the latter are of high conservation value, as many have a very disjunct global distribution and are highly restricted within Europe, being considered rare or threatened on both an Irish and European scale [39], [40].

Currently, conservation strategies rarely consider climate change [41], [42]. Yet, most plant and vertebrate species have been projected to lose climatically suitable space within protected European conservation areas [42], suggesting that the current protected areas network will be less effective under future climate change, as species ranges may shift to non-protected areas [43], [44], [45]. This will require a range of new conservation approaches and conservation plans that accommodate shifts in species' range by incorporating potential future species ranges [42], [45], [46]. To this aim, knowledge of the potential species' response to future environmental change, including range contractions, will help target limited conservation resources to those species and populations most at risk. Therefore, the projected potential range changes for key montane oceanic heath species in this study will be spatially related to existing protected areas in order to assess the adequacy of these for species conservation under future climate change.

We used a state-of-the-art ensemble species distribution modelling (SDM) approach, which is a conservation planning tool that statistically correlates current species distributions with climatic and other environmental variables and enables projections of future potential distributions. Because of the hepatic mat species' very specific topographical requirements and the restricted occurrence of vascular arctic-montane species in the region, we used topography as a key modelling variable to add precision to the projections. Our aims were to:

- project changes in the future potential distribution of upland bryophyte and vascular plant species under a changing climate and assess the potential implications for oceanic montane plant communities
- investigate the effect of the addition of topographical variables on model predictive capacity for species distribution, since many upland bryophyte and vascular plant species are confined to areas with specific topographical characteristics
- assess the implications of climate change-induced shifts in species range for future conservation management of European protected montane habitats

Materials and Methods

Study area

The study area consists of the whole island of Ireland (both the Republic and Northern Ireland), on the north-western fringe of Europe (Figure 1A) between 51°25′N and 55°25′N; 5°26′W and 10°37′W. Ireland has a temperate oceanic climate, with mean monthly temperatures ranging from 6–6.5°C in January to 15–15.5°C in July [47] and relatively high (750 mm to >2500 mm per annum) and frequent (150 to >220 days per annum) rainfall [48]. Rising to 1041 m, most of the mountains are coastal and, in the western part of the country have a hyperoceanic climate, which is highly suitable for the growth of oceanic montane vegetation [34] (Figure 1B). Global climate change is projected to alter the climate of Ireland; temperatures are predicted to rise by 3–4°C by the end of this century, while an increased seasonal variation in rainfall amounts is projected, leading to lower summer and higher winter precipitation [49], [50].

Species data and selection criteria

The species in this study are either confined to montane habitats or form an important component of montane vegetation. Thirty species (14 bryophyte and 16 vascular plant species) were selected (Table S1), representing four groups: characteristic of, or restricted

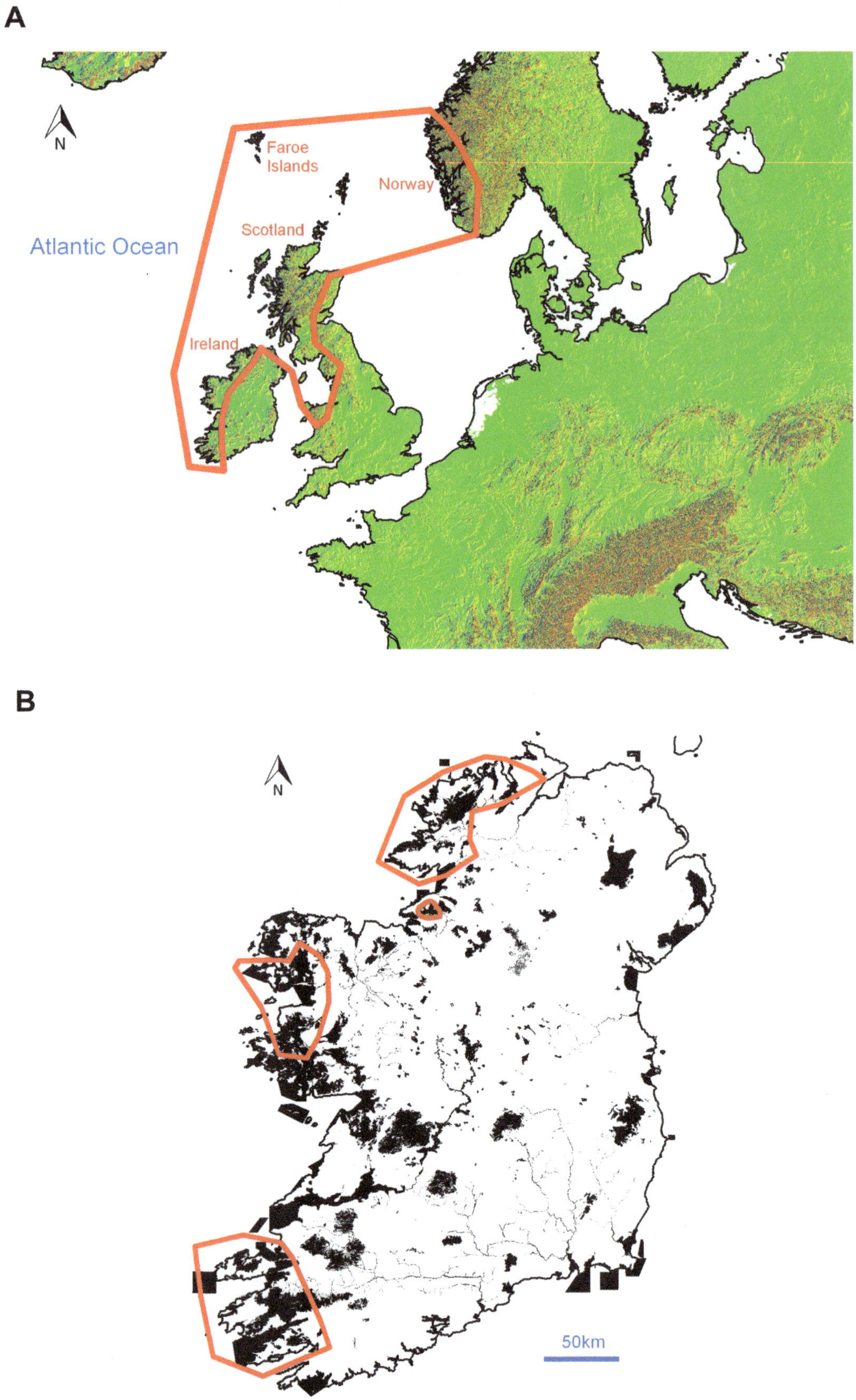

Figure 1. Maps showing the distribution of oceanic montane vegetation in Europe and Ireland. (A) Distribution range (red line) of temperate oceanic montane vegetation in Europe. (B) Main areas (red lines) where conditions are currently optimal for the occurrence of oceanic montane vegetation in Ireland, i.e. areas of altitude over 500 m, that are within the area defined as hyperoceanic by [94]; shaded areas are sites

designated as the Natura 2000 network of protected areas (including marine sites) on the island of Ireland.

to, montane heath vegetation; characteristic of montane cliffs; characteristic of mixed northern hepatic mat ('hepatic mat') and oceanic bryophyte species that frequently grow in association with hepatic mats, but also occur in a wider range of upland habitats, henceforth referred to as 'oceanic montane bryophytes'. The species selected for use in these models were either species that were considered to be key components of Irish montane vegetation, based on analysis of ecological data collected from montane vegetation in western Ireland [51], or species that are restricted only to montane vegetation within the study area, selected using a combination of expert knowledge and analysis of the current known distribution of these species [35], [39], [52].

Species were also allocated to specific groups: 1) having a wide, narrow or disjunct distribution; 2) being associated with a particular biogeographical element, divided into biome and eastern limit categories (after [53] for vascular plants and [54] for bryophytes); 3) the maximum and minimum altitude of occurrence (following [52] for vascular plants; [55] for mosses and [56] for liverworts). The eastern limit category is a measure of how far east a species' distribution stretches, using Britain and Ireland as a reference point, ranging from hyperoceanic to circumpolar.

The data for most species were obtained from the National Biodiversity Network (NBN) Gateway [57] and the data for a number of rare oceanic bryophyte species were obtained from the National Parks and Wildlife Service (NPWS), Dublin [58]. These data represent high quality presence-absence distribution data on a 10 km×10 km grid cell basis (Table S2), which is the smallest available grid size for the area studied. No information was available on the abundance of each species within each grid cell, but the majority of species selected form a significant component of the vegetation in which they occur and are likely to occur at multiple locations across each 10 km×10 km grid cell in which they are present.

Climate data

Climate data from the 1961–1990 baseline period were obtained for the whole island of Ireland [50], [59]. These 10 km×10 km (Irish National Grid) resolution data were derived from daily climate data from 560 precipitation stations and 70 temperature stations spatially interpolated using a polynomial regression method with an inbuilt adjustment for elevation. Variables used include mean, minimum and maximum monthly temperatures, and mean monthly precipitation as well as various derived bioclimatic variables (e.g. net annual rainfall, mean winter temperature, continentality index). The climate change data used in the current study incorporated the mean values of A2 and B2 scenarios obtained from statistically downscaled outputs from the HadCM3 Global Climate Model [59].

Topographic data

A range of topographical data were extracted from the GTOPO30 digital elevation model (DEM) [60] for each 10 km×10 km grid cell, including mean, maximum and minimum elevation, area of land surface >350 m, area of land surface > 500 m, mean slope, area of land surface occupied by aspects facing north-west, north and north-east. Based on the known ecological and topographical conditions required by the species [52], [55], [56], we quantified the area of land surface in each grid cell occupied by those specific topographical variables for inclusion in the models. For example, for the montane heath species we quantified the area of land surface in each grid cell occupied by elevations >350 m, with an aspect facing north-west, north and north-east. Although these species are not restricted exclusively to areas where these conditions are met within the study area, they would very rarely grow at lower altitudes or with other aspects, and reach their optimum occurrence at these aspects and altitudes. Data processing was undertaken in ArcGIS v9.3 (ESRI, Redlands, CA, USA). Hawth's Tools, an extension to ArcGIS, was used to carry out the polygons in polygon analysis [61].

Variable selection

Sixty climatic and fifteen topographic variables were considered for analysis. All were tested for collinearity, then selected to avoid this, using variance inflation factors (VIFs). These were calculated between the variables and those with the highest VIFs were eliminated until the VIFs for all variables were below a value of 5, sufficiently low to avoid collinearity [62]. Knowledge of the species' ecological requirements informed the final selection of the most ecologically important variables.

The variables used for species of montane heaths and cliffs, were: December minimum temperature, July maximum temperature, mean winter precipitation (December – February), mean slope and area of the land surface in each grid cell >350 m, with an aspect of north to north-east. The variables used for hepatic mat and oceanic bryophyte species were: December minimum temperature, May maximum temperature, mean summer precipitation (June – August), area of cell with slope >10% and area of the land surface in each grid cell >500 m, with an aspect of north to north-east (Table 1).

Species distribution modelling

We predicted the distribution of the 30 species using an ensemble of species distribution modelling (SDM) techniques appropriate for the presence-absence species data collated, within the BIOMOD2 framework; Generalized Linear Models (GLM), Random Forests (RF), Generalized Boosting Models (GBM), Artificial Neural Networks (ANN), Flexible Discriminant Analysis (FDA) and Generalized Additive Models (GAM) [63], [64]. Models were created using climatic variables only, or climatic with topographic variables, to establish the effects of topography on species distribution and to investigate the effects on predictions of disregarding topography [65]. The predictive performance of these models was measured using Cohen's Kappa statistic (K), True Skill Statistic (TSS), the Area Under the Curve (AUC) of the Receiver Operating Characteristic, sensitivity and specificity scores. Thresholds used to assess model predictive performance are outlined in Table S3. Only those models obtaining AUC, TSS and Kappa scores above 0.8, 0.6 and 0.6, respectively (i.e. with minimum 'Good' rating), were used to build the ensemble models. Model evaluation was carried out by splitting the data; 80% for calibration and 20% for validation. All models were calibrated using the R environment software [66] and followed the default settings of BIOMOD2 [64]. The simulated current and future distributions of the species were compared for the models using both climate and topography variables and the % decrease or increase in range for each species calculated. Central to possible range changes of species under changing climate, is the ability of those species to colonise new potentially suitable areas. For vascular plants and bryophytes, this may depend on species' dispersal ability. However, detailed dispersal distances are

Table 1. Current mean, minimum, maximum and range values for climatic and topographic variables and projected climate variables for 2055.

Climate variables	Units	Used for	Mean	Min	Max	Range
May maximum temperature	°C	Oceanic bryophytes	14.12	12.17	15.09	2.92
July maximum temperature	°C	Montane heath	18.36	15.77	19.63	3.85
December minimum temperature	°C	All species	2.39	−0.57	4.87	5.44
Mean winter precipitation (December-February)	Mm	Montane heath	124.10	63.82	224.99	161.17
Mean summer precipitation (June-August)	Mm	Oceanic bryophytes	83.84	49.64	127.90	78.27
Climate variables - 2055			**Mean 2055**	**Min 2055**	**Max 2055**	**Range 2055**
May maximum temperature	°C	Oceanic bryophytes	15.75	13.11	15.09	3.43
July maximum temperature	°C	Montane heath	22.61	16.37	19.63	7.85
December minimum temperature	°C	All species	3.48	0.12	4.87	5.81
Mean winter precipitation (December-February)	Mm	Montane heath	137.67	62.34	224.99	181.99
Mean summer precipitation (June-August)	Mm	Oceanic bryophytes	61.10	26.11	127.90	97.77
Topography variables			**Mean**	**Min**	**Max**	**Range**
Area with aspect northeast to northwest, above 500 m altitude	Km^2	Oceanic bryophytes	0.14	0.00	20.92	20.92
Area with aspect northeast to northwest, above 350 m altitude	Km^2	Montane heath	0.46	0.00	24.30	24.30
Area of slope >10%	Km^2	Oceanic bryophytes	1.66	0.00	40.50	40.50
Mean slope	°	Montane heath	1.18	0.00	5.02	5.02

Which variable is used for oceanic bryophyte, montane heath, cliff, or all species is also indicated.

unavailable for most species; therefore, we examined two extreme scenarios:

1. Unlimited dispersal; where the entire projected future range of the species is taken to be the actual future distribution.
2. Limited (i.e. no) dispersal; where the future distribution results solely from the overlap between current and projected future range of the species.

Overlap with protected areas

Gap analysis is a protocol for assessing the extent to which valued biodiversity attributes are represented within protected areas [67], [68], and was used here to calculate the overlap between protected areas and species' current and projected potential future distributions. The species studied here belong to habitats mostly protected in Special Areas of Conservation (SAC) designated under the EU Habitats Directive, typically found in blanket and raised bogs, wet, dry, alpine and subalpine heaths, and siliceous and calcareous rocky slopes and scree [69]. The analysis was performed using current gridded presence/absence data, our future projected distributions for the species and the SAC spatial extent data [70] in which the proportion of each grid cell area occupied by SAC was computed. As the species data used were at a coarser resolution to the SAC data, thresholds were devised to match the datasets. Grid cells were considered protected if their proportion of protected area equalled or exceeded 10%. Comparisons of a range of thresholds (≥2%, ≥5%, ≥10%, ≥20%, ≥ 30%, ≥50%) were made to rule out threshold effects [67]. The analysis was undertaken using the R environment software [66].

Statistical analyses

Statistically significant differences in model performance statistics, range changes and change in overlap with protected areas for the various species groups were tested using the Wilcoxon signed-rank test and the significance of between-group relationships was calculated using Wilcoxon rank-sum test, as the data were not normally distributed [66].

Results

Model performance

A comparison of the predictive performance statistics of models using climate variables only and those produced using both climate and topography, revealed that AUC ($p<0.01$), Kappa ($p< 0.001$) and TSS ($p<0.001$) were significantly higher for most species, across all modelling techniques, when topography was added to the models (Figure 2; values for individual species in Table S4a). Mean AUC for all species increased from 0.92 (± standard error 0.01) to 0.95 (±0.01); mean Kappa from 0.46 (±0.02) to 0.58 (±0.02) and mean TSS from 0.7 (±0.02) to 0.8 (±0.02) upon addition of topographic variables. Species of disjunct distributions showed a significantly greater increase in performance ($p<0.01$ for AUC and Kappa; $p<0.001$ for TSS) than widely distributed species, as did species which are restricted only to montane habitats as opposed to those that are not habitat specific ($p<0.01$ for all measures of performance). Overall, the models for most species were shown to be very good or excellent at correctly predicting true presences and true absences. Mean (min, max) sensitivity and specificity for all modelled species using both climate and topography were 84.50 (67.40, 100) and 90.25 (76.75, 96.31), respectively (Table S4a). Highly significant correlations were found between all predictive performance statistics, except for Kappa (Table S4b). In general, the models tended to over-predict the current distribution of these species, but the areas identified as suitable were similar to the actual recorded distribution.

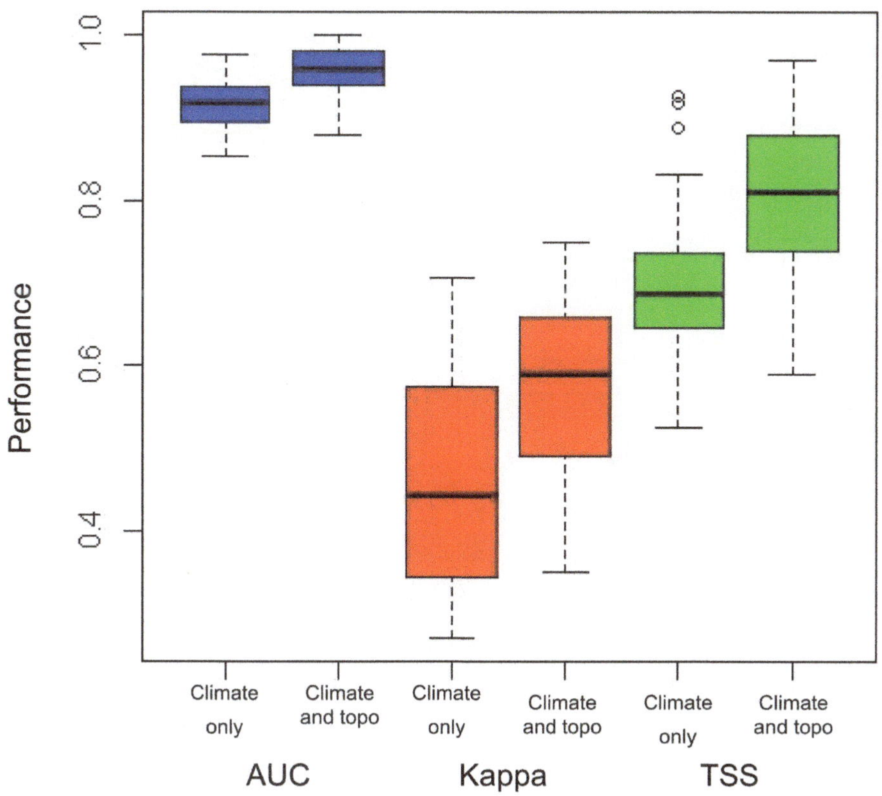

Figure 2. Box plots of the predictive performance of species distribution models for 30 montane species, comparing models with only climate variables to those containing both climate and ecologically relevant topographic variables. The methods of evaluation used were AUC, Kappa and TSS. The difference in performance between models created using climate variables only and climate and topographic variables was significant in all cases ($p<0.001$).

Projected species range changes

A variety of range change responses (i.e. changes in areas of suitable climate) were shown for the modelled species under both unlimited and limited dispersal scenarios (Table S5). Comparing bryophyte and vascular plant species under a scenario of unlimited dispersal, they show a significant difference ($p<0.001$) in their range changes (Figure 3A; means for all categories shown in Figure 3 are listed in Table S6). Bryophytes are projected to show a mean increase in suitable climate space of +12.91% (±4.47), whereas vascular species show a mean decrease of −13.64% (±5.42). Under a limited dispersal scenario (species unable to move to new areas), both bryophyte and vascular plant species decrease in range; bryophytes are projected to show a mean decrease of −12.96% (±2.0), not significantly different from the potential decrease (−23.92%±5.20) of vascular plant species (Figure 3B).

When grouped by community (Figures 3C and 3D), the difference in range change between groups was also significant under unlimited dispersal ($p<0.001$) but not under conditions of limited dispersal. Only the bryophyte communities (groups 3 and 4), particularly the hepatic mat species (group 4: range change of + 25.0% ±6.41, Table S6), show a potential increase in range under unlimited dispersal, but show a notable decrease in range with limited dispersal. In contrast, the species of exposed ridges and montane cliffs (groups 1 & 2) are projected to decrease in range under both unlimited and limited dispersal scenarios, especially under the latter.

When species were grouped by biome and eastern limit category, the differences between the grouping categories were less significant ($p<0.01$) under an unlimited dispersal scenario and not significant if no dispersal is possible. Under an unlimited dispersal scenario, species within the arctic-montane and boreo-arctic-montane biomes (groups 1 & 2) show a decrease in suitable range (−18.21±7.39 and −13.53±5.37 respectively; Table S6), while the species of the other two biomes show a mean increase (Figure 3E). The range change for the species of the most northerly biomes (1 & 2) is more negative than for the other two groups under a limited dispersal scenario (Figure 3F).

In terms of the eastern limit category, species with circumpolar (group 4; mean change of −13.49%±4.42; Table S6) and easterly (group 3) distributions generally show a mean decrease in range under conditions of unlimited dispersal, while species of suboceanic, oceanic and hyperoceanic distribution show a mean increase in range (Figure 3G); all but one of these are bryophytes. Under conditions of limited dispersal, the range change is broadly similar and negative for all groups (Figure 3H).

A number of potential trends can be identified from the spatially extrapolated model outputs shown as species distribution maps (Figure 4). Oceanic bryophytes (Figure 4A–4C), of hepatic mat and oceanic montane species groups, such as *Anastrepta orcadensis* and *Scapania ornithopodioides*, are projected to lose space to the south and gain space in northerly areas, suggesting a northward shift in distribution of these species (assuming unlimited dispersal). However, species of montane heath and montane cliffs (Figure 4D–4F), many of which are of circumpolar arctic-montane distribution, e.g. *Salix herbacea* and *Sedum rosea*, show a projected general contraction in range to core areas of high altitude, which are centred in the south-west, east and centre-west of Ireland.

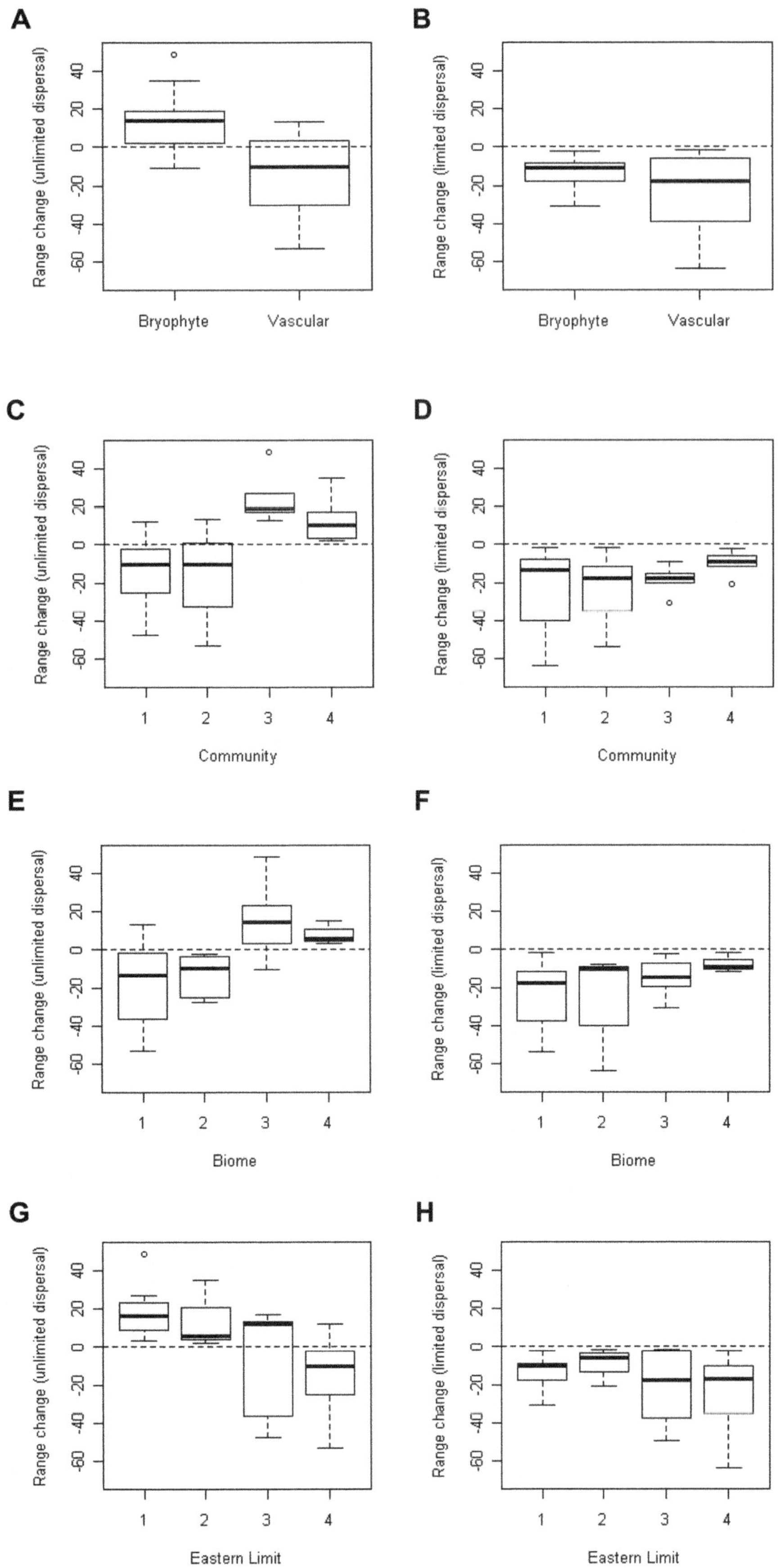
A
Range change (unlimited dispersal)
-60
-40
-20
0
20
40
Bryophyte
Vascular
B
Range change (limited dispersal)
-60
-40
-20
0
20
40
Bryophyte
Vascular
C
Range change (unlimited dispersal)
1
2
3
4
Community
D
Range change (limited dispersal)
1
2
3
4
Community
E
Range change (unlimited dispersal)
1
2
3
4
Biome
F
Range change (limited dispersal)
1
2
3
4
Biome
G
Range change (unlimited dispersal)
1
2
3
4
Eastern Limit
H
Range change (limited dispersal)
1
2
3
4
Eastern Limit

Figure 3. Boxplots of the change in range (%) of species under conditions of unlimited and limited dispersal, grouped by a number of categories. (A)–(B) are grouped by plant type (bryophyte (n = 14) or vascular plant (n = 16)); (C)–(D) are grouped by community (1: montane heath (n = 10), 2: montane cliff (n = 9), 3: hepatic mat (n = 5), 4: oceanic montane bryophyte (n = 6)); (E)–(F) are grouped by biome [53], [54] (1: Arctic-montane (n = 10), 2: Boreo-arctic-montane (n = 5), 3: Boreal montane (n = 12), 4: Temperate and Southern Temperate (n = 3)); (G)–(H) are grouped by eastern limit category [53], [54] (1: hyperoceanic and oceanic (n = 8), 2: suboceanic (n = 3), 3: European, Eurosiberian and Eurasian (n = 5), 4: circumpolar (n = 14)). Range changes calculated using an ensemble of models produced by BIOMOD2.

Overlap with protected areas

The results of the gap analysis indicate that there is projected to be less overlap between the distribution of the majority of species modelled and protected areas in the future (Figure 5; Table S7); in some cases even the current overlap is very low and for some, especially *Vaccinium vitis-idaea*, the future overlap may be extremely low. When grouped by plant community (Figure 6A), the percent overlap with protected areas of species of hepatic mat (92.14%±3.97: current projections; 84.73%±4.53: future projections) and montane cliffs (90.13%±2.19: current projections; 77.53%±2.69: future projections) are significantly ($p<0.01$) greater than for species of montane heath (59.07%±6.18: current projections; 52.44%±5.68: future projections) under both current and future predictions. Only the distribution of species of montane cliffs (mean decrease of −13.86%±2.73) are projected to show a significant ($p<0.01$) decrease in overlap with protected areas between the present and future. When grouped by biome (Figure 6B), species of arctic-montane distribution, which under current projections show a high level of overlap with protected areas (89.11%±2.62), are projected to significantly ($p<0.01$) decrease overlap with protected areas under future climate change scenarios (mean decrease of −12.61%±2.78). Species of Temperate and Southern Temperate affinity show the least projected change in overlap with protected areas.

Discussion

The results of our study indicate that the montane oceanic vegetation of western Europe is likely to face a number of threats to their future conservation in response to climate change. Arctic-montane species, in particular, are projected to see areas of suitable climate either shift northwards or contract to higher elevations, while internationally important bryophyte species and communities are also under threat. As the effectiveness of the current protected area network is projected to be reduced for these species, climate change may pose a major challenge to the effective conservation of montane oceanic species and plant communities in the future.

Model performance

This study demonstrates the need to include ecologically relevant topographic variables when projecting species distributions in relation to climate change on a regional scale, especially for species strongly affected by topography and the resultant microclimate. Species distributions are governed by a range of biotic and abiotic factors at different spatial scales [71]. At large continental scales, species distributions are shaped by macroclimate, whereas at smaller local or regional scales, factors such as topography modify this to produce a complex pattern of species distribution and diversity [72]. Species were modelled here at a regional scale and the inclusion of ecologically relevant topographic variables along with climatic variables improved model performance for nearly all species and all modelling techniques. This is important particularly when predicting the distribution of montane plant species [65], while the addition of further autecological information would facilitate the determination of appropriate conservation measures for montane species and habitats [73].

Our study also indicated that the extent of the distribution of a species and the degree to which it is restricted to areas with specific microclimatic conditions influence how well the species model performs when climate variables only are used. Therefore, topographically-dependent oceanic bryophytes, especially hepatic mat species that have a very narrow, often disjunct distribution [33], perform much better when topography is included in the model. Similarly, many arctic-montane species do not model well with only climatic variables, as they are strongly dependent on altitude and other topographic factors to create the required conditions for their growth. The species that exhibit the highest performance statistics when only climate variables are used are generally widely distributed, and also occur outside of montane areas. Therefore, they are generally not restricted by altitude, aspect or other topographic variables and their distribution is more strongly controlled by climatic factors.

The data-splitting cross-validation approach used in our study relies on use of the same data-set, though it has been suggested that this may give an overly optimistic model performance assessment [74]. However, our models were shown to be very good at correctly predicting true presences and true absences, and in the absence of directly relevant independent data-sets and in the knowledge that no predictive modelling will be 100% accurate, this methodology still gives useful results for interpretation [75].

We are fully aware that the relatively coarse-scale resolution for both climate and species distribution data may lead to erroneous predictions, especially in montane regions [76], [77]. Some previous studies have shown greater areas of suitable climate/habitat space predicted using coarser resolution data [71], [77], [78], while others show greater persistence of suitable climate/habitat under finer resolution data [76]. It is possible that by modelling at the 10 km×10 km resolution in the current study we are losing the potential to identify local climate refugia for species persistence and rapid migration [77], particularly in areas such as the west coast of Ireland where there is large local scale variations both in elevation and climate. The averaging of climate variables over broader areas (e.g. 10 km×10 km grids), is likely to result in the loss of the fine-scale variability (e.g. broader ranges of temperature and precipitation values) relevant to the species physiological limitations [77], [78]. However, our knowledge of the species and their ecology has enabled us to infer that the overall projections presented here are entirely probable. While finer-resolution climate and environmental data are available, species data at equivalent resolutions will be an on-going challenge for many regions of the world, including Europe and Ireland. These results, despite their limitations and pending finer-scale data, are important to publicise, since one purpose of this paper is to influence policy-makers in advance of any major range changes occurring.

Range Changes

This study shows how bryophyte and vascular plant species may respond differently to climate change and oceanic bryophytes may retain a higher proportion of suitable climate space than montane

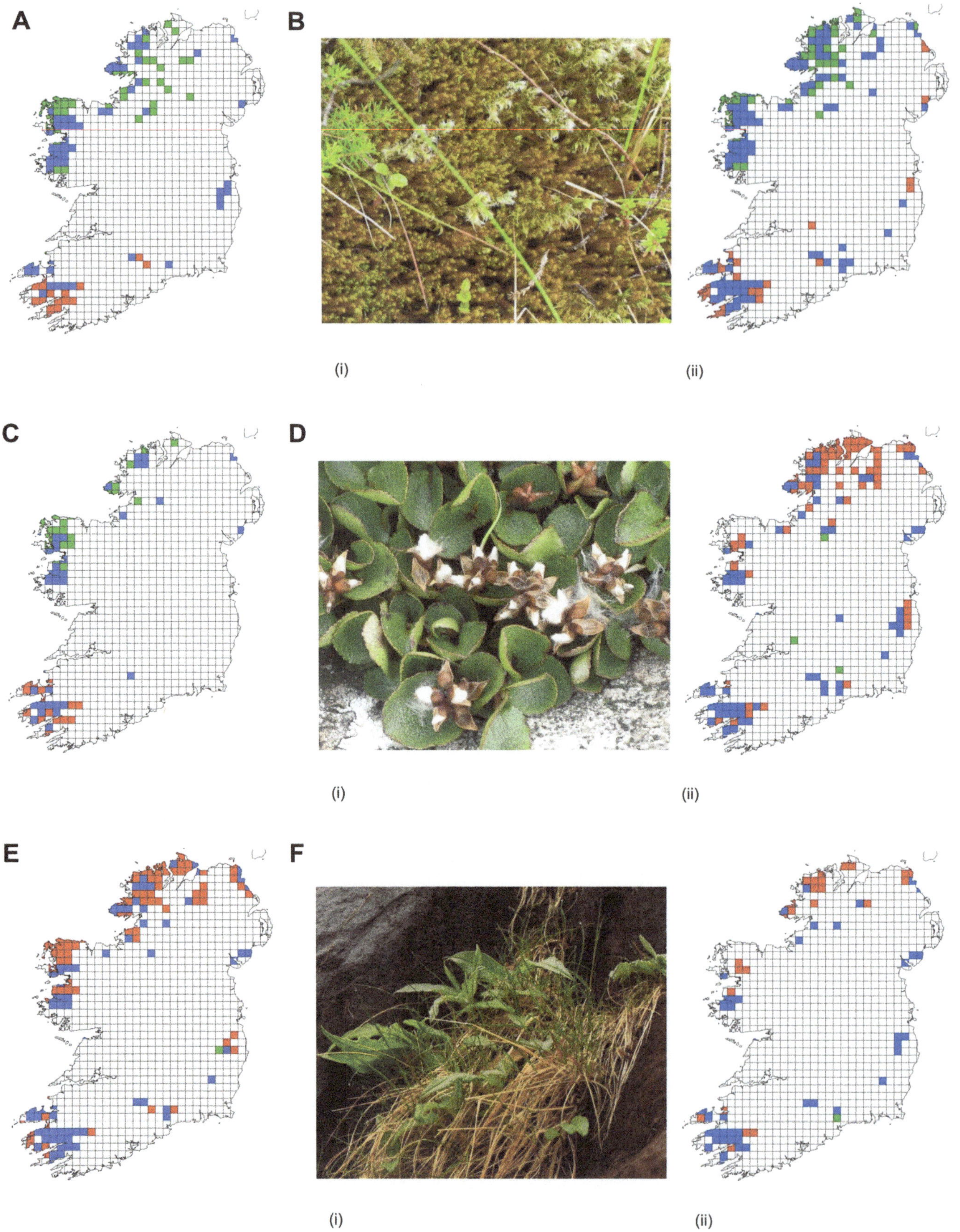
A
B
(i)
(ii)
C
D
(i)
(ii)
E
F
(i)
(ii)

Figure 4. Species distribution maps, showing the projected change in spatial distribution of 6 montane species under predicted climate change scenarios for 2055, representative of the two primary patterns of range change displayed by the modelled species. Species (A) *Anastrepta orcadensis*, (B) *Herbertus aduncus* subsp. *hutchinsiae* ((i) photograph of *H. aduncus* subsp. *hutchinsiae* from Co. Donegal, Ireland; (ii) distribution map) and (C) *Scapania ornithopodioides* are oceanic bryophytes, and show a potential northward shift in range. Species (D) *Salix herbacea* ((i) photograph of *S. herbacea* from Co. Donegal, Ireland; (ii) distribution map), (E) *Sedum rosea* and (F) *Saussurea alpina* ((i) photograph of *S. alpina* from Co. Kerry, Ireland; (ii) distribution map) are of arctic montane distribution, and will potentially contract in range to grid cells of higher altitude. Green grid cells = Gain; Blue grid cells = Stable; Red grid cells = Loss.

vascular plant species. However, the majority of the oceanic bryophytes are projected to lose a large proportion of their climatically suitable area in the south of their range and gain this in the north, so would only be able to migrate successfully if there were no barriers to their dispersal. It is also possible that some species may adapt *in situ* and therefore may not need to migrate for the species to survive [12], [14]. In contrast, many of the vascular plants of montane heaths and cliff ledges are projected to lose space in areas of lower altitude and become restricted to the areas of highest altitude, with little or no gain of potential climate space. Most species that showed this contraction in range are of arctic-montane distribution, while species of temperate and southern distribution generally showed a projected increase, or minor decrease, in suitable space. Both of these patterns are likely to be in response to temperature changes.

These patterns are also likely to be influenced by rainfall changes, particularly for the widespread oceanic and hepatic mat bryophytes. These results suggest that the degree of oceanicity of a species' distribution is likely to influence its response to climate change. Species of hyperoceanic, oceanic and suboceanic distribution mainly show a pattern of northward shift in range. The specific cause of these shifts in range is not clear.

When considering species' potential range changes, it is essential to take dispersal capacity into account [79]. Neither dispersal scenario used in this study is likely to reflect the species' true dispersal capacity [80], as they are unlikely to be able to colonise every area that may become suitable for their growth in the future, but may shift their range to some extent in response to climate changes. The limited dispersal scenario is likely to be closer to their true response, as most species studied have poor dispersal capacity, are of disjunct distribution and are limited to relatively isolated mountain ranges, creating a further barrier to their dispersal. However, little is known about their true dispersal capacity, particularly of the bryophytes. Liverworts of the hepatic mat community are not known to produce sporophytes in Ireland, spreading only vegetatively, primarily by fragmentation [35].

Biotic interactions are likely to have a major influence on future composition of montane vegetation in the face of climate change. Temperature rises are likely to lead to increased competition in montane habitats [7], [20], with less cold-tolerant species

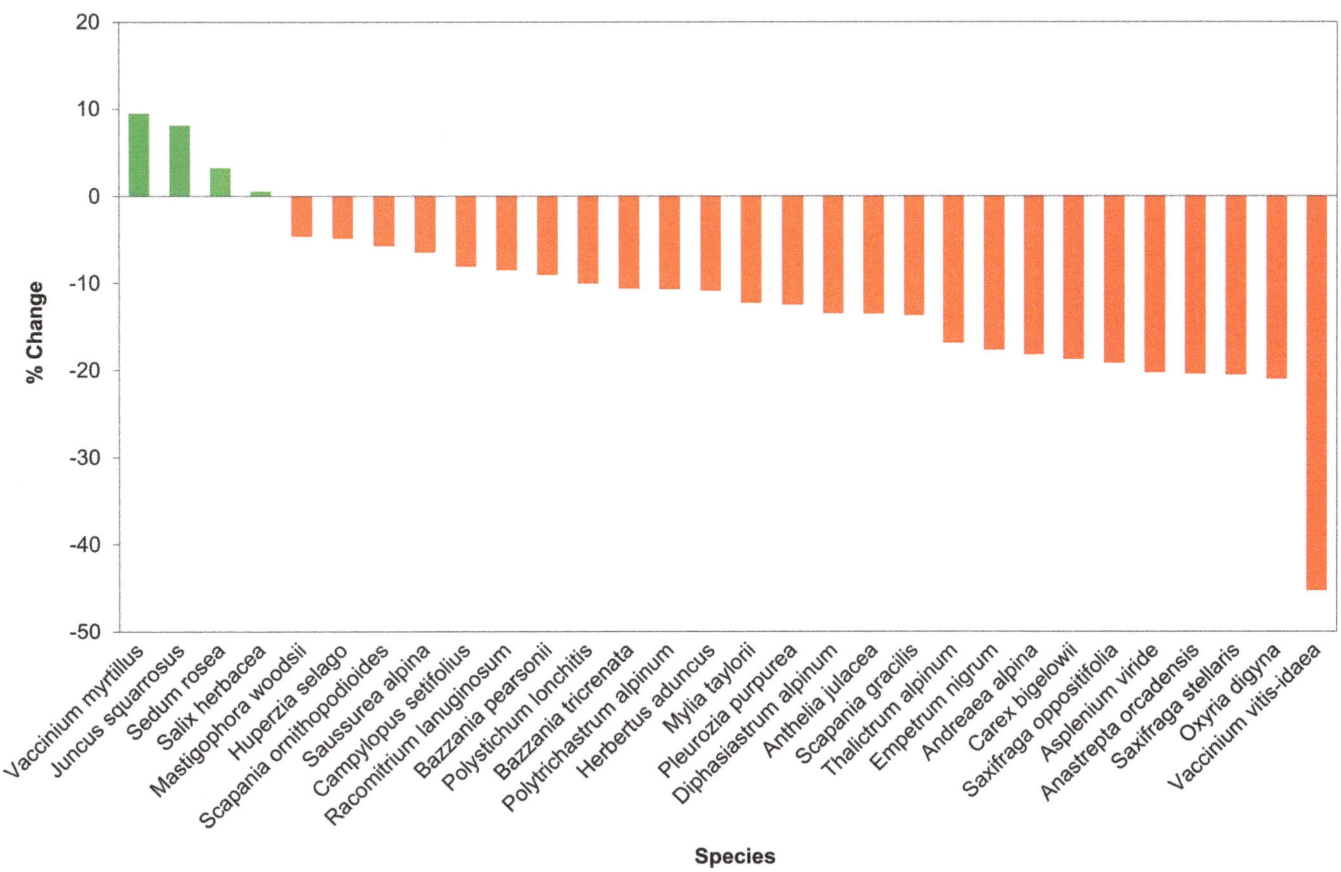

Figure 5. Projected percentage change in overlap with protected areas from current to future distribution, for all species modelled.

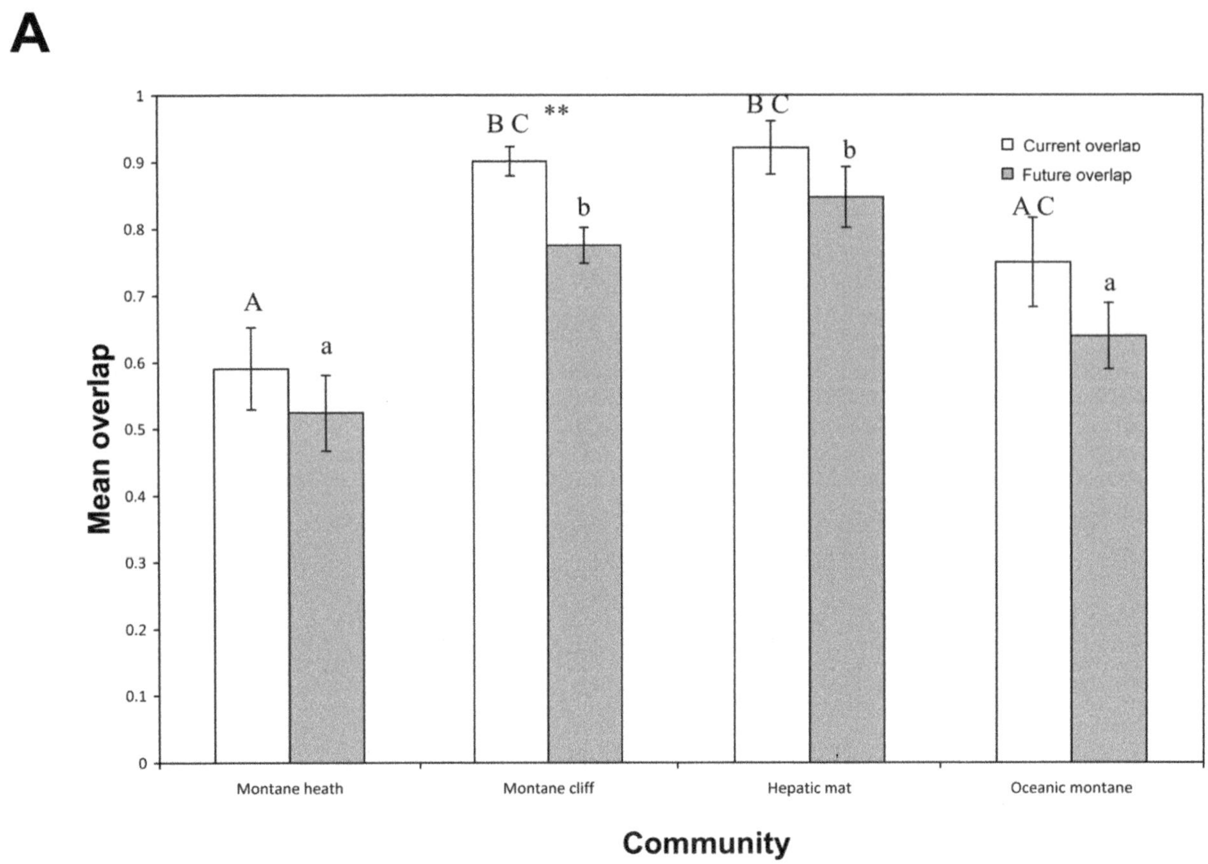
A
B C
**
B C
b
b
A C
A
a
a
Current overlap
Future overlap
Mean overlap
1
0.9
0.8
0.7
0.6
0.5
0.4
0.3
0.2
0.1
0
Montane heath
Montane cliff
Hepatic mat
Oceanic montane
Community

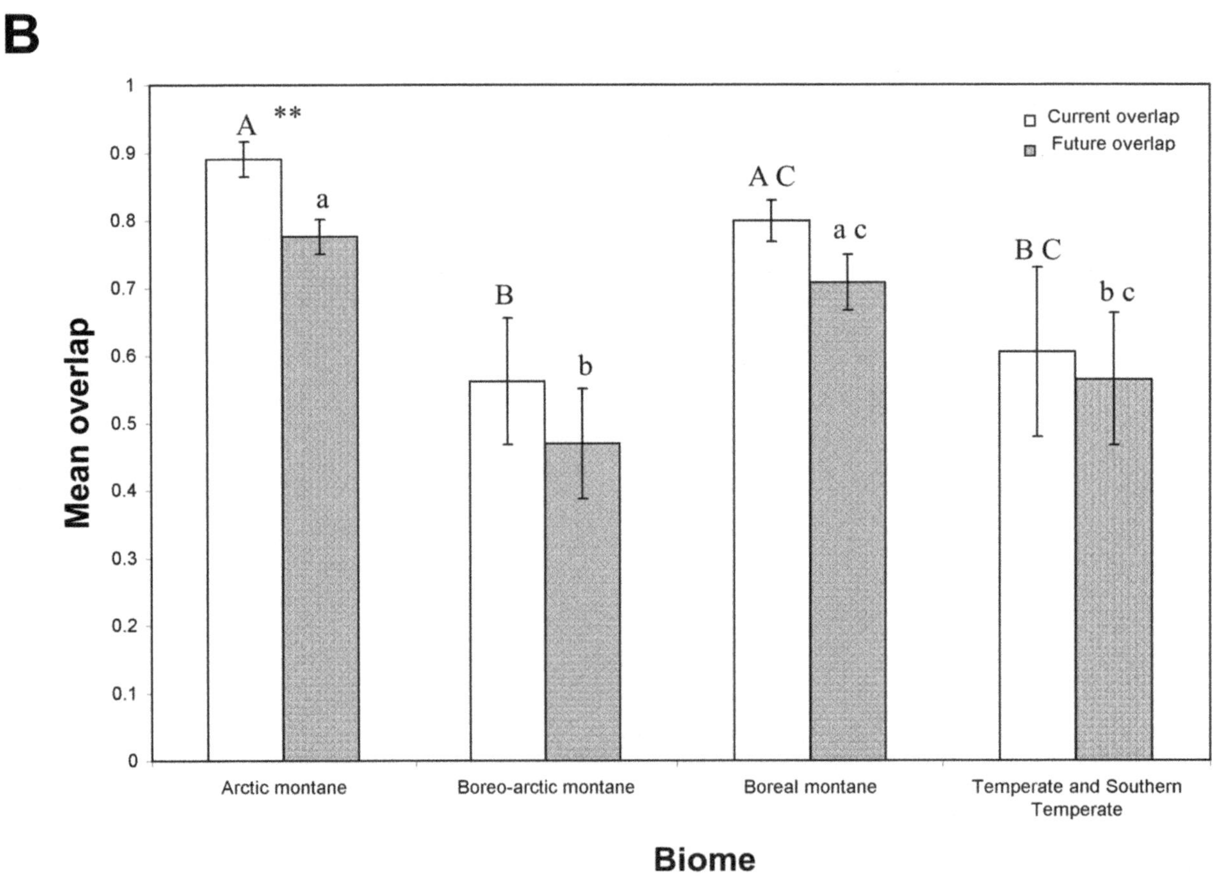
B
A
**
a
A C
a c
B C
b c
B
b
Current overlap
Future overlap
Mean overlap
1
0.9
0.8
0.7
0.6
0.5
0.4
0.3
0.2
0.1
0
Arctic montane
Boreo-arctic montane
Boreal montane
Temperate and Southern Temperate
Biome

Figure 6. Proportion of overlap between current and future distribution of species groups and protected areas. (A) Grouped by plant community; a different letter signifies a significant difference between categories (Wilcoxon ranked sum; $p<0.05$; capitals for present, lower case for future scenarios), categories that show a significant change in overlap between current and future (Wilcoxon signed rank; $p<0.01$) are denoted by **. (B) Grouped by biome; letters signify significant differences between categories (Wilcoxon ranked sum; $p<0.05$), categories that show a significant change in overlap between current and future (Wilcoxon signed rank; $p<0.01$) are denoted by **.

expanding their niche upward in altitude and others, already present in these habitats, growing more efficiently to outcompete specialised montane species [24]. This indirect effect of temperature rises may have a greater negative impact on the range of montane species than other consequences of climate change [81], and may lead to a homogenisation of montane vegetation, as has been detected in Scotland [17]. However, competition may not be as great in exposed montane heath, as exposure and wind strength may limit the growth of many species, negating the impact of rising temperatures, and enabling the survival of adapted arctic-montane species [82]. In contrast, oceanic bryophytes of sheltered habitats, particularly those of the mixed northern hepatic mat, may be more threatened by competition. A combination of decreased rainfall at key periods of the year and temperature rises will likely favour the growth of large acrocarpous and pleurocarpous mosses and vascular plant species, to the possible exclusion of hepatic mat liverworts [35].

Despite the fact that climate change might be too rapid for vulnerable species to migrate to new regions of suitable climate space [83] and may cause a loss in genetic diversity in northern species [84], plants living in marginal areas, such as mountain-tops, have an inherent genetic variability and adaptability at the metapopulation level that may impose a greater degree of resilience on the component species than in more stable habitats [85]. Indeed, arctic-alpine vascular plants can be extremely long-lived [86] and, with some recruitment and genetic variability, species have survived many climatic oscillations [87]. Little is yet known of bryophyte genetic variability [88]. However, when addressing conservation aims in the face of climate change, all potential range changes must be considered and the extension of protected areas examined, as a basic safety-net method to prevent possible regional species extinctions.

Conservation of montane vegetation in response to climate change

Hyperoceanic plant communities are highly restricted in Europe to the western fringes of Ireland, Britain, Norway and the Faroes; those communities dominated by bryophytes exhibit an even more restricted distribution [24], [89]. However, there is little specific protection of these bryophyte-dominated communities at a European level. As Ireland is at the south-western limit of the range of occurrence of many of these communities and their constituent species, climate change poses a greater threat to their future survival than in other parts of their range. Therefore, Ireland has a major role to play in conserving the European populations of these species. If their range were to move northwards, it would also increase the role of Scotland, Norway and the Faroe Islands in the conservation of this vegetation (*sensu* [90]).

These results suggest there will be a reduction in area of all true montane habitats in Ireland, as defined by their characteristic species. Many are listed under Annex 1 of the EU Habitats Directive [91], including Alpine and boreal heaths (4060), siliceous Alpine and boreal grassland (6150), as well as significant elements of siliceous rocky slope vegetation (8220), calcareous rocky slope vegetation (8210) and hydrophilous tall herb vegetation (6430). Other Annex habitats, currently integral parts of montane vegetation in European oceanic areas, including wet heath (4010), dry heath (4030) and active blanket bog (*7130), may also come under considerable threat with climate change.

A number of the modelled species, as well as others of these habitats that could not be modelled, as they occur in too few grid cells, are listed in the Irish Red Data Books [40], [92], [93] and are already highly threatened in Ireland. It is likely that the further threat of climate change may preclude the future survival of these species.

The results of the gap analysis indicate that the current protected area network in Ireland may not adequately conserve many montane species under future climate change, particularly arctic-montane and boreo-arctic montane species, which are likely to be the most vulnerable species to climate change. Although the modelled oceanic bryophyte species are also projected to lose climate space in protected areas, a large proportion of their range still overlaps with protected areas, in contrast to species of montane heath. This is likely to be due to the fact that the distribution of the oceanic species, by definition, is restricted to western, coastal areas, which is where Irish protected areas are concentrated [34]. These results can inform future strategies for managing and designating protected areas in montane oceanic areas.

Conclusions

The marginal position of European oceanic regions results in their indigenous species and plant communities being of high conservation value, with a richness and diversity of oceanic bryophytes not seen elsewhere in Europe. The threat to their conservation from climate change is particularly high. The relatively small extent and low altitude of Ireland's mountain areas means that the constituent species of these communities have very little chance of shifting their range to areas of suitable climate in response to climate change. The challenge facing policy makers and conservation organisations is to recognise the uniqueness and value of oceanic vegetation and ensure the highest conservation status of this vegetation in response to the effects of climate change and other threats. By incorporating the results of studies of species distributional changes under future climate change, the robustness and relevance of protected areas and conservation strategies can be ensured into the future.

Supporting Information

Table S1 List of species modelled and their main attributes (threat status for bryophytes after [40]**; threat status for vascular species after** [93]**); biome, eastern limit category and altitude range after** [53] **and** [54]**).**

Table S2 Presence/absence data for all species for all Irish 10 km x 10 km grid cells. Data obtained from National Biodiversity Network Gateway [57] and National Parks and Wildlife Service, Dublin [58].

Table S3 Threshold values for classifying model predictive accuracy [95], [96]**, where AUC is Area Under the**

Curve and Kappa/TSS is Cohen's Kappa statistic (K) and True Skill Statistic (TSS) respectively.

Table S4 A: AUC, Kappa and TSS values for all species using an ensemble modelling approach in BIOMOD2, from models created using both climate and topography variables and climate variables only. Model predictive accuracy is acceptable when AUC>0.7 and Kappa/TSS>0.4 (Table S3). Sensitivity and specificity presented only for models created using both climatic and topographic variables. B: Correlation coefficients between the five model predictive performance statistics: Cohen's Kappa statistic (K), True Skill Statistic (TSS), Area Under the Curve (AUC) of the Receiver Operating Characteristic, Sensitivity and Specificity scores.

Table S5 Range changes (%), derived from ensemble of models produced in BIOMOD2, for all species under conditions of unlimited and limited dispersal.

Table S6 Mean range changes (%) ± standard error of the mean, under conditions of unlimited and limited dispersal, under the categories of plant type, community, biome and eastern limit.

Table S7 Outputs from gap analysis, showing the proportional overlap of both current and future distribution of all species with protected areas, using a range of threshold values (≥2%, ≥5%, ≥10%, ≥20%, ≥30%, ≥50% of grid cell occupied by protected areas).

Acknowledgments

We would like to thank Prof John Sweeney at the Irish Climate Analysis and Research Unit (ICARUS), at NUI Maynooth for providing climate and climate change data. Thanks also to Dr Neil Lockhart (National Parks and Wildlife Service, Dublin) for providing data on bryophytes. We also wish to thank Dr Wilfried Thuiller, Laure Gallien and Damien Georges at Laboratoire d'Écologie Alpine, CNRS and Université Joseph Fourier, Grenoble, France for their advice on the application of BIOMOD to undertake the climate change and biodiversity modelling. Thanks also to Laura Meller of the University of Helsinki for her advice on gap analysis. We are also grateful to two anonymous referees for very helpful comments and suggestions on the draft paper.

Author Contributions

Conceived and designed the experiments: DB RH MSS. Performed the experiments: DB RH. Analyzed the data: DB RH MSS. Contributed reagents/materials/analysis tools: DB RH MSS. Wrote the paper: RH DB MSS.

References

1. Diaz HF, Grosjean M, Graumlich L (2003) Climate variability and change in high elevation regions: past, present and future. Climate Change 59: 1–4.
2. Burt TP, Holden J (2010) Changing temperature and rainfall gradients in the uplands. Climate Research 45: 57–70.
3. Coll J, Gibb SW, Price MF, McClatchey J, Harrison J (2010) Developing site scale projections of climate change in the Scottish Highlands. Climate Research 45: 71–85.
4. Parmesan C, Yohe G (2003) A globally coherent fingerprint of climate change impacts across natural systems. Nature 421: 37–42.
5. Walther G-R, Post E, Convey P, Menzel A, Parmesan C, et al. (2002) Ecological responses to recent climate change. Nature 416: 389–395.
6. Grabherr G, Gottfried M, Pauli H (1994) Climate effects on mountain plants. Nature 369: 448.
7. Parolo G, Rossi G (2008) Upward migration of vascular plants following a climate warming trend in the Alps. Basic and Applied Ecology 9: 100–107.
8. Araújo MB, Thuiller W, Pearson RG (2006) Climate warming and the decline of amphibians and reptiles in Europe. Journal of Biogeography 33: 1712–1728.
9. Lenoir J, Gégout JC, Marquet PA, de Ruffray P, Brisse H (2008) A significant upward elevation shift in plant species optimum elevation during the 20th Century. Science 320: 1768–1770.
10. Pearson RG, Dawson TE (2003) Predicting the impacts of climate change on the distribution of species: are bioclimatic envelope models useful? Global Ecology and Biogeography 12: 361–371.
11. Berry PM, Harrison PA, Dawson TP, Walmsley CA (Eds.) (2005) Modelling Natural Resource Responses to Climate Change (MONARCH): a local approach. UKCIP technical report, Oxford, UK.
12. Geml J, Kauff F, Brochmann C, Taylor DL (2010) Surviving climate changes: high genetic diversity and transoceanic gene flow in two arctic-alpine lichens, *Flavocetraria cucullata* and *F. nivalis* (Parmeliaceae, Ascomycota). Journal of Biogeography 37: 1529–1542.
13. Alsos IG, Eidesen PB, Ehrich D, Skrede I, Westergaard K, et al. (2007) Frequent long-distance plant colonization in the changing Arctic. Science 316: 1606–1609.
14. Abbot RJ, Chapman HM, Crawford RMM, Forbes DG (1995) Molecular diversity and derivation of populations of *Silene acaulis* and *Saxifraga oppositifolia* from the high Arctic and more southerly latitudes. Molecular Ecology 4: 199–208.
15. Engler R, Randin CF, Thuiller W, Dullinger S, Zimmerman NE et al. (2011) 21st century climate change threatens mountain flora unequally across Europe. Global Change Biology 17: 2330–2341.
16. Thuiller W, Lavorel S, Araújo MB, Sykes MT, Prentice IC (2005) Climate change threats to plant diversity in Europe. Proceedings of the National Academy of Sciences 102: 8245–8250.
17. Britton AJ, Beale CM, Towers W, Hewison RL (2009) Biodiversity gains and losses: evidence for homogenisation of Scottish alpine vegetation. Biological conservation 142: 1728–1739.
18. Guisan A, Theurillat J-P (2000) Assessing alpine plant vulnerability to climate change: a modelling perspective. Integrated Assessment 1: 307–320.
19. Klanderud K, Birks HJB (2003) Recent increases in species richness and shifts in altitudinal distributions of Norwegian mountain plants. The Holocene 13: 1–6.
20. Pauli H, Gottfried M, Reiter K, Klettner K, Grabherr G (2007) Signals of range expansions and contractions of vascular plants in the high Alps: observations (1994-2004) at the GLORIA master site Schrankogel, Tyrol, Austria. Global Change Biology 13: 147–156.
21. Trivedi MR, Morecroft MD, Berry PM, Dawson TP (2008) Potential effects of climate change on plant communities in 3 montane nature reserves in Scotland, UK. Biological Conservation 141: 1665–1675.
22. Körner C (2003) Alpine Plant Life: functional plant ecology of high mountain ecosystems, second ed. Springer, Berlin, Germany.
23. Bassler C, Müller J, Hothorn T, Kneib T, Badeck F, et al. (2009) Estimation of the extinction risk for high-montane species as a consequence of global warming and assessment of their suitability as cross-taxon indicators. Ecological Indicators 10: 341–352.
24. Crawford RMM (2000) Tansley Review no. 114. Ecological hazards of oceanic environments. New Phytologist 147: 257–281.
25. Crawford RMM (2001) Plant community responses to Scotland's changing environment. Botanical Journal of Scotland 53: 77–105.
26. Fosaa AM, Sykes MT, Lawesson JE, Gaard M (2004) Potential effects of climate change on plant species in the Faroe Islands. Global Ecology and Biogeography 13: 427–437.
27. Gignac LD (2001) Bryophytes as indicators of climate change. The Bryologist 104: 410–420.
28. Gignac LD (2011) Bryophytes as predictors of climate change, in: Tuba Z, Slack NG and Stark LR (Eds.), Bryophyte Ecology and Climate Change. Cambridge University Press, Cambridge, UK, pp. 461–482.
29. Frahm J-P, Klaus D (2001) Bryophytes as indicators of recent climate fluctuations in Central Europe. Lindbergia 26: 97–104.
30. Bergamini A, Ungricht S, Hofmann H (2009) An elevational shift of cryophilous bryophytes in the last century – an effect of climate warming? Diversity and Distribution 15: 871–879.
31. Pócs T (2011) Signs of climate change in the bryoflora of Hungary, in: Tuba Z, Slack NG, Stark LR (Eds.), Bryophyte Ecology and Climate Change. Cambridge University Press, Cambridge, UK, pp. 359–370.
32. Bates JW, Preston CD (2011) Can the effects of climate change on British bryophytes be distinguished from those resulting from other environmental changes?, in: Tuba Z, Slack NG, Stark LR (Eds.), Bryophyte Ecology and Climate Change. Cambridge University Press, Cambridge, UK, pp. 371–407.
33. Ratcliffe DA (1968) An ecological account of Atlantic bryophytes in the British Isles. New Phytologist 67: 365–439.
34. Hodd RL, Sheehy Skeffington MJ (2011) Climate change and oceanic montane vegetation: a case study of the montane heath and associated plant communities in western Irish mountains, in: Hodkinson T, Jones M, Waldren S, Parnell J

(Eds.), Climate Change, Ecology and Systematics. Cambridge University Press, Cambridge, UK, pp. 490–515.
35. Hodd RL, Sheehy Skeffington MJ (2011) Mixed northern hepatic mat: a threatened and unique bryophyte community. Field Bryology 104: 2–11.
36. Frahm J-P (2008) Diversity, dispersal and biogeography of bryophytes (mosses). Biodiversity and Conservation 17: 277–284.
37. Rothero G (2003) Bryophyte conservation in Scotland. Botanical Journal of Scotland 55: 17–26.
38. Holden J, Shotbolt L, Bonn A, Burt TP, Chapman PJ, et al. (2007) Environmental change in moorland landscapes. Earth Science Reviews 82: 75–100.
39. Lockhart ND, Hodgetts NG, Holyoak DT (2012) Rare and Threatened Bryophytes of Ireland. National Museum of Northern Ireland, Hollywood, Co. Down, UK.
40. Lockhart ND, Hodgetts NG, Holyoak DT (2012) Ireland Red List No. 8: Bryophytes. National Park and Wildlife Service, Department of Arts, Heritage and the Gaeltacht, Dublin, Ireland.
41. Hannah L, Midgley GF, Millar D (2002) Climate change-integrated conservation strategies. Global Ecology and Biogeography 11: 485–495.
42. Araújo MB, Alagador D, Cabeza M, Nogués-Bravo D, Thuiller W (2011) Climate change threatens European conservation areas. Ecology Letters 14: 484–492.
43. Carroll C, Dunk JR, Moilanen A (2010) Optimizing resilience of reserve networks to climate change: multispecies conservation planning in the Pacific Northwest, USA. Global Change Biology 16: 891–904.
44. D'Amen MD, Bombi P, Pearman PB, Schmatz DR, Zimmermann NE, et al. (2011) Will climate change reduce the efficacy of protected areas for amphibian conservation in Italy. Biological Conservation 144: 989–997.
45. Faleiro FV, Machado RB, Loyola RD (2013) Defining spatial conservation priorities in the face of land-use and climate change. Biological Conservation 158: 248–257.
46. Kujala H, Moilanen A, Araújo MB, Cabeza M (2013) Conservation planning with uncertain climate change projections. PLOS ONE 8: 1–12.
47. Rohan PK (1986) The Climate of Ireland. The Stationery Office, Dublin.
48. Met Éireann (2011) The Irish Meterological Service Online. Available: www.met.ie Accessed 2011 Jan.
49. McGrath R, Lynch P (Eds.) (2008) Ireland in a Warmer World: scientific predictions of the Irish climate in the 21st Century. Community Climate Change Consortium for Ireland (C4I), Dublin, Ireland.
50. Sweeney J, Fealy R (2002) A preliminary investigation of future climate scenarios for Ireland. Proceedings of the Royal Irish Academy 102B: 121–128.
51. Hodd RL (2012) A study of the ecology of the oceanic montane vegetation of western Ireland and its potential response to climate change. Unpublished PhD thesis, NUI Galway, Ireland.
52. Preston CD, Pearman DA, Dines TD (Eds.) (2002) New Atlas of the British and Irish Flora. Oxford University Press, Oxford, UK.
53. Hill MO, Preston CD, Roy DB (2004) PLANTATT: Attributes of British and Irish plants: status, size, life history, geography and habitats. Centre for Ecology and Hydrology, Cambridgeshire, UK.
54. Hill MO, Preston CD, Bosanquet SDS, Roy DB (2007) BRYOATT: Attributes of British and Irish mosses, liverworts and hornworts. Centre for Ecology and Hydrology, Cambridgeshire, UK.
55. Smith AJE (2004) The Moss Flora of Britain and Ireland, second ed. Cambridge University Press, Cambridge, UK.
56. Paton JA (1999) The Liverwort Flora of Britain and Ireland. Harley Books, Colchester, UK.
57. (2008) NBN Gateway. Available: http://data.nbn.org.uk Accessed 2008 Dec.
58. NPWS (2010) Draft Database for the Red List for Bryophytes in Ireland. National Parks and Wildlife Service, Dublin, Ireland.
59. Sweeney J, Fealy R (2003) Establishing reference climate scenarios, in: Sweeney J. (Ed.), Climate Change Scenarios and Impacts for Ireland. Environmental Protection Agency, Dublin. ERTRI Report 15.
60. U.S. Geological Survey (2008) GTOPO30 Global Digital Elevation Mod. U.S. Geological Survey's EROS Data Center, Sioux Falls, South Dakota, USA. Available: http://www.temis.nl/data/gtopo30.html Accessed 2008 Sep.
61. Beyer HL (2004) Hawth's Analysis Tools for ArcGIS. (http://www.spatialecology.com /htools) (accessed January 2009).
62. Zuur AF, Ieno EN, Smith GM (2007) Analysing Ecological Data. Springer, New York, USA. Series: Statistics for Biology and Health.
63. Thuiller W, Lafourcade B, Engler R, Araújo MB (2009) BIOMOD – A platform for ensemble forecasting of species distributions. Ecography 32: 369–373.
64. Thuiller W, Georges D, Engler R (2012) Biomod2: ensemble platform for species distribution modelling. Available: http://cran.r-project.org/web/packages/biomod2/index.html.
65. Virkkala R, Marmion M, Heikkinen RK, Thuiller W, Luoto M (2010) Predicting range change shifts of northern bird species: influence of modelling techniques and topography. Acta Oecologica 36: 269–281.
66. R Development Core Team (2010) R, A language and environment for statistical computing. R Foundation for Statistical Computing, Vienna, Austria. Available: http://www.R-project.org/.
67. Alagador D, Martins MJ, Cerdeira JO, Cabeza M, Araújo MB (2011). A probabilistic-based approach to match species with reserves when data are at different resolutions. Biological Conservation 144: 811–820.
68. Scott JM, Csuti B, Caicco S (1991) Gap Analysis: assessing protection needs, in: Hudson, WE (Ed.), Landscape linkages and biodiversity. Island Press, Washington, DC, USA, pp. 15–26.
69. NPWS (2013) The Status of EU Protected Habitats and Species in Ireland. Habitat Assessments Volume 2. Version 1.0. Unpublished Report, National Parks & Wildlife Service, Department of Arts, Heritage and the Gaeltacht, Dublin, Ireland.
70. (2013) NPWS Available: http://www.npws.ie/protectedsites/specialareasofconservationsac/
71. Trivedi MR, Berry PM, Morecroft MD, Dawson TP (2008) Spatial scale affects bioclimatic model projections of climate change impacts on mountain plants. Global Change Biology 14: 1089–1103.
72. Bruun HH, Moen J, Virtanen R, Grytnes J-A, Oksanen L, et al. (2006) Effects of altitude and topography on species richness of vascular plants, bryophytes and lichens in alpine communities. Journal of Vegetation Science 17: 37–46.
73. Ellis NE, Good JEG (2005) Climate change and effects on Scottish and Welsh montane ecosystems: a conservation perspective, in: Thompson DBA, Price MF Galbraith CA (Eds.), Mountains of northern Europe: conservation, management, people and nature. The Stationery Office, Edinburgh, UK, pp.99–102.
74. Roberts DR, Hamann A (2012) Method selection for species distribution modelling: are temporally or spatially independent evaluation necessary? Ecography 35: 792–802.
75. Araújo MB, Pearson RG, Thuiller W, Erhard M (2005) Validation of species-climate impact models under climate change. Global Change Biology 11: 1504–1513.
76. Randin CF, Engler R, Normand S, Zappa M, Zimmermann NE, et al. (2009) Climate change and plant distribution: local models predict high-elevation persistence. Global Change Biology 15: 1557–1569.
77. Franklin J, Davis FW, Ikegami M, Syphard AD, Flint LE, et al. (2013) Modeling plant species distributions under future climates: how fine scale do climate projections need to be? Global Change Biology 19: 473–483.
78. Seo C, Thorne JH, Hannah L, Thuiller W (2009) Scale effect in species distribution models: implications for conservation planning under climate change. Biology Letters 5: 39–43.
79. Brooker RW, Travis JMJ, Clarke EJ, Dytham C (2007) Modelling species' range shifts in a changing climate: the impacts of biotic interactions, dispersal distance and the rate of climate change. Journal of Theoretical Biology 245: 59–65.
80. Engler R, Randin CF, Vittoz P, Czáka T, Beniston M et al. (2009) Predicting future distributions of mountain plants under climate change: does dispersal capacity matter? Ecography 32: 34–45.
81. Sætersdal M, Birks HJB (1997) A comparative ecological study of Norwegian mountain plants in relation to possible future climate change. Journal of Biogeography 24: 127–152.
82. Crabtree D, Ellis CJ (2010) Species interaction and response to windspeed alter the impact of projected temperature change in a montane ecosystem. Journal of Vegetation Science 21: 744–760.
83. Jump AS, Peñuelas J (2005) Running to stand still: adaptation and the response of plants to rapid climate change. Ecology Letters 8: 1010–1020.
84. Alsos IG, Ehrich D, Thuiller W, Bronken Eidesen P, Tribsch A, et al. (2012) Genetic consequences of climate change for northern plants. Proceedings of the Royal Society B: 279, 2042–205.
85. Crawford RMM (2008) Plants at the Margin –Ecological Limits and Climate Change. Cambridge University Press, Cambridge, UK.
86. de Witte LC, Armbruster GFJ, Gielly L, Taberlet P, Stöcklin J (2012) AFLP markers reveal high clonal diversity and extreme longevity in four key arctic-alpine species. Molecular Ecology 21: 1081–1097.
87. Crawford RMM (2014) Tundra-Taiga Biology. Human, Plant and Animal Survival in the Arctic. Oxford University Press, Oxford, UK.
88. Baczkiewicz A (2012) Genetic diversity of leafy liverwort species (Jungermanniidae, Marchantiophyta) in Poland: regional genetic differentiation of leafy liverworts. Biodiversity research and Conservation 27: 55–76.
89. Averis AM (1994) The Ecology of an Atlantic Liverwort Community. Unpublished PhD thesis, University of Edinburgh, UK.
90. Harrison PA, Berry PM, Butt N, New M (2006) Modelling climate change impacts on species distributions at the European scale: implication for climate change policy. Environmental Science and Policy 9: 116–128.
91. European Communities (1992) Council Directive 92/43/EEC on the conservation of natural habitats and of wild flora and fauna. Official Journal L 206, 22/07/1992.
92. Curtis TGF, McGough HN (1988) The Irish Red Data Book. 1. Vascular Plants. Government Publications, Dublin, Ireland.
93. Kingston N (2012) *Checklist of protected & rare species in Ireland.* Unpublished Report. National Parks & Wildlife Service, Dublin, Ireland.
94. Averis AM, Averis ABG, Birks HJB, Horsfield D, Thompson DBA, et al. (2004) An Illustrated Guide to British Upland Vegetation. Joint Nature Conservation Committee, Peterborough, UK.
95. Swets JA (1988) Measuring the accuracy of diagnostic systems. Science 240: 1285–1293.
96. Allouche O, Tsoar A, Kadmon R (2006) Assessing the accuracy of species distribution models: prevalence, kappa and the true skill statistic (TSS). Journal of Applied Ecology 43: 1223–1232.

Marine Protected Area Networks: Assessing Whether the Whole Is Greater than the Sum of Its Parts

Kirsten Grorud-Colvert[1]*, Joachim Claudet[2,3], Brian N. Tissot[4], Jennifer E. Caselle[5], Mark H. Carr[6], Jon C. Day[7], Alan M. Friedlander[8], Sarah E. Lester[9], Thierry Lison de Loma[10,3], Daniel Malone[11], William J. Walsh[12]

1 Department of Integrative Biology, Oregon State University, Corvallis, Oregon, United States of America, **2** National Center for Scientific Research (CNRS), University of Perpignan, Perpignan cedex, France, **3** Laboratoire d'Excellence 'CORAIL', Perpignan cedex, France, **4** Humboldt State University Marine Laboratory, Trinidad, California, United States of America, **5** Marine Science Institute, University of California Santa Barbara, Santa Barbara, California, United States of America, **6** Department of Ecology and Evolutionary Biology, Long Marine Lab, University of California Santa Cruz, Santa Cruz, California, United States of America, **7** Great Barrier Reef Marine Park Authority, Townsville, Queensland, Australia, **8** Department of Biology, University of Hawai'i, Honolulu, Hawai'i, United States of America, **9** Marine Science Institute, University of California Santa Barbara, Santa Barbara, California, United States of America, **10** Centre de Recherches Insulaires et Observatoire de l'Environnement (CRIOBE), Moorea, French Polynesia, **11** Institute of Marine Sciences, Long Marine Lab, University of California Santa Cruz, Santa Cruz, California, United States of America, **12** Hawai'i Division of Aquatic Resources, Kailua-Kona, Hawai'i, United States of America

Abstract

Anthropogenic impacts are increasingly affecting the world's oceans. Networks of marine protected areas (MPAs) provide an option for increasing the ecological and economic benefits often provided by single MPAs. It is vital to empirically assess the effects of MPA networks and to prioritize the monitoring data necessary to explain those effects. We summarize the types of MPA networks based on their intended management outcomes and illustrate a framework for evaluating whether a connectivity network is providing an outcome greater than the sum of individual MPA effects. We use an analysis of an MPA network in Hawai'i to compare networked MPAs to non-networked MPAs to demonstrate results consistent with a network effect. We assert that planning processes for MPA networks should identify their intended outcomes while also employing coupled field monitoring-simulation modeling approaches, a powerful way to prioritize the most relevant monitoring data for empirically assessing MPA network performance.

Editor: Jesus Gomez-Gardenes, Universidad de Zarazoga, Spain

Funding: Funding was provided by the David and Lucille Packard Foundation (http://www.packard.org/), Gordon and Betty Moore Foundation (http://www.moore.org/), EU FP7 COCONET (http://www.coconet-fp7.eu/), biodivERsA BUFFER (http://www.biodiversa.org/518), Hawai'i Coral Reef Initiative Research Program (HCRI-RP, http://www.cop.noaa.gov/ecosystems/coralreefs/current/hcri-factsheet-cr.aspx) under the National Oceanic and Atmospheric Administration (NOAA), and Coral Reef Monitoring Grants under NOAA's Coral Reef Conservation Program (http://coralreef.noaa.gov/aboutcrcp/workwithus/funding/). The funders had no role in study design, data collection and analysis, decision to publish, or preparation of the manuscript.

Competing Interests: The authors have declared that no competing interests exist.

* Email: grorudck@science.oregonstate.edu

Introduction

Anthropogenic impacts are increasingly modifying our oceans [1], subjecting marine ecosystems to threats ranging from climate change to pollution to overfishing [2,3]. As a result, no-take marine reserves and other types of marine protected areas (MPAs) have been recommended as one tool to conserve marine biodiversity, ecosystem function, and the goods and services provided by healthy ecosystems [4,5,6]. Growing scientific information has shown that no-take marine reserves can provide benefits for adjoining fished areas [7,8,9] and serve as experimental controls for evaluating the impact of extractive activities on marine ecosystems and for distinguishing such effects from a changing global climate [10]. Full protection inside marine reserves has often led to consistent increases in species density, biomass, size, and diversity, with these results spanning diverse regions and reserves of varying sizes and ages (e.g., [11,12,13,14] but see [15]). However, most of these data are from individual marine reserves, or groups of reserves [7,8], which are each compared separately. To date, there is little evidence that MPAs in a given network are performing synergistically.

Properly designed networks of MPAs can theoretically outperform single marine reserves for a variety of ecological, economic, and social management goals. In theory, MPA networks can minimize the potential negative economic, social, and cultural impacts of a single large no-take reserve while producing similar or even greater ecological and economic returns from fishing outside the no-take areas (e.g., [16,17,18]). The International Union for Conservation of Nature's Marine Program defines a network as "a collection of individual marine protected areas (MPAs) or reserves operating co-operatively and synergistically, at various spatial scales and with a range of protection levels that are designed to meet objectives that a single reserve cannot achieve" [19]. This definition is clearly open to interpretation. How can we accurately assess whether an MPA network is fulfilling its specific objectives, and how does this compare to the methods used to assess whether a single MPA is effective?

It can be difficult to identify attainable management goals for MPA networks—and to design a process for evaluating whether they achieve those goals—without a clear understanding of their objectives or potential functions. Different types of networks exist based on varying management needs and goals. Theoretical studies address optimal size and spacing of protected areas to integrate population connectivity into the design of MPA networks (e.g., [17,20,21]), but there is no empirical evidence for the predicted outcomes of either existing networks or those under development. Recently, we proposed an analytical framework for assessing whether ecological effects across an entire network are greater than the sum of the ecological effects that occur within each MPA in the network [22]. In this paper we provide (1) a brief review of this analytical framework for network effects along with the definitions of different network types to clarify what can realistically be expected from each; and 2) to our knowledge, the first analysis that evaluates an MPA network effect, using monitoring data from a MPA network in Hawai'i.

Monitoring Networks Based on Expected Outcomes

MPA networks can vary in their objectives, which should be explicitly considered to evaluate whether management goals are being met [23,24]. However, complex scenarios are not conducive to the development of a single, one-size-fits-all assessment. Appropriate targets can vary based on situation-dependent criteria such as marine habitat distribution, the life-history traits of species targeted by management, and the socio-economic and cultural context in which the network is established. How can we assess the effectiveness of an MPA network at achieving a set of specific objectives? First, we must understand the different types of networks and the potential outcomes and limitations of each type.

We have defined five different types of "networks" that represent different goals and intended outcomes from collections of marine reserves and other MPAs (Table 1), [22]. Briefly, a network of MPAs could be an *ad-hoc* or *regional network*, a grouping of MPAs that are in proximity to each other but were not planned as a synergistic network; a *conservation network*, designed to have strict conservation goals in order to conserve the representative ecological characteristics of an area or ecosystem by protecting replicated sites that encompass habitats or species of interest; a *management network*, which manages and facilitates the economic uses of marine resources at a broader scale than a single-MPA approach would have afforded; a *social network* based on human interactions across groups of people including MPA managers, stakeholders, decision-makers, and scientists who transfer knowledge, share best practices, and build capacity; or a *connectivity network*, a set of multiple marine reserves and other MPAs designed *a priori* to be connected by the dispersal of larvae and/or movement of juveniles and adults, whose general goal is to maximize conservation and/or fisheries benefits from no-take areas. Examples of each of these network types can be found worldwide (Table 1). However, population connectivity is integral for effectively achieving the goals of protecting an adequate portion of a region (regional network), a particular species, group of taxa, or habitat (conservation network), and an assemblage of fished species that are harvested in areas outside the network to benefit local fisheries and the communities they support (management network). Thus, we assert that population connectivity should be a fundamental goal of network design and establishment to meet ecological goals.

A properly designed connectivity network should ensure that it is not merely establishing a disconnected collection of single reserves and other types of MPAs, but that it instead protects a set of sites that allow connections among populations within protected habitats and ecosystems. Population connectivity includes not only larval dispersal, but also movement of juveniles and adults, which can augment the increased benefits provided by MPA networks as long as fishing mortality encountered while moving between MPAs does not negate the benefits provided by individual reserves [25,26]. Key considerations for a connectivity network also include appropriate coverage across a geographical gradient and expected economic outcomes for managed fisheries in the surrounding waters (e.g., [27,28,29]). Below, we summarize an analytical framework for evaluating whether a connectivity network is effectively meeting the goal of increasing the density, production, or fishery yield of targeted species.

An Analytical Framework for Monitoring Networks

The literature includes many papers on the proper design and evaluation of marine reserves and other types of MPAs (e.g., [28,30]) as well as design criteria for incorporating connectivity into network design (e.g., [20,31,32,33]) but there are very few papers that provide guidance on the evaluation of an MPA network (e.g., [34]). Those tasked with monitoring a network of MPAs must know how to measure whether that network is effective in meeting its goals. When focusing on connectivity networks, the key question is whether there is a significant overall network effect that is greater than the sum of the individual MPA effects. We test if the magnitude (or effect size) of a response (e.g., greater biomass of a target species inside MPAs relative to outside the MPAs) at the scale of the network (ES_{network}, i.e., both inside and outside MPAs) is greater than the sum of the magnitudes of change occurring in individual MPAs (ES_{MPA}). In other words, there is a synergistic interaction occurring among protected areas, such that:

$$\text{ES}_{\text{network}} = \sum \text{ES}_{\text{MPA}} + \text{ES}_{\text{interaction}}$$

where $ES_{interaction}$ is the magnitude of the interaction effect for the MPAs within a network, and the hypothesis to be tested is:

$$\text{H} : \text{ES}_{\text{interaction}} > 0$$

The evaluation of this effect would require both consideration of the overall goals of the network, which vary with the type of network [22], and data from a rigorously designed monitoring program at the appropriate scale with suitable controls at both local and regional levels. Based on this mathematical framework, we use data from the islands of Hawai'i and Maui to illustrate one of the first MPA network-wide analyses.

Network Analysis: Hawai'i

We tested whether a planned network of MPAs, established in 1999 on the west coast of the island of Hawai'i (hereafter West Hawai'i), exhibited a management response consistent with a network effect. The nine MPAs in West Hawai'i, combined with eight relatively small preexisting protected areas, collectively protect 35% of the coast from fishing by the aquarium trade [35]. The Hawai'i aquarium trade most heavily exploits the yellow tang (*Zebrasoma flavescens*) [36], a common fish on Hawaiian reefs and an ecologically important herbivore [37]. Although yellow tang can live over 40 years, the prime target size range for aquarium-trade yellow tang corresponds to fish ≤2 years [38]. Reports of illegal fishing in the network are rare; MPAs are close

Table 1. Definitions, goals, and examples for each type of marine protected area (MPA) network.

MPA network type	Definition	General network goals	Example network(s)
Ad-hoc or Regional	An unplanned collection of MPAs in a given area, not established with a cohesive goal	To meet international conservation targets, serve as potential foundation for a planned network	North-western Mediterreanean, Hawai'i, Caribbean
Conservation	A collection of MPAs in a given area aimed at protecting conservation priority sites	To protect replicates of representative ecosystems, critical areas, damaged habitats	Great Barrier Reef, Chile, Australian Commonwealth MPA networks, Florida Keys
Management	A collection of MPAs in a given area established to manage a marine resource and multiple human uses	To protect targeted species, increase reproductive capacity, increase yield, optimize coastal uses while meeting conservation targets, avoid conflicts	West Hawai'i, US West Coast Rockfish Conservation Areas, US Essential Fish Habitat Closures
Social	A collection of MPAs whose managers, practitioners, stakeholders, decision-makers, scientists, and others interact and transfer knowledge	To promote interaction among participants to effectively plan, manage, implement, or monitor area-based management of marine resources and associated uses	Mediterreanean Protected Areas Network (MedPAN), Caribbean Marine Protected Area Managers (CaMPAM)
Connectivity	A set of multiple MPAs connected by the movement and dispersal of larvae, juveniles, or adults	To maximize conservation benefits but minimize no-take area by establishing multiple, interconnected MPAs	Papua New Guinea, Gulf of California, California coast, Moorea, West Hawai'i

to shore and poachers' activity is easily spotted [37]. Due to protection from fishing, increases in yellow tang abundance are evident in these MPAs [39]. In addition, there is large-scale population connectivity of yellow tang within the West Hawai'i MPA network and evidence of larval seeding to unprotected areas [40], demonstrating that these MPAs function as a connectivity network.

We compared yellow tang abundance in the planned West Hawai'i network to yellow tang abundance in two marine reserves on Maui ('Ahihi-Kina'u, established in 1973, and Molokini, established in 1977). These two reserves were implemented based on separate management processes and were not planned to be ecologically connected. Although similar connectivity studies have not been performed on Maui, the lack of replicated MPAs across the dispersal distance of yellow tang illustrates the lack of network connectivity planning for the two separate Maui reserves. In contrast to the West Hawai'i MPAs, which only limit aquarium collecting and mostly protect herbivores, no-take reserves on Maui provide protection for all species. However, the small size of most of Hawai'i's protected areas limits their effectiveness for larger, more mobile predators [41]. Thus, there are likely few significant trophic cascades that can cause major differences between MPAs in West Hawai'i and marine reserves on Maui in terms of yellow tang abundance. Poaching is considered to be rare in Maui marine reserves [42].

Although Maui has one additional marine reserve (Honolua-Mokulēia Bay, located ~65 km from the nearest marine reserve at Molokini), which contributes to the total protected area representing <1% of the coast, only 'Ahihi-Kina'u and Molokini were monitored during time frames that overlapped with the West Hawai'i MPAs. On both islands, study sites were selected for the analysis based on similar monitoring methods during 2000–2003, except for Molokini, which included data from 1996–2004 [39]. After 2004, monitoring of the 'Ahihi-Kina'u and Molokini reserves ended, while sampling of the West Hawai'i MPAs is ongoing. At each island "control" study sites open to all fishing activities (nine in West Hawai'i and three on Maui) were monitored during the same time period (Figure 1).

Sampling methods are detailed in Tissot et al. [39], but briefly, fish densities for both yellow tang and potential predators were estimated via visual search by a pair of divers along four 25×4 m strip transects per site. We estimated densities of top and mid-level predators using predator groups previously identified for Maui and West Hawai'i [43], excluding those that are known not to prey on Acanthurids such as yellow tang (i.e. the introduced snapper, *Lutjanus kasmira* or taape; [44,45]. All sites were surveyed bimonthly, weather permitting, for a total of six surveys per site per year with the exception of logistical constraints in the summer of 2002. To be consistent among islands, we used 1999–2000 as a "before" MPA survey period (1996–2000 for Molokini) and 2001–2003 (2001–2004 for Molokini) as an "after" MPA implementation survey period.

To test for a network effect, we first conducted a three-way repeated-measure ANOVA with nested study sites after diagnostic tests confirmed the assumption of equal variances. The statistical model for the analysis is:

$$\mu = \mathrm{Net} + \mathrm{MPA} + \mathrm{BA} + \mathrm{Net} * \mathrm{MPA} + \mathrm{Net} * \mathrm{BA} + \mathrm{MPA} * \mathrm{BA} + \mathrm{Net} * \mathrm{MPA} * \mathrm{BA} + \mathrm{Site(MPA)} + \mathrm{Time} + \mathrm{error}$$

Where, $\boldsymbol{\mu}$ denotes the population mean, **Net** the partitioned variability between network and non-network locations, **MPA** the partitioned variability between MPAs and open areas, **BA** the partitioned variability between before and after time periods, **Site** the partitioned variability among replicated study sites nested within MPAs, and **Time** the surveys as a random repeated measure. The random repeated measures accounted for the yearly averages of yellow tang density at each site and corrected for the non-independence in yearly sampling by site. We accounted for differences in the total area protected by the West Hawai'i sites and Maui sites by using ratios of before-after, outside-inside, and networked and non-networked yellow tang densities.

We predict that a significant network effect would be indicated by a greater increase (before versus after) in abundance in MPAs (and perhaps open areas) around an island with a network than on

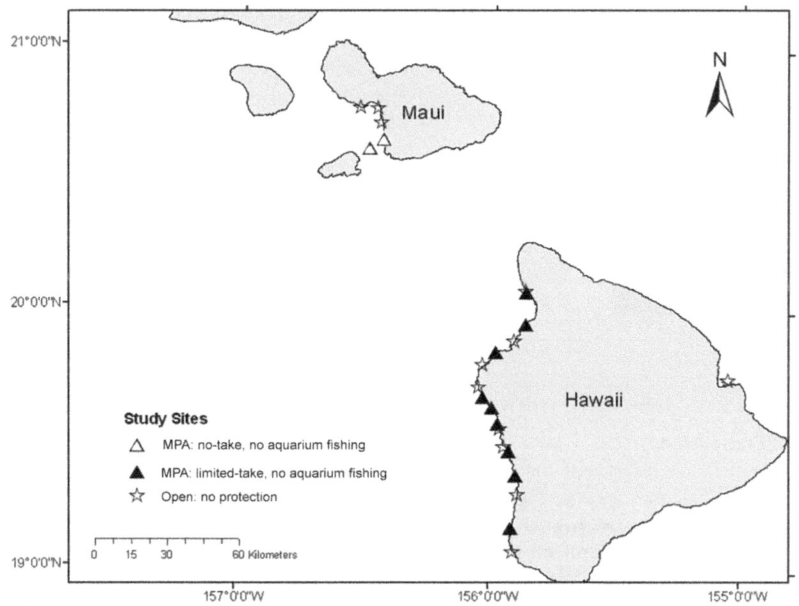

Figure 1. Map of study sites inside and outside protected areas in West Hawai'i and Maui.

one without a network. This result would be indicated statistically by any significant first or second order interaction involving the factors **Net** and **BA**, and a significant effect size of the network. An effect size of a network could be estimated as:

$$ES = \ln\left(\frac{\dfrac{\bar{X}_{A,P,Network}}{\bar{X}_{B,P,Network}} \Big/ \dfrac{\bar{X}_{A,C,Network}}{\bar{X}_{B,C,Network}}}{\dfrac{\bar{X}_{A,P,non-Network}}{\bar{X}_{B,P,non-Network}} \Big/ \dfrac{\bar{X}_{A,C,non-Network}}{\bar{X}_{B,C,non-Network}}}\right)$$

where $\bar{X}$ is the average density of yellow tang before (B) or after (A) the MPAs were established in the protected (P) or control (C) locations within the *Network* (e.g., West Hawai'i) or *non-Network* (e.g., Maui) conditions, which is tested for a statistical difference from zero using a z-test.

The results of these analyses for the West Hawai'i and Maui data show significant **Net*BA** and **Net*MPA** interactions (Table 2). These results support the hypothesis that MPAs on West Hawai'i had significantly greater percent change in density within MPAs and open areas before vs. after network establishment compared to the Maui non-network sites during the same period (Figure 2). Moreover, the network effect size was significantly greater than zero (Mean network effectiveness $= 1.63 \pm 0.69$ [95% C.I.], $n = 18$, z-test, $p < 0.001$). Overall, yellow tangs within marine reserves and open areas on Maui declined during the study period, while both MPAs and open areas in West Hawai'i increased during the same time period. Although both marine reserves on Maui declined overall, abundances at 'Ahihi-Kina'u during the study period were high but variable and yellow tang were relatively less common at Molokini.

Our data illustrate results consistent with a network effect that is greater than the sum of individual MPA effects. Although the West Hawai'i MPAs account for a greater area of protected coastline (35%) than the Maui marine reserves (<1%), a network effect is still indicated when the total area protected is standardized by response ratio (e.g. outside-inside comparisons). We compared protected sites that were both no-take for yellow tang, however the Maui sites were completely no-take marine reserves while some fishing for non-aquarium trade species is allowed in the West Hawai'i MPAs. To investigate whether differences in yellow tang predators may influence the densities of yellow tang across protection levels, we compared the top and mid-level predators by protection level (MPA, Open) and by island (Maui, West Hawai'i) (Table 3). For mid-level predators, the significant interaction term describing the effects of both island and protection level shows that there are in fact higher densities of these predators in open vs. MPA sites on Maui (Table 4). Overall there were no significant differences in densities of top predators when collectively comparing Maui and West Hawai'i. Thus, lower densities of yellow tang in Maui MPAs are not due to higher predator densities than the West Hawai'i MPAs. The length of protection for Maui marine reserves is greater than that of the West Hawai'i MPAs (established in 1973, 1977, vs. 2000, respectively). If the densities of yellow tang were greater in Maui reserves due to the longer timeline to achieve a response, we would expect our network effect size to be lower as a result. Yet we did not see a higher yellow tang response in the older Maui reserves but instead a higher response in the Hawai'i network. In addition, more recent data from West Hawai'i demonstrate that in 2012 yellow tang responses to MPA network protection were still as high or higher than those observed during the first five years after reserve establishment [46], indicating that the network effect we observed may in fact be a conservative estimate. Collectively, these data show that the West Hawai'i MPAs have functioned as a network based on the established management goals [39].

The ideal analysis for the network effect would involve similar types of comparisons to those made here but among MPAs that were created over the same time period and followed for longer lengths of time. These data gaps illustrate the importance of prioritizing and standardizing monitoring efforts given the financial and logistical resources at hand [47]. To that end, field monitoring coupled with simulation modeling is a powerful way to both generate data-driven hypotheses and target the monitoring data most useful for detecting a network effect [48].

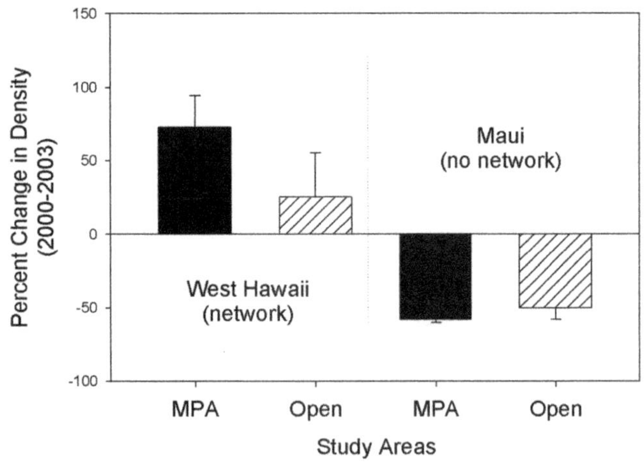

Figure 2. The mean percent change (±1 SE) in density of yellow tang (*Zebrasoma flavescens*) before versus after MPA network establishment in West Hawai'i within limited-take and no-take MPAs and control (open to fishing) sites on West Hawai'i and Maui.

The Value of Interactive Empirical-Modeling Approaches

From a practical point of view, a monitoring program that provides the full complement of data to evaluate an MPA network effect can be difficult and cost-prohibitive. Quantitative monitoring of algae, fishes, and invertebrate densities, fishing mortality, socio-economic indicators, and measures of larval retention within and connectivity among MPAs [49] should be evaluated both before and after MPA establishment, from areas inside MPAs and in unprotected areas *inside* and *outside* the network—all while incorporating sufficient sample replication, avoidance of spatial confounding, and appropriate temporal replication [22,50]. Given the time and energy required to monitor a single MPA, let alone an MPA network, the logistical and financial considerations associated with such an endeavor would be high. Further, the reality of having access to a full network that is designed to function as a connectivity network, coupled with access to another comparable control "non-network" not designed for connectivity, is unlikely.

To prioritize data collection given limited resources, and to further refine the empirical approach discussed above, predictions generated from spatially explicit models can serve as model experiments to generate hypotheses and test the effectiveness of an MPA network [16,20,48]. One can ask how the ecosystem and fisheries responses would change if one or more protected areas were poorly connected to the rest of the network and then compare how both the individual MPAs and the network as a whole respond to different levels of connectivity. These predictions can be vital for estimating the timing and magnitude of expected MPA effects by quantifying the separate contributions of reduced fishing mortality and potential network effects, identifying potential spillover effects in references areas, characterizing the relative importance of multiple factors affecting network success, and ultimately refining future monitoring plans to inform adaptive management of the network. For example, during the process to design an MPA network mandated by California's Marine Life Protection Act (MLPA), three-dimensional ocean circulation models were coupled with spatially explicit models of the dynamics of fished species and fishing effort to evaluate the projected impact on abundance and fishery yield of all submitted MPA designs [6,51,52]. These spatially explicit metapopulation models use estimated larval dispersal kernels, habitat distribution, and fishing mortality (inside and outside reserves) to predict spatial and temporal patterns of larval production and replenishment, which in turn predict resulting geographic patterns of adult abundance, size structure, biomass and spawning potential, and levels of population sustainability [6,53]. Empirical estimates of abundance, size structure, and recruitment rates derived from monitoring studies, such as those illustrated in our Hawai'i example above, are critical for both parameterizing these population models and evaluating the model predictions.

The design of monitoring programs can also benefit from these model predictions. For example, models can predict not only the rate and magnitude of change in response variables but also the likely degree of associated spatial and temporal variability [46,47,54,55]. Given this information, monitoring can optimally allocate sampling effort to sampling designs that increase the

Table 2. Three-way analysis of variance of yellow tang abundance on West Hawai'i versus Maui (factor Network), within and outside protected areas (MPA), before (1999–2000) vs. after (2001–2003) network establishment on West Hawai'i (BA), and interactions among all fixed factors (see text for details).

Source	DF	SS	MS	F	P
Network	1	1275	1275	6.90	0.009
MPA	1	4572	4572	14.01	0.001
BA	1	38	38	0.21	0.649
Network*MPA	1	908	908	4.91	0.028
Network*BA	1	1275	1275	6.90	0.009
MPA*BA	1	301	301	1.63	0.203
Network*MPA*BA	1	148	148	0.80	0.373
Site(MPA)	16	9212	57	3.11	<0.001
Time	5	1945	389	2.10	0.066
Error	235	43442	185		
Total	263	67390			

Replicated study sites for all treatments were nested within MPA and time was treated as a random, repeated measure factor.

Table 3. The mean density (±1 SE) of top and mid-level predators on Maui and West Hawai'i under different management regimes.

	Mean density (100 m^{-2})	
	Open sites	MPA sites
Top predators		
West Hawai'i	0.01 (0.01)	0.04 (0.03)
Maui	0.39 (0.10)	0.08 (0.05)
Mid-level predators		
West Hawai'i	0.36 (0.02)	0.41 (0.02)
Maui	0.68 (0.11)	0.39(0.10)

Open = open to fishing. MPA= marine protected area. Top predators: Carangidae, Carcharhinidae, Sphyraenidae; Mid-level predators: Aulostomidae, Lutjanidae, Muraeidae, Scorpaenidae, Serranidae, Synodontidae.

likelihood of detecting responses empirically (i.e., by identifying the timing, number, and distribution of spatial and temporal replicates). This iterative approach provides a powerful tool for adaptive management when empirical data from monitoring informs models and, in turn, model results further refine monitoring designs [46].

Conclusions

As more MPA networks are established, it becomes increasingly important to implement thoughtful and rigorous monitoring plans to assess their effectiveness [24]. The benefits of connectivity underscore many network management goals, yet the other types of networks defined here can accomplish their own set of protective objectives (Table 1). Thus, it is essential to carefully consider which network type(s) is most in line with the stated goals. Thorough monitoring designs should be outlined prior to network establishment, accounting for outside, inside, control, and experimental MPA sampling across an appropriate spatial and temporal scale and prioritizing the data types most critical for assessing network effects. Our example of MPAs in West Hawai'i and Maui illustrates the need for well-balanced data in order to effectively determine whether a network effect is occurring. In reality, there are many logistical constraints on MPA monitoring, which highlight the value of coupling models with empirical data to generate network-level hypotheses and evaluate the responses to protection in a network. We encourage regional management bodies, scientists, and stakeholders to discuss all of these key components as a regular part of the network planning process and to work collectively to develop optimal monitoring strategies given the available resources.

Acknowledgments

We thank S. Gaines and C. Costello for formative discussions about the assessment of marine reserve networks as well as M. Gomei for useful conversations about social networks. Some of the Maui data were generously provided by R. Sparks and S. Hau. This paper is an outcome of conversations that began at a symposium organized by the Communication Partnership for Science and the Sea (COMPASS) and the Partnership for Interdisciplinary Studies of Coastal Oceans (PISCO) at the 2009 International Marine Conservation Congress in Washington, D.C. JC would like to thank EU FP7 COCONET and biodivERsA BUFFER for funding. This is contribution # 446 from PISCO, the Partnership for Interdisciplinary Studies of Coastal Oceans.

Author Contributions

Conceived and designed the experiments: KGC JC BNT JEC MHC JCD AMF SEL TL DM WJW. Performed the experiments: BNT WJW. Analyzed the data: BNT. Wrote the paper: KGC JC BNT JEC MHC JCD AMF SEL TL DM WJW.

Table 4. Two-way analysis of variance of top predator and mid-level predator densities on West Hawai'i and Maui (factor Island), within and outside protected areas (MPA), and for the interaction among these fixed factors (see text for details).

Predator level	Source	DF	SS	MS	F	P
Top	Island	1	1.59	1.59	0.62	0.43
	MPA	1	0.78	0.78	0.31	0.58
	Interaction	1	1.12	1.12	0.44	0.51
	Error	117	299.6	2.56		
	Total	120	304.8			
Mid-level	Island	1	0.56	0.56	3.59	0.06
	MPA	1	0.35	0.35	2.24	0.14
	Interaction	1	0.72	0.72	4.57	0.04
	Error	117	18.4	0.16		
	Total	120	20.7			

Top predators: Carangidae, Carcharhinidae, Sphyraenidae; Mid-level predators: Aulostomidae, Lutjanidae, Muraeidae, Scorpaenidae, Serranidae, Synodontidae.

References

1. Halpern BS, Walbridge S, Selkoe KA, Kappel CV, Micheli F, et al. (2008) A global map of human impact on marine ecosystems. Science 319: 948–952.
2. Jackson JBC (2008) Ecological extinction and evolution in the brave new ocean. Proc Natl Acad Sci U S A 105: 11458–11465.
3. Lotze HK, Lenihan HS, Bourque BJ, Bradbury RH, Cooke RG, et al. (2006) Depletion, degradation, and recovery potential of estuaries and coastal seas. Science 312: 1806–1809.
4. Browman HI, Stergiou KI (2004) Marine protected areas as a central element of ecosystem-based management: defining their location, size and number. Mar Ecol Prog Ser 274: 271–272.
5. Halpern BS, Lester SE, McLeod KL (2010) Placing marine protected areas onto the ecosystem-based management seascape. Proc Natl Acad Sci U S A 107: 18312–18317.
6. Rassweiler A, Costello C, Siegel DA (2012) Marine protected areas and the value of spatially optimized fishery management. Proc Natl Acad Sci U S A.
7. Harrison Hugo B, Williamson David H, Evans Richard D, Almany Glenn R, Thorrold Simon R, et al. (2012) Larval export from marine reserves and the recruitment benefit for fish and fisheries. Curr Biol 22: 1023–1028.
8. Miller I, Cheal AJ, Emslie MJ, Logan M, Sweatman H (2012) Ongoing effects of no-take marine reserves on commercially exploited coral trout populations on the Great Barrier Reef. Mar Environ Res 79: 167–170.
9. Abesamis RA, Russ GR, Alcala AC (2006) Gradients of abundance of fish across no-take marine reserve boundaries: Evidence from Philippine coral reefs. Aquatic Conservation-Marine and Freshwater Ecosystems 16: 349–371.
10. Carr MH, Woodson CB, Cheriton OM, Malone D, McManus MA, et al. (2011) Knowledge through partnerships: integrating marine protected area monitoring and ocean observing systems. Front Ecol Environ 9: 342–350.
11. Claudet J, Osenberg CW, Benedetti-Cecchi L, Domenici P, Garcia-Charton JA, et al. (2008) Marine reserves: size and age do matter. Ecol Lett 11: 481–489.
12. Lester SE, Halpern BS, Grorud-Colvert K, Lubchenco J, Ruttenberg BI, et al. (2009) Biological effects within no-take marine reserves: A global synthesis. Mar Ecol Prog Ser 384: 33–46.
13. Molloy PP, McLean IB, Cote IM (2009) Effects of marine reserve age on fish populations: a global meta-analysis. J Appl Ecol 46: 743–751.
14. Fenberg P, Caselle JE, Claudet J, Gaines SD, Garcia-Charton JA, et al. (2012) The science of European marine reserves: Status, efficacy, and future needs. Mar Pol 36: 1012–1021.
15. Edgar GJ, Bustamante RH, Farina JM, Calvopina M, Martinez C, et al. (2004) Bias in evaluating the effects of marine protected areas: the importance of baseline data for the Galapagos Marine Reserve. Environ Conserv 31: 212–218.
16. Gaines SD, White C, Carr MH, Palumbi S (2010) Designing marine reserve networks for both conservation and fisheries management. Proc Natl Acad Sci U S A doi:10.1073/pnas.0906473107.
17. Almany GR, Connolly SR, Heath DD, Hogan JD, Jones GP, et al. (2009) Connectivity, biodiversity conservation and the design of marine reserve networks for coral reefs. Coral Reefs 28: 339–351.
18. Murray SN, Ambrose RF, Bohnsack JA, Botsford LW, Carr MH, et al. (1999) No-take reserve networks: sustaining fishery populations and marine ecosystems. Fisheries 24: 11–25.
19. IUCN-WCPA (2008) Establishing marine protected area networks–Making it happen. Washington, D.C.: IUCN World Commission on Protected Areas, National Oceanic and Atmospheric Administration, and The Nature Conservancy. 118 p.
20. White JW, Botsford LW, Hastings A, Largier JL (2010) Population persistence in marine reserve networks: incorporating spatial heterogeneities in larval dispersal. Mar Ecol Prog Ser 398: 49–67.
21. Shanks AL, Grantham BA, Carr MH (2003) Propagule dispersal distance and the size and spacing of marine reserves. Ecol Appl 13: S159–S169.
22. Grorud-Colvert K, Claudet J, Carr MH, Caselle JE, Day J, et al. (2011) The assessment of marine reserve networks: Guidelines for ecological evaluation. In: Claudet J, editor. Marine Protected Areas–A Multidisciplinary Approach. Cambridge, U.K.: Cambridge University Press. pp. 293–321.
23. Claudet J, Pelletier D (2004) Marine protected areas and artificial reefs: A review of the interactions between management and scientific studies. Aquat Living Resour 17: 129–138.
24. Day J, Hockings M, Jones G (2002) Measuring effectiveness in marine protected areas - principles and practice. In: Beumer JP, Grant A, Smith DC, editors. World Congress on Aquatic Protected Areas: what works best and how do we know? Cairns, Queensland. pp. 401–414.
25. Botsford LW, Brumbaugh DR, Grimes C, Kellner JB, Largier J, et al. (2009) Connectivity, sustainability, and yield: bridging the gap between conventional fisheries management and marine protected areas. Rev Fish Biol Fish 19: 69–95.
26. Russ GR, Alcala AC (1996) Do marine reserves export adult fish biomass? Evidence from Apo Island, central Philippines. Mar Ecol Prog Ser 132: 1–9.
27. Gleason M, Fox E, Ashcraft S, Vasques J, Whiteman E, et al. (2013) Designing a network of marine protected areas in California: Achievements, costs, lessons learned, and challenges ahead. Ocean Coast Manage 74: 90–101.
28. Fernandes L, Day J, Lewis A, Slegers S, Kerrigan B, et al. (2005) Establishing representative no-take areas in the Great Barrier Reef: Large-scale implementation of theory on marine protected areas. Conserv Biol 19: 1733–1744.
29. White JW, Botsford LW, Hastings A, Baskett ML, Kaplan DM, et al. (2013) Transient responses of fished populations to marine reserve establishment. Conservation Letters 6: 180–191.
30. NRC (2001) Marine protected areas: tools for sustaining oceans ecosystems. Washington, D. C.: National Academy Press. 288 p.
31. McCook LJ, Almany GR, Berumen ML, Day JC, Green AL, et al. (2009) Management under uncertainty: guide-lines for incorporating connectivity into the protection of coral reefs. Coral Reefs 28: 353–366.
32. McInerney CE, Allcock AL, Johnson MP, Prodohl PA (2012) Ecological coherence in marine reserve network design: An empirical evaluation of sequential site selection using genetic structure. Biol Conserv 152: 262–270.
33. GBRMPA (2002) Biophysical Operational Principles as recommended by the Scientific Steering Committee for the Representative Areas Program. Great Barrier Reef Marine Park Authority 6p.
34. Hamilton SL, Caselle JE, Malone D, Carr MH (2010) Incorporating biogeography into evaluations of the Channel Islands marine reserve network. Proc Natl Acad Sci U S A doi:10.1073/pnas.0908091107.
35. Tissot BN, Walsh WJ, Hixon MA (2009) Hawaiian islands marine ecosystem case study: Ecosystem and community-based management in West Hawaii. Coast Manage 37: 255–273.
36. Stevenson TC, Tissot BN, Dierking J (2011) Fisher behaviour influences catch productivity and selectivity in West Hawaii's aquarium fishery. ICES J Mar Sci 68: 813–822.
37. Williams ID, Walsh WJ, Claisse JT, Tissot BN, Stamoulis KA (2009) Impacts of a Hawaiian marine protected area network on the abundance and fishery sustainability of the yellow tang, Zebrasoma flavescens. Biol Conserv 142: 1066–1073.
38. Pillay D, Branch GM, Griffiths CL, Williams C, Prinsloo A (2010) Ecosystem change in a South African marine reserve (1960–2009): role of seagrass loss and anthropogenic disturbance. Marine Ecology-Progress Series 415: 35–48.
39. Tissot B, Walsh W, Hallacher L (2004) Evaluating effectiveness of a marine protected area network in west Hawai'i to increase productivity of an aquarium fishery. Pac Sci 58: 175–188.
40. Christie MR, Tissot BN, Albins MA, Beets JP, Jia YL, et al. (2010) Larval connectivity in an effective network of marine protected areas. Plos One 5.
41. Friedlander AM, Brown EK, Monaco ME (2007) Coupling ecology and GIS to evaluate efficacy of marine protected areas in Hawaii. Ecol Appl 17: 715–730.
42. Stevenson TC, Tissot BN (2013) Evaluating marine protected areas for managing marine resource conflict in Hawaii. Mar Pol 39: 215–223.
43. Friedlander AM, DeMartini EE (2002) Contrasts in density, size, and biomass of reef fishes between the northwestern and the main Hawaiian islands: the effects of fishing down apex predators. Mar Ecol Prog Ser 230: 253–264.
44. Schumacher BD (2011) Habitat use and trophic ecology of the introduced snapper *Lutjanus kasmira* and native goatfishes in Hawai'i. PhD. Dissertation, University of Hawaii. 230 p.
45. Parrish JD, Aeby GC, Conklin EJ, Ivey GL, Schumacher BD (2000) Interactions of nonindigenous blueline snapper (taape) with native deep-water fishery species. Project completion report, Hawaii Department of Land and Natural Resources. 76 pp.
46. Walsh W, Cotton SP, Barnett C, Couch C, Preskitt L, et al. (2013) Long-term monitoring of coral reefs of the main Hawaiian Islands. NOAA Coral Reef Conservation Program, Hawai'i Division of Aquatic Resources. 97 p.
47. Sweatman H, Cheal A, Coleman G, Delean S, Fitzpatrick B, et al. (2001) Long-term monitoring of the Great Barrier Reef, Status Report Number 5. Australian Institute of Marine Science. 106 p.
48. White JW, Botsford LW, Baskett ML, Barnett LAK, Barr RJ, et al. (2011) Linking models with monitoring data for assessing performance of no-take marine reserves. Front Ecol Environ 9: 390–399.
49. Jones GP, Almany GR, Russ GR, Sale PF, Steneck RS, et al. (2009) Larval retention and connectivity among populations of corals and reef fishes: history, advances and challenges. Coral Reefs 28: 307–325.
50. Willis TJ, Millar RB, Babcock RC, Tolimieri N (2003) Burdens of evidence and the benefits of marine reserves: putting Descartes before des horse? Environ Conserv 30: 97–103.
51. CDFG (2008) California Marine Life Protection Act: Master plan for marine protected areas, revised draft January 2008.
52. Saarman E, Gleason M, Ugoretz J, Airame S, Carr MH, et al. (2013) The role of science in supporting marine protected area network planning and design in California. Ocean Coast Manage 74: 45–56.
53. White C, Kendall BE, Gaines S, Siegel DA, Costello C (2008) Marine reserve effects on fishery profit. Ecol Lett 11: 370–379.
54. Joseph LN, Elkin C, Martin TG, Possingham HP (2009) Modeling abundance using N-mixture models: the importance of considering ecological mechanisms. Ecol Appl 19: 631–642.
55. Pelletier D, Claudet J, Ferraris J, Benedetti-Cecchi L, Garcia-Charton JA (2008) Models and indicators for assessing conservation and fisheries-related effects of marine protected areas. Can J Fish Aquat Sci 65: 765–779.

18

Incorporating Conservation Zone Effectiveness for Protecting Biodiversity in Marine Planning

Azusa Makino[1]*, Carissa J. Klein[1], Maria Beger[1], Stacy D. Jupiter[2], Hugh P. Possingham[1,3]

1 Australian Research Council Centre of Excellence for Environmental Decisions, School of Biological Sciences, The University of Queensland, Brisbane, Queensland, Australia, **2** Fiji Country Program, Wildlife Conservation Society, Suva, Fiji, **3** Department of Life Sciences, Imperial College London, Ascot, Berkshire, United Kingdom

Abstract

Establishing different types of conservation zones is becoming commonplace. However, spatial prioritization methods that can accommodate multiple zones are poorly understood in theory and application. It is typically assumed that management regulations across zones have differential levels of effectiveness ("zone effectiveness") for biodiversity protection, but the influence of zone effectiveness on achieving conservation targets has not yet been explored. Here, we consider the zone effectiveness of three zones: permanent closure, partial protection, and open, for planning for the protection of five different marine habitats in the Vatu-i-Ra Seascape, Fiji. We explore the impact of differential zone effectiveness on the location and costs of conservation priorities. We assume that permanent closure zones are fully effective at protecting all habitats, open zones do not contribute towards the conservation targets and partial protection zones lie between these two extremes. We use four different estimates for zone effectiveness and three different estimates for zone cost of the partial protection zone. To enhance the practical utility of the approach, we also explore how much of each traditional fishing ground can remain open for fishing while still achieving conservation targets. Our results show that all of the high priority areas for permanent closure zones would not be a high priority when the zone effectiveness of the partial protection zone is equal to that of permanent closure zones. When differential zone effectiveness and costs are considered, the resulting marine protected area network consequently increases in size, with more area allocated to permanent closure zones to meet conservation targets. By distributing the loss of fishing opportunity equitably among local communities, we find that 84–88% of each traditional fishing ground can be left open while still meeting conservation targets. Finally, we summarize the steps for developing marine zoning that accounts for zone effectiveness.

Editor: Sebastian C. A. Ferse, Leibniz Center for Tropical Marine Ecology, Germany

Funding: The authors are thankful for support from the National Center for Ecological Analysis and Synthesis, as a part of the working group "Conservation Decision Making in The Coral Triangle", funded by the David and Lucile Packard Foundation, and a grant from the John D. and Catherine T. MacArthur Foundation (0-94985-000-GSS) to the Wildlife Conservation Society. Additional support was provided by the Australian Research Council Centre of Excellence for Environmental Decisions and the Australian Research Council Centre of Excellence for Coral Reef Studies. AM is funded by the by ITO Foundation, Japan, by the Australian Research Council Centre of Excellence for Environmental Decisions, and the University of Queensland. CJK is supported by an Australian Research Council Postdoctoral Fellowship (Project number DP110102153). MB is supported by a Discovery Early Career Research Award to the ARC Centre of Excellence for Environmental Decisions (CE110001014). The funders had no role in study design, data collection and analysis, decision to publish, or preparation of the manuscript.

Competing Interests: The authors have declared that no competing interests exist.

* E-mail: azusamakino@gmail.com

Introduction

Marine spatial planning that incorporates multiple conservation zones in a marine protected area (MPA) network provides planners and policy-makers with more flexibility to accommodate the objectives of multiple users compared with a traditional two-zone planning process (*i.e.* reserve zones versus non-reserve zones) [1]. Different types of zones within a management scheme afford varying degrees of protection for biodiversity depending on the degree of restriction of human use. For example, in the Great Barrier Reef Marine Park in Australia, the "General Use Zone" is the least restricted zone and allows activities such as mining, tourism, fishing, mariculture, and shipping, whereas the "Preservation Zone" is completely no access [2]. Establishing multiple zones in a MPA network may improve the overall effectiveness of the MPA network by minimizing costs to, and conflict between, different activities (*e.g.* conservation versus fishing) [3,4]. Furthermore, well- designed and managed MPA networks will benefit both biodiversity and fisheries by providing many direct (*e.g.* restoring fish populations) and indirect benefits (*e.g.* abating threats from overfishing that can cause coral degradation, spill over benefit for fishery) [5–8].

Marine zoning approaches have been developed and used around the world. In the Great Barrier Reef Marine Park in Australia, managers decided on the location of different zones on the basis of extensive stakeholder consultation, and the spatial allocation of no-take reserves were informed using Marxan, a systematic conservation planning tool [2,9]. A more advanced version of this tool, Marxan with Zones, has been used to inform the allocation of zones for fishery and biodiversity objectives in California, U.S.A [10], the Caribbean [11], and Raja Ampat, Indonesia [12]. Other methods for systematic marine zoning are to use multi-criteria decision analysis frameworks to incorporate opinions of different stakeholders (*e.g.* Villa *et al.* [13] and Portman [14]), develop a GIS- based model (*e.g.* Bruce and Eliot [15]), or using a web-based zoning platform [16]. Although ecological effects were mentioned in the use of multi-criteria method by Lahdelma et al. [17], none of these studies on marine zoning incorporate how different kinds of zones differentially contribute to the conservation of various aspects of biodiversity (henceforward "zone effectiveness"). We focus only on the ecological effectiveness defined as the relative contribution of actions to realizing conservation objectives and not the management effectiveness that is based on both human behavior and species' ecology [18]. While most managers implicitly realize that less restrictive zones will be less effective for biodiversity conservation which is also quantified [19], planners typically assume that all zones are either completely effective or completely ineffective when calculating performance against biodiversity targets.

Incorporating zone effectiveness into the design of marine zoning for a MPA network requires quantitative data on the effectiveness of each zone at protecting each species or habitats. Obtaining such information is challenging as we know very little about zone effectiveness *in situ* because data are costly and time-consuming to acquire [20]. In general, permanent closure zones are recognized as ecologically effective because they prohibit extractive activities, which provide well known benefits to biodiversity [21]. Other types of zones contribute differently to the protection of biodiversity, depending on the kinds of activities allowed and their intensity [22,23]. For example, high fishing pressure threatens marine biodiversity [24]. Different zones regulating fishing activities include temporal closures, gear prohibitions, species-specific bans, and quotas [2]. These actions for conservation are not equally effective at protecting biodiversity [25]. Furthermore taxonomic groups with different types of life histories or ecological characteristics will be affected differently by different kinds of zones [21]. However, ecological effects of different types of conservation zone on different taxa are poorly understood and quantified relative to permanent closure zones.

Quantifying the relative costs of each zone is also a critical component to zoning as it helps planners produce plans that minimise negative socioeconomic impacts (*e.g.* regulating fishing activities in a way that would reduce the opportunity cost of foregone fishing). In addition to zone effectiveness, the relative cost of different conservation zones can impact the location and size of each zone [10]. Although there are many possible conservation costs (*e.g.* acquisition costs, management costs, transection costs, etc. Neidoo et al. [26]) most marine planners only consider opportunity costs [27], which can differ between zones in a given place, depending on which fishing activities are prohibited in the zone [10,12]. For example, the opportunity cost of a permanent closure zone that prohibits all fishing activities will be higher than in a partial protection zone that allows some kinds of fishing. Data on the opportunity cost of different zones are rarely available for marine planning (but see Klein *et al.* [10]), and can be costly to collect, especially over large planning regions [28]. To compensate, planners often develop surrogates to represent the opportunity cost of implementing a no-take area based on population [29], area [30], threats [31], and etc. However, it is unclear how these cost surrogates can be applied to other types of conservation zones. Given this uncertainty, we explore how the relative cost difference of permanent closure zones and partial protection zones impacts conservation plans. The consideration of opportunity costs is just one way to address fishing interests in marine planning. Another way is to ensure that a designated or even proportion of fishing grounds for different fisheries [10] or communities [12] are represented in a fishing zone [32].

We test our methodology in a spatial prioritization that considers two different conservation zones (permanent closure and partial protection) and one fishing zone (open zone) in Fiji under five different planning scenarios, where we varied the effectiveness of the partial protection zone, relative zone costs, and targets for fishing communities. We specifically selected Fiji as a case study because zone effectiveness score were previously developed by local experts [18]. Zone effectiveness was designed specifically for looking at the effectiveness of Fiji's MPA network at achieving the Fiji government biodiversity target to effectively protect 30% of marine areas by 2020, as well as the Aichi biodiversity target 11. We evaluate the results from the scenarios to determine: (1) the effects on spatial priorities using different levels of zone effectiveness and different zone costs; and (2) the maximum proportions of each fishing ground that can be left open while achieving conservation targets for a minimum cost.

Methods

Study region

We focused our study on the traditional fishing grounds in the Vatu-i-Ra Seascape in Fiji, which includes the inshore marine and coastal habitats of the four provinces of Ra, Tailevu, Lomaiviti and Bua (Fig. 1). Fiji's traditional fishing grounds (*i qoliqoli*) are legally demarcated by the iTaukei Land and Fisheries Commission (the traditional fishing grounds areas of the four provinces are shown in Fig. 1). The traditional fishing grounds extend to the edge of the furthest reef crest (fringing, barrier or patch) away from districts that have traditional fishing rights in the area (Fig. 1). Within these areas, indigenous Fijian maintain customary fishing access rights and have some ability to designate management zones, with support from co-management partners in the Fiji Locally Managed Marine Area network [33]. The Fiji locally managed marine areas are equivalent to MPAs although they aim for the sustainable use of the marine areas and not for conservation *per se*. They are governed by a group of local resource management and conservation practitioners [18]. In this study, we used a broader definition of MPAs that includes locally managed marine areas, as they also contribute to the protection of biodiversity. In Fiji, there are permanent closure zones where all extractive activities are prohibited and periodically harvested zones [18]. Some of the periodically harvested closure zones have a controlled harvest once per year, but many of them do not have a well-developed management plan for harvesting that specifies frequency and duration [18,34].

Spatial prioritization

We used the systematic planning tool Marxan with Zones, which is a decision-support tool for selecting priority areas for zoning marine or terrestrial areas to achieve conservation and/or fishing targets for minimum cost. Marxan with Zones minimizes the total cost of a zoning system, which is the sum of the costs of placing each planning unit into a particular zone subject to meeting target amounts of conservation and/or fishing features [35]. Marxan with Zones allows users to set how much each zone will contribute towards meeting conservation and/or fishing targets - zone effectiveness [35]. Specifically, Marxan with Zones solves the mathematical problem:

$$\text{minimize} \sum_{i=1}^{m} \sum_{k=1}^{p} c_{ik} x_{ik},$$

$$\text{subject to} \sum_{i=1}^{m} \sum_{k=1}^{p} a_{ij} e_{jk} x_{ik} \geq s_j \forall j \text{ and } \sum_{i=1}^{m} \sum_{k=1}^{p} a_{ih} e_{hk} x_{ik} \geq t_h \forall h,$$

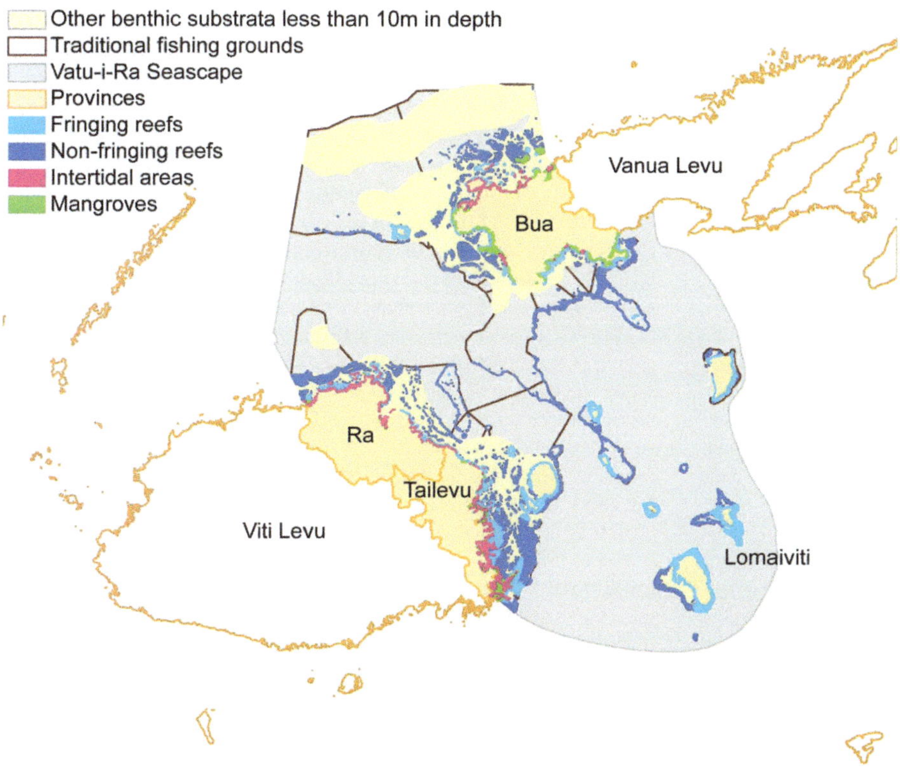

Figure 1. Study region and habitat maps of conservation features. Our study region is the traditional fishing grounds of four provinces (*i.e.* Ra, Tailevu, Lomaiviti and Bua) along the coastline of the Vatu-i-Ra Seascape, Fiji. Conservation features were (1) fringing reefs, (2) non-fringing reefs, (3) mangroves, (4) intertidal areas, and (5) other benthic substrata less than 10 m in depth.

where m is the total number of planning units ($i = 1, \ldots, m$), p is the total number of zones ($k = 1, \ldots, p$), and c_{ik} is the cost of allocating planning unit i to zone k. For all planning units ($i = 1, \ldots, m$) and zones k ($k = 1, \ldots, p$), x_{ik} is a member of $\{0, 1\}$ and if planning unit i is allocated as zone k, $x_{ik} = 1$ and if not $x_{ik} = 0$. A planning unit i cannot be allocated to more than one zone, thus $\sum_{k=1}^{p} x_{ik} \leq 1$. The first constraint is to meet the conservation target while applying zone effectiveness: a_{ij} is the amount of conservation feature j ($i = 1, \ldots, n$) in planning unit i, and e_{jk} is the zone effectiveness of zone k to conservation feature j. The target amount for conservation feature j is defined as s_j. The second constraint is to meet the target amount for the fishing feature h ($h = 1, \ldots, q$), defined as t_h. The amount of the fishing feature h in planning unit i is a_{ih} and e_{hk} is the zone effectiveness of zone k to meet the target of the fishing feature h.

Conservation targets and planning units

We considered five different marine habitats in the region as conservation features: fringing reefs, non-fringing reefs, mangroves, intertidal areas, and other benthic substrata less than 10 m in depth (Fig. 1) [18]. Intertidal areas include mudflats but exclude the mangroves. Areas less than 10 m in depth where there are no fringing reefs, non-fringing reefs, mangroves or intertidal areas are classified as other benthic substrata less than 10 m in depth. In addition to conservation features, we also had fishing features, which are the traditional fishing grounds in the study region (n = 27). For the conservation features, we aimed to protect 30% of the distribution of each feature in either a permanent closure or partial protection zone [18]. The conservation targets were in accordance with national targets and outputs from an expert workshop in Fiji [18]. For fishing features, we aimed to evenly include as much of each traditional fishing ground in the open zone without compromising the conservation targets. We did this by gradually increasing the target for fishing features until a conservation target was not achieved. Not all features were targeted for all planning scenarios, described in the subsection of "5. Planning scenarios" below.

We overlaid 1 km^2 hexagonal planning units on the study region (n = 10044). We only included traditional fishing grounds in our planning region that were larger than 20 km^2 and had no more than 10% of their total area in closures from fishing. We excluded small traditional fishing grounds since they do not always contain conservation features (*i.e.* habitats) and/or were too small to allocate multiple zones. Each planning unit could be allocated to one of the three zones to create a MPA network except for existing permanent closure zones and periodically harvested closure zones. Existing permanent closure zones and periodically harvested closure zones in the study region were assigned to the permanent closure zone and the partial protection zone, respectively, for all analyses. Conservation features within those existing conservation areas contributed towards the overall conservation targets in every analysis.

Zones effectiveness and costs

We considered three zones: (1) a permanent closure zone, which offers the highest level of protection; (2) a partial protection zone, which may be periodically harvested for fisheries resources; and (3) an open zone, under which all activities are permissible and we assume it offers no protection for biodiversity. A partial protection zone has no predetermined harvest frequency, intensity or duration (referred to as "periodic closure with uncontrolled harvesting" zones in Mills *et al.* [18]). Experts determined the effectiveness of zones at protecting selected species groups (*e.g.* corals, targeted invertebrates, non-targeted invertebrates, targeted fish, non-targeted fish, and coralline algae in fringing reef habitat) that they considered of national importance in five marine habitats in our study region, Fiji, working under the assumptions that permanent closure zones offer the highest benefit for biodiversity, while periodic harvests reduce zone effectiveness [18]. We chose to use the more conservative measures of zone effectiveness for the partial protection zones by adopting the measures defined in Mills *et al.* [18] for periodically harvested areas in an uncontrolled manner. The effectiveness of any zone is a real number between 0 and 1 inclusive, where 0 is completely ineffective and 1 is 100% effective in meeting targets in that zone. Thus, the zone effectiveness, e_{jk}, for permanent closure zones at protecting conservation feature j was $e_{jk} = 1$ (maximum effectiveness), and for open zones at protecting conservation feature j was $e_{jk} = 0$(completely ineffective) for every conservation feature. For partial protection zones (referred to as "periodic closure with uncontrolled harvesting" zones in Mills *et al.* [18]) at protecting conservation feature j, the highest, average, and lowest zone effectiveness values determined by local experts in Fiji were used (Table 1) [18]. For fishing feature h, the zone effectiveness for the open zone was $e_{hk} = 1$, and $e_{hk} = 0$ for the permanent closure and partial protection zone.

We estimated the opportunity costs of fishing using the fishing pressure data modeled by Klein *et al.* [29] for the permanent closure zone. The approach assumed that fishing is allowed on all coral reefs. This assumption was reasonable as all reefs are open access for subsistence use under the present Fiji Fisheries Act (1942). They estimated fishing pressure to be proportional to the number of people living within 35 km of each planning unit on the basis that fishing pressure is roughly correlated with the coastal population in Fiji [29,36]. This was used as there is no existing comprehensive fishing information available for Fiji and the global datasets did not adequately represent fishing pressure in the study region [24,37]. We assigned the relative costs of partial protection zones as 75%, 50% and 25% of the cost of the permanent closure zones since there was no information how the opportunity costs of fishing would scale among conservation zones in Fiji. The opportunity cost for an open zone was zero since there were no restrictions on fishing.

Table 1. Zone effectiveness for the partial protection zone at achieving conservation targets, as determined by experts in Fiji (referred to as "periodic closure with uncontrolled harvesting" zone in Mills *et al.* [18]).

	Scenario			
Conservation Features	**(1) Equal**	**(2) Highest**	**(3) Average**	**(4) Lowest**
Fringing reefs	1	0.6	0.39	0.1
Non-fringing reefs	1	0.8	0.46	0.1
Mangrove	1	0.85	0.56	0.1
Intertidal	1	0.8	0.48	0.1
Other benthic less than 10 m	1	0.8	0.6	0.3

Zone effectiveness at meeting conservation targets for the permanent closure zone was always 1 (fully effective), and for the open zone was always 0 (completely ineffective) for all scenarios. Scenario 5 used the same average zone effectiveness as scenario 2.

Planning scenarios

We considered five different planning scenarios, where we varied the effectiveness of the partial protection zone, relative zone costs, and targets for fishing communities. In scenario 1, we considered partial protection zones to be as effective as permanent closure zones (equal zone effectiveness) at meeting conservation targets to represent cases where locally managed partial protection zones are highly effective in protecting biodiversity [38,39] (Table 2). In scenario 2-4, we applied the highest (scenario 2), average (scenario 3) and lowest (scenario 4) zone effectiveness scores to the partial protection zones (Table 1 and 2) at meeting conservation targets using the consensus values from the expert workshop [18]. In scenario 5, we treated each traditional fishing ground as a fishing feature to include in a zone and targeted them the same as we targeted conservation features. We increased the targeted amount of fishing features until conservation target became impossible to achieve, to find the highest amount of traditional fishing grounds that can remain open while still meeting the conservation targets. In our study region, the traditional fishing grounds of four provinces include existing permanent closure or partial protection zones but we only allowed the fishing feature to be achieved in the open zone. The five scenarios were analyzed for each of the three different zone costs for partial protection zones (75%, 50% and 25% of the cost of the permanent closure zones) as described in subsection 2.4. For each scenario, we ran Marxan with Zones 100 times.

In order to test whether solutions for the 4 scenarios that used different degrees of zone effectiveness and costs differ statistically, we conducted model-based clustering. We combined all solutions of scenarios 1–4 and assigned an unknown structure. We then classified solutions into clusters that are cohesive and separated from other clusters in multivariate space. Each solution contained information of the planning units that were allocated to the permanent closure and partial protection zones. We used multivariate normal mixture modeling that evaluated alternative numbers of clusters as most likely models to determine the statistically appropriate classification of the solutions among scenarios 1 to 4 (*i.e.* 100 solutions per scenario). The function "mclustBIC" in R package mclust was applied to implement the modeling [40,41]. We tested these models with the Bayesian Information Criterion (BIC). BIC provides an approximation to the Bayes factor, allowing comparison of a model with differing numbers of clusters [41]. BIC is defined as twice the log likelihood in mclust, unlike as for ordinal BIC. Thus, a high value of BIC indicates stronger evidence for a certain number of clusters [41]. We used a sub-model of VEV: volumes of all clusters as varying (V), shapes of all clusters as equal (E), and orientation of all clusters as varying (V) to identify the number of clusters across all solutions between scenarios (n = 400) with the highest BIC.

We compared the best available solution, which was the least costly solution that met the targets out of the 100 runs, from scenario to scenario as well as the "selection frequency," which was how many times each planning unit was selected out of 100 runs. We used the best ten solutions to get average values for the total costs.

Results

The highest BIC was seen when the number of clusters was four, showing that the statistically best number of the clusters is four, which is the number of scenarios (1–4). This suggested

Table 2. Description of scenarios using different zone effectiveness values, e_{jk}, zone effectiveness of zone k ($k=3$) for conservation feature j ($j=5$) and e_{hk} for fishing feature h ($h=27$).

Scenario	Description
1	Partial protection zones are as effective as permanent closure zone for all conservation features to achieve conservation targets. $e_{jk}=1$ for the permanent closure zone, $e_{jk}=0$ for the open zone, $e_{jk}=1$ for the partial protection zone.
2	Uses the highest zone effectiveness for partial protection zone. $e_{jk}=1$ for the permanent closure zone, $e_{jk}=0$ for the open zone, $e_{jk}=$highest for the partial protection zone (see Table1).
3	Uses the average zone effectiveness for the partial protection zone. $e_{jk}=1$ for the permanent closure zone, $e_{jk}=0$ for the open zone, $e_{jk}=$average for the partial protection zone (see Table1).
4	Uses the lowest zone effectiveness for the partial protection zone. $e_{jk}=1$ for the permanent closure zone, $e_{jk}=0$ for the open zone, $e_{jk}=$lowest for the partial protection zone (see Table1).
5	Uses the average zone effectiveness and aims an equitable amount of fishing grounds remained open for fishing. $e_{jk}=1$ for the permanent closure zone, $e_{jk}=0$ for the open zone, $e_{jk}=$average for the partial protection zone (see Table1). For fishing feature h, $e_{hk}=0$ for the permanent closure zone, $e_{hk}=1$ for the open zone, $e_{hk}=0$ for the partial protection zone

Zone effectiveness was the same for all conservation features as well as for all fishing features. Fishing features were targeted only in scenario 5.

solutions between scenarios differed statistically (Fig. 2). Regardless of zone costs, all selected planning units were allocated to the partial protection zone when the zone effectiveness of the partial protection zone equaled that of the permanent closure zone (Fig. 3). As zone effectiveness for the partial protection zone decreased, an increasing number of planning units were allocated to the permanent closure zone. Furthermore, almost all of the selected planning units were allocated to the permanent closure zone when the zone effectiveness for the partial protection zone was lowest (*i.e.* the difference between the zone effectiveness between zones was greatest).

When the cost of partial protection zones relative to permanent closure zones decreased (*i.e.* zone cost of a partial protection zone went from 75% to 25% of zone cost of a permanent closure zone), the number of planning units selected into a conservation zone (either partial protection or permanent closure zone) increased, except for when the zone effectiveness was equal for each zone (Fig. 3). The difference between zone costs influenced the proportion of planning units allocated to each zone in all cases except when the partial protection zone effectiveness was equal to the permanent closure zone effectiveness. When zone effectiveness of the partial protection zone was equal (scenario 1) or lowest (scenario 4), the relative zone cost difference did not substantially influence the number of planning units allocated to the permanent closure zone.

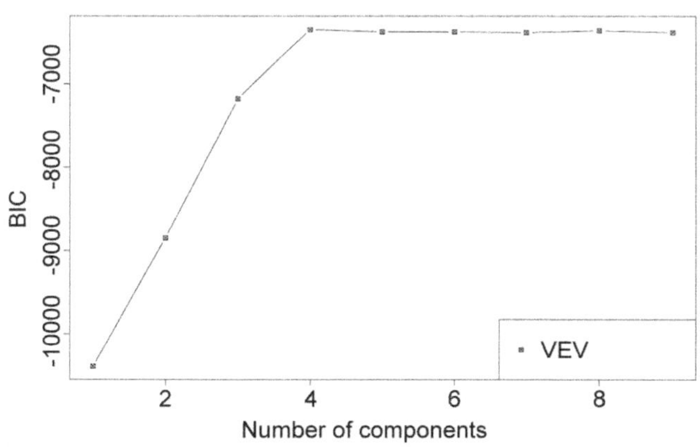

Figure 2. Result of classification of the solutions using multivariate normal mixture modeling. We used the sub-model of VEV (volumes of all clusters as varying (V), shapes of all clusters as equal (E), and orientation of all clusters as varying (V)) to identify the number of clusters across all solutions between scenarios (n = 400) with the highest Bayesian Information Criterion (BIC). BIC is defined as twice the log likelihood in this method [41], unlike as for ordinal BIC.

These trends were also seen in selection frequency of planning units. We defined planning units that were selected more than 50 times out of 100 Marxan runs as "high priority" planning units so we could compare the solutions among scenarios (Table 3). As the zone effectiveness of partial protection zones decreased, the number of high priority planning units for permanent closure zones increased and that for partial protection zones decreased for all zone costs (Table 3). When the zone effectiveness of partial protection zones at achieving conservation targets was $e_{jk}=1$ (equal to that of permanent closure zone, scenario 1), there were no high priority planning units allocated to the permanent closure zone. There were also no high priority planning units allocated to partial protection zone when the partial protection zone effectiveness was the lowest (Table 3). As the zone cost of partial protection zone varied from 75% to 25% of zone cost of permanent closure zone, the number of planning units selected into a conservation zone increased, including when the zone effectiveness was equal for each zone (Table 3).

The total opportunity cost almost doubled when zone effectiveness of the permanent closure zone and the partial protection zone were not equal (Fig. 4). The most costly solutions for marine zoning were produced when the zone effectiveness of the partial protection zone was low (scenario 4). The cost declined as the zone cost of the partial protection zone decreased relative to the permanent closure zone (Fig. 4).

When we explored the maximum equitable amount of fishing grounds remaining open for fishing while still meeting the conservation targets (scenario 5), we found 84%, 87%, 88% (for relative zone costs of partial protection zones to permanent closure zones of 25%, 50%, and 75%, respectively) of the areas of every traditional fishing ground can remain open for fishing. In contrast, in scenario 3 (average partial protection zone effectiveness), using the same zone effectiveness as in scenario 5, some traditional fishing grounds were excessively impacted because open zones for fishing were distributed inequitably. For example, without targets for fishing features (*i.e.* traditional fishing grounds), one traditional fishing ground had less than 40% of its area open for fishing while another traditional fishing ground remained entirely open for fishing. For scenario 5 where an equitable amount of fishing

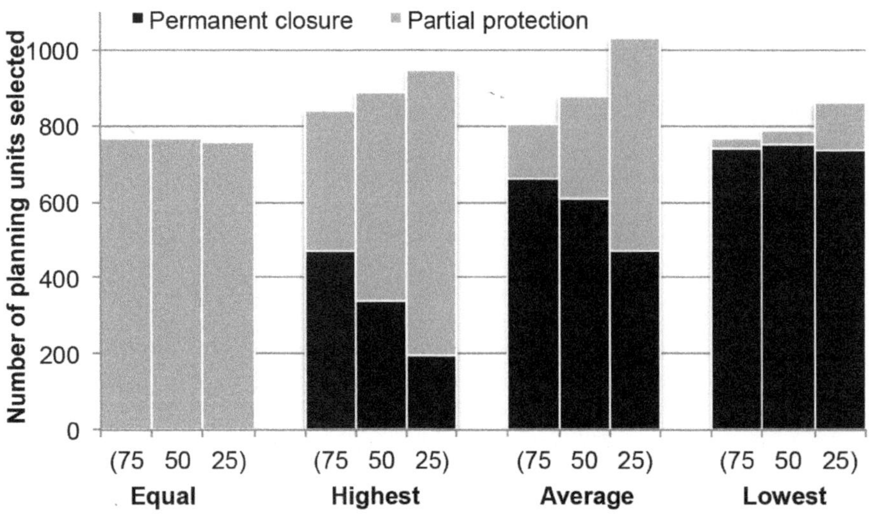

Figure 3. The number of planning units selected as a permanent closure and partial protection zone in the best solution. The allocation of selected planning units in the best solution (*i.e.* one solution that had the minimum score out of 100 runs) for each scenario. Scenarios used different zone effectiveness values (equal, highest, average, lowest) and relative zone costs. The numbers on the x-axis indicate the zone cost of the partial protection zone relative to the permanent closure zone (*i.e.* "75" means that the zone cost of partial protection zone is 75% of that of permanent closure zone).

grounds remained open for fishing, the overall reserve system costs were 14%, 15%, and 21% (relative zone costs of partial protection zones to permanent closure zones of 25%, 50%, and 75%, respectively) greater than scenario 3 (average partial protection zone effectiveness) that used the same zone effectiveness but did not set the fishing targets.

Discussion

Spatial prioritization traditionally assumes that a site is either inside, or outside, a protected area system. However, the ability of Marxan with Zones to place any planning unit in one of many zones opens up many options for more advanced fine-scale conservation planning [10,12]. Here, we focused on advancing conservation zoning by exploring the impact of differential zone effectiveness and zone costs on zoning plans at a MPA network scale. We found that differential zone effectiveness and zone costs influenced the location of priorities for different zones and the allocation of planning units into a particular zone. For example, highly effective and low cost zones were more likely to be selected. As zone effectiveness and relative zone cost changed, a trade-off between zone effectiveness and cost was seen in the number of planning units allocated to permanent closure or partial protection zones. This trade-off was not seen when zone effectiveness of the partial protection zone was equal, or low, relative to the permanent closure zone because zone effectiveness became the driver for priority selection.

It is important to consider differential effectiveness between different types of conservation zones to avoid missing priority areas that need to be prudently allocated to a permanent closure zone. When we assumed that the two conservation zones were equally effective at achieving conservation targets, planning units were

Table 3. The number of planning units that were selected more than 50 times out of 100 Marxan run in scenarios (1: equal, 2: highest, 3: average, 4: lowest zone effectiveness) with the different relative zone costs of partial protection zones to permanent closure zones of 25%, 50%, and 75%.

		Scenario			
Zone	**Relative zone cost**	**(1) Equal**	**(2) Highest**	**(3) Average**	**(4) Lowest**
Permanent closure	25%	0	161	457	690
Partial protection	25%	711	688	414	0
Open	25%	9306	9130	9009	9183
Permanent closure	50%	0	341	594	722
Partial protection	50%	692	409	23	0
Open	50%	9318	9158	9152	9255
Permanent closure	75%	0	448	663	722
Partial protection	75%	683	91	0	0
Open	75%	9329	9221	9244	9273

This table shows the selection frequency of a planning unit out of 100 Marxan runs.

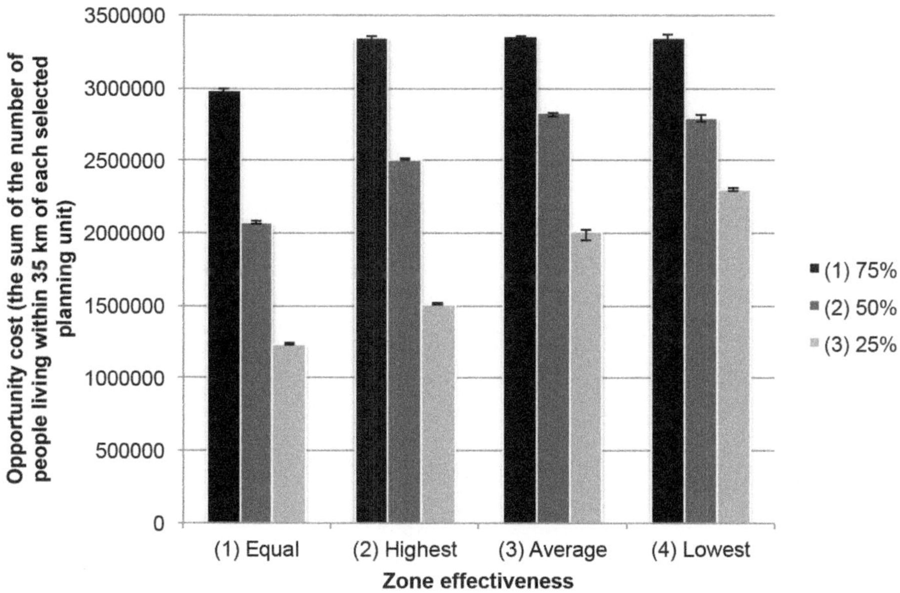

Figure 4. The average opportunity costs of ten best solutions of scenarios. Scenarios used different zone effectiveness values (1: equal, 2: highest, 3: average, 4: lowest zone effectiveness) with the different relative zone costs (75%, 50%, and 25%). The opportunity costs of fishing were obtained using the fishing pressure data and are the sum of the number of people living within 35 km of each planning unit.

only selected in the most inexpensive zone, the partial protection zone. Although this could be avoided by adding a zone-specific target for the permanent closure zone (see Klein et al. [10]), it does not accurately represent the contribution of different zones to conserving biodiversity.

Differential zone effectiveness and zone costs also influenced the opportunity costs of zoning plans. As we did not have spatially explicit opportunity cost data for our analysis, we developed a surrogate to represent fishing pressure (*i.e.* opportunity cost), but acknowledge that our model may not accurately represent fishing in Fiji and that planners in other regions should seek to use actual fishing data (or other human use data) to estimate the opportunity cost of a zone. However, the consideration of opportunity costs in planning is challenging, as spatially explicit data on opportunity costs is rarely available, especially at large spatial scales although the importance of including the opportunity costs of conservation when planning for MPA is well established [27,42]. This challenge is magnified when opportunity costs for more than one type of conservation zone are required. However, we demonstrated an approach to overcome this challenge when planning for multiple conservation zones by considering the relative cost between zones. Thus, planners can zone for multiple conservation zones with just two types of information on costs: (1) spatially explicit cost data for one zone; (2) relative difference in costs between conservation zones.

If the adverse impacts of MPAs are equitably spread amongst communities then we believe that a conservation plan is more likely to be implemented [43,44]. However, it is important to note that the efficiency of any plan is reduced when we try to increase social and/or economic equity (*i.e.* the total cost increased when an equitable amount of fishing grounds remained open for fishing while still meeting conservation targets) [44]. This is because the inclusion of conservation zones is shared equally across communities. We do not expect communities in Fiji to exactly adopt any of our zoning plans; instead, our zoning plans can be used in conjunction with local community expertise and/or other planning processes as supporting information when developing conservation plans. For example, there currently are projects underway in Fiji to develop provincial-level zoning plans for Ra and Bua provinces, for which our technique would be appropriate when combined with initial stakeholder consultation to achieve consensus on zone type, zone effectiveness and relative costs, as well as follow-up consultations to refine placement of spatial boundaries of zones.

We demonstrated a simple approach for considering differential levels of ecological zone effectiveness when designing marine protected areas with different zones. Further, we showed how consideration of zone effectiveness affects the location and cost of spatial priorities for different zones. This approach will be informative to other places where more types of zones and human activities are being planed for, such as mining, wind farming, tourism, and other types of fisheries management zones. Such an approach would require information about the contribution of each zone to achieving conservation and industry targets, as well as the costs associated with designation of each zone. Although zone effectiveness values from experts, like those used here, are valuable, we recognize that an evaluation using empirical data would improve the robustness of values [45,46]. In addition, quantifying the relationship between the zone effectiveness and the zone size is needed as the size of a managed area may change its overall effectiveness [47,48]. For example, a permanent no-take area smaller than the size of the home range of targeted fish species will not provide full protection for those species [49]. Further, we did not take into account that zone effectiveness for a given zone can differ from place to place [22,38,39] as data were unavailable; however, given available data, this could be considered in our zoning approach. Finally, applying not only ecological effectiveness but also management effectiveness of existing closures in addition to the ecological effectiveness would improve our results.

We suggest that application of our approach should consider more realistic cost estimates for each zone. We estimated the zone cost based on a surrogate measure for foregone profit from fishing (*i.e.*, opportunity cost). Opportunity cost estimates would be improved if supported by empirical data gathered through interviews with fishermen or other fishing reports [28,50]. Further, there are other types of financial costs associated with establishing and managing MPAs, including management and transaction costs [26,51]. We did not consider MPA management costs because MPAs generally depend on voluntary compliance in Fiji and the relative zone costs are uncertain. Including cost information not only for fishers but also for different stakeholders would enhance our study to reflect the reality of zoning. We arbitrarily set three different relative zone costs. Assessing the zone costs using opinions from experts, empirical data or traditional ecological knowledge could improve our results [52]. Finally, our approach was static in that it did not consider the spillover effects of MPAs or the redistribution of fishing effort after protection. Incorporating these dynamic process into MPA planning is important yet complex, highlighting areas of further research.

Marine zoning requires identifying and involving stakeholders and deciding conservation objectives, the same as traditional conservation planning [53]. Recommendations and principles for designing MPAs on size, shape, connectivity and target amounts exist (*e.g.* McLeod et al. [54] and Fernandes et al. [55]). Additional guidelines are required for marine zoning. The first additional step is to list activities allowed in each zone. The activities guide table for the Great Barrier Reef Marine Park is a useful example for this stage (see http://www.gbrmpa.gov.au/zoning-permits-and-plans/zoning/zoning-maps. Accessed 30 September 2013.). The second step it to decide which zones should be included, which will depend on the conservation context. This includes clarifying the purpose of the zone, *i.e.* is it for biodiversity, fishery, customs/tradition, or industries. The third step is to assess the zone effectiveness by empirical means or expert elicitation. The fourth step is to decide how much of each feature should be represented in each type of zone. The final step is to calculate or estimate the cost of zones (*e.g.* relative cost compared with permanent zone). Although effective solutions can be produced using a systematic conservation tool, like Marxan with Zones, stakeholders should also be a part of zoning decisions [56].

Marine zoning has the ability to accommodate objectives of multiple users and minimize conflict between different resource users and stakeholders [3,4]. However, the effects of differential effectiveness between zones had not been investigated in systematic conservation planning. In this study, we explored the impact of differential zone effectiveness on the location and costs of conservation priorities to enable more reliable conservation planning and help achieve conservation targets more effectively in Fiji. Our approach is applicable to any country, especially in places where strong traditional management practices with partial or temporary closure systems exist and permanent closure zones are unlikely to be accepted because of concerns for food security and cultural and social factors [57,58].

Acknowledgments

We are grateful for support from the Australian Research Council Centre of Excellence for Environmental Decisions and the Wildlife Conservation Society Fiji Country Program. We are grateful to the Fiji Locally Managed Marine Area network for sharing data on the location of existing community-managed fishing closures. We thank Dr. Naoki H. Kumagai for his statistical support.

Author Contributions

Conceived and designed the experiments: AM CJK MB SDJ HPP. Performed the experiments: AM. Analyzed the data: AM. Contributed reagents/materials/analysis tools: AM CJK MB SDJ HPP. Wrote the paper: AM CJK MB SDJ HPP. Obtained permission for use of the data: AM SDJ.

References

1. Crowder LB, Osherenko G, Young OR, Airamé S, Norse EA, et al. (2006) Resolving mismatches in U.S. ocean governance. Science 313: 617–618.
2. Day JC (2002) Zoning—lessons from the Great Barrier Reef Marine Park. Ocean Coast Manag 45: 139–156.
3. Douvere F (2008) The importance of marine spatial planning in advancing ecosystem-based sea use management. Marine Policy 32: 762–771.
4. Brown K, Adger WN, Tompkins E, Bacon P, Shim D, et al. (2001) Trade-off analysis for marine protected area management. Ecol Econ 37: 417–434.
5. Babcock RC, Shears NT, Alcala AC, Barrett NS, Edgar GJ, et al. (2010) Decadal trends in marine reserves reveal differential rates of change in direct and indirect effects. Proc Natl Acad Sci U S A 107: 18256–18261.
6. Halpern BS (2003) The impact of marine reserves: do reserves work and does reserve size matter? Ecological Applications 13: 117–137.
7. Halpern BS, Warner RR (2002) Marine reserves have rapid and lasting effects. Ecol Lett 5: 361–366.
8. Mumby PJ, Dahlgren CP, Harborne AR, Kappel CV, Micheli F, et al. (2006) Fishing, trophic cascades, and the process of grazing on coral reefs. Science 311: 98–101.
9. Fernandes L, Day JON, Lewis A, Slegers S, Kerrigan B, et al. (2005) Establishing representative no-take areas in the Great Barrier Reef: Large-scale implementation of theory on marine protected areas. Conserv Biol 19: 1733–1744.
10. Klein CJ, Steinback C, Watts M, Scholz AJ, Possingham HP (2010) Spatial marine zoning for fisheries and conservation. Front Ecol Environ 8: 349–353.
11. Agostini VN, Margules SW, Schill SR, Knowles JE, Blyther RJ (2010) Marine zoning in Saint Kitts and Nevis: A path towards sustainable management of marine resources. The Nature Conservancy, Florida, USA. 288p. Available: http://www.marineplanning.org/pdf/StKitts_Nevis_Full_Report.pdf Accessed 30 September 2013.
12. Grantham HS, Agostini VN, Wilson J, Mangubhai S, Hidayat N, et al. (2013) A comparison of zoning analyses to inform the planning of a marine protected area network in Raja Ampat, Indonesia. Marine Policy 38: 184–194.
13. Villa F, Tunesi L, Agardy T (2002) Zoning marine protected areas through spatial multiple-criteria analysis: the case of the Asinara island national marine reserve of Italy. Conserv Biol 16: 515–526.
14. Portman ME (2007) Zoning design for cross-border marine protected areas: The Red Sea Marine Peace Park case study. Ocean Coast Manag 50: 499–522.
15. Bruce EM, Eliot IG (2006) A spatial model for marine park zoning. Coast Manage 34: 17–38.
16. Merrifield MS, McClintock W, Burt C, Fox E, Serpa P, et al. (2013) MarineMap: A web-based platform for collaborative marine protected area planning. Ocean Coast Manag 74: 67–76.
17. Lahdelma R, Salminen P, Hokkanen J (2000) Using multicriteria methods in environmental planning and management. Environ Manage 26: 595–605.
18. Mills M, Jupiter SD, Pressey RL, Ban NC, Comley J (2011) Incorporating effectiveness of community-based management in a national marine gap analysis for Fiji. Conserv Biol 25: 1155–1164.
19. Sciberras M, Jenkins S, Kaiser M, Hawkins S, Pullin A (2013) Evaluating the biological effectiveness of fully and partially protected marine areas. Environmental Evidence 2: 4.
20. Chape S, Harrison J, Spalding M, Lysenko I (2005) Measuring the extent and effectiveness of protected areas as an indicator for meeting global biodiversity targets. Philos Trans R Soc Lond B Biol Sci 360: 443–455.
21. Lester SE, Halpern BS, Grorud-Colvert K, Lubchenco J, Ruttenberg BI, et al. (2009) Biological effects within no-take marine reserves: a global synthesis. Mar Ecol Prog Ser 384: 33–46.
22. Lester SE, Halpern BS (2008) Biological responses in marine no-take reserves versus partially protected areas. Mar Ecol Prog Ser 367: 49–56.
23. Mora C, Andréfouët S, Costello MJ, Kranenburg C, Rollo A, et al. (2006) Coral reefs and the global network of marine protected areas. Science 312: 1750–1751.
24. Halpern BS, Walbridge S, Selkoe KA, Kappel CV, Micheli F, et al. (2008) A global map of human impact on marine ecosystems. Science 319: 948–952.
25. Shahabuddin G, Rao M (2010) Do community-conserved areas effectively conserve biological diversity? Global insights and the Indian context. Biol Conserv 143: 2926–2936.
26. Naidoo R, Balmford A, Ferraro PJ, Polasky S, Ricketts TH, et al. (2006) Integrating economic costs into conservation planning. Trends Ecol Evol 21: 681–687.
27. Ban NC, Klein CJ (2009) Spatial socioeconomic data as a cost in systematic marine conservation planning. Conservation Letters 2: 206–215.
28. Scholz AJ, Steinback C, Kruse SA, Mertens M, Silverman H (2011) Incorporation of spatial and economic analyses of human-use data in the design of marine protected areas. Conserv Biol 25: 485–492.

29. Klein CJ, Jupiter SD, Selig ER, Watts ME, Halpern BS, et al. (2012) Forest conservation delivers highly variable coral reef conservation outcomes. Ecol Appl 22: 1246–1256.
30. Leslie H, Ruckelshaus M, Ball IR, Andelman S, Possingham HP (2003) Using siting algorithms in the design of marine reserve networks. Ecol Appl 13: 185–198.
31. Tallis H, FerdaÑA Z, Gray E (2008) Linking terrestrial and marine conservation planning and threats analysis. Conserv Biol 22: 120–130.
32. Halpern BS, Klein CJ, Brown CJ, Beger M, Grantham HS, et al. (2013) Achieving the triple bottom line in the face of inherent trade-offs among social equity, economic return, and conservation. Proc Natl Acad Sci U S A published ahead of print March 25 2013, doi:10.1073/pnas.1217689110
33. Govan H, Tawake A, Tabunakawai K, Jenkins A, Lasgorceix A, et al. (2009) Status and potential of locally-managed marine areas in the Pacific Island Region: meeting nature conservation and sustainable livelihood targets through wide-spread implementation of LMMAs. SPREP/WWF/WorldFish-Reefbase/CRISP, Suva, Fiji. 148p. Available: http://www.spc.int/DigitalLibrary/Doc/FAME/Reports/Govan_09_Status_LMMAs.pdf Accessed 30 September 2013.
34. Cohen PJ, Foale SJ (2013) Sustaining small-scale fisheries with periodically harvested marine reserves. Marine Policy 37: 278–287.
35. Watts ME, Ball IR, Stewart RS, Klein CJ, Wilson K, et al. (2009) Marxan with Zones: Software for optimal conservation based land- and sea-use zoning. Environ Model Softw 24: 1513–1521.
36. Teh LCL, Teh LSL, Starkhouse B, Sumaila UR (2009) An overview of socio-economic and ecological perspectives of Fiji's inshore reef fisheries. Marine Policy 33: 807–817.
37. Burke L, K Reytar, M Spalding, Perry A (2011) Reefs at risk revisited. World Resources Institute,Washington, D C, USA 114p. Available: http://pdf.wri.org/reefs_at_risk_revisited.pdf Accessed 30 September 2013.
38. Bartlett CY, Manua C, Cinner J, Sutton S, Jimmy R, et al. (2009) Comparison of outcomes of permanently closed and periodically harvested coral reef reserves. Conserv Biol 23: 1475–1484.
39. Cinner J, Marnane MJ, McClanahan TR, Almany GR (2006) Periodic closures as adaptive coral reef management in the Indo-Pacific. Ecology and Society 11: 31.
40. Fraley C, Raftery AE (2002) Model-based clustering, discriminant analysis, and density estimation. Journal of the American Statistical Association 97: 611–631.
41. Fraley C, Raftery AE, Murphy TB, Scrucca L (2012) mclust version 4 for R: normal mixture modeling for model-based clustering, classification, and density estimation. Technical Report No. 597 Department of Statistics, University of Washington.
42. Adams VM, Mills M, Jupiter SD, Pressey RL (2011) Improving social acceptability of marine protected area networks: A method for estimating opportunity costs to multiple gear types in both fished and currently unfished areas. Biol Conserv 144: 350–361.
43. Alcala AC, Russ GR (2006) No-take marine reserves and reef fisheries management in the Philippines: A new people power revolution. Ambio 35: 245–254.
44. Halpern BS, Klein CJ, Brown CJ, Beger M, Grantham HS, et al. (2013) Achieving the triple bottom line in the face of inherent tradeoffs among social equity, economic return and conservation. Proc Natl Acad Sci U S A 110: 6229–6234.
45. Edwards HJ, Elliott IA, Pressey RL, Mumby PJ (2010) Incorporating ontogenetic dispersal, ecological processes and conservation zoning into reserve design. Biol Conserv 143: 457–470.
46. Agardy T, Bridgewater P, Crosby MP, Day J, Dayton PK, et al. (2003) Dangerous targets? Unresolved issues and ideological clashes around marine protected areas. Aquat Conserv 13: 353–367.
47. Claudet J, Osenberg CW, Benedetti-Cecchi L, Domenici P, García-Charton J-A, et al. (2008) Marine reserves: size and age do matter. Ecol Lett 11: 481–489.
48. Weeks R, Russ GR, Alcala AC, White AT (2010) Effectiveness of marine protected areas in the Philippines for biodiversity conservation. Conserv Biol 24: 531–540.
49. Gaines SD, White C, Carr MH, Palumbi SR (2010) Designing marine reserve networks for both conservation and fisheries management. Proc Natl Acad Sci U S A 107: 18286–18293.
50. Yates KL, Schoeman DS (2013) Spatial access priority mapping (SAPM) with fishers: a quantitative GIS method for participatory planning. PLoS One 8: e68424.
51. Balmford A, Gravestock P, Hockley N, McClean CJ, Roberts CM (2004) The worldwide costs of marine protected areas. Proc Natl Acad Sci U S A 101: 9694–9697.
52. Berkes F, Colding J, Folke C (2000) Rediscovery of traditional ecological knowledge as adaptive management. Ecol Appl 10: 1251–1262.
53. Pressey RL, Bottrill MC (2009) Approaches to landscape- and seascape-scale conservation planning: convergence, contrasts and challenges. Oryx 43: 464–475.
54. McLeod E, Salm R, Green A, Almany J (2009) Designing marine protected area networks to address the impacts of climate change. Front Ecol Environ 7: 362–370.
55. Fernandes L, Green A, Tanzer J, White A, Alino P, et al. (2012) Biophysical principles for designing resilient networks of marine protected areas to integrate fisheries, biodiversity, and climate change objectives in the Coral Triangle. A Project of the Coral Triangle Support Partnership. 152p. Available: http://www.coraltriangleinitiative.org/sites/default/files/resources/MPA%20Network%20Design_Principles_Full%20Report_FINAL_CTSP%20Jan%2023%202012.pdf Accessed 30 September 2013.
56. Pomeroy R, Douvere F (2008) The engagement of stakeholders in the marine spatial planning process. Marine Policy 32: 816–822.
57. Foale S, Cohen P, Januchowski-Hartley S, Wenger A, Macintyre M (2011) Tenure and taboos: origins and implications for fisheries in the Pacific. Fish Fish 12: 357–369.
58. McClanahan TR, Cinner JE (2008) A framework for adaptive gear and ecosystem-based management in the artisanal coral reef fishery of Papua New Guinea. Aquat Conserv 18: 493–507.

19

The Use of a Predictive Habitat Model and a Fuzzy Logic Approach for Marine Management and Planning

Tarek Hattab[1,2]*, Frida Ben Rais Lasram[1], Camille Albouy[3], Chérif Sammari[4], Mohamed Salah Romdhane[1], Philippe Cury[2], Fabien Leprieur[5], François Le Loc'h[2]

1 UR 03AGRO1 Ecosystèmes et Ressources Aquatiques, INAT (Institut National Agronomique de Tunisie), Tunis, Tunisia, **2** UMR 212 Ecosystèmes Marins Exploités, IRD (Institut de Recherche pour le Développement), Sète, France, **3** Département de biologie, Chimie et Géographie, UQAR (Université du Québec à Rimouski), Québec, Canada, **4** Laboratoire du milieu marin, INSTM (Institut National des Sciences et Technologies de la Mer), Salammbô, Tunisia, **5** Laboratoire Ecologie des Systèmes Marins Côtiers UMR 5119, UM2 (Université de Montpellier 2), Montpellier, France

Abstract

Bottom trawl survey data are commonly used as a sampling technique to assess the spatial distribution of commercial species. However, this sampling technique does not always correctly detect a species even when it is present, and this can create significant limitations when fitting species distribution models. In this study, we aim to test the relevance of a mixed methodological approach that combines presence-only and presence-absence distribution models. We illustrate this approach using bottom trawl survey data to model the spatial distributions of 27 commercially targeted marine species. We use an environmentally- and geographically-weighted method to simulate pseudo-absence data. The species distributions are modelled using regression kriging, a technique that explicitly incorporates spatial dependence into predictions. Model outputs are then used to identify areas that met the conservation targets for the deployment of artificial anti-trawling reefs. To achieve this, we propose the use of a fuzzy logic framework that accounts for the uncertainty associated with different model predictions. For each species, the predictive accuracy of the model is classified as 'high'. A better result is observed when a large number of occurrences are used to develop the model. The map resulting from the fuzzy overlay shows that three main areas have a high level of agreement with the conservation criteria. These results align with expert opinion, confirming the relevance of the proposed methodology in this study.

Editor: Joshua S. Madin, Macquarie University, Australia

Funding: This work was partly funded by the Institut de Recherche pour le Développement(http://www.ird.fr/) and by the 'Fondation pour la Recherche sur la Biodiversité' (http://www.fondationbiodiversite.fr/). The funders had no role in study design, data collection and analysis, decision to publish, or preparation of the manuscript.

Competing Interests: The authors have declared that no competing interests exist.

* E-mail: hattab.tarek@gmail.com

Introduction

Understanding species distributions is essential for conservation planning and forecasting [1], particularly in the present context of stock depletion and species extinction. Species distribution maps play an important role in developing spatial management measures such as the identification of Essential Fish Habitats or the establishment Marine Protected Areas (MPAs) [2,3], which in turn contribute to sustainable ecosystem-based marine management [3–5]. However, information on the true distribution of many marine organisms remains limited, particularly for species that are difficult to detect [6]. Modelling species distributions based on data samples is one solution to address this lack of knowledge. For instance, Species Distribution Models (SDMs) relate species' distributions based on data samples with the associated environmental and geographical characteristics of the surveyed locations [7].

Nevertheless, the perception remains that the distributions of marine species are uncertain and dependent on the sampling process used to generate the models. Bottom trawls are commonly used as a sampling technique to assess the spatial distribution of commercial species and to obtain fisheries-independent abundance data [8]. However, there are some limitations in the species detections associated with this technique due to a range of factors, e.g., catchability, gear efficiency, and gear-specific selectivity [8]. A common problem in recording species' distribution results from a false absence, which occurs when a species is not available for capture despite occupying the site, or when a species occurs at a site but is simply not captured. Such data can severely limit the fit of many SDMs [9] and can decrease the reliability of prediction models (see [10,11]). The issue of false absences is further complicated by locations with favourable environmental conditions but where species are absent due to biotic interactions, dispersal limitations or fishing pressure. This latter case is particularly critical when modelling commercial exploited species [12]. Yet the problem of imperfect detection when modelling marine species distributions has been rarely mentioned in the literature [13].

Confirmed absences are very difficult to obtain, especially from bottom trawl survey data and for mobile species. A much higher level of sampling effort is required to ensure their reliability relative to presence data [14]. To cope with the lack of confirmed absence data, presence-only models, or profile techniques, have often been used [12]. These models differ from the group discrimination approaches that require presence–absence or abundance data [15]. Some well-known examples of profile-type models are the Ecological Niche Factor Analysis (ENFA; [16]), the Genetic Algorithm for Rule-Set Prediction [17], and the maximum entropy method [18]. Comparisons between the various SDMs reveal that group discrimination approaches tend to perform better than the profile-type, or presence-only models [19–21]. A common problem of the profile-type models is that their predictions are often overly optimistic, i.e., they predict the species occurring at too many locations [22].

In order to use group discrimination approaches, artificial absence data are increasingly being used in situations where no confirmed absence data are available. Typically called pseudo-absences, artificial absences are generated and inserted into the selected model in lieu of confirmed absence data. The method selected to generate the pseudo-absences is particularly important because it can influence the final quality of the model [21,22]. Several approaches have been suggested for the generation of pseudo-absences: *(i)* a random selection of absence points across the entire available area (e.g., [21,23]; *(ii)* a random selection with a geographically-weighted exclusion (e.g., [24]; and *(iii)* a selection of sampled locations (i.e., occurrence locations for other species) at which the target species has not been recorded (e.g., [5]). However, there are several issues with these approaches. The first two may produce false absences, even in environmentally-favourable areas for the species [25,26]. The latter approach is unsuitable for bottom trawl survey data because it is unlikely that an entire target group of species caught by a particular type of fishing gear will share a similar sampling bias.

To deal with presence-only records, Engler *et al.* [22] suggested an intermediate methodology between presence-only and presence-absence distribution models. This approach proposed the use of a habitat suitability map as a way of selecting weighted pseudo-absence data points. These points are added to the original presence-only data and used to improve the logistic regression procedure. This mixed method could be more suitable for bottom trawl survey data than the other approaches outlined above.

Distribution models often result in a predicted probability surface that is then translated into a presence-absence classification map for use in different conservation applications [27]. For instance, these presence-absence maps are commonly aggregated (the 'predict first, assemble later' strategy from [28]) to identify areas that will experience the greatest changes in species composition due to climate change [29]. To convert a probability surface into a binary map, a number of threshold selection methods have been proposed [27]. Given the variety of approaches available to generate a dataset, the method chosen can have a dramatic effect on a model's accuracy and its predictions [27,30], as well as the subsequent conservation planning decisions and outcomes [31]. To address the uncertainty in the presence-absence classification map, we propose the use of a fuzzy logic approach that can be applied in situations where vagueness and uncertainty exist. The fuzzy logic approach has several advantages in situations where: *(i)* there are no clear cut definitions; and *(ii)* results cannot be categorised as either 0 or 1 [32]. Given its current application in a range of scientific disciplines [33], it is logical to extend the use of fuzzy logic to the threshold selection procedure in the context of SDMs.

Predicting the distributions of commercially targeted marine species is particularly urgent in over-exploited and damaged ecosystems, such as the Gulf of Gabes. With a soft bottom, shallow slope and a high diversity of fishes, the Gulf of Gabes is the most important fishing ground in Tunisia [34,35]. This coastal area supports 60% of the national fishing fleet and contributes 42% of the national annual fish and crustacean production [36]. This intense fishing activity began in the early 1980s, after which fisheries experienced large fluctuations in landings until the late 1990s when catches began to progressively decline [35]. This was mainly due to the decreasing of demersal stocks caused by intense bottom trawling activities. Depletion rates are so alarming that in the near future the Gulf of Gabes will be subject to a habitat conservation management plan that excludes trawling activities.

In this study, we initially assess the reliability of the mixed approach proposed by Hengl *et al.* [37]. We combine ENFA predictions for generating pseudo-absences and regression-kriging (RK) for modelling the spatial distribution of 27 commercially-targeted species in the Gulf of Gabes based on bottom trawls survey data. We then propose a fuzzy logic framework to transform modelled probabilities of occurrence into binary predictions of species presence and absence. We illustrate this framework by applying it to the task of identifying areas that meet the conservation targets for the deployment of artificial anti-trawling reefs (AARs) in the Gulf of Gabes.

Materials and Methods

Study Area

The Gulf of Gabes is located in the southern Mediterranean Sea and covers the second widest continental shelf area of this semi-enclosed sea. The Gulf of Gabes has high fisheries productivity and it serves as a feeding and reproduction area for numerous populations of fishes and crustaceans [34]. Indeed, this ecosystem supports one of the most extensive biocenosis of seagrass (*Posidonia oceanica*; [38]), which constitutes a major nursery site for several marine species [39]. Accordingly, the Gulf of Gabes is one of the most productive ecosystems in the Mediterranean Sea and has great economic and ecological importance [38].

Bottom trawling is the predominant fishing activity in this area and the gear type that has the largest impact on the target demersal fishes [34]. The regular incursions of trawlers into areas that are shallower than their regulated depth have led to the extensive degradation of *P. oceanica* meadows [40]. Due to the lack of monitoring and surveillance activities carried out by the marine police and fishery authorities, illegal fishing still takes place. Consequently, effective management measures are required to prevent illegal fishing activities.

The proposed fisheries management plan for the Gulf of Gabes includes a perimeter of AARs that combine anti-trawling structures with artificial reefs. The major functions of these structures will be to protect the coastal zone marine ecosystems and species from the mechanical impacts of trawling. This measure will be especially important for high diversity communities, such as the *P. oceanica* seagrass beds and associated fauna of biological interest [41]. In addition, AARs aim to reduce the fishing mortality of commercial species, to protect nursery areas and juvenile fish, to create new fishing grounds and/or improve existing grounds, and to increase natural productivity [42]. The AARs will be deployed in areas that are chosen by considering both the presence of favourable habitats of commercial species and areas of high-density *P. oceanica* seagrass beds.

Bottom Trawl Survey Data and Model Variables

Species occurrence data were collected from the Tunisian bottom trawl survey database gathered by the National Institute of Marine Sciences and Technologies (INSTM, Tunisia) onboard the R/V Hannibal. The sampling net used (vertical opening trawl, GOV: 42/55) had a 20 mm diamond stretched mesh at the cod end and a 15 m horizontal opening. The trawl was towed at a speed of 2.9 knots for one hour and sampling areas were fixed

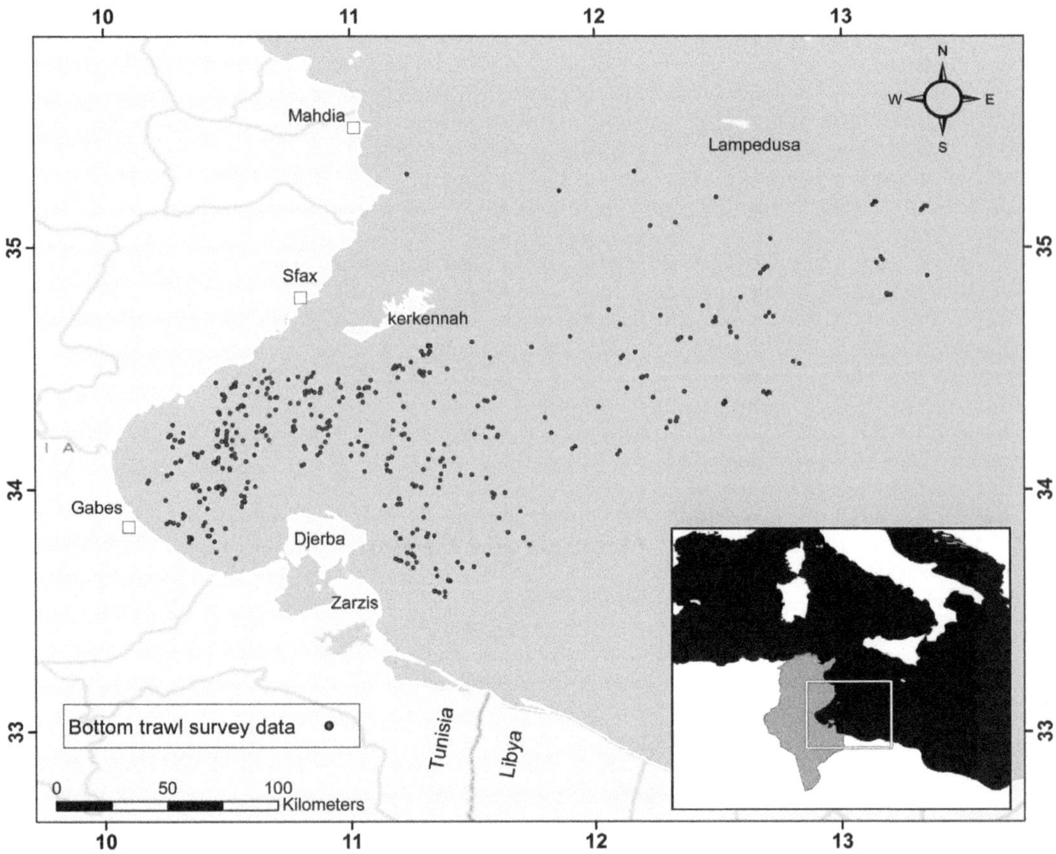

Figure 1. The Gulf of Gabes study area with black points indicating the location of the benthic trawls used in this study.

according to a stratified random sampling based on three depth strata (0–25 m, 25–65 m, and 65–200 m). A total of 360 trawl hauls were completed between 1998 and 2005. Central point geo-referenced position data for each trawl haul and the associated catch contents were extracted from the database (Figure 1).

To build the SDMs, we selected four local habitat variables (depth, slope, aspect, and seafloor type) and one spatial predictor (distance to shore). Previous studies undertaken at a similar spatial extent showed that these variables had a strong influence on species distributions in coastal environments (Figure 2) (e.g., slope [43,44], depth [5,45], aspect [43,44,46], distance to shore [47,48], and seafloor type [5]).

The bathymetry of the Gulf of Gabes was extracted as a digital gridded depth data set from a digital elevation model with a 90 m resolution. The bathymetric elevation data was derived from source soundings collected by the INSTM and referenced to the local tidal datum. The seafloor bathymetric slope and aspect were derived from the bathymetric base map using a 3×3 cell neighbourhood window around the processing cell. Respectively, these represent the rate of change in bathymetry and the azimuthal direction of the steepest slope over the analysis window. The aspect was transformed into two derived variables: Eastness (values close to 1 represent an eastward aspect, while values close to −1 represent a westward aspect) and Northness (values close to 1 represent a northward aspect, while values close to −1 represent a southward aspect).

To develop the map of the distance to shore, Euclidean distances were calculated from the shorelines and islands throughout the study area based on a gridded map with a 90 m resolution. Release and recapture locations were then sampled using ArcGIS 10 to obtain distance values. Digital seafloor type data were obtained from the INSTM. The original 15 seafloor types were grouped into eight broader seafloor types that represent the relatively distinct physical environments thought to influence the distributions of demersal marine species. We mapped the seafloor type by attributing a seafloor category to the center of each 0.0081 km^2 grid cell.

Selected Species

Since the advent of bottom trawl surveys in the Gulf of Gabes, 152 different species of fishes, cephalopods, and crustaceans have been identified. As small sample sizes pose challenges to any statistical analyses and result in decreased predictive potential, we decided to concentrate on the relatively common species (defined here as species present in >10% of trawl hauls). Having applied this criterion we retained a total of 27 species: 20 fishes, four cephalopods, two decapods, and one stomatopod (Table 1). Collectively, these species constituted 60% of total biomass of the landings and 98% of bottom trawl landings in the Gulf of Gabes [36].

The Modelling Framework

Confirmed absence data were not available for the 27 species selected for this study. As group discrimination approaches usually perform better than presence-only models (as discussed above), we selected a hybrid approach. This approach combines group discrimination and profile techniques to model species distributions (Figure 3) [22,25,37,49].

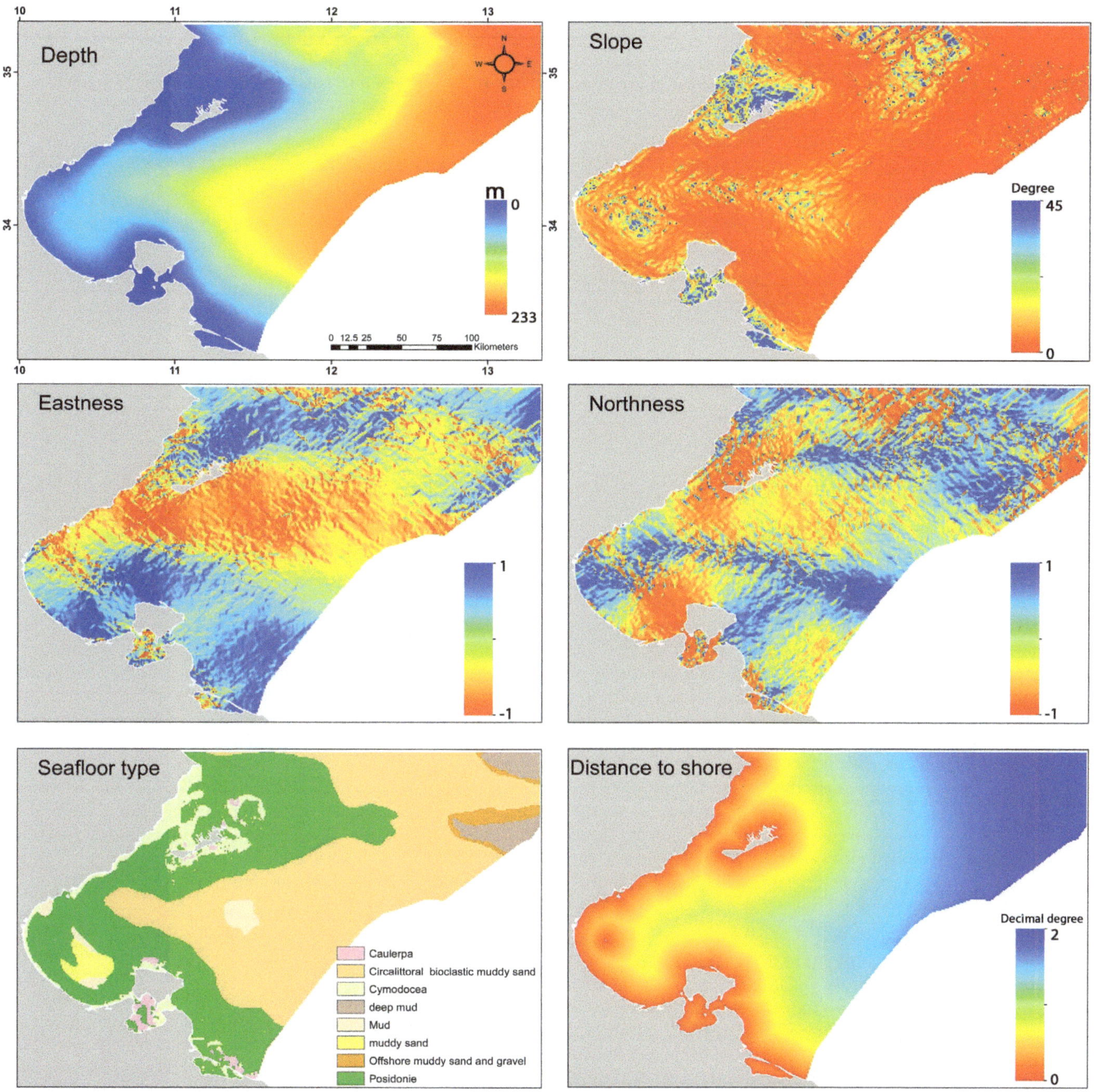

Figure 2. Maps of the six habitat variables (including two derived variables) used to generate the predicted distribution maps.

Presence-only models. ENFA [16] was used to create an habitat suitability map that depicts areas where species are unlikely to occur. ENFA is a specific ordination technique that compares a species' environmental niche and the environmental characteristics of the study area and assigns a degree of suitability to each point on a map (typically from 0 to 100). Thus, it quantifies the dissimilarity between an ecological niche and the ecological space. The first component of the technique, called the marginality factor, is defined as the standardised difference between the centroids of the ecological space and the ecological niche. The second component, the specialisation factors, are successively extracted from the n–1 residual dimensions and represent the narrowness of the ecological niche relative to the ecological space [16]. ENFA was preferred to another widely used presence-only model, namely maximum entropy method, as several studies recently showed that this technique may lead to spurious inferences (e.g., [50]). Then, ENFA is recognized as one of the best presence-only methods to model habitat suitability for marine species (see [6,51]).

To evaluate the accuracy of the ENFA, we performed a Monte-Carlo randomisation test with 100 permutations. This test for the significance of the marginality factor by randomising the locations of selected species within the study area. At each permutation the ENFA was performed on the random locations and the results

Table 1. Code, scientific name and the number of occurrences of each of the 27 species modelled in this study.

Code	Scientific name	Number of occurrences	Code	Scientific name	Number of occurrences
BALICAR	*Balistes carolinensi*	87	OCTOVUL	*Octopus vulgaris*	77
BOOPBOO	*Boops boops*	299	PAGEERY	*Pagellus erythrinus*	223
DIPLANN	*Diplodus annularis*	190	PAGRCAE	*Pagrus caeruleosticutus*	85
ELEDMOS	*Eledone moschata*	64	PENAKER	*Penaeus kerathurus*	219
ENGRENC	*Engraulis encrasicolus*	47	POMTSAL	*Pomatomus saltator*	51
GOBINIG	*Gobius niger*	61	SCOMSCO	*Scomber scombrus*	95
LITHMOR	*Lithognathus mormyrus*	111	SEPIOFF	*Sepia officinalis*	270
LOLIVUL	*Loligo vulgaris*	134	SERAHEP	*Serranus hepatus*	34
MERLMER	*Merluccius merluccius*	164	SOLEAEG	*Solea aegyptica*	104
METAMON	*Metapenaus monoceros*	46	SPARAUR	*Sparus aurata*	34
MUGICEP	*Mugil cephalus*	38	SPICMAE	*Spicara maena*	62
MULLBAR	*Mullus barbatus*	191	SQUIMAN	*Squilla mantis*	43
MULLSUR	*Mullus surmuletus*	69	TRACTRA	*Trachurus trachurus*	169
MUSTMUS	*Mustelus mustelus*	62			

were then evaluated against the observed locations. We ran the ENFA using the *R* adehabitat package [52,53].

Simulation of pseudo-absences. To generate pseudo-absence data, two methods were used to choose geographic coordinates: (*i*) at random across the Gulf of Gabes; and (*ii*) weighted by ENFA predictions and the geographical location of presence-only records [37]. We undertook both methods with the aim of assessing the level of improvement gained by using the environmentally- and geographically-weighted method as compared with the random method. The weighted method proposed by Hengl *et al.* [37] is based on both the Habitat Suitability Index (HSI, derived through ENFA) and the distance from the observations that are subsequently used to weight pseudo-absence points (Figure 3). Given that HSI values are scaled between 0 and 100, Hengl *et al.* [37] defined the probability distribution (τ) used to generate the pseudo-absence locations as:

$$\tau(x*)=\left[\frac{d_R(x)+(100-HSI(x))}{2}\right]^2 \quad (1)$$

where d_R is the normalised distance in the range [0, 100%], i.e., the distance from the observation points divided by the maximum distance. The square term is used to ensure that there are progressively more pseudo-absences at the edge of low HSI and large distances will approximately follow a Poisson distribution. In this way, pseudo-absences are located both in areas of low HSI (unsuitable habitat) and further away from the occurrence locations.

Based on Equation 1, the HSI and the map with buffers around the occurrences were combined to create a weighted map. We then performed the random generation of points with a probability density proportional to the values of the weighted map. To account for the variability arising from this weighted selection, 10 groups of pseudo-absences were generated. This allowed us to assess the stability of the final predictions for different simulations. For each set, the number of simulated pseudo-absences was equal to the number of presences. This is supported by the statistical theory of model-based designs, also known as "D-designs" [37]. According to this theory, the optimal design to minimise prediction variance is when an equal number of observations are at opposite value extremes [37,54] and there is a higher spreading in the feature space.

A total of 11 groups of pseudo-absences were obtained. One group was generated entirely at random and the remaining 10 groups were weighted by ENFA predictions and the geographical locations of the presence-only records (Figure 3).

Regression modelling. Once the pseudo-absences were simulated, they could then be combined with the occurrence locations to build a regression model to predict the probability distribution of occurrences. Prior to running the regression analysis, the six original habitat predictors were converted to principal components (to reduce their dimensions and the multicollinearity effect) using the Hill-Smith ordination method [55] that deals with mixed variable types (i.e., quantitative and factor).

We used a generalised linear model (GLM) [56] for the regression analysis, assuming a binomial error in the response variable. Model residuals were then analysed by fitting a variogram to assess their level of spatial dependence (Figure 3). Model residuals exhibited no spatial dependence for six species (i.e., the squid, *Loligo vulgaris*; the stomatopod, *Squilla mantis*; the demersal fishes, *Merlucius merlucius*, *Mugil cephalus*, *Mullus barbatus*, and *Mullus surmuletus*; and the pelagic fish, *Trachurus trachurus*). For the remaining species ($n = 21$), we used logistic Regression Kriging (RK) models that explicitly incorporate spatial dependence into predictions (Figure 3). This method assumes that the model residuals have a spatial structure resulting from either 'model' factors such as incorrectly specified or inadequate predictor variables, or 'real' factors such as biotic processes that cause spatial patterns [57]. It combines the predictions from a regression model along with the resulting kriged residuals [58]. Specifically, the regression modelling was supplemented with the use of variograms to assess the level of spatial dependence among residuals (Figure 3). Regression residuals were then interpolated and added back to the regression estimate (see [58] for more details). Finally, to select the most parsimonious model for each of the selected species, we applied an automatic stepwise model selection using the Akaike Information Criterion [59].

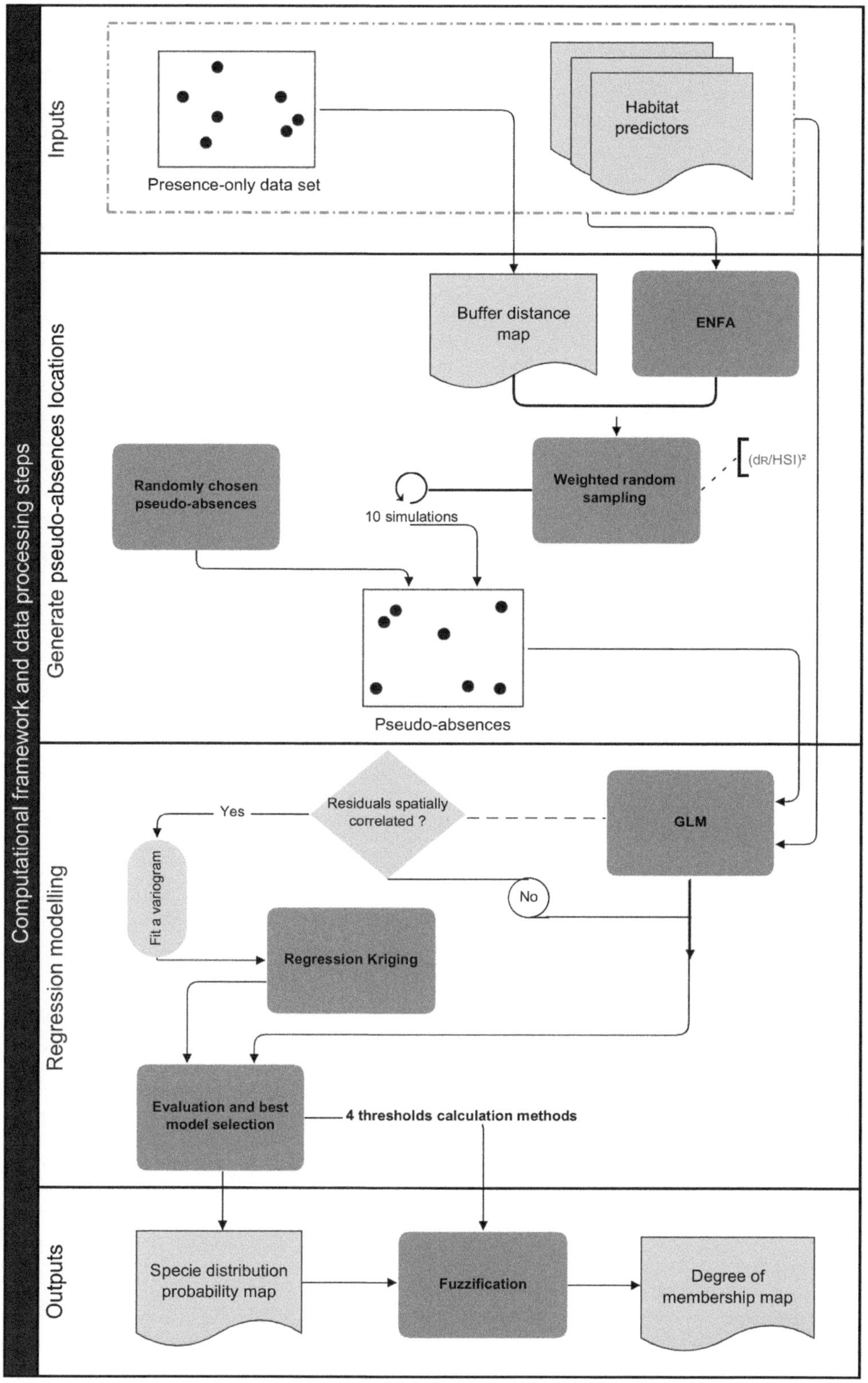

Figure 3. Computational framework and data processing steps.

Model accuracy. For every species, the predictive accuracy of the model was evaluated by a 10-fold cross validation [56]. The receiver operating characteristic curve method was then applied to derive the area under the curve (AUC) index [60] to measure the model's performance.

The simulation of pseudo-absences may generate absences far from the environmental conditions of presences, which may artificially increases the rate of well-predicted absences and hence the AUC scores [24,25,61]. In addition, the AUC test statistic may not always reflect a model's ability to prioritise areas in terms of their habitat suitability relative to alternative models (e.g., [61,62]). Model assessment was therefore supplemented with the Point Biserial Coefficient (PBC) [20,24,63], the sensitivity (presences correctly predicted as presences), and the specificity metrics (absences correctly predicted as absences) [60]. The PBC was calculated as a Pearson's correlation coefficient between the observation in the occurrence dataset (presence (1) or pseudoabsence (0)) and the prediction and therefore takes into account how far the prediction varies from the observation. An independent examination of the percentage of presence and absence errors was recommended by Lobo *et al* [61] to help in the model selection process according to the researcher's goals, rather than the use of a synthetic measure such as the AUC. Predictions were further inspected visually and compared to plotted occurrence data in order to assess their plausibility.

Finally, we used the Pearson's correlation coefficient to calculate the pairwise correlation between the final predictions maps derived from the 10 groups of pseudo-absences. This allowed us to assess the stability of predictions for each of the different weighted simulations of pseudo-absences.

Each data processing step was completed in *R*, drawing on code developed by Hengl *et al.* [37], and automating the calculation for several species simultaneously. For each modelled species, the regression models, 10-fold cross validation, and evaluation procedures were carried out for the 11 simulated groups of pseudo-absences.

Conservation Planning Procedure

Probability of occurrence maps can be generated using the most accurate model selected for each species from the 11 pseudo-absences groups. Once these maps have been converted to presence-absence data, they can be used to identify areas that meet the conservation targets for selection of AAR deployment sites. These targets specify the inclusion of both favourable habitats for commercially-targeted species and areas of high density *P. oceanica* seagrass beds.

Threshold approaches and fuzzy modelling. The most common method used to convert probabilities of occurrence to presence-absence data is the use of an optimum probability threshold [64]. Different methods have been proposed to select a probability threshold [27]. Among these, the most widely used threshold optimisation criteria are:

(1) *Sens = Spec* - The threshold where sensitivity equals specificity (i.e., where positive observations are equally as likely to be wrong as negative observations).

(2) *Max (Sens+Spec)* - The threshold that maximises the sum of sensitivity and specificity (i.e., it minimises the mean error rates for both positive and negative observations). This threshold is equivalent to finding the point on the receiver operating characteristic curve whose tangent has a slope of one.

(3) *MaxKappa* - The threshold that results in the maximum value of Kappa statistic.

(4) *MaxPCC* - The threshold that results in the maximum percent of correctly classified observations.

For each species, a probability threshold was determined using each of the four optimisation criteria outlined above. As these methods do not provide similar threshold values, the use of one method over another can influence conservation planning outcomes, e.g., modifying areas that are expected to be suitable for a given species [27]. To avoid the subjective selection of a particular threshold, predicted distributions of selected species were defined by fuzzy sets theory [65]. Fuzzy logic is useful in circumstances that involve uncertainty, imprecision, and vagueness by replacing the sharp boundary between the suitable and non-suitable classes with the concept of a degree of truth (membership). In the fuzzification process, crisp attribute values (probability of presence) are transformed linearly into a common suitability scale (0 to 1) using the fuzzy linear membership function. A membership value of 0 is assigned to the lowest probability threshold value (as calculated using the four criteria outlined above) and a value of 1 to the highest threshold value (see example in Figure 4). Each map cell was assigned a fuzzy membership value resulting from the fuzzy linear membership function.

Fuzzy overlay of conservation criteria. For each species, the favourability value of an area is defined as the degree of membership of that area to the fuzzy set of favourable areas for the species. Fuzzified inputs can be combined together to identify the most favourable area for the majority of commercial species by using a fuzzy operator. The fuzzy AND operator was applied to return the minimum of the fuzzy memberships from the fuzzy input maps. The result of this aggregation is a final fuzzy set expressing the site suitability for all key species (see example of the method in Figure 4). To identify areas with a high density of seagrass beds, a recent map showing seagrass recovery rates was obtained from the INSTM. This map was fuzzified using a fuzzy linear membership function that assigned a membership value of 0 for recovery rates less than 30% and a value of 1 for recovery rates greater than 60%. Finally, the maps that express the fuzzy memberships of site suitability for all key species and seagrass recovery rates were combined together using the fuzzy AND operator. For example, "IF the favourability value of an area for the species 1 IS high, AND the favourability value of an area for the species 2 IS high, etc…, AND the recovery rate of seagrass IS high THEN the area has high conservation criteria".

Results and Discussion

The Monte-Carlo randomisation tests show that the ENFA marginality factor is highly significant for each modelled species (all $p<0.001$). This implies that the habitat occupied by the species modelled differ unequivocally from the average environmental conditions found in the broader study area. This indicates that a species-specific habitat selection process takes place. The ENFA also show that the potential distributions of species are much larger than their realised distributions, based on sampling locations (see examples of three species in Figure 5). In contrast, incorporating pseudo-absence data into logistic RK models results in predicted distributions that are closer to the realised distributions (Figure 5).

When the pseudo-absence data are randomly selected, the potential distribution of the species are less extensive in comparison with those obtained using a weighted selection. This difference can be explained by the fact that a random simulation can select absences between observed occurrences points, therefore generating pseudo-absences in favourable areas. This can subsequently lead to an underestimation in the realised distributions [25,66].

For each method generating pseudo-absences, the distribution of the AUC, PBC, the sensitivity and the specificity metrics is shown in Figure 6. For each species, model accuracy differs according to the method used to generate pseudo-absences. All the models based on the environmentally- and geographically-

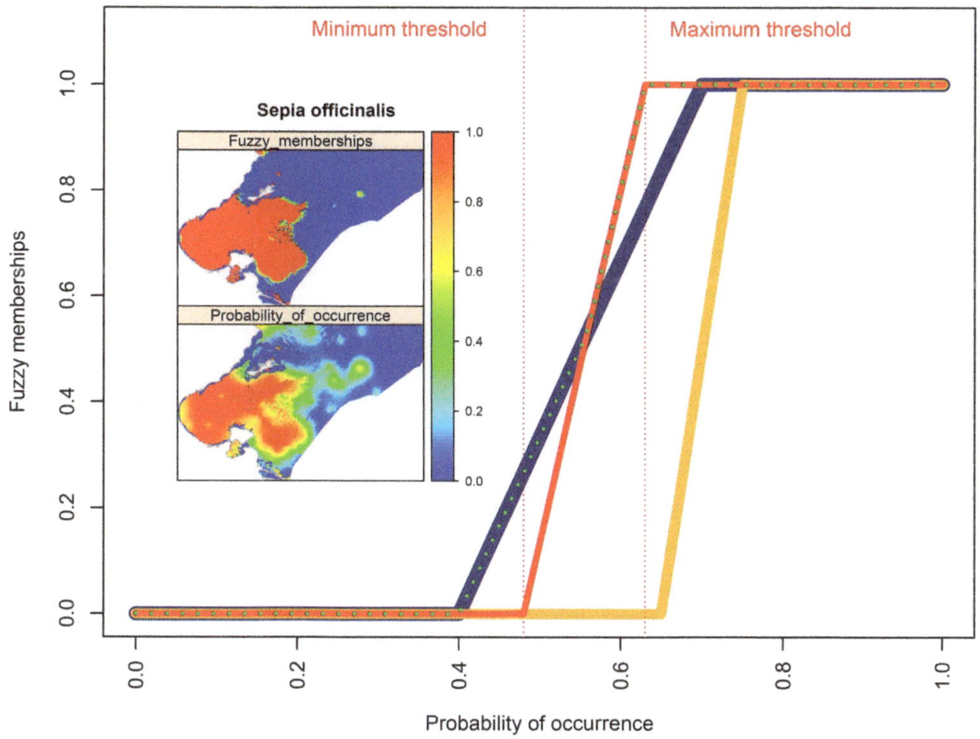

Figure 4. Threshold approaches and fuzzy modelling. An illustration of the fuzzification process performed using a fuzzy linear membership function. The red, orange, and blue curves represent the fuzzy membership sets of three different species. The dotted green line represents the fuzzy AND overlay outcomes. The inset maps display an example of the fuzzification process for the cuttlefish *(Sepia officinalis)* which corresponds with the red curve.

weighted method achieved a high level of accuracy as indicated by the four measures of model accuracy (Mean±standard deviation: 0.9±0.07, 0.65±0.23, 0.87±0.08 and 0.85±0.09 respectively for the AUC, PBC, sensitivity and specificity) (Figure 6). These values were significantly higher (Wilcoxon rank test: $p<0.001$) than those obtained by models based on the random selection method (Mean±standard deviation: 0.78±0.1, 0.36±0.26, 0.77±0.11 and 0.69±0.1 for the AUC, PBC, sensitivity and specificity, respectively).

Since all the resulting SDMs are based on pseudo-absences, both specificity and AUC scores estimate the degree of accuracy of the absence information used in the model training process. Thus, a high specificity score only implies that most of the data considered as absence data are correctly predicted and does not imply a high performance in the prediction of the unknown true absences. However, the sensitivity and PBC values were higher when the weighted method of generating pseudo-absence was used, implying a high performance of this method in the predictions of the known true presences as compared with the random method.

Models based on the weighted method of generating pseudo-absence data provide significantly better results on average, which aligns with the results of Engler *et al.*, Chefaoui & Lobo and Hengl *et al.* [22,25,37]. It contrasts, however, with the findings of Wisz & Guisan and Barbet-Massin *et al.* [19,26]. Using virtual species, these authors found that randomly-selected pseudo-absences yielded the most reliable species distribution models. However, this may be explained by the fact that both studies used a large number of pseudo-absences (e.g., 10.000 data points), whereas this study used the same number of simulated pseudo-absences as the number of occurrences. Currently, there is no consensus on the number of pseudo-absences that are required to optimise model predictions. Some authors suggest a ratio of 10:1 (e.g., [25,49]) while others recommend using large numbers of pseudo-absences when they are randomly selected (e.g., [20,26]). Intuitively, it makes sense to generate an equal number of pseudo-absence data points as occurrence data points [58,67] to avoid the bias caused by a presence-absence ratio that is too low [68,69]. Indeed this was the outcome in McPherson *et al.* [70]who found SDMs had the best predictive accuracy when prevalence values (the proportion of data points representing a species' presence) were around 0.5. Therefore, several authors recommend resampling the training data to balance presence and absence data points [70,71].

For each species, the range of pairwise correlations between the final habitat prediction maps (derived from each of the 10 groups of pseudo-absences) varied according to the number of occurrences used to develop the models (Figure 7). The lowest correlation occurs when the number of occurrences is less than 61. The array of evaluation measures, based on all the replicated runs, does not show a clear trend in relation to the number of occurrences used to fit the models. Even with a small number of occurrences, values of AUC and PBC indicate excellent predictive accuracy of the models. However, large fluctuations in the predictive accuracy and the models' subsequent predictions are observed when the models are based on a low number of occurrences. Indeed, the selection of pseudo-absences can induce variability in model predictions when several runs are made with a small set of occurrence data, each run having its own dataset for calibration.

Williams [72] found that the predictive ability of some ecological modelling approaches varies with a species' detectability. While presence-absence approaches generally have higher

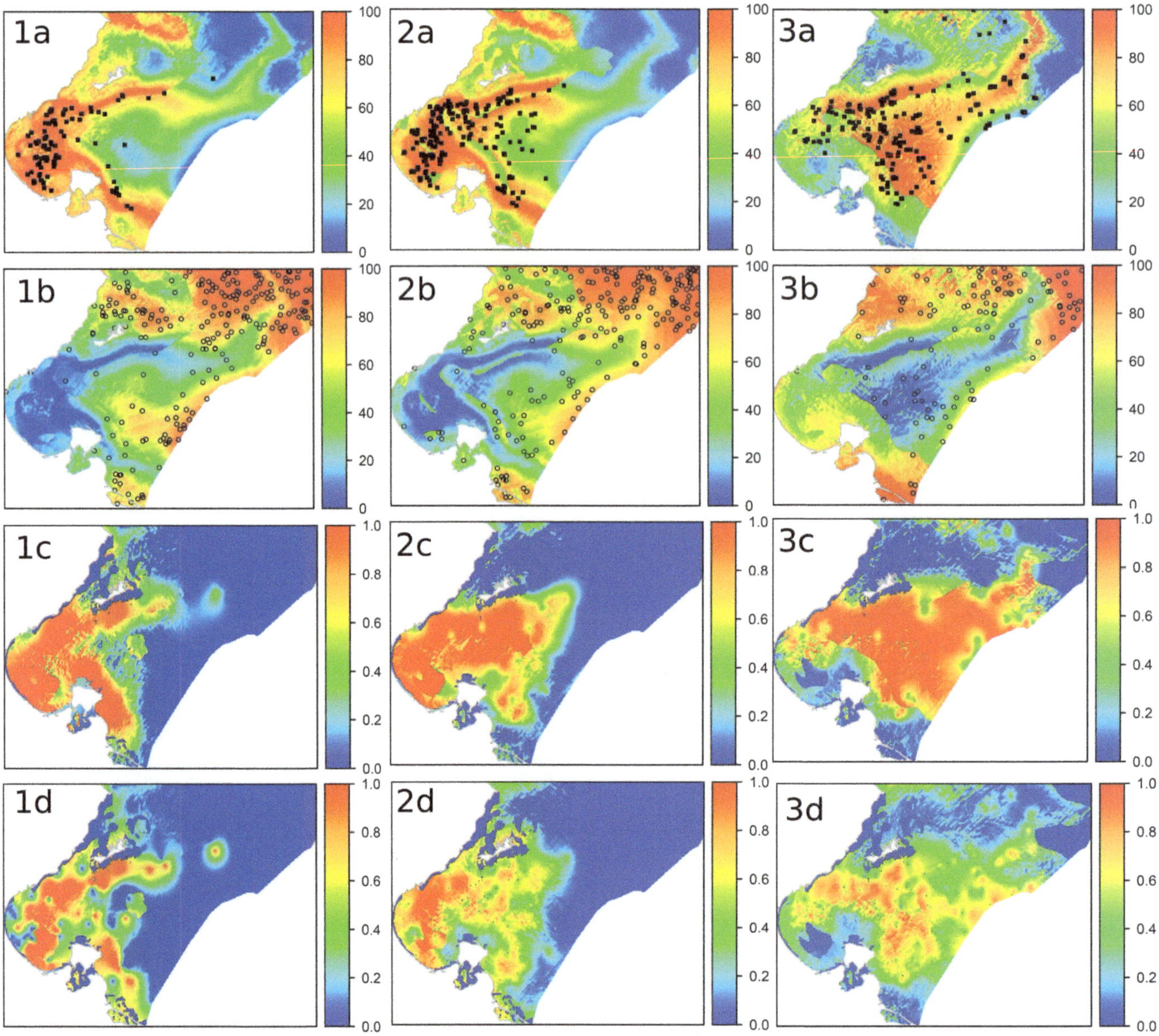

Figure 5. Spatial prediction maps in the Gulf of Gabes for three example species. (1) *Lithognathus mormyrus,* (2) *Penaeus kerathurus,* (3) *Pagellus erythrinus* (a) the habitat suitability index map with presence-only data; (b) the weighted map and the randomly-generated pseudo-absences using the Equation 1; (c) probabilities predicted using the binomial regression-kriging (RK) with a weighted selection of pseudo-absences; and (d) probabilities predicted using the binomial RK with a random selection of pseudo-absences.

predictive abilities for species with a high detectability, they do not perform as well as presence-only approaches when detectability is low. However, for some species such as the pelagic fish species, *Trachurus trachurus*, the fitted model is found to be moderately accurate though it is based on a relatively high number of data points (338 presences/pseudo-absences). This result may be related to the fact that the selected predictors (e.g., depth, seafloor type) have greater ecological significance to model the distribution of demersal and benthic species. [66]. This implies that the performance of SDMs strongly depends on the environmental predictors, which in turn depend of the type of organism that is studied [62]. In addition, the realism and the robustness of models may have been influenced by our automatic variable selection procedure [73]. Indeed, ecologically important variables may have been excluded from the stepwise models and/or non-meaningful variables may have been incorporated into models. This was particularly the case for some of the studied pelagic species (*Trachurus trachurus, Spicara maena, Scomber scombrus*). These species were therefore excluded from the conservation planning procedure.

Overall, our results suggest that simulating pseudo-absences with an environmentally and geographically weighted method rather than a purely heuristic approach enhances the accuracy of predictions. This method provides a robust result when a relatively large number of occurrence data points with good spatial coverage are used. The RK method also shows great potential as an approach to incorporate spatial dependence in SDMs, by combining information on species-habitat relationships (i.e., through the deterministic model) and error components. We believe that the combined ENFA and RK method has several

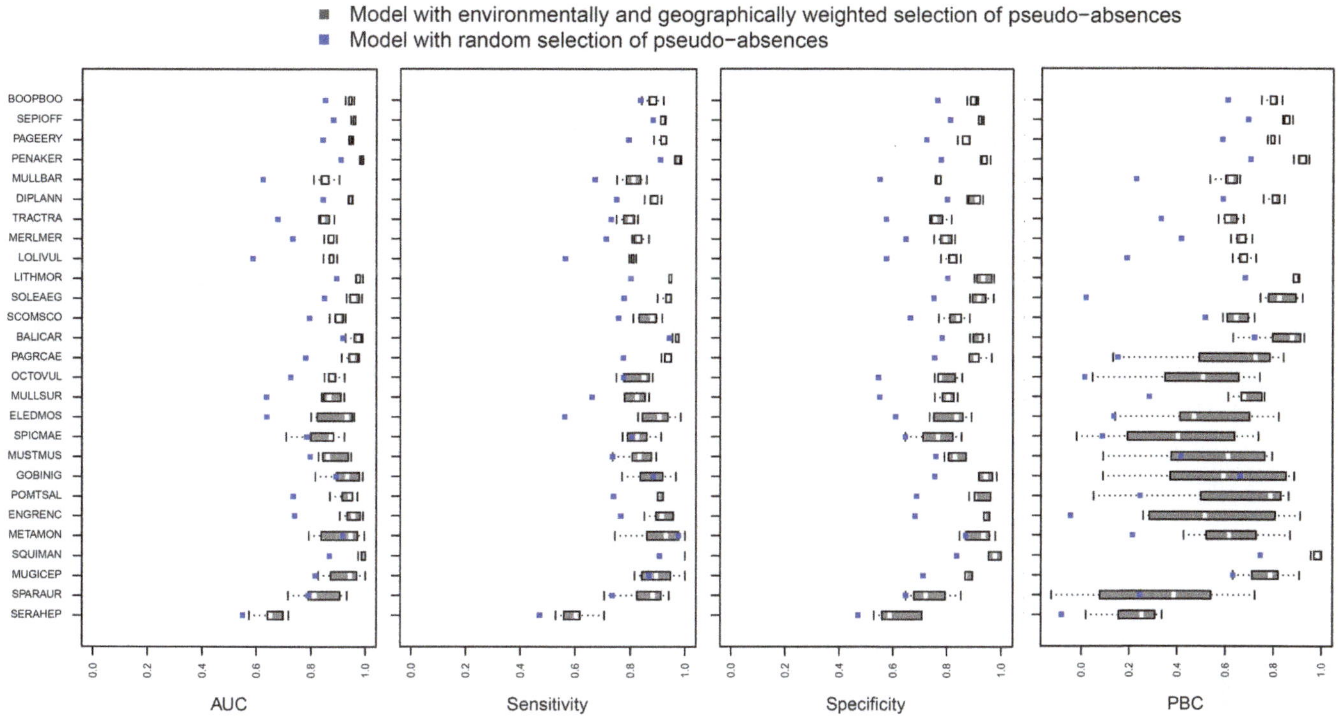

Figure 6. Distributions of the area under the curve (AUC), PBC, sensitivity and specificity values. Calculated AUC, PBC, sensitivity and specificity values for the 27 modelled species (species codes are listed in Table 1). Species are sorted (top to bottom) by decreasing number of occurrences (used to develop the models).

advantages when applied to trawl survey data, especially when addressing their imperfect ability to detect a species. Furthermore, this method applies both the spatial auto-correlation structure and the trend component of the spatial variation to make spatial predictions of species' distributions. This method can be applied to other areas where survey data are available, such as the Medits (International bottom trawl surveys in the Mediterranean) dataset.

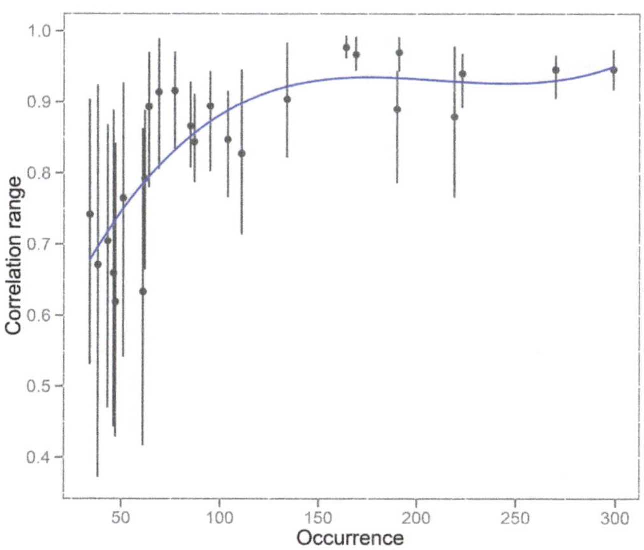

Figure 7. Stability of predictions. The range of Pearson's pairwise correlation coefficients between the final prediction maps according to the number of occurrences used to develop the models (red line: smooth curve fitted by Loess function).

The second purpose of this study is to identify the areas required to meet the conservation targets of AARs based on probabilistic predicted maps. In this study we propose a novel method to develop these maps that uses fuzzy sets to address the uncertainty associated with the selection of probability threshold optimisation criteria. Fuzzy logic is interested in capturing partial truths, that is, how to reason about things that are not wholly true or false; while probability is concerned with making predictions about events based on a partial state of knowledge [65]. The fuzzy sets theory is used to transform the probability of presence into a membership degree using not only a single threshold value but several values obtained with different cut-off threshold optimization criteria.

Of the 27 species modelled in this study, an excellent or high predictive accuracy is only found for 12 benthic and demersal species. It was these species that were selected for the fuzzy overlay of conservation criteria. Of these, the 2004–2005 stock assessment results for the Gulf of Gabes report that *Solea aegyptiaca*, *Octopus vulgaris*, and *Sepia officinalis* were fully exploited (Othman Jarboui, personal communication: INSTM) and *Mullus barbatus*, *Pagrus caeruleostictis*, and *Pagellus erythrinus* were overexploited.

The resultant fuzzy overlay map (Figure 8) highlights three main areas that meet the conservation criterion to a high level. The largest area is located south of Kerkennah Island while the remaining two areas are in the coastal region off Mahres Harbour and north of Jerba Island. As well as meeting the AAR conservation criteria, the Kerkennah Island area has already been reported as a biodiversity hotspot for megabenthic fauna [74] and is currently proposed as a potential MPA [75]. The area north of Jerba Island was also proposed as suitable for MPA establishment by Ben Mustapha & Afli [75]. In addition to its dense seagrass

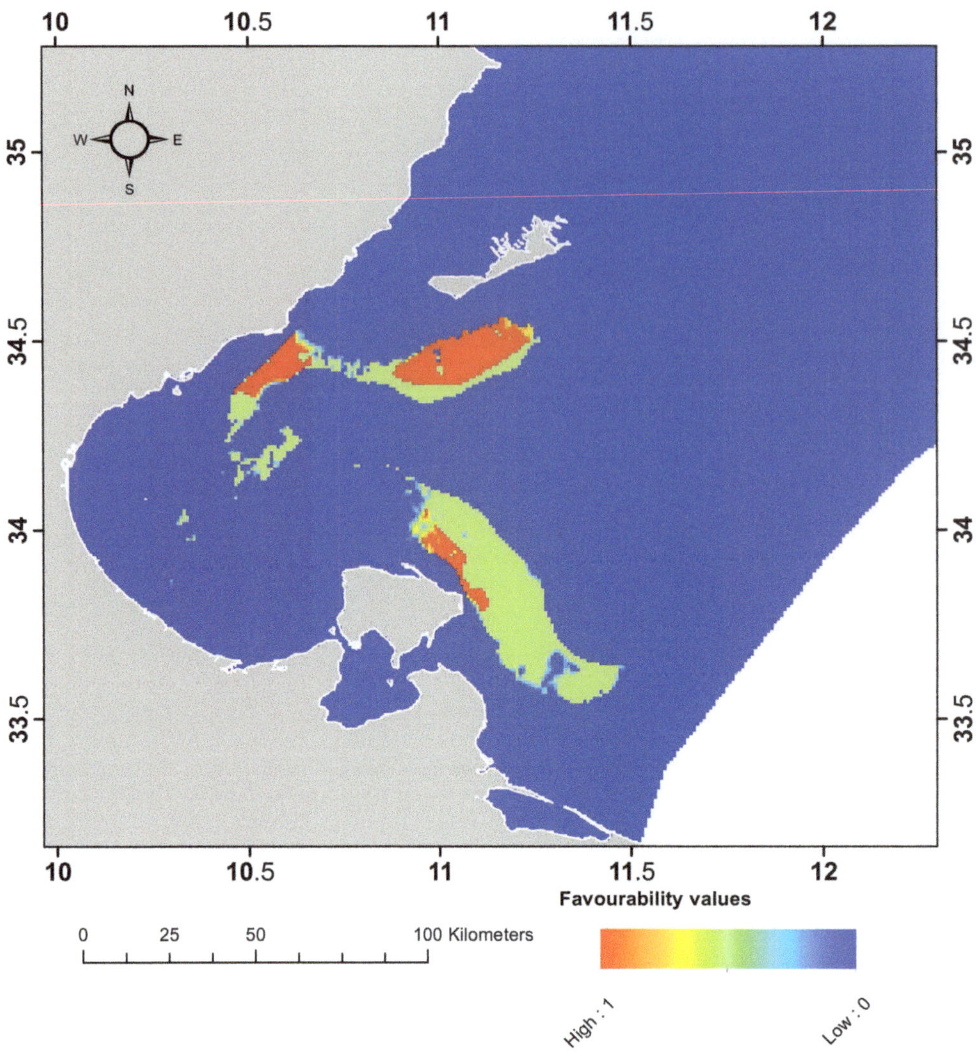

Figure 8. Map of favourability values for the deployment of AARs.

beds, the area is also characterised by coralligenous assemblages and is a recognised nursery site for several juvenile of commercially-targeted species [75]. These expert opinions corroborate our findings and confirm the relevance of the established methodology for the selection of AAR deployment sites.

Conclusions

The comparison between the two pseudo-absence data generation methods reveals that the environmentally- and geographically-weighted method has significant potential to reduce prediction errors in SDMs. When confirmed absence data are not available, we recommend using the method proposed by Hengl *et al.* [37], which combines ENFA and RK predictions to deal with an imperfect ability to detect a species and incorporate spatial dependence in predictions. This study proposes a novel method to developing predicted maps that uses fuzzy sets to address the uncertainty associated with selecting probability threshold criteria. In the context of conservation planning, we illustrate the advantages of this novel method, using it to identify areas within the Gulf of Gabes that meet the conservation targets for AARs. Three key areas that met the conservation criteria to a high level were identified, and these areas are recommended as deployment sites for AARs. The location and spatial arrangement of reef units must be carefully planned. Reef characteristics such as the number of modules, the distance between modules as a function of trawl parameters, and the weight of modules as a function of exposure to currents should all be informed by scientific studies. Additional factors that should be considered in the deployment of AARs are economic costs and socio-economic effects.

The areas with a nonzero favourability value for AAR deployment cover 1578 km^2 and encompass 30% of seagrass beds in good condition. However, these areas represent biodiversity targets without regard to cost. In this context this study can contribute to effective conservation planning in a broader prioritization process, which must be optimized by accounting for the cost of conservation. For instance, future works should focus on systematic conservation planning [76,77] that attempts to solve a cost-effectiveness problem (i.e. how to achieve the most conservation given limited resources). This is particularly important in the Gulf of Gabes where the diverse stakeholder group often holds conflicting values and opinions (e.g., conflicts between professional and local artisanal fishers).

Acknowledgments

The authors would like to acknowledge the National Institute of Marine Sciences and Technologies (INSTM, Tunisia) that contributed unpublished data. We are very grateful to Jane Alpine for English editing. We also thank the two anonymous referees for their helpful comments on the manuscript.

Author Contributions

Conceived and designed the experiments: TH. Performed the experiments: TH. Analyzed the data: TH FLL FBRL. Contributed reagents/materials/analysis tools: TH CS MSR PC. Wrote the paper: TH FLL FBRL FL CA PC CS MSR.

References

1. Harris PT, Whiteway T (2009) High seas marine protected areas: Benthic environmental conservation priorities from a GIS analysis of global ocean biophysical data. Ocean & Coastal Management 52: 22–38. doi:10.1016/j.ocecoaman.2008.09.009.
2. Valavanis VD, Pierce GJ, Zuur AF, Palialexis A, et al (2008) Modelling of essential fish habitat based on remote sensing, spatial analysis and GIS. Hydrobiologia 612: 5–20. doi:10.1007/s10750-008-9493-y.
3. Johnson AF, Jenkins SR, Hiddink JG, Hinz H (2012) Linking temperate demersal fish species to habitat: scales, patterns and future directions. Fish and Fisheries. doi:10.1111/j.1467-2979.2012.00466.x.
4. Degraer S, Verfaillie E, Willems W, Adriaens E, Vincx M, et al (2008) Habitat suitability modelling as a mapping tool for macrobenthic communities: An example from the Belgian part of the North Sea. Continental Shelf Research 28: 369–379. doi:10.1016/j.csr.2007.09.001.
5. Maxwell DL, Stelzenmüller V, Eastwood PD, Rogers SI (2009) Modelling the spatial distribution of plaice (Pleuronectes platessa), sole (Solea solea) and thornback ray (Raja clavata) in UK waters for marine management and planning. Journal of Sea Research 61: 258–267. doi:10.1016/j.seares.2008.11.008.
6. MacLeod CD, Mandleberg L, Schweder C, Bannon SM, Pierce GJ (2008) A comparison of approaches for modelling the occurrence of marine animals. Hydrobiologia 612: 21–32. doi:10.1007/s10750-008-9491-0.
7. Elith J, Leathwick JR (2009) Species Distribution Models: Ecological explanation and prediction across space and time. Annual Review of Ecology, Evolution, and Systematics 40: 677–697. doi:10.1146/annurev.ecolsys.110308.120159.
8. Hoffman JC, Bonzek CF, Latour RJ (2009) Estimation of Bottom Trawl Catch Efficiency for Two Demersal Fishes, the Atlantic Croaker and White Perch, in Chesapeake Bay. Marine and Coastal Fisheries 1: 255–269. doi:10.1577/C08-048.1.
9. Guisan A, Edwards TC, Hastie T (2002) Generalized linear and generalized additive models in studies of species distributions: setting the scene. Ecological Modelling 157: 89–100.
10. Anderson R. (2003) Real vs. artefactual absences in species distributions: tests for Oryzomys albigularis (Rodentia: Muridae) in Venezuela. Journal of Biogeography 30: 591–605.
11. Loiselle B., Howell C., Graham C., Goerck J., Brooks T, et al (2003) Evitando Dificultades Resultantes del Uso de Modelos de Distribución de Especies en Planeación de Conservación. Conservation Biology 17: 1591–1600. doi:10.1111/j.1523-1739.2003.00233.x.
12. Robinson L, Elith J, Hobday A, Pearson R, Kendall B, et al (2011) Pushing the limits in marine species distribution modelling: lessons from the land present challenges and opportunities. Global Ecology and Biogeography 20: 789–802.
13. Monk J (2013) How long should we ignore imperfect detection of species in the marine environment when modelling their distribution? Fish and Fisheries. doi:10.1111/faf.12039.
14. Mackenzie DI, Royle JA (2005) Designing occupancy studies: general advice and allocating survey effort. Journal of Applied Ecology 42: 1105–1114. doi:10.1111/j.1365-2664.2005.01098.x.
15. Anderson R., Dudík M, Ferrier S, Guisan A, J Hijmans R, et al (2006) Novel methods improve prediction of species' distributions from occurrence data. Ecography 29: 129–151.
16. Hirzel AH, Hausser J, Chessel D, Perrin N (2002) Ecological-niche factor analysis: how to compute habitat-suitability maps without absence data? Ecology 83: 2027–2036.
17. Stockwell D (1999) The GARP modelling system: problems and solutions to automated spatial prediction. International Journal of Geographical Information Science 13: 143–158. doi:10.1080/136588199241391.
18. Phillips SJ, Anderson RP, Schapire RE (2006) Maximum entropy modeling of species geographic distributions. Ecological Modelling 190: 231–259. doi:10.1016/j.ecolmodel.2005.03.026.
19. Segurado P, Araújo MB (2004) An evaluation of methods for modelling species distributions. Journal of Biogeography 31: 1555–1568. doi:10.1111/j.1365-2699.2004.01076.x.
20. Elith J, H. Graham C, P. Anderson R, Dudík M, Ferrier S, et al (2006) Novel methods improve prediction of species' distributions from occurrence data. Ecography 29: 129–151. doi:10.1111/j.2006.0906-7590.04596.x.
21. Barbet-Massin M, Jiguet F, Albert CH, Thuiller W (2012) Selecting pseudo-absences for species distribution models: how, where and how many? Methods in Ecology and Evolution 3: 327–338. doi:10.1111/j.2041-210X.2011.00172.x.
22. Engler R, Guisan A, Rechsteiner L (2004) An improved approach for predicting the distribution of rare and endangered species from occurrence and pseudo-absence data. Journal of Applied Ecology 41: 263–274.
23. Torres LG, Read AJ, Halpin P (2008) Fine-scale habitat modeling of a top marine predator: do prey data improve predictive capacity. Ecological Applications 18: 1702–1717.
24. Jones MC, Dye SR, Pinnegar JK, Warren R, Cheung WW (2012) Modelling commercial fish distributions: Prediction and assessment using different approaches. Ecological Modelling 225: 133–145.
25. Chefaoui RM, Lobo JM (2008) Assessing the effects of pseudo-absences on predictive distribution model performance. Ecological Modelling 210: 478–486. doi:10.1016/j.ecolmodel.2007.08.010.
26. Wisz MS, Guisan A (2009) Do pseudo-absence selection strategies influence species distribution models and their predictions? An information-theoretic approach based on simulated data. BMC Ecology 9: 8. doi:10.1186/1472-6785-9-8.
27. Freeman EA, Moisen GG (2008) A comparison of the performance of threshold criteria for binary classification in terms of predicted prevalence and kappa. Ecological Modelling 217: 48–58. doi:10.1016/j.ecolmodel.2008.05.015.
28. Ferrier S, Guisan A (2006) Spatial modelling of biodiversity at the community level. Journal of Applied Ecology 43: 393–404. doi:10.1111/j.1365-2664.2006.01149.x.
29. Albouy C, Guilhaumon F, Araújo MB, Mouillot D, Leprieur F (2012) Combining projected changes in species richness and composition reveals climate change impacts on coastal Mediterranean fish assemblages. Global Change Biology 18: 2995–3003. doi:10.1111/j.1365-2486.2012.02772.x.
30. Nenzén HK, Araújo M (2011) Choice of threshold alters projections of species range shifts under climate change. Ecological Modelling 222: 3346–3354.
31. Wilson KA, Westphal MI, Possingham HP, Elith J (2005) Sensitivity of conservation planning to different approaches to using predicted species distribution data. Biological Conservation 122: 99–112. doi:10.1016/j.biocon.2004.07.004.
32. Teh LCL, Teh LSL (2011) A fuzzy logic approach to marine spatial management. Environmental Management 47: 536–545.
33. Montero J (2009) Fuzzy Logic and Science. In: Seising R, editor. Views on Fuzzy Sets and Systems from Different Perspectives. Studies in Fuzziness and Soft Computing. Springer Berlin Heidelberg, Vol. 243. 67–77.
34. Hattour A (1991) Le chalutage dans les eaux Tunisiennes réalités et considérations législatives particulièrement dans les Golfe de Tunis et de Gabès. Note de l'Institut National Scientifique et Technique d'Océanographie et de Pêche de Salammbô: 13p.
35. Najar B, Ben Mariem S, Hadj Ali M (2010) Évolution des profils des débarquements de poissons dans la région de Gabes, Tunisie. Commission International pour l'Exploration Scientifique de la Mer Méditerranée 39: 601p.
36. Direction Générale de la Pêche et de l'Aquaculture (DGPA) (2010) Annuaire des statistiques des pêches en Tunisie. Ministère de l'Agriculture, Tunisie.
37. Hengl T, Sierdsema H, Radović A, Dilo A (2009) Spatial prediction of species' distributions from occurrence-only records: combining point pattern analysis, ENFA and regression-kriging. Ecological Modelling 220: 3499–3511. doi:10.1016/j.ecolmodel.2009.06.038.
38. Batisse M, De Grissa AJ (1995) Marine Region 3: Mediterranean. In G Kelleher, C Bleakley and S Wells (eds) A Global Representative System of Marine Protected Areas Vol I, The Great Barrier Reef Marine Park Authority, The World Bank and IUCN, Washington, DC: 77–104.
39. Francour P (1997) Fish Assemblages of Posidonia oceanica Beds at Port-Cros (France, NW Mediterranean): Assessment of Composition and Long-Term Fluctuations by Visual Census. Marine Ecology 18: 157–173. doi:10.1111/j.1439-0485.1997.tb00434.x.
40. Ben Mustapha K, Hattour A, Mhetli M, El Abed A, Tritar B (1999) Etat de la Bionomie Benthique des Etages Infra et Circalittoral du Golfe de Gabès. Bulletin de l'Institut National des Sciences et Technologies de la Mer 26: 5–48.
41. Jensen A, Collins K, Lockwood AP (2000) Artificial Reefs in European Seas. Lockwood, A.P. (Eds.). Springer. 508 p.
42. Munoz-Perez JJ, Mas G, Jose M, Naranjo JM, Torres E, et al (2000) Position and monitoring of anti-trawling reefs in the Cape of Trafalgar (Gulf of Cadiz, SW Spain). Bulletin of Marine Science 67: 761–772.
43. Moore C., Harvey E., Niel K (2010) The application of predicted habitat models to investigate the spatial ecology of demersal fish assemblages. Marine Biology 157: 2717–2729. doi:10.1007/s00227-010-1531-4.
44. Monk J, Ierodiaconou D, Bellgrove A, Harvey E, Laurenson L (2011) Remotely sensed hydroacoustics and observation data for predicting fish habitat suitability. Continental Shelf Research 31: 17–27. doi:10.1016/j.csr.2010.02.012.
45. Katsanevakis S, Maravelias CD (2008) Bathymetric distribution of demersal fish in the Aegean and Ionian Seas based on generalized additive modeling. Fisheries Science 75: 13–23. doi:10.1007/s12562-008-0033-5.

46. Pittman SJ, Brown KA (2011) Multi-Scale Approach for Predicting Fish Species Distributions across Coral Reef Seascapes. PLoS ONE 6: e20583. doi:10.1371/journal.pone.0020583.
47. Palialexis A, Georgakarakos S, Karakassis I, Lika K, Valavanis VD (2011) Prediction of marine species distribution from presence–absence acoustic data: comparing the fitting efficiency and the predictive capacity of conventional and novel distribution models. Hydrobiologia 670: 241–266. doi:10.1007/s10750-011-0673-9.
48. Monk J, Ierodiaconou D, Harvey E, Rattray A, Versace VL (2012) Are We Predicting the Actual or Apparent Distribution of Temperate Marine Fishes? PLoS ONE 7: e34558. doi:10.1371/journal.pone.0034558.
49. Jiménez-Valverde A, Gómez JF, Lobo JM, Baselga A, Hortal J (2008) Challenging species distribution models: the case of Maculinea nausithous in the Iberian Peninsula. Annales Zoologici Fennici. Vol. 45. 200–210.
50. Yackulic CB, Chandler R, Zipkin EF, Royle JA, Nichols JD, Campbell Grant E, Veran S (2012) Presence-only modelling using MAXENT: when can we trust the inferences? Methods in Ecology and Evolution: 236–243. doi:10.1111/2041-210x12004.
51. Monk J, Ierodiaconou D, Versace VL, Bellgrove A, Harvey E, et al (2010) Habitat suitability for marine fishes using presence-only modelling and multibeam sonar. Marine Ecology Progress Series 420: 157–174.
52. Calenge C (2006) The package "adehabitat" for the R software: A tool for the analysis of space and habitat use by animals. Ecological Modelling 197: 516–519. doi:10.1016/j.ecolmodel.2006.03.017.
53. R Development Core Team (2011) R: A Language and Environment for Statistical Computing. Vienna, Austria. Available: http://www.R-project.org/.
54. Montgomery DC (2007) Design and Analysis of Experiments, 6th Edition Set. New York: Wiley. 752 p.
55. Hill M, Smith A (1976) Principal component analysis of taxonomic data with multi-state discrete characters. Taxon 25: 249–255.
56. McCullagh P, Nelder JA (1989) Generalized Linear Models, Second Edition. 2nd edition London. Chapman and Hall. 512 p.
57. Miller J, Franklin J, Aspinall R (2007) Incorporating spatial dependence in predictive vegetation models. Ecological Modelling 202: 225–242. doi:10.1016/j.ecolmodel.2006.12.012.
58. Hengl T, Heuvelink GBM, Rossiter DG (2007) About regression-kriging: From equations to case studies. Computers & Geosciences 33: 1301–1315. doi:10.1016/j.cageo.2007.05.001.
59. Chambers JM, Hastie T, editors (1991) Statistical models in S. London: Chapman & Hall. 608 p.
60. Fielding AH, Bell JF (1997) A review of methods for the assessment of prediction errors in conservation presence-absence models. Environmental Conservation 24: 38–49.
61. Lobo JM, Jiménez-Valverde A, Real R (2008) AUC: a misleading measure of the performance of predictive distribution models. Global Ecology and Biogeography 17: 145–151.
62. Austin M (2007) Species distribution models and ecological theory: A critical assessment and some possible new approaches. Ecological Modelling 200: 1–19. doi:10.1016/j.ecolmodel.2006.07.005.
63. Zheng B, Agresti A (2000) Summarizing the predictive power of a generalized linear model. Statistics in medicine 19: 1771–1781.
64. Manel S, Williams HC, Ormerod SJ (2001) Evaluating presence–absence models in ecology: the need to account for prevalence. Journal of Applied Ecology 38: 921–931.
65. Zadeh L (1965) Fuzzy sets. Information and Control 8: 338–353. doi:10.1016/S0019-9958(65)90241-X.
66. Hanberry B., He H., Palik BJ (2012) Pseudo-absence generation strategies for Species Distribution Models. PloS one 7: e44486.
67. Zarnetske PL, Edwards Jr TC, Moisen GG (2007) Habitat classification modeling with incomplete data: Pushing the habitat envelope. Ecological Applications 17: 1714–1726.
68. King G, Zeng L (2001) Logistic Regression in Rare Events Data. Political Analysis 9: 137–163.
69. Dixon PM, Ellison AM, Gotelli NJ (2005) Improving the precision of estimates of the frequency of rare events. Ecology 86: 1114–1123. doi:10.1890/04-0601.
70. McPherson JM, Jetz W, Rogers D (2004) The effects of species' range sizes on the accuracy of distribution models: ecological phenomenon or statistical artefact? Journal of Applied Ecology 41: 811–823. doi:10.1111/j.0021-8901.2004.00943.x.
71. Liu C, Berry PM, Dawson TP, Pearson RG (2005) Selecting thresholds of occurrence in the prediction of species distributions. Ecography 28: 385–393. doi:10.1111/j.0906-7590.2005.03957.x.
72. Williams AK (2003) The influence of probability of detection when modeling species occurrence using GIS and survey data [PhD thesis]. Virginia Polytechnic Institute and State University. Available: http://vtechworks.lib.vt.edu/handle/10919/11129. Accessed 1 October 2012.
73. Guisan A, Thuiller W (2005) Predicting species distribution: offering more than simple habitat models. Ecology letters 8: 993–1009.
74. El Lakhrach H, Hattour A, Jarboui O, Elhasni K, Ramos-Espla A (2012) Spatial distribution and abundance of the megabenthic fauna community in Gabes gulf (Tunisia, eastern Mediterranean Sea). Mediterranean Marine Science 13: 12–29.
75. Ben Mustapha K, Afli A (2007) Quelques traits de la biodiversité marine de Tunisie: Proposition d'aires de conservation et de gestion. Report of the MedSudMed Expert Consultation on Marine Protected Areas and Fisheries Management. MedSudMed Technical Documents. Rome (Italy). 32–55.
76. Margules CR, Pressey RL (2000) Systematic conservation planning. Nature 405: 243–253.
77. Watson JE, Grantham H, Wilson KA, Possingham HP (2011) Systematic conservation planning: Past, present and future. Wiley-Blackwell: 136–160.

PERMISSIONS

All chapters in this book were first published in PLOS ONE, by The Public Library of Science; hereby published with permission under the Creative Commons Attribution License or equivalent. Every chapter published in this book has been scrutinized by our experts. Their significance has been extensively debated. The topics covered herein carry significant findings which will fuel the growth of the discipline. They may even be implemented as practical applications or may be referred to as a beginning point for another development.

The contributors of this book come from diverse backgrounds, making this book a truly international effort. This book will bring forth new frontiers with its revolutionizing research information and detailed analysis of the nascent developments around the world.

We would like to thank all the contributing authors for lending their expertise to make the book truly unique. They have played a crucial role in the development of this book. Without their invaluable contributions this book wouldn't have been possible. They have made vital efforts to compile up to date information on the varied aspects of this subject to make this book a valuable addition to the collection of many professionals and students.

This book was conceptualized with the vision of imparting up-to-date information and advanced data in this field. To ensure the same, a matchless editorial board was set up. Every individual on the board went through rigorous rounds of assessment to prove their worth. After which they invested a large part of their time researching and compiling the most relevant data for our readers.

The editorial board has been involved in producing this book since its inception. They have spent rigorous hours researching and exploring the diverse topics which have resulted in the successful publishing of this book. They have passed on their knowledge of decades through this book. To expedite this challenging task, the publisher supported the team at every step. A small team of assistant editors was also appointed to further simplify the editing procedure and attain best results for the readers.

Apart from the editorial board, the designing team has also invested a significant amount of their time in understanding the subject and creating the most relevant covers. They scrutinized every image to scout for the most suitable representation of the subject and create an appropriate cover for the book.

The publishing team has been an ardent support to the editorial, designing and production team. Their endless efforts to recruit the best for this project, has resulted in the accomplishment of this book. They are a veteran in the field of academics and their pool of knowledge is as vast as their experience in printing. Their expertise and guidance has proved useful at every step. Their uncompromising quality standards have made this book an exceptional effort. Their encouragement from time to time has been an inspiration for everyone.

The publisher and the editorial board hope that this book will prove to be a valuable piece of knowledge for researchers, students, practitioners and scholars across the globe.

LIST OF CONTRIBUTORS

Carlos Werner Hackradt, José Antonio García-Charton and Á ngel Pérez-Ruzafa
Departamento de Ecología e Hidrología, Universidad de Murcia, Murcia, Spain

Mireille Harmelin-Vivien
Aix-Marseille Université , Université du Sud Toulon-Var, CNRS/INSU/IRD, Institut Méditerranéen d'Océanologie (MIO), Marseille, France

Laurence Le Diréach
GIS Posidonie, Institut Méditerranéen d'Océanologie (MIO), Marseille, France

Just Bayle-Sempere, Pablo Sanchez-Jerez and Carlos Valle
Departamento de Ciencias del Mar y Biología Aplicada, Universidad de Alicante, Alicante, Spain

Eric Charbonnel
Parc Marin de la Côte Bleue, Observatoire, Carry-le-Rouet, France

Denis Ody
WWF-France, Marseille, France

Olga Reñones
Instituto Español de Oceanografía, Centro Oceanográfico de Baleares, Palma de Mallorca, Spain

Steve Rocliffe and Julie P. Hawkins
Environment Department, University of York, York, United Kingdom

Shawn Peabody
Blue Ventures Conservation, London, United Kingdom

Melita Samoilys
CORDIO East Africa, Mombasa, Kenya

Carolyn K. Robb
School of Environment and Management, Royal Roads University, Victoria, British Columbia, Canada

Nicholas A. Farmer
National Oceanographic and Atmospheric Administration, National Marine Fisheries Service, Southeast Regional Office, St. Petersburg, Florida, United States of America

Mandy Karnauskas
National Oceanographic and Atmospheric Administration, National Marine Fisheries Service, Southeast Fisheries Science Center, Miami, Florida, United States of America

Iliana Chollett and Peter J. Mumby
Marine Spatial Ecology Lab, College of Life and Environmental Sciences, University of Exeter, Exeter, United Kingdom
Marine Spatial Ecology Lab, School of Biological Sciences, Goddard Building, University of Queensland, Brisbane, Australia

Susana Enríquez
Instituto de Ciencias del Mar y Limnología, Universidad Nacional Autónoma de México, Puerto Morelos, México

Takahiro Shimada
School of Earth and Environmental Sciences, James Cook University, Townsville, Queensland, Australia

Christian Gredzens
School of Earth and Environmental Sciences, James Cook University, Townsville, Queensland, Australia
School of Marine and Tropical Biology, James Cook University, Townsville, Queensland, Australia

Helene Marsh and Mark Hamann
School of Earth and Environmental Sciences, James Cook University, Townsville, Queensland, Australia
Centre for Tropical Water and Aquatic Ecosystem Research, James Cook University, Townsville, Queensland, Australia

Mariana M. P. B. Fuentes
ARC Centre of Excellence for Coral Reef Studies, James Cook University, Townsville, Queensland, Australia

Colin J. Limpus
Aquatic Threatened Species Unit, Queensland Department of Environment and Heritage Protection, Brisbane, Queensland, Australia

Simonetta Fraschetti, Giuseppe Guarnieri, Stanislao Bevilacqua and Antonio Terlizzi
Department of Biological and Environmental Sciences and Technologies, University of Salento, Consorzio Nazionale Interuniversitario per le Scienze del Mare, Lecce, Italy

Ferdinando Boero
Department of Biological and Environmental Sciences and Technologies, University of Salento, Consorzio Nazionale Interuniversitario per le Scienze del Mare, Lecce, Italy
Institute of Marine Sciences of National Research Council, Genova, Italy

David Harasti
Fisheries Research, Marine Ecosystems, NSW Department of Primary Industries, Nelson Bay, New South Wales, Australia

Keith Martin-Smith
School of Zoology, University of Tasmania, Hobart, Tasmania

William Gladstone
School of the Environment, University of Technology, Sydney, New South Wales, Australia

Tessa Mazor and Salit Kark
ARC Centre of Excellence for Environmental Decisions, School of Biological Sciences, The University of Queensland, Brisbane, Queensland, Australia

Hugh P. Possingham
ARC Centre of Excellence for Environmental Decisions, School of Biological Sciences, The University of Queensland, Brisbane, Queensland, Australia
Grand Challenges in Ecosystems and the Environment, Silwood Park, Imperial College, London, United Kingdom

Dori Edelist
Leon Recanati Institute for Maritime Studies, Department of Maritime Civilizations, University of Haifa, Mount Carmel, Haifa, Israel

Eran Brokovich
Department of Geography, The Hebrew University of Jerusalem, Mount Scopus, Jerusalem, Israel

Howard Peters, Bethan C. O'Leary, Julie P. Hawkins and Callum M. Roberts
Environment Department, University of York, York, United Kingdom

Kent E. Carpenter
International Union for Conservation of Nature, Global Marine Species Assessment, Biological Sciences, Old Dominion University, Norfolk, Virginia, United States of America

Caroline E. Whidden, Katrina G. DeZeeuw, Jackie K. Zorz and Amanda M. Cockshutt
Department of Chemistry & Biochemistry, Mount Allison University, Sackville, NB, Canada

Andrew P. Joy and David A. Barnett
Atlantic Cancer Research Institute, Moncton, NB, Canada

Milo S. Johnson,
Department of Biological Sciences, Dartmouth College, Hanover, New Hampshire, United States of America

Olga Zhaxybayeva
Department of Biological Sciences, Dartmouth College, Hanover, New Hampshire, United States of America
Department of Computer Science, Dartmouth College, Hanover, New Hampshire, United States of America

Lucinda L. Douglass and Daniel Beaver
Centre for Biodiversity and Conservation Science, School of Biological Sciences, The University of Queensland, Brisbane, Queensland, Australia
Centre for Conservation Geography, Sydney, New South Wales, Australia

Joel Turner
Centre for Conservation Geography, Sydney, New South Wales, Australia

Hedley S. Grantham
Centre for Biodiversity and Conservation Science, School of Biological Sciences, The University of Queensland, Brisbane, Queensland, Australia
Betty and Gordon Moore Centre for Science and Oceans, Conservation International, Arlington, Virginia, United States of America

Angelika Brandt
Biocentre Grindel and Zoological Museum, University of Hamburg, Hamburg, Germany

Stefanie Kaiser
Biocentre Grindel and Zoological Museum, University of Hamburg, Hamburg, Germany
German Centre for Marine Biodiversity Research, Wilhelmshaven, Germany

Andrew Constable and Ben Raymond
Australian Antarctic Division, Department of the Environment, Australian Government, Kingston, Tasmania, Australia
Antarctic Climate and Ecosystems Cooperative Research Centre, University of Tasmania, Hobart, Tasmania, Australia

Rob Nicoll
WWF Australia, Ultimo, New South Wales, Australia

Alexandra Post
Marine and Coastal Environment Group, Geoscience Australia, Canberra, Australian Capital Territory, Australia

Anne Molcard
Université du Sud Toulon-Var, Aix-Marseille Université ,CNRS/INSU/IRD, Mediterranean Institute of Oceanography (MIO), La Garde, France

Anna-Maria Rammou, Andrea Doglioli and Anne Petrenko
Aix-Marseille Université , CNRS/INSU/IRD, Mediterranean Institute of Oceanography (MIO), Marseille, France

Léo Berline
Université du Sud Toulon-Var, Aix-Marseille Université ,CNRS/INSU/IRD, Mediterranean Institute of Oceanography (MIO), La Garde, France
CNRS, Laboratoire d'Océanographie de Villefranche, Villefranche-sur-Mer, France
Université Pierre et Marie Curie, Paris 6, Laboratoire d'Océanographie de Villefranche, Villefranche-sur-Mer, France

Wing Yee Lui and Alice Wong
School of Biological Sciences, Kadoorie Biological Sciences Building, The University of Hong Kong, Hong Kong SAR, China

Karen Lau, Keng Po Lai, Anna Tse and Rudolf Shiu Sun Wu
School of Biological Sciences, Kadoorie Biological Sciences Building, The University of Hong Kong, Hong Kong SAR, China
The State Key Laboratory in Marine Pollution, Hong Kong, Hong Kong SAR, China

Richard Yuen Chong Kong
The State Key Laboratory in Marine Pollution, Hong Kong, Hong Kong SAR, China

Jessie Yun Juan Bao, Na Zhang, Amy Tong and Si Lok
Genome Research Centre, The Hong Kong Jockey Club Building for Interdisciplinary Research, The University of Hong Kong, Pokfulam, Hong Kong SAR, China

Jing Woei Li
School of Life Sciences, Hong Kong Bioinformatics Centre, The Chinese University of Hong Kong, Hong Kong SAR, China

Ursula Witte
Oceanlab, University of Aberdeen, Aberdeen, United Kingdom

Niels Jobstvogt
Oceanlab, University of Aberdeen, Aberdeen, United Kingdom
Aberdeen Centre for Environmental Sustainability (ACES), University of Aberdeen, Aberdeen, United Kingdom

Michael Townsend
National Institute of Water and Atmospheric Research (NIWA), Hamilton, New Zealand

Nick Hanley
Department of Geography and Sustainable Development, University of St Andrews, St Andrews, United Kingdom

Rory L. Hodd and Micheline Sheehy Skeffington
Plant Ecology Research Unit (PERU), Botany and Plant Science, National University of Ireland Galway, Galway, Ireland

David Bourke
Department of Botany, and Trinity Centre for Biodiversity Research, School of Natural Sciences, Trinity College, Dublin, Dublin, Ireland

Kirsten Grorud-Colvert
Department of Integrative Biology, Oregon State University, Corvallis, Oregon, United States of America

Joachim Claudet
National Center for Scientific Research (CNRS), University of Perpignan, Perpignan cedex, France
Laboratoire d'Excellence 'CORAIL', Perpignan cedex, France

Brian N. Tissot
Humboldt State University Marine Laboratory, Trinidad, California, United States of America

Jennifer E. Caselle
Marine Science Institute, University of California Santa Barbara, Santa Barbara, California, United States of America

Mark H. Carr
Department of Ecology and Evolutionary Biology, Long Marine Lab, University of California Santa Cruz, Santa Cruz, California, United States of America

Jon C. Day
Great Barrier Reef Marine Park Authority, Townsville, Queensland, Australia

Alan M. Friedlander
Department of Biology, University of Hawai'i, Honolulu, Hawai'i, United States of America

Sarah E. Lester
Marine Science Institute, University of California

Santa Barbara, Santa Barbara, California, United States of America

Thierry Lison de Loma
Laboratoire d'Excellence 'CORAIL', Perpignan cedex, France
Centre de Recherches Insulaires et Observatoire de l'Environnement (CRIOBE), Moorea, French Polynesia

Daniel Malone
Institute of Marine Sciences, Long Marine Lab, University of California Santa Cruz, Santa Cruz, California, United States of America

William J. Walsh
Hawai'i Division of Aquatic Resources, Kailua-Kona, Hawai'i, United States of America

Azusa Makino, Carissa J. Klein and Maria Beger
Australian Research Council Centre of Excellence for Environmental Decisions, School of Biological Sciences, The University of Queensland, Brisbane, Queensland, Australia

Stacy D. Jupiter
Fiji Country Program, Wildlife Conservation Society, Suva, Fiji

Hugh P. Possingham
Australian Research Council Centre of Excellence for Environmental Decisions, School of Biological Sciences, The University of Queensland, Brisbane, Queensland, Australia
Department of Life Sciences, Imperial College London, Ascot, Berkshire, United Kingdom

Frida Ben Rais Lasram and Mohamed Salah Romdhane
UR 03AGRO1 Ecosystèmes et Ressources Aquatiques, INAT (Institut National Agronomique de Tunisie), Tunis, Tunisia

Tarek Hattab
UR 03AGRO1 Ecosystèmes et Ressources Aquatiques, INAT (Institut National Agronomique de Tunisie), Tunis, Tunisia
UMR 212 Ecosystèmes Marins Exploités, IRD (Institut de Recherche pour le Développement), Sète, France

Philippe Cury and François Le Loc'h
UMR 212 Ecosystèmes Marins Exploités, IRD (Institut de Recherche pour le Développement), Sète, France

Camille Albouy
Département de biologie, Chimie et Géographie, UQAR (Université du Québec à Rimouski), Québec, Canada

Chérif Sammari
Laboratoire du milieu marin, INSTM (Institut National des Sciences et Technologies de la Mer), Salammbô, Tunisia

Fabien Leprieur
Laboratoire Ecologie des Systémes Marins Côtiers UMR 5119, UM2 (Université de Montpellier 2), Montpellier, France

Index